Maxwell's Enduring Legacy
A Scientific History of the Cavendish Laboratory

The Cavendish Laboratory is arguably the most famous physics laboratory in the world. Founded in 1874, it rapidly gained a leading international reputation through the researches of the Cavendish professors, beginning with Maxwell, Rayleigh, J.J. Thomson, Rutherford and Bragg. Its name will always be associated with the discoveries of the electron, the neutron, the structure of the DNA molecule and pulsars, but these are simply the tip of the iceberg of outstanding science. The physics carried out in the laboratory is the central theme of the book and this is explained in reasonably non-technical terms. The research activities are set in their international context. Generously illustrated, with many pictures of the apparatus used and diagrams from the original papers, the story is brought right up to date with descriptions of the science carried out under the leadership of the very different personalities of Mott, Pippard and Edwards.

Malcolm Longair is Jacksonian Professor Emeritus of Natural Philosophy and Director of Development at the Cavendish Laboratory. He was appointed the ninth Astronomer Royal of Scotland in 1980, as well as the Regius Professor of Astronomy, University of Edinburgh, and the director of the Royal Observatory, Edinburgh. He was head of the Cavendish Laboratory from 1997 to 2005 and he has served on and chaired many international committees, boards and panels, working with both NASA and the European Space Agency. He has received much recognition for his work over the years, including a CBE in the Millennium Honours list for his services to astronomy and cosmology.

Maxwell's Enduring Legacy

A Scientific History of the Cavendish Laboratory

MALCOLM LONGAIR

University of Cambridge

CAMBRIDGE
UNIVERSITY PRESS

CAMBRIDGE
UNIVERSITY PRESS

University Printing House, Cambridge CB2 8BS, United Kingdom

Cambridge University Press is part of the University of Cambridge.

It furthers the University's mission by disseminating knowledge in the pursuit of education, learning and research at the highest international levels of excellence.

www.cambridge.org
Information on this title: www.cambridge.org/9781107083691

© Cambridge University Press 2016

First published 2016

Printed in the United Kingdom by Clays, St Ives plc

A catalogue record for this publication is available from the British Library

Library of Congress Cataloguing in Publication data

ISBN 978-1-107-08369-1 Hardback

For Deborah

Contents

Preface

The title and subtitle of this book deserve some explanation. To take the subtitle first, this book is a *scientific history*, rather than a history in the conventional sense – my interest and qualifications for writing it are as a scientist rather than as a historian. Specifically, my principal interest is in the *content* of the physics and how it came about, rather than a history of personalities, politics, administrative structures and so on. The latter topics cannot be ignored and obviously have a very significant bearing in framing the whole story, but that is not my prime goal. My objective is to set the scientific achievements in the context of the development of physics as a whole. This book is not a panegyric about the Laboratory, but an attempt to understand the areas in which it has been successful and what influenced the directions the research programme took.

The reason for the main title will become apparent as the story unfolds. James Clerk Maxwell was the first Cavendish Professor of Experimental Physics. His epochal scientific achievements in essentially all branches of physics need little emphasis here, but what is less well known is the profound effect his personality and research style had upon the early direction of the Laboratory. The achievement is all the more remarkable granted the rather barren field in which the seeds of future success were sown. Perhaps most significant of all is the remarkable agenda which Maxwell set for the Laboratory in his inaugural lecture of 1871. The subsequent history can be regarded as the working out of Maxwell's vision over succeeding generations of research workers. Tragically, Maxwell died in 1879 before the full impact of his agenda and reforms could be appreciated, but they were brought to fruition by his successors, Rayleigh, J.J. Thomson and Rutherford. By the time of Rutherford's death in 1937, the Laboratory had more than 50 years of history behind it and had made revolutionary contributions to many areas of experimental physics. As the Laboratory continued to grow after the Second World War, the organisation of research had to change, but the essential Maxwellian philosophy was maintained at the research group level. At the same time, the Laboratory had to cope with the demands of 'Big Science' and to exploit the opportunities offered by large national and international facilities.

At the time of writing, there are about 1000 members of the Laboratory spanning essentially the whole of physics. It is almost certainly foolhardy, to put it at its politest, for one person to attempt to do justice to the vast panorama of physics which has been carried out in the Laboratory. To keep the book within finite bounds, I have had to be selective. Maintaining a balance among the amazing diversity of topics is in the end a matter of subjective judgement and I hope I have not disappointed too many of my colleagues by what has been left out. My apologia for attempting this history is that, having spent much of my career working in the Laboratory and been its head for eight years from 1997 to 2005, I have at least some understanding of most of the research activities which have taken place in the Laboratory. It was enormously gratifying to be exposed to the wonderful range of science being undertaken by my colleagues – writing this book has made me admire their achievements all the more.

Malcolm Longair

Acknowledgements

This book has benefitted enormously from the insights provided by so many of my colleagues over the years since I became a graduate student in the Laboratory in 1963. I joined the staff as a demonstrator in 1970 and then progressed to a lectureship in 1975. I was in Edinburgh from 1980 to 1990 before returning to the Laboratory as Jacksonian Professor of Natural Philosophy in 1991. Thus, I had experience of working in the Laboratory as a relatively junior staff member during the Mott and Pippard eras, but missed much of the Edwards years. But working under and with successive heads of the Laboratory – Nevill Mott, Brian Pippard, Alan Cook and Archie Howie, until my own spell as Head of Department – was a unique education in how physics works. I am profoundly indebted to all of them for the support and opportunities they provided.

There is no doubt that the strongest Cambridge influences upon my thinking were Martin Ryle and Peter Scheuer. Martin was a truly inspirational, and often frightening, personality whose insight, dedication and sheer hard work were an inspiration. From Peter, I learned what it really means to understand physics – he had as deep an intellectual grasp of the essence of physics as any scientist I have met. He strongly influenced my own thinking. The strongest influence outside Cambridge was undoubtedly Yakov Borisevich Zeldovich, with whom I had the privilege of working in Moscow, with Rashid Sunyaev, for the academic year 1968–69. Encountering how Soviet physicists thought on their home territory was a mind-opening experience and one which taught me new ways of thinking about physics. My debt to these friends and mentors is incalculable.

In my last year of service to the University before my retirement in 2008, I was appointed Director of Development by Peter Littlewood with the aim of creating a Development Portfolio, among the many aims being the reconstruction of the whole Laboratory. On retirement, I was rehired half-time to continue in that role. In 2008, in discussion with Jeffrey Hughes, distinguished historian of science and technology at the University of Manchester, it became apparent that a new history of the Laboratory was badly needed, but there was no point in pretending that this was not a major challenge. We considered a multi-author approach to the history, but it would have been very difficult and time-consuming to bring off in a reasonable time. I am most grateful to Jeffrey for his insights. In the end, I decided that I was probably in as good a position as anyone to make the attempt. Once my book *Quantum Concepts in Physics* (Longair, 2013) was completed at the end of 2012, I decided to make the Cavendish history my priority writing task. I have been most gratified by the support and enthusiasm of so many of my colleagues for this endeavour.

Some friends and colleagues deserve special mention. First and foremost, I am very grateful to Isobel Falconer who has kindly read the whole book. With her in-depth knowledge of the early history of the Cavendish Laboratory from the perspective of a professional historian of science, she provided invaluable advice and insight into these and later parts of the story. John Waldram went far beyond the call of duty in ensuring that my description of low-temperature physics was adequate and accurate, as well as providing insights into large sections of the book. He consulted widely with the low-temperature physicists in the Quantum Matter Group, the contributions of John Adkins and Gil Lonzarich being particularly appreciated. Archie Howie, with contributions from Mick Brown, performed the same role in ensuring that my descriptions of microstructural physics were not too wide of the mark and kindly provided helpful 'tutorials'. These resulted in a much improved display of electron microscopy in the Cavendish Collection of Historic Scientific Instruments, as well as much invaluable content for this book. Volker Heine kindly made available his personal history of the Solid State Theory/Theory of Condensed Matter Group, which is now available on the TCM website and which many will enjoy.

As Director of Development, I have been the editor of the Laboratory's alumnus magazine, *CavMag*, from its inaugural edition in 2009. The articles in this twice-yearly publication provide a record of highlights of the Laboratory's current research programme. I am most grateful to the many colleagues who have provided excellent and stimulating articles, which have made my task of bringing the history right up to date all the easier. In parallel with these activities, I have been overseeing the refurbishment of the displays in the Cavendish Collection of Historic Scientific Instruments, supported by summer students who played a full role in refreshing the displays. Amy Howard and Rosie Davies did outstanding jobs in producing excellent new diagrams and text, a number of examples of their work appearing in the diagrams in this book. I am most grateful to both of them.

As in my earlier books, I have greatly benefitted from the advice of David Green on the subtleties of LaTeX coding – his expertise has greatly improved the appearance of the text. I am most grateful to the librarians in the Rayleigh Library, Nevenka Huntic and Helen Suddaby, the latter now retired, who were unfailingly helpful in tracking down the rarer books and papers referred to in the text. Kelvin Fagan, the Laboratory's photographer, kindly provided images from the photographic archives, as well as taking many new photographs which have been used extensively throughout this book. I am most grateful for his photographic skills and expertise in producing some wonderful images.

One final remark about the writing of this history may help the reader understand my approach. While the colleagues mentioned above have been most generous with their thoughts, I have not attempted to undertake a systematic set of interviews with the many remarkable physicists who have worked in the Laboratory. Rather, I have concentrated on using the written word as much as possible. This involved going back to most of the original papers, and that proved to be an essential and illuminating experience. In addition to the published biographies of many of the protagonists in this story, I gained excellent information and insights from the *Biographical Memoirs of Fellows of the Royal Society* for those deceased staff members who were Fellows of the Society. The approach of the authors of these memoirs, who are all physicists familiar with the work of those commemorated, matched exactly my own aspirations. I can strongly recommend these volumes for further information and enlightenment.

The usual disclaimer, that I am solely responsible for errors of content and judgement, is most apposite for this book. The story could be told in many different ways, with different emphases according to the inclination and experience of the writer. I have tried to create a balanced story, in the full knowledge that in the end it has involved personal choices.

As usual, the love, support, encouragement and understanding of my family, Deborah, Mark and Sarah, mean more to me than can ever be expressed by the written word.

Figure credits

I am most grateful to the publishers who have given permission for the use of the figures from journals, books and other media for which they hold the copyright. The following list includes the original publishers of the figures and the publishers who now hold the copyright, as well as the specific forms of acknowledgement requested by them. Many of the images were created by research groups in the Laboratory, and I am most grateful to them for permission to use them in this book. I also had the benefit of the efforts of the summer students who created many new illustrations for the displays in the Cavendish Collection of Historic Scientific Instruments and which I have used in this book. Any images which are not credited in the figure captions themselves are the property of the Cavendish Laboratory.

AAS Publications. Reproduced by kind permission of AAS Publications. Box 8.1(*b*).

Acta Crystallographica. Reproduced by kind permission of Acta Crystallographica. Figure 12.6.

Adam Hilger. Reproduced by kind permission of Adam Hilger. Figure 9.3.

Advances in Physics. Reproduced by kind permission of Advances in Physics. Box 20.8(*b*).

American Journal of Science. Reproduced by kind permission of the American Journal of Science. Figure 6.1(*a*) and (*b*).

Anderson, Rae. Reproduced by kind permission of Rae Anderson, University of San Diego. Box 18.5.

Angewandte Chemie. Reproduced by kind permission of Angewandte Chemie. Box 22.4(*c*).

Astronomical Society of the Pacific. Reproduced by kind permission of Astronomical Society of the Pacific. Figure 17.2(*b*).

Astronomy and Astrophysics. Reproduced by kind permission of Astronomy and Astrophysics. Box 22.1(*c*) and (*e*).

Astrophysical Journal Letters. Reproduced by kind permission of Astrophysical Journal Letters. Figure 21.5(*b*).

Bickerstaff, R., Manchester. Reproduced by kind permission of R. Bickerstaff, Manchester. Figure 1.10.

Building Design Partnership (BDP). Reproduced by kind permission of Building Design Partnership (BDP). Box 22.17(*d*).

Cambridge Collection in the Cambridge City Library. Reproduced by kind permission of Cambridge Collection in the Cambridge City Library. Figures A.1 through A.9.

Cambridge Display Technologies (CDT). Reproduced by kind permission of Cambridge Display Technologies (CDT). Figure 20.6(*b*).

Cambridge University Press. Reproduced by kind permission of Cambridge University Press. Box 7.1, Figures 8.1(*b*) and 9.14(*a*) and (*b*).

Cavendish Laboratory. Reproduced by kind permission of the Cavendish Laboratory. Boxes 3.1, 4.1, 6.2, 6.4, 7.4(*b*), 8.3(*a*), 10.3(*a*) and (*b*), 15.1(*a*), (*b*) and (*c*), 15.6, 15.8(*a*), (*b*) and (*c*), 15.10(*a*), (*b*) and (*c*), 15.11, 15.13(*b*), 18.1(*a*) and (*b*), 18.2(*a*) and (*b*), 20.1(*a*) and (*b*), 22.3(*a*), 22.7(*a*), (*b*) and (*c*), 22.8(*a*), 22.9(*b*), (*c*) and (*d*), 22.10(*a*), 22.11(*a*) and (*b*), 22.12(*a*) and (*c*), 22.13, 22.14, 22.17(*a*), (*b*) and (*c*).

Cavendish Laboratory. Reproduced by kind permission of the Cavendish Laboratory. Figures 2.2(*b*), 3.2(*a*) and (*b*), 3.3(*a*) and (*b*), 4.1(*a*) and (*b*), 4.3, 5.2, 5.3, 6.3(*a*) and (*b*), 6.5, 6.6(*b*), 6.7(*b*), 7.2(*a*) and (*b*), 7.3, 7.4, 7.5, 7.8(*b*), 8.2(*b*), 8.4, 8.5, 9.2(*a*), 9.4(*a*), 9.6(*a*), 9.8, 9.9, 9.10(*b*), 9.11(*b*), 9.12, 9.13, 9.16, 10.1(*a*) and (*b*), 10.2(*a*) and (*b*), 12.1, 12.3(*a*) and (*b*), 12.4(*a*) and (*b*), 12.5(*a*) and (*b*), 12.8(*b*), 12.9, 12.12(*a*) and (*b*), 12.13(*a*) and (*b*), 12.15(*b*), 12.16(*a*), 12.18, 13.1, 13.3, 14.1(*a*), 14.2(*a*) and (*b*), 14.5(*a*), 14.7(*a*) and (*b*), 14.9, 14.13, 14.14, 15.6, 16.4, 17.4, 18.3(*a*) and (*b*), 18.4(*a*) and (*b*), 18.5(*a*) and (*b*), 20.3(*a*) and (*b*), 20.9, 20.11(*a*) and (*b*), 20.13(*b*), 20.15(*a*) and (*b*), 20.17, 21.4(*a*), 22.2(*b*).

Cavendish Laboratory. Reproduced by kind permission of the following groups:

Atomic, Mesoscopic and Optical Physics Group. Boxes 22.5(*a*), (*b*) and (*c*).
Astrophysics Group. Figure 21.7(*a*) and Box 22.1(*a*).
Biological and Soft Systems Group. Figures 20.4(*a*) and (*b*) and Box 22.4(*a*).
High Energy Physics Group. Box 21.1 and Figures 17.6(*a*) and (*b*), 17.7(*a*) and (*b*).
Microelectronics Group. Box 22.6(*a*) and (*b*).
Optoelectroncs Group. Box 20.3 and Figure 20.6(*a*).
Quantum Matter Group. Boxes 20.8(*c*), 22.8(*b*) and (*c*).
Scientific Computing Group. Box 22.16.
Semiconductor Physics Group. Box 18.3(*a*) and (*b*).
Surface, Microstructure and Fracture Group. Boxes 15.13(*a*) and 15.9(*a*) and (*b*) and Figure 20.18(*a*) and (*b*).
Thin Film Magnetism Group. Boxes 20.9(*a*) and (*b*), 22.15(*a*) and (*b*).

Cell. Reproduced by kind permission of Cell. Box 22.4(*d*).

CERN. Reproduced by kind permission of CERN. Boxes 21.1, 21.2, 22.2(*b*), (*c*) and (*d*), Figures 17.5, 17.7(*a*) and (*b*), 21.1, 21.2.

Churchill College. Reproduced by kind permission of Antony Hewish and Churchill College Archives. Figure 14.10.

Creative Commons. Reproduced by kind permission of Creative Commons. Box 10.3(*b*) (in public domain, http://commons.wikimedia.org/wiki/File:Vector_de_Burgers.PNG), Box 12.1(*a*) (PDB: 1GZX Proteopedia Hemoglobin) and (*b*) (in public domain) and Box 18.4 (placed in public domain). Figure 6.4(*a*) (in public domain, Wikimedia Commons) and (*b*) (in public domain, http://en.wikipedia.org/wiki/Image:Becquerel_plate.jpg).

Deutsches Museum, Munich. Reproduced by kind permission of the Deutsches Museum, Munich. Figure 1.2.

Edinburgh Philosophical Journal. Reproduced by kind permission of Edinburgh Philosophical Journal. Figure 1.1.

European Space Agency. Reproduced by kind permission of European Space Agency. Figure 21.4(b).

European Southern Observatory. Reproduced by kind permission of European Southern Observatory. Figure 21.8.

Fermilab. Reproduced by kind permission of Fermilab. Box 22.2(a).

Fourth Estate Publishers. Reproduced by kind permission of Fourth Estate Publishers. Box 22.10.

Getty Images. Reproduced by kind permission of Getty Images. Figure 2.2(a).

Glasgow University. Reproduced by kind permission of Glasgow University Library, Special Collections Department. Figure 2.1.

Historical Studies in the Physical Sciences. Reproduced by kind permission of Historical Studies in the Physical Sciences. Figure 5.1(a).

International Union of Crystallography. Reproduced by kind permission of the International Union of Crystallography. Figure 12.6.

Jacques Gabay Publisher, Paris. Reproduced by kind permission of Jacques Gabay Publisher, Paris. Figure 1.9.

Jestico and Whiles. Model and photographs reproduced by kind permission of Jestico and Whiles. Figure 22.4.

John Campbell. Reproduced by kind permission of John Campbell. Box 8.1(b).

Journal of Investigative Dermatology. Reproduced by kind permission of Journal of Investigative Dermatology. Box 20.2.

Journal of Physics C: Solid State Physics. Reproduced by kind permission of Journal of Physics C: Solid State Physics. Figure 20.1(b) and 20.2.

Journal of Scientific Instruments. Reproduced by kind permission of Journal of Scientific Instruments. Figures 11.1 and 11.3(a) and (b).

Journal of Superconductivity. Reproduced by kind permission of Journal of Superconductivity. Figure 20.10(a).

Journal of the American Chemical Society. Reproduced by kind permission of Journal of the American Chemical Society. Box 22.3(c).

Journal of the Royal Institution. Reproduced by kind permission of Journal of the Royal Institution. Figure 6.6(a).

Macmillan. Reproduced by kind permission of Macmillan. Box 6.3, Figures 3.1(a) and (b), 16.1, 16.2, 16.3.

Monthly Notices of the Royal Astronomical Society. Reproduced by kind permission of Monthly Notices of the Royal Astronomical Society. Box 22.1(b) and (d) and Figures 12.10(a) and (b), 14.4(a) and (b), 14.7(c), 17.1, 17.2(a), 17.3(b), 21.3 and 21.5(a).

MRC Laboratory of Molecular Biology. Reproduced by kind permission of MRC Laboratory of Molecular Biology. Figures 13.1, 15.4(a) and (b).

NanoLetters. Reproduced by kind permission of NanoLetters. Box 22.12(b).

NASA, ESA and the Space Telescope Science Institute. Reproduced by kind permission of NASA, ESA and the Space Telescope Science Institute. Figures 14.3 and 21.9.

Nature. Reproduced by kind permission of Nature. Boxes 10.2(*a*) and (*b*), 20.7 and 20.8(*a*). Figures 11.2, 12.11, 14.5(*b*), 14.8, 21.6(*b*).

Observatory. Reproduced by kind permission of The Observatory. Figure 14.1(*b*).

Oxford University Press. Reproduced by kind permission of Oxford University Press. Box 22.3(*b*). Figures 1.7, 1.8 and 4.2.

Philosophical Magazine. Reproduced by kind permission of the Philosophical Magazine. Box 7.4(*a*), 8.1(*a*) and 8.2. Figures 6.7(*a*), 6.8, 6.9(*a*) and (*b*), 7.9, 7.10, 8.2(*a*), 8.3, 9.1, 9.2(*b*), 15.3(*a*) and (*b*).

Philosophical Transactions of the Royal Society of London. Reproduced by kind permission of the Royal Society of London. Boxes 7.5, 15.2, 15.3 and 15.4. Figures 1.3, 7.6(*a*) and (*b*), 10.3(*a*), (*b*) and (*c*), 18.6.

Physical Review. Reproduced by kind permission of the American Physical Society. Figure 20.12. Copyright 1999 by the American Physical Society.

Physical Review Letters. Reproduced by kind permission of the American Physical Society. Box 22.4(*b*). Figures 20.1(*a*), 20.10(*b*) and 20.14. Copyright 1986, 1993, 2003 and 2004 by the American Physical Society.

Proceedings of the Cambridge Philosophical Society. Reproduced by kind permission of Proceedings of the Cambridge Philosophical Society. Figure 7.7(*b*).

Proceedings of the National Academy of Sciences of the United States of America. Reproduced by kind permission of National Academy of Sciences of the United States of America. Figure 20.7.

Proceedings of the Royal Society of London. Reproduced by kind permission of the Royal Society of London. Boxes 8.3(*b*) and 15.12. Figures 1.6, 8.1(*a*), 9.6(*b*), 9.7, 9.10(*a*), 9.15, 11.4(*a*) and (*b*), 12.16(*b*), 15.1, 15.2(*a*) and (*b*), 15.5.

Progress in Surface Science. Reproduced by kind permission of Journal of Physics: Condensed Matter. Figure 20.17.

Royal Institution of Great Britain. Reproduced by kind permission of Royal Institution of Great Britain. Figures 1.4 and 1.5(*a*) and (*b*).

Royal Observatory, Edinburgh. Reproduced by kind permission of Royal Observatory, Edinburgh. Figures 21.6(*a*) and 21.7(*b*).

Science. Reproduced by kind permission of Science. Boxes 20.5 and 22.10(*b*).

Springer Science and Business Media. Reproduced by kind permission of Springer Science and Business Media. Box 14.1 and Figure 14.6.

Vieweg, Braunschweig. Reproduced by kind permission of Vieweg, Braunschweig. Figure 7.8(*a*).

UIT Publications. Reproduced by kind permission of UIT Publications. Box 22.9(*a*).

World Wide Web. Reproduced by kind permission of World Wide Web. Box 12.3(*b*), diagram by Jonathan Atteberry. Box 22.10(*c*), courtesy of Toshiba Ltd and Paul Townsend.

University of Cambridge Computer Laboratory. Reproduced by kind permission of University of Cambridge Computer Laboratory. Figure 12.2.

Zeitschrift für Physik. Reproduced by kind permission of Zeitschrift für Physik. Figures 9.11(*a*) and 9.5. Also, with kind permission from Springer Science and Business Media.

Zhurnal Russkoe Fiziko-Khimicheskoe Obshchestvo. Reproduced by kind permission of the Zhurnal Russkoe Fiziko-Khimicheskoe Obshchestvo. Figure 1.11.

The following diagrams were drawn by the author and his colleagues, some of them for this publication and the others for the displays in the Cavendish Collection of Historic Scientific Instruments. Boxes 11.1, 11.2, 12.3(*a*), 12.14, 12.15(*a*), 12.17, and 15.7. Figures 5.1(*b*), 7.1, 9.4(*b*), 12.7, 12.8(*a*), 12.14, 12.15(*a*), 12.17, 13.2(*a*) and (*b*), 14.11, 14.12, 16.5(*a*) and (*b*), 17.3(*a*), 18.1(*a*), (*b*) and (*c*), 18.2(*a*), (*b*) and (*c*), 19.1, 20.13(*a*), 20.8, 22.1 and 22.3. Note 8.1.

PART I

To 1874

William Cavendish (1808–1891), 7th Duke of Devonshire
Chancellor of Cambridge University (1861–1891)
Oil painting: copy after George Frederic Watts
by Katharine Maude Humphrey

Physics in the nineteenth century 1

Where better to start than with the definition of *physics* from the *Oxford English Dictionary: The Definitive Record of the English Language*:

> **physics, n.** The branch of science concerned with the nature and properties of non-living matter and energy, in so far as they are not dealt with by chemistry or biology; the science whose subject matter includes mechanics, heat, light and other radiation, sound, electricity, magnetism, gravity, the structure of atoms, the nature of subatomic particles, and the fundamental laws of the material universe.

Most twenty-first-century physicists would find this definition sufficiently all-encompassing to describe what physics is about. Many, including the present author, would argue that the extensive nature of the subject and its emphasis upon model-building and problem-solving enables it to embrace a vastly wider range of cognate fields, including chemistry, biology, medicine and economics. But the mid-nineteenth-century scientist, and in particular one associated with Cambridge University, would have formed a very different view.

1.1 Discoveries in physics, 1687 to 1874

It is enlightening to study a time-line of major discoveries and advances in what we would now call physics, as well as selected topics from cognate disciplines such as chemistry, mathematics and particularly technology. The items in Table 1.1 span the period from the date of publication of Isaac Newton's *Philosophiæ Naturalis Principia Mathematica* (Newton, 1687) to the opening of the Cavendish Laboratory in 1874. The contents of the table are necessarily subjective but provide a broad-brush picture of the standard history of physics. The country and location where the advances were made, as well as those which originated in Cambridge, are also indicated.

There are a number of striking features of Table 1.1. There is no question about the overwhelming impact of Newton's laws of motion and of gravity of 1687. Newton and his followers promoted vigorously the corpus of Newton's great works in the sciences of mechanics, statics, dynamics, gravity and optics. Nowhere was this more assiduously practiced than in his alma mater, Cambridge University, where, well into the nineteenth century, Newton's approach to the laws of motion were taught to the exclusion of the more powerful mathematical methods of analysis which had been developed in continental Europe.

Table 1.1 Selected advances in physics and cognate disciplines, 1687–1874

Date	Discovery	UK involvement	Cambridge involvement
1687	Newton's laws of motion and of gravity	Newton	Newton
1690	Huygens' *Treatise on Light* (Netherlands)		
1704	Newton's *Opticks*	Newton	Newton
1776	James Watt's first commercial steam engine	Watt (Birmingham)	
1782	Chemistry, conservation of mass: Lavoisier (France)		
1785	Coulomb's inverse square laws of electrostatics and magnetostatics (France)		
1788	Lagrange's equations of motions (France)		
1801	Young's wave theory of light	Young	Young
1800–10	Gas laws: Gay-Lussac (France), Charles (France), Dalton	Dalton (Manchester)	
1808	Dalton's atomic theory of matter	Dalton (Manchester)	
1811	Avogadro's hypothesis (Italy)		
1820	Electric and magnetic fields and current electricity Ørsted (Denmark), Ampère (France), Biot (France), Savart (France)		
1822	Fourier's *Théorie Analytique de la Chaleur* (France)		
1822	Babbage's design of his difference engine	Babbage (London)	
1824	The Carnot cycle and ideal gas engines (France)		
1827	Hamilton's equations	Hamilton (Ireland)	
1827	Electrical resistance: Ohm's law (Germany)		
1831	Faraday's lines of force and the law of electromagnetic induction	Faraday (RI London)[a]	
1842–43	Conservation of energy: Mayer (Germany), Kelvin[b]	Kelvin[b] (Glasgow)	
1845	Faraday on the rotation of polarised light	Faraday (RI London)[a]	
1845–50	Conservation of energy: Joule, Helmholtz (Germany)	Joule (Salford)	
1850–51	Second law of thermodynamics: Clausius (Germany), Kelvin[b]	Kelvin[b] (Glasgow)	
1854–56	Kelvin[b] and the transatlantic telegraph cable	Kelvin[a] (Glasgow)	
1857–59	Kinetic theory: Clausius (Germany), Maxwell	Maxwell (Aberdeen)	
1859	Black-body radiation and Kirchhoff's laws (Germany)		
1863	Entropy: Clausius (Germany), Maxwell (London), Kelvin[b] (Glasgow)	Maxwell (London), Kelvin[b] (Glasgow)	
1864–65	Maxwell's dynamical theory of the electromagnetic field	Maxwell (London)	
1867	Maxwell's dynamical theory of gases	Maxwell (Glenlair)	
1869	Mendeleyev (Russia) and the periodic table		

[a] 'RI London' refers to the Royal Institution of Great Britain in Albemarle Street, London.
[b] In this table, 'Kelvin' means William Thomson, who took the title 1st Baron Kelvin of Largs.

From roughly the turn of the nineteenth century onwards, the emerging sciences of electricity, magnetism, heat and thermodynamics were added to the edifice of Newtonian physics, but these could not be readily incorporated into the Newtonian world view. Equally significantly, the new sciences were to find immediate practical application in the industrial revolution which was in full swing by the beginning of the nineteenth century – indeed, the boot was very much on the other foot, with industry driving research in the basic sciences. In parallel with the physical and mathematical elaboration of new areas of science, there was an irresistible drive by engineers, industrialists and entrepreneurs to exploit these discoveries for the benefit of industry, for military advantage and for the improvement of the well-being of society. It is revealing to review in parallel developments in the physical sciences and the corresponding needs of industry and their contributions to society at large.

1.2 Precise measurement and the determination of time

Precise measurement has always had a key role to play in astronomical observations, which were essential for keeping an accurate track of time and for navigation at sea. Pride of place must go to Tycho Brahe, who was inspired to begin his great series of astronomical observations of the stars and planets by the inaccuracies he found in the Alphonsine Tables. The predicted positions of the Sun, Moon, stars and the planets in the tables were based upon a complex Ptolemaic model of the solar system which was used to keep track of time and to establish the dates of religious festivals. With sponsorship by Frederick II to the tune of about 1% of the gross national product of Denmark for about 20 years, Tycho constructed his great observatory on the island of Hven. In fact, Tycho built two independent observatories, Uraniborg and Stjerneborg. He employed numerous clocks to ensure the precise measurement of the times of transit of the stars and planets. The instruments were the most advanced of their time and the whole enterprise resulted in an improvement in the precision with which the positions of the celestial bodies were known by a factor of 10 over previous measurements. This achievement resulted from a number of technical innovations:

- understanding of the importance of *systematic* and *random* errors
- understanding of the effects of bending of the instruments on astronomical observations
- understanding of the effects of atmospheric refraction
- maintaining the independence of the astronomical data by observers at his two observatories.

Over a period of 20 years, he measured positions to a precision of about 1–2 arcmin for the Sun, Moon, planets and 777 stars. In 1600, the year before his death, he employed Johannes Kepler to analyse his observations of the motion of Mars. From Tycho's observations, Kepler deduced his laws of planetary motion, which in turn led to Newton's laws of motion and of gravity.[1]

The qualities of skill, craftsmanship and invention available in eighteenth-century England are well illustrated by the achievements of John Harrison in designing and building clocks which could maintain accurate timekeeping despite the rigours of life on the high seas. While the determination of latitude at sea was straightforward, by measuring the angle from the celestial pole to the horizon, the determination of longitude was much trickier. With a sufficiently stable clock, longitude could be found by comparing a clock on board ship, which kept track of Greenwich Mean Time, with the local standard of time, found when noon was observed at the location of the ship. The sea-clocks had to be able to withstand severe temperature, pressure and humidity changes, as well as the pitching and rolling motions of the ship, particularly during storms and bad weather. The importance to a maritime nation such as the United Kingdom of being able to measure longitude accurately was recognised by a prize of £20,000 offered by the British government for the construction of a chronometer which could precisely maintain time at sea. In 1761, Harrison's first marine watch achieved an accuracy of about 1 km on the transatlantic voyage to Jamaica.

1.3 Fraunhofer, Kirchhoff and the development of optical spectroscopy

In the early years of the nineteenth century, Joseph Fraunhofer opened up the new science of precision spectroscopy. The breakthrough can be traced to Thomas Young's pioneering experiments and theoretical understanding of the laws of interference and diffraction of waves. In his Bakerian Lecture of 1801 to the Royal Society of London, *On the Theory of Light and Colours*, he used the wave theory of light of Christiaan Huygens to account for the results of interference experiments such as his double-slit experiment (Young, 1802). Among the most striking achievements of this paper was the measurement of the wavelengths of light of different colours using a diffraction grating with 500 grooves per inch. From this time onwards, wavelengths were used to characterise the colours in the spectrum.

In 1802, William Wollaston, the Jacksonian Professor of Natural Philosophy at Cambridge University, made spectroscopic observations of sunlight and discovered five strong dark lines, as well as two fainter lines (Wollaston, 1802). The full significance of these observations only began to be appreciated following the remarkable experiments of Fraunhofer. He was the son of a glazier and became one of the two directors of the Benediktbeuern glassworks in Bavaria in 1814. The firm manufactured high-quality optical glass for military and surveying instruments. Fraunhofer's motivation for studying the solar spectrum was his realisation that accurate measurements of the refractive indices of glasses should be made using monochromatic light. His visual observations were made by placing a prism in front of a 25 mm aperture telescope. In his spectroscopic observations of the Sun, he discovered that the dark lines were narrow and provided precisely defined wavelength standards. He labelled the ten strongest lines in the solar spectrum A, a, B, C, D,

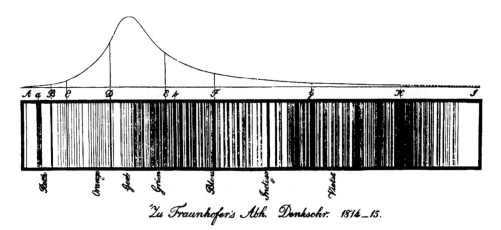

'Zu Fraunhofer's Abh. Denkschr. 1814_15.

Fraunhofer's solar spectrum of 1814, showing the vast numbers of dark absorption lines. The colours of the various regions of the spectrum are labelled, with the letters A, a, B, C, D, E, b, F, G and H indicating the most prominent absorption lines. The continuous line above the spectrum shows the approximate solar continuum intensity, as estimated by Fraunhofer (1817a,b).

Fig. 1.1

E, b, F, G and H and recorded 574 fainter lines between the B and H lines (Figure 1.1) (Fraunhofer, 1817a,b). A major advance was the invention of the spectroscope, with which the deflection of light passing through the prism could be measured precisely. To achieve this, Fraunhofer placed a theodolite on its side and observed the spectrum through a telescope mounted on the rotating ring (Figure 1.2).

In a second paper, Fraunhofer measured the wavelengths of what are now referred to as the *Fraunhofer lines* in the solar spectrum using a diffraction grating, which consisted of a large number of equally spaced thin wires (Fraunhofer, 1821); he was one of the early pioneers in the production of diffraction gratings. He found that the wavelengths of these lines were stable and so provided accurate wavelength standards.

From the perspective of the glass industry, Fraunhofer was then able to characterise the chromatic properties of glasses and lenses quantitatively and precisely. These developments led to much superior glasses, as well as to much improved polishing and testing methods for glasses and lenses. These technical improvements also resulted in the best astronomical telescopes then available. Fraunhofer's masterpiece was the 24 cm Dorpat Telescope, built for Wilhelm Struve at the Dorpat (now Tartu) Observatory in Estonia. In addition, he built a heliometer for Friedrich Bessel at Königsberg, which enabled the first parallax of a star to be measured.

The understanding of the dark lines in the solar spectrum had to await developments in laboratory spectroscopy. In his papers of 1817, Fraunhofer noted that the dark D lines coincided with the bright double line seen in lamplight. In 1849, Léon Foucault in Paris performed a key experiment in which sunlight was passed through a sodium arc so that the two spectra could be compared precisely. To his surprise, the solar spectrum displayed even darker D lines when passed through the arc than without the arc present (Foucault, 1849). He followed up this observation with an experiment in which the continuum spectrum of

Fig. 1.2 A portrait of Fraunhofer with his spectroscope (courtesy of the Deutsches Museum, Munich). This portrait is located in the museum's Hall of Fame.

light from glowing charcoal was passed through the arc and the dark D lines of sodium were found to be imprinted on the transmitted spectrum.

Ten years later, the experiment was repeated by Gustav Kirchhoff at his laboratory at the University of Heidelberg. He made the further crucial observation that, for the observation of an absorption feature, the source of the light had to be hotter than the absorbing flame. These results were immediately followed up in 1859 by his understanding of the relation between the emissive and absorptive properties of any substance, what is now known as *Kirchhoff's law of the emission and absorption of radiation* (Kirchhoff, 1859).[2] This states that, in thermal equilibrium, the radiant energy emitted by a body at any frequency is precisely equal to the radiant energy absorbed at the same wavelength. He showed that in thermodynamic equilibrium

$$\alpha_\nu B_\nu(T) = j_\nu \tag{1.1}$$

(Box 1.1). In other words, the emission and absorption coefficients, j_ν and α_ν respectively, for any physical process are related by the unknown spectrum of equilibrium radiation $B_\nu(T)$, the spectrum of black-body radiation. In 1859, very little was known about the

Kirchhoff's law of the emission and absorption of radiation

Box 1.1

For isotropic radiation, the *monochromatic emission coefficient* j_ν is defined such that the increment in intensity dI_ν radiated into the solid angle $d\Omega$ from the cylindrical volume dV of area dA and length dl is

$$dI_\nu \, dA \, d\Omega = j_\nu \, dV \, d\Omega,$$

where j_ν has units $W \, m^{-3} \, Hz^{-1} \, sr^{-1}$. Since the emission is assumed to be isotropic, the *volume emissivity* of the medium is $\varepsilon_\nu = 4\pi j_\nu$. We can write the volume of the cylinder as $dV = dA \, dl$ and so

$$dI_\nu = j_\nu \, dl.$$

The *monochromatic absorption coefficient* is defined by the relation

$$dI_\nu \, dA \, d\Omega = -\alpha_\nu I_\nu \, dA \, d\Omega \, dl.$$

In thermodynamic equilibrium, the magnitudes of the left- and right-hand sides of this equation must be equal:

$$\alpha_\nu B_\nu(T) = j_\nu,$$

where $I_\nu = B_\nu(T)$ is the unknown spectrum of equilibrium radiation $B_\nu(T)$ relating the emission and absorption coefficients for any physical process. This expression enabled Kirchhoff to understand the relation between the emission and absorption properties of flames, arcs, sparks and the solar atmosphere.

experimental or theoretical form of $B_\nu(T)$. As Kirchhoff remarked, 'It is a highly important task to find this function.' This was to remain one of the great experimental challenges for the remaining decades of the nineteenth century. Kirchhoff's profound insight was the beginning of a long and tortuous story which was to lead to Planck's discovery of the formula for black-body radiation over 40 years later.

Throughout the 1850s, there was considerable effort in Europe and in the USA aimed at identifying the emission and absorption lines produced by different substances in flame, spark and arc spectra. Different elements and compounds possess distinctive patterns of spectral lines, and attempts were made to relate these to the lines observed in the solar spectrum. In 1859, for example, Julius Plücker identified the Fraunhofer F line with the bright Hβ line of hydrogen and the C line was more or less coincident with Hα, demonstrating the presence of hydrogen in the solar atmosphere.

The most important work resulted from the studies of Robert Bunsen and Kirchhoff. In Kirchhoff's great papers of 1861 to 1863, entitled *Investigations of the Solar Spectrum and the Spectra of the Chemical Elements*, the solar spectrum was compared with the spark spectra of 30 elements using a four-prism arrangement with which it was possible to simultaneously view the spectrum of the element and the solar spectrum (Kirchhoff, 1861, 1862, 1863). Kirchhoff concluded that the cool outer regions of the solar atmosphere contained iron, calcium, magnesium, sodium, nickel and chromium and probably cobalt, barium, copper and zinc as well.

1.4 Electricity and magnetism

The disciplines which were to drive the new physics of the second half of the nineteenth century were electricity, magnetism and heat. These advances not only opened up new fields of research in physics but were to have immediate social, industrial and commercial impact. From disciplines which had been the province of natural philosophers and learned societies, new industries were created which depended upon the understanding of the laws of physics at a fundamental level. In this section, we trace the revolutionary impact of electricity and magnetism and, in the next section, thermodynamics and the theory of heat.

1.4.1 Electrostatics and magnetostatics

By the end of the eighteenth century, many of the basic experimental features of electrostatics and magnetostatics had been established. In the 1770s and 1780s, Charles-Augustin Coulomb performed sensitive electrostatic experiments which established the inverse-square laws of electrostatics and magnetostatics. In SI notation,[3] these can be written

$$f = \frac{q_1 q_2}{4\pi \epsilon_0 r^2} \, i_r \quad \text{and} \quad f = \frac{\mu_0 p_1 p_2}{4\pi r^2} \, i_r, \tag{1.2}$$

where i_r is the unit vector directed radially *away* from either electrostatic charge q_1, q_2 or magnetostatic monopole with pole strengths p_1, p_2 in the direction of the other.

As will be discussed in Chapter 2, the late eighteenth and early nineteenth centuries was a period of extraordinary brilliance in French mathematics, and included personalities such as Siméon-Denis Poisson, who was a pupil of Pierre-Simon Laplace, and Joseph-Louis Lagrange. In 1812, Poisson published his famous *Mémoire sur la Distribution de l'Électricité à la Surface des Corps Conducteurs* (Poisson, 1812), in which he demonstrated that many of the problems of electrostatics can be simplified by the introduction of the electrostatic potential V which is the solution of Poisson's equation,

$$\frac{\partial^2 V}{\partial x^2} + \frac{\partial^2 V}{\partial y^2} + \frac{\partial^2 V}{\partial z^2} = -\frac{\rho_e}{\epsilon_0}, \quad E = -\text{grad}\, V, \tag{1.3}$$

where ρ_e is the electric charge density distribution and E the electric field strength. In 1826, he published the corresponding expressions for the magnetic flux density B in terms of the magnetostatic potential V_{mag}:

$$\frac{\partial^2 V_{\text{mag}}}{\partial x^2} + \frac{\partial^2 V_{\text{mag}}}{\partial y^2} + \frac{\partial^2 V_{\text{mag}}}{\partial z^2} = 0, \quad B = -\mu_0 \, \text{grad}\, V_{\text{mag}}. \tag{1.4}$$

1.4.2 Current electricity

Until 1820, electrostatics and magnetostatics appeared to be quite separate phenomena, but this changed dramatically with the development of the science of *current electricity*. In the latter years of the eighteenth century, the Italian anatomist Luigi Galvani discovered

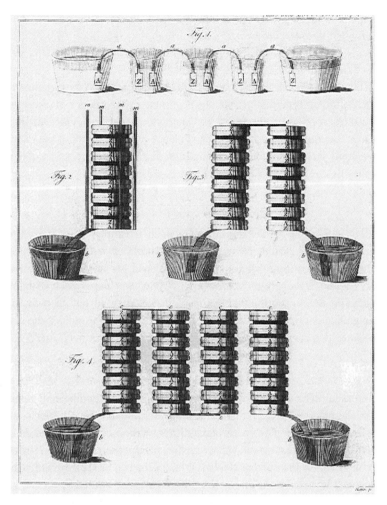

Diagrams from Alessandro Volta's letter of 20 March 1800 to Joseph Banks, President of the Royal Society of London, showing Volta's crown of cups (upper panel) and examples of combinations of voltaic piles (lower panels) (Volta, 1800).

Fig. 1.3

that electrical effects could stimulate the muscular contraction of frogs' legs. In 1791, he showed that, when two dissimilar metals were used to make the connection between nerve and muscle, the same form of muscular contraction was observed, and he announced the discovery of *animal electricity* (Galvani, 1791). Alessandro Volta suspected that the electric current was associated with the presence of different metals in contact with a moist body. He described his inventions of the *crown of cups* and of the *voltaic pile* in a letter dated 20 March 1800 to Joseph Banks, President of the Royal Society of London (Volta, 1800). The voltaic pile consisted of interleaved layers of copper and zinc separated by layers of pasteboard soaked in a conducting liquid (Figure 1.3). With this pile, Volta was able to demonstrate all the phenomena of electrostatics – the production of electric discharges, electric shocks and so on. The most important aspect of Volta's experiments was that he had invented a *controllable sources* of electric current. A problem with the voltaic cell was that it had a relatively short lifetime because the pasteboard dried out, but this was not a problem with his *crown of cups*, in which the electrodes were placed in series in glass vessels, resulting in a steady source of direct-current electricity. This arrangement was described by Volta in the same letter to Banks (Figure 1.3).

In 1820, Hans-Christian Ørsted demonstrated that there is always a magnetic field associated with an electric current; this marked the beginning of the science of *electromagnetism* (Ørsted, 1820). As soon as his discovery was announced, the physicists Jean-Baptiste Biot and Félix Savart set out to determine the dependence of the strength of the magnetic field at distance r from a current element of length $\mathrm{d}l$ in which a current I is flowing. In the same year, they discovered the *Biot–Savart law*,

$$\mathrm{d}\boldsymbol{B} = \frac{\mu_0 I \,(\mathrm{d}\boldsymbol{l} \times \boldsymbol{r})}{4\pi r^3}. \tag{1.5}$$

The term $\mathrm{d}\boldsymbol{l}$ is the length of the current element in the direction of the current I and $\boldsymbol{r}$ is measured from the current element $\mathrm{d}\boldsymbol{l}$ to the point at vector distance $\boldsymbol{r}$ (Biot and Savart, 1820). In 1826, André-Marie Ampère published his famous treatise *Théorie des Phénomènes Électro-Dynamique, Uniquement Déduite de l'Expérience* (Ampère, 1826), which included the demonstration that the magnetic field of a current loop could be represented by an equivalent magnetic shell. In the treatise, he also formulated the equation for the force between two current elements $\mathrm{d}\boldsymbol{l}_1$ and $\mathrm{d}\boldsymbol{l}_2$ carrying currents I_1 and I_2:

$$\mathrm{d}\boldsymbol{F}_2 = \frac{\mu_0 I_1 I_2 \,[\mathrm{d}\boldsymbol{l}_1 \times (\mathrm{d}\boldsymbol{l}_2 \times \boldsymbol{r})]}{4\pi r^3}. \tag{1.6}$$

$\mathrm{d}\boldsymbol{F}_2$ is the force acting on the current element $\mathrm{d}\boldsymbol{l}_2$, $\boldsymbol{r}$ being measured from $\mathrm{d}\boldsymbol{l}_1$. Ampère also demonstrated the relation between this law and the Biot–Savart law.

In 1826–27, Georg Simon Ohm formulated the relation between potential difference V and the current I, now known as *Ohm's law*, $V = RI$, where R is the resistance of the material through which the current flows (Ohm, 1826a,b, 1827)). Sadly, this pioneering work was not well received by Ohm's colleagues in Cologne and he resigned from his post in disappointment. The relation was, however, confirmed by the experiments of Gustav Theodor Fechner and Charles Wheatstone in the 1830s and 1840s. The central importance of Ohm's work was subsequently recognised, for example, with the award of the Copley Medal of the Royal Society of London in 1841. The status of the law remained somewhat insecure until the careful experiments of James Clerk Maxwell and George Chrystal in 1876, one of the first research results of the newly founded Cavendish Laboratory.

1.4.3 Electromagnetic induction

The results described in Sections 1.4.1 and 1.4.2 were known by 1830 and comprise the whole of *static electricity*, namely, the forces between *stationary* charges, magnets and currents. Over the succeeding 20 years, the basic experimental features of time-varying electric and magnetic fields were established, the hero of this story being Michael Faraday. He carried out his pioneering researches at the Royal Institution of Great Britain (RI) in London. The Institution was founded in 1799 by 58 influential figures in society, each of whom contributed 50 guineas to found an

... institution for diffusing the knowledge, and facilitating the general introduction, of useful mechanical inventions and improvements; and for teaching, by courses of philosophical lectures and experiments, the application of science to the common purposes of life. (James, 2000)

Among the few bona fide men of science was Henry Cavendish. The prospectus was drawn up by Benjamin Thompson, Count Rumford, who will reappear later in the story. Humphry Davy was appointed Professor of Chemistry in 1802 and vigorously promoted the objectives of the Institution through popular lectures on topics such as chemical agriculture and through his invention in 1815–16 of the miners' safety lamp. With the assistance of his influential supporters, Davy's laboratory became the best equipped in Britain. During the first decade of the nineteenth century, Davy pioneered the use of electrochemistry to isolate the chemical elements sodium, potassium, barium, calcium, magnesium and strontium.

Ørsted's announcement of the discovery of a connection between electricity and magnetism in 1820 caused a flurry of scientific activity. Many articles were submitted to the scientific journals describing other electromagnetic effects and attempting to explain them. Faraday had been appointed as chemical assistant to Davy in 1813 and quickly established his reputation as a brilliant experimental scientist. He was asked by the editor of the *Philosophical Magazine* to survey the mass of experiment and speculation about electricity and magnetism and, as a result, began his systematic study of electromagnetic phenomena. He soon invented the first electric motors using the force between a current-carrying wire and a dipolar magnetic field (Figure 1.4). These experiments led Faraday to the key concept of *magnetic lines of force*, which sprang from the observation of the patterns which iron filings take up about a magnet.

The great advance occurred in 1831. Believing firmly in the symmetry of nature, Faraday conjectured that, since an electric current could produce a magnetic field, it must also be possible to generate an electric current from a magnetic field. In 1831, he learned of Joseph Henry's experiments in Albany, New York, in which very powerful electromagnets were used. Faraday immediately had the idea of observing the strain in the material of a strong electromagnet caused by the lines of force. He built a strong electromagnet by winding an insulating wire, through which a current could be passed, onto a thick iron ring, thus creating a magnetic field within the ring. The effects of the strain were to be detected by another winding on the ring, which was attached to a galvanometer to measure the amount of electric current produced (Figure 1.5(*a*)).

The experiment was carried out on 29 August 1931 but the result was not at all what Faraday had expected. When the primary circuit was closed, there was a displacement of the galvanometer needle in the secondary winding – an electric current had been induced in the secondary wire through the medium of the iron ring. Deflections of the galvanometer were *only* observed when the current in the electromagnet was switched on and off, that is, with changing currents and, consequently, changing magnetic fields. This was the discovery of *electromagnetic induction*.

Over the next few weeks, there followed a series of remarkable experiments in which the nature of electromagnetic induction was elucidated. Faraday improved the sensitivity of his apparatus and observed that the sense of the electric current produced in the second circuit

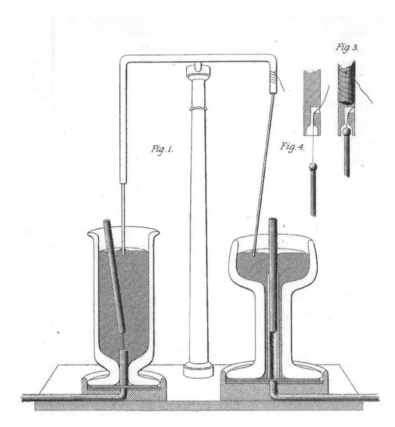

Fig. 1.4 Faraday's experiments illustrating the forces acting between a current-carrying wire and a magnet. To complete the electrical circuits, the vessels were filled with mercury. In the left side of the diagram, the current-carrying wire is fixed and the magnet rotates about the vertical axis; in the right side of the diagram, the magnet is fixed and the current-carrying wire rotates about the vertical axis. These were the first electric motors to be constructed (courtesy of the Royal Institution of Great Britian).

when the electric current was switched off was the opposite of that when it was switched on. Next, he tried coils of different shapes and sizes and discovered that the iron bar was not needed to create the effect. On 17 October 1831, a new experiment was carried out in which an electric current was created by sliding a cylindrical bar magnet into a long coil (or solenoid) connected to a galvanometer. Then, in a famous experiment demonstrated at the Royal Society of London on 28 October 1831, he showed how a continuous electric current could be generated by rotating a copper disc between the poles of the 'great horse-shoe magnet' which belonged to the Society. The axis and the edge of the disc were connected to a galvanometer and, as the disc rotated, the needle was deflected. A slightly later version of a similar experiment (1832) is shown in Figure 1.5(*b*). On 4 November 1831, Faraday found that simply moving a copper wire between the poles of the magnet could create an electric current. Thus, within a period of four months, he had discovered the *transformer* and the *dynamo* for generating electric currents. These were to prove crucial for industry and society within a generation.

Fig. 1.5

(*a*) The apparatus with which Faraday first demonstrated electromagnetic induction. Two insulated coils of copper wire were wound on an iron ring. Passing a current through one wire induced an electromotive force in the other by electromagnetic induction only when current was switched on and off. (*b*) Faraday's magnetic spark generator consisted of an insulated copper coil wound around a bobbin which was held between the poles of stacked permanent magnets. Rotating the bobbin rapidly produced a large enough voltage for a spark to be generated across the small gap between the open ends of the wire in air (both images courtesy of the Royal Institution and the Science Photo Library).

Faraday established the qualitative form of his law of induction in terms of the concept of lines of force: *the electromotive force induced in a current loop is directly related to the rate at which magnetic field lines are cut*, adding that

> By magnetic curves, I mean lines of magnetic force which would be depicted by iron filings.[4]

In 1834, Emil Lenz showed that the induced electromotive force always gives rise to a current, the magnetic field associated with which opposes the original change in magnetic flux (Lenz, 1834).

Faraday could not formulate his theoretical ideas mathematically, but the law of induction was put into mathematical form by Wilhelm Weber, who wrote the electromotive force $\mathcal{E}$ induced in a closed circuit as

$$\mathcal{E} = -\frac{\partial \Phi}{\partial t}, \tag{1.7}$$

where Φ is the magnetic flux enclosed by the circuit and the minus sign reflects Lenz's law. Faraday was convinced that the concept of lines of force provided the key to understanding electromagnetic phenomena. In 1846, he speculated in a discourse to the Royal Institution that light might be some form of disturbance propagating along the field lines. Two quotes from the lecture (Faraday, 1846) are revealing of his thinking:

> All I can say is, that I do not perceive in any part of space, whether (to use the common phrase) vacant or filled with matter, anything but forces and the lines in which they are exerted.

> My view . . . considers therefore radiation as a high species of vibration in the lines of force which are known to connect particles and also masses of matter together. It endeavours to dismiss the aether, but not the vibrations.

He published these ideas in a paper entitled *Thoughts on Ray Vibrations*, but they were received with considerable scepticism. Nonetheless, James Clerk Maxwell showed in his papers of 1861–62 and 1865 that light is indeed a form of electromagnetic radiation. With his outstanding physical intuition and mathematical ability, he was able to put Faraday's discoveries into mathematical form and then show that any electromagnetic wave propagating in a vacuum travels at the speed of light. In Maxwell's words,[5]

> As I proceeded with the study of Faraday, I perceived that his method of conceiving of phenomena was also a mathematical one, though not exhibited in the conventional form of mathematical symbols . . . I found, also, that several of the most fertile methods of research discovered by the mathematicians could be expressed much better in terms of ideas derived from Faraday than in their original form.

In a series of experiments carried out towards the end of 1845, Faraday attempted to find out if the polarisation of light could be influenced by the presence of a strong electric field, but these were unsuccessful. Turning instead to magnetism, his initial experiments were also unsuccessful, until he passed polarised light through lead borate glass, which has a large refractive index, in the presence of a strong magnetic field. He demonstrated the phenomenon now known as *Faraday rotation*, in which the plane of polarisation of linearly polarised light is rotated when the light rays travel along the magnetic field direction in the presence of a transparent dielectric. The phenomenon turned out to be closely related to the phenomenon of birefringence in crystals, in which the ordinary and extraordinary rays have different phase velocities, but now it is the right- and left-handed circularly polarised components, into which the linear polarisation can be decomposed, which have different phase velocities.[6] William Thomson interpreted this phenomenon as evidence that the magnetic field caused the rotational motion of the electric charges in molecules. Following an earlier suggestion by Ampère, he envisaged magnetism as being essentially rotational in nature, and this was to influence strongly Maxwell's model for the magnetic field in free space.

Maxwell's analysis of the empirical laws of electromagnetism while he was Professor of Natural Philosophy at Aberdeen and then at King's College, London led to his formulation of the equations for the electromagnetic field in 1861–62 and 1865 (Maxwell, 1861a,b, 1862a,b, 1865), one of the great triumphs of nineteenth-century physics.[7] In Maxwell's great paper of 1865, the equations were written out as 20 equations in 20 unknowns.[8] Maxwell's equations were reorganised into their more familiar compact form by Heaviside,

Table 1.2 Maxwell's comparison of measurements of the velocity of light and the ratio of electric units[a]

Velocity of light (m s^{-1})		Ratio of electric units (m s^{-1})	
Fizeau	314,000,000	Weber	310,740,000
Aberration, Sun's parallax	308,000,000	Maxwell	288,000,000
Foucault	298,360,000	Thomson	282,000,000

[a] Maxwell, 1873.

Helmholtz and Hertz in subsequent years:

$$\operatorname{curl} \boldsymbol{E} = -\frac{\partial \boldsymbol{B}}{\partial t}, \tag{1.8}$$

$$\operatorname{curl} \boldsymbol{H} = \boldsymbol{J} + \frac{\partial \boldsymbol{D}}{\partial t}, \tag{1.9}$$

$$\operatorname{div} \boldsymbol{D} = \rho_{\mathrm{e}}, \tag{1.10}$$

$$\operatorname{div} \boldsymbol{B} = 0. \tag{1.11}$$

$\boldsymbol{D}$ and $\boldsymbol{B}$ are the electric and magnetic flux densities, $\boldsymbol{E}$ and $\boldsymbol{H}$ are the electric and magnetic field strengths and $\boldsymbol{J}$ is the current density. The term $\partial \boldsymbol{D}/\partial t$ is the famous *displacement current*, originally introduced on the basis of Maxwell's mechanical model for material media, or vacua (Maxwell, 1861a,b, 1862a,b). All mention of the mechanical origins of his model disappeared from the final version of the theory in his great paper, *A Dynamical Theory of the Electromagnetic Field* (Maxwell, 1865).

In his papers of 1861–62, Maxwell demonstrated that electromagnetic waves are propagated at the speed of light, $c = (\epsilon_0 \mu_0)^{-1/2}$ *in vacuo*. This relation between the equivalents of the quantities ϵ_0 and μ_0 and the speed of light had already been noted empirically by Weber and Kohlrausch (1856) in their pioneering paper on the accurate determination of the relation between electrostatic and electromagnetic units of electric charge, and this was well known to Maxwell. A proof of this relation is contained in the notes.[9] Furthermore, Weber had already shown that electromagnetic waves propagate along wires at the speed of light, although his equations did not include Maxwell's displacement current term.

Table 1.2, taken from Maxwell's *Treatise on Electricity and Magnetism*, shows a comparison of various determinations of the speed of light, the first two columns showing the direct measurements of Fizeau, Foucault and others and the last two from the ratio of the electromagnetic to electrostatic units of electric charge (Maxwell, 1873). Maxwell was fully aware of the importance of the accurate determinations of the speed of light. Immediately following this table in the *Treatise*, he writes:

> It is manifest that the velocity of light and the ratio of the units are quantities of the same order of magnitude. Neither of them can be said to be determined as yet with such a degree of accuracy as to enable us to assert that the one is greater or less than the other.

It is to be hoped that, by further experiment, the relation between the magnitudes of the two quantities may be more accurately determined. (Maxwell, 1873)

This was to become one of the major themes of experimental research of the newly founded Cavendish Laboratory during the Maxwell and Rayleigh years. But this was of more than just theoretical importance. The requirements of the burgeoning electrical industries needed precise measurements of the physical properties of all types of materials. A crucial example is William Thomson's involvement in the laying of the first transatlantic telegraph cables.

1.4.4 William Thomson and the transatlantic telegraph cable

The developing relation between the needs of industry and commerce and the application of basic physics is splendidly illustrated by William Thomson's involvement in the laying of transatlantic telegraph cables.[10] These endeavours were to have far-reaching consequences for the researches of the Cavendish Laboratory under Maxwell and Rayleigh in the 1870s and 1880s. The story also illuminates Thomson's expertise in the techniques of continental mathematical analysis, which he had fostered in his courses in natural philosophy at Glasgow University.

Electric telegraphy became feasible once the ability to maintain currents in electric cables over long distances was established. By breaking the circuit, electric pulses could be transmitted down the cable and so coded signals could be sent electrically. The most famous means of coding messages was the Morse code invented by Samuel Morse in 1837. In Great Britain, the creation of what was to become a vast network of telegraph lines began in the same year. The first successful submarine cable between France and England was established in 1851, the year of the great International Exhibition at the Crystal Palace. The great challenge was the laying of the first transatlantic cables, which were of great commercial and economic significance but which required very large investments, involving the laying of 2000 miles of underwater cable.

From our perspective, the key issue concerned the time delay in the propagation of powerful enough signals over such large distances. It was known that there was a delay in the arrival of the signal and its dependence on the length and properties of the cable were estimated by experiment. The underwater cables were constructed with a central copper core surrounded by a sheath of gutta percha, which had been introduced about 1848 as an excellent and flexible insulator. In 1854–55, Thomson worked out the dependence of the delay time on the properties of the cable in an exchange of letters with George Stokes. Thomson used his formidable skills in mathematical physics to show that the delay is proportional to the square of the length of the cable. His analysis is full of interest and is reproduced here in its simplest form (Thomson, 1855).

Thomson recognised that a copper wire insulated by gutta percha is no more than what we would now call a coaxial cable, for which an elementary calculation shows that the capacitance per unit length c is

$$c = \frac{2\pi\epsilon\epsilon_0}{\ln\left(\dfrac{r_o}{r_i}\right)} \text{ (SI units)}, \quad \text{or} \quad c = \frac{\epsilon}{2\ln\left(\dfrac{r_o}{r_i}\right)} \text{ (esu units)}, \qquad (1.12)$$

where ϵ is the relative permittivity, or dielectric constant, of gutta percha and r_o and r_i are the outer radius of the gutta percha sheath and the radius of the copper core, respectively.

Next, Thomson wrote down the expression for the time evolution of the current in the wire. The rate of change of charge entering the element dx of the cable is

$$\frac{dQ}{dt}\,dx = -\sigma\frac{dI}{dx}\,dx, \qquad (1.13)$$

where the factor σ is the ratio of electrostatic to electromagnetic units in the cgs system (see Note 9 of Chapter 1). This term arises because, prior to the SI system, the unit of charge was defined by Coulomb's law of electrostatic attraction, whereas the current was defined by the force between parallel current elements, which were defined in electromagnetic units. In the modern SI system $\sigma = 1$ by definition, but in 1855 the values of unit electrostatic and electromagnetic charges had to be determined experimentally. The current I can be written in terms of the potential difference over the element of cable length dx by the relation $dV = -rI\,dx$, where r is the resistance per unit length of the cable, this relation being written in electromagnetic units. Therefore,

$$I = -\frac{1}{r}\frac{dV}{dx}. \qquad (1.14)$$

Substituting this result into (1.13), we obtain the partial differential equation for the rate of change of electrostatic potential along the wire,

$$c\frac{\partial V}{\partial t} = \sigma\frac{1}{r}\frac{\partial^2 V}{\partial x^2}, \qquad (1.15)$$

where we have used the relation $Q = cV$. Thomson immediately recognised that this equation is identical to the one-dimensional *heat diffusion equation* for the temperature T,

$$\frac{\partial T}{\partial t} = \frac{K}{\rho C}\frac{\partial^2 T}{\partial x^2}, \qquad (1.16)$$

where K is the thermal conductivity of the medium, ρ its density and C the specific heat capacity. In addition, he noted that the insulation of the wire would not be perfect and so he added a 'loss term' proportional to $-V$ to (1.15), which could then be written in the form

$$\frac{cr}{\sigma}\frac{\partial V}{\partial t} = \frac{\partial^2 V}{\partial x^2} - hV. \qquad (1.17)$$

By the method of separation of variables, this equation can be restored to the form of a diffusion equation by writing

$$V = \exp\left(-\frac{h\sigma}{cr}t\right)\phi, \qquad \frac{cr}{\sigma}\frac{\partial\phi}{\partial t} = \frac{\partial^2\phi}{\partial x^2}. \qquad (1.18)$$

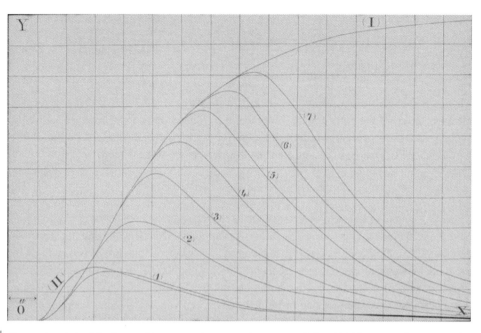

Fig. 1.6 Thomson's analysis of the change of the measured current at the remote end of a cable if the potential is maintained for a time na, where n takes the integral values shown in brackets beside each line (Thomson, 1855). The line marked (II) is the derivative of (I) with respect to x and corresponds to response of a very short pulse applied to the wire.

Solutions of this equation were well known to Thomson from Fourier's *Analytic Theory of Heat*, with which he was thoroughly familiar from his studies of the works of Fourier and the French school of mathematics.

First, we can use the method of dimensions to determine from the diffusion equation the dependence of the time T it would take the signal to travel a distance L. Writing $\partial/\partial t \equiv 1/T$ and $\partial/\partial x^2 \equiv 1/L^2$, we find

$$T \propto rcL^2/\sigma. \tag{1.19}$$

This is Thomson's demonstration that the time delay of an electrostatic pulse varies as the square of the length of the cable.

In his paper of 1855, he used the full power of Fourier's analytic techniques to determine not only the delay time of the signal, but also to work out how the current builds up at the far end of the cable. The solutions of Thomson's analysis are shown in Figure 1.6, in which the abscissa x is time measured in units of

$$a = \frac{rcl^2}{\sigma \pi^2} \ln\left(\frac{4}{3}\right). \tag{1.20}$$

The lines on the graph show the time dependence of the current received at the remote end of the cable if a steady driving voltage is maintained for a time na, where n is the figure in brackets beside each curve, (1), (2), (3), ... If the current were maintained for an infinite time, the curve labelled (I) would be obtained. The response to a very short pulse, a

δ-function input, is found from the derivative of (I) with respect to x and is labelled (II). After switching off, there is a roughly exponential decay of the signal. The current at the remote end of the cable increases as the duration of the input voltage pulse increases. Notice that the detected current at the remote end is negligibly small until after time a and that, after about $5a$, the current reaches roughly half its final strength.

Thomson obtained from the Astronomer Royal, George Airy, retardation times and the dimensions for the cables from Greenwich to Edinburgh, Greenwich to Brussels and London to Manchester and found that his scaling relation was in good agreement with the measured delay times. The ratio of the electrostatic to electromagnetic units σ was, however, quite poorly known. In his paper of 1855, Thomson estimated that σ should lie between 419,000,000 and 104,000,000 m s^{-1}. He was well aware of the commercial importance of understanding the theory of the telegraph cable, since these considerations determined the costs involved. As he wrote in his paper,

> We may infer that the time required to reach a stated fraction of the maximum strength of current at the remote end will be proportional to (rcl^2/σ). We may be *sure* beforehand that the American telegraph will succeed, with a battery sufficient to give a sensible current at the remote end, when kept long enough in action; but the time required for each deflection will be sixteen times as long as would be with a wire a quarter of the length, such, for instance, as in the French submarine telegraph to Sardinia and Africa . . .
>
> It will be an economical problem, easily solved by the ordinary analytic method of maxima and minima, to determine the dimensions of wire and covering which, with stated prices of copper, gutta percha, and iron, will give the stated rapidity of action with the smallest initial expense. (Thomson, 1855)

Prior to the laying of the first cable, a dispute arose between Thomson and Wildman Whitehouse, a surgeon turned electrical engineer, who was to be given the role of electrician for the laying of the transatlantic cable. Whitehouse was a practical engineer who relied upon somewhat rudimentary data to infer that the delay in the signal was simply proportional to the length of the cable. A capital investment of £350,000 was raised for the laying of the cable by the Atlantic Telegraph Company, which was registered in 1856. Thomson became a director of the company.

On the basis of his analysis which led to (1.12) and (1.19), Thomson recommended that, to reduce the delay time, the thickness of the copper core should be increased, which would have the beneficial effect of reducing both the resistance r and the capacitance c per unit length. But this would have increased both the cost and mass of the cable. These recommendations ran counter to the recommendations of Whitehouse, who overruled Thomson and went ahead with the manufacture and laying of 2500 miles of cable, about 1 cm in diameter and weighing about one ton per nautical mile. After numerous tribulations, the cable was successfully laid from Valencia Island off the south-west coast of Ireland to Trinity Bay in Newfoundland, but it took 20 minutes to relay a simple sentence and was only detectable thanks to a very sensitive galvanometer developed by Thomson. The cable deteriorated rapidly and was no longer useable in a matter of weeks.

At the subsequent inquest set up by the Board of Trade and the Atlantic Telegraph Company in 1859, Whitehouse became the scapegoat for the disaster. The cause of the failure was partly faulty insulation, partly the difficulty of deploying the cables, but mostly because Whitehouse had used very high voltages of 2000 V, which would have caused sparks a quarter of an inch long in air, to make the messages detectable at the far end. Smith and Wise write that Whitehouse essentially used a brute-force method to make the signals detectable, and comment that

> Whitehouse's rejection of theory – or at least his failure to grasp theory – thus led him to adopt an empirical approach which, though successful enough with small-scale projects, proved disastrously expensive when used on the Atlantic cable. To ignore theory on such a project was a luxury which simply could no longer be afforded. The 2000 mile cable was not a cheap, expendable piece of apparatus, but an extremely valuable commercial property. (Smith and Wise, 1989)

Thomson's analysis of the disaster revealed a number of major deficiencies in the processes of manufacture of the cable and its deployment. Among the problems were the fact that the conductivity of samples of commercial copper wire could differ by a factor two, with a corresponding impact upon the rate of transmission of information. In August 1863, the Committee recommended a new manufacturer for the cable and a capital of £1 million was secured for its manufacture and laying. The new cable weighed 6000 tons and the only vessel capable of laying it was the gigantic but laid-up monster, the *Great Eastern*, built by Isambard Kingdom Brunel and John Scott Russell. The laying of the cable was again not without its vicissitudes, but in 1866 the project was successfully completed. In recognition of his efforts, Thomson was knighted at Windsor Castle on 10 November 1866. As Sviedrys (1976) has written,

> [The knighthood] was an unusual distinction for a University professor, and it enhanced his stature in the eyes of University officials who were at the time drawing up plans to transfer the University to a new site. Realising that Thomson had placed the University of Glasgow in the forefront of British science, they acknowledged their debt by providing him with excellent experimental facilities in their new building and thus conferred official status on his laboratory.

In 1870, Thomson moved into his new laboratory, consisting of six specially adapted and well-lighted rooms. Equally significant was the fact that Thomson acquired considerable wealth from his involvement in these and other commercial enterprises, building a mansion at Largs on the River Clyde where he moored his 126-ton schooner, the *Lalla Rookh*.

The history of these events has been told in some detail since they were to have an important bearing on the decision to found the Cavendish Laboratory and on the nature of its research programme under Maxwell and Rayleigh. Let us list some of the more important of these.

- Perhaps the most significant impact is best summarised in the words of Smith and Wise:

> Science-based industry had here made an explicit appearance, complete with the emphatic priority of science education over practical experience in the training of

technical personnel, well-paid employment as the reward for such training, and finance from industry to fund the research laboratories of university professors. (Smith and Wise, 1989)

As they remark earlier,

> Economic historians have long spoken of a second industrial revolution in the last third of the nineteenth century, referring to the emergence of science-based industries, notably those producing and utilising chemical fertilisers, synthetic dyes, and electric power. (Smith and Wise, 1989)

This was the background to the foundation of a number of physics laboratories throughout the UK and Europe. For example, Werner von Siemens founded his telegraph company with Johann Halske in 1847 and this developed into a major international telegraph company. In 1866, he applied his experimental skills to the commercial manufacture of electric dynamos, leading to the establishment of the discipline of power electrical engineering. His entrepreneurial skills extended to electric lighting and telephone networks.

- Thomson's papers demonstrated the power of theoretical analysis in supporting the needs of industry. At the same time, the story reveals how poorly the fundamental constants of nature, such as ϵ_0 and μ_0, as well as the properties of materials, were known at the time. These needs were to be reflected in the institution of science laboratories and national institutions dedicated to the characterisation of materials and the establishment and maintenance of standards.

- Thomson made imaginative and fluent application of advanced continental mathematics in his elucidation of the scaling laws and the solutions of the equations which were to be at the heart of the application of theory to research and industry.

- Thomson applied the discipline of natural philosophy – what we would now call physics – to the understanding of practical problems. In the courses of instruction which he developed at Glasgow University and which Peter Guthrie Tait instituted in Edinburgh, the students acquired skills in both theoretical and experimental physics, although the practical element was not nearly as systematic as it was to become in the last decades of the century. It is no surprise that Thomson and Tait's *Treatise on Natural Philosophy* (1867) and Maxwell's *Treatise on Electricity and Magnetism* (Maxwell, 1873), both having their methodological and philosophical roots in the influential Scottish Enlightenment, should have attained a special position in the education of physicists in the last third of the nineteenth century. Nor is it a coincidence that so many inventions and discoveries were made by Scottish engineers, inventors and scientists in the nineteenth century.

- Finally, the dates should be noted. Maxwell was just about to begin work on his epochal papers on the theory of the electromagnetic field, which would eventually lead to his equations in the period 1861–65. Thomson's analysis did not include the displacement current and, indeed, he argued strongly against its inclusion in the electromagnetic field equations since he regarded it as a purely theoretical construct.

1.5 The laws of thermodynamics

Towards the end of the eighteenth century, a controversy developed concerning the nature of heat. The idea of heat as a form of motion can be traced back to the atomic theory of the Greek philosopher Democritus in the fifth century BC, who considered matter to be made up of indivisible particles whose motions gave rise to the myriad of different forms in which matter is found in nature. These ideas resonated with the thoughts of natural philosophers during the seventeenth and eighteenth centuries, advocates including Francis Bacon, Robert Boyle, Thomas Hooke, Isaac Newton, Gottfried Liebniz, Henry Cavendish, Thomas Young and Humphry Davy, as well as the seventeenth-century philosophers Thomas Hobbes and John Locke.

During the eighteenth century, an alternative picture gained currency, its origin being derived from the concept of *phlogiston*. The idea was that this hypothetical element could be combined with a body to make it combustible. There was a body of opinion that heat should be considered a physical substance. Lavoisier and others named this 'matter of heat' *caloric*, which was to be thought of as an 'imponderable fluid', that is, a massless fluid, which was conserved when heat flowed from one body to another. Its flow was conceived as being similar to the flow of water in a waterwheel, in which mass is conserved. There were, however, problems with the caloric theory – for example, when a warm body is brought into contact with ice, caloric flows from the warm body into the ice but, although ice is converted into water, the temperature of the ice–water mixture remains the same. It had to be supposed that caloric could combine with ice to form water. The theory attracted the support of a number of distinguished scientists including Joseph Black, Pierre-Simon Laplace, John Dalton, John Leslie, Claude Berthollet and Johan Berzelius.

The two theories came into conflict at the end of the eighteenth century. Among the most important evidence against the caloric theory were the experiments of Benjamin Thompson, Count Rumford. His great contribution to thermodynamics was the realisation that heat could be created in unlimited quantities by the force of friction. This phenomenon was mostly vividly demonstrated by his experiments of 1798 in which cannons were bored with a blunt drill. In his famous experiment, he heated about 19 pounds of water in a box surrounding the cylinder of the cannon and after two and a half hours, the water boiled. The summit of Rumford's achievement was the first determination of the mechanical equivalent of heat, which he found from the amount of work done by the horses in rotating the drilling piece. In modern terms, he found a value of 5.60 J cal^{-1}. It was to be more than 40 years before Mayer and Joule returned to this problem.

Among the key contributions in the understanding of the nature of heat was Jean Baptiste Joseph Fourier's treatise *Théorie Analytique de la Chaleur*, published in 1822 (Fourier, 1822). Fourier worked out the mathematical theory of heat transfer in terms of partial differential equations which did not require the construction of any specific model for the physical nature of heat. Fourier's methods were firmly based on the French tradition of rational mechanics and gave mathematical expression to the *effects* of heat without enquiring into its *causes*.

1.5.1 The first law of thermodynamics

The idea of conservation laws for mechanical systems had been discussed in eighteenth-century treatises, in particular in the work of Liebniz. He argued that 'living force' or *vis viva*, which we now call kinetic energy, is conserved in mechanical processes. By the 1820s, the relation of kinetic energy to the work done was clarified. In the 1840s, a number of scientists independently came to the correct conclusion about the interconvertibility of heat and work, following the path pioneered by Rumford. One of the first was Julius Mayer, who proposed the equivalence of work and energy in 1842 on the basis of experiments on the work done in compressing and heating a gas. He elevated his results to the principle of the conservation of energy and from the adiabatic expansion of gases derived an estimate for the mechanical equivalent of heat of 3.58 J cal^{-1}. His pioneering works were somewhat overshadowed by those of James Joule, whose experiments were carried out in a laboratory which his father had built for him next to the family brewery. Joule's genius as a meticulous experimenter was to put the science of heat on an exact experimental basis.

The results of Joule's paddle-wheel experiments greatly excited the young William Thomson in 1847. At the Oxford meeting of the British Association for the Advancement of Science, Thomson heard Joule describe the results of his recent experiments on the mechanical equivalent of heat. Up until that time, Thomson had adopted the assumption made by Carnot that heat (or caloric) is conserved in the operation of heat engines. He found Joule's results astonishing and, as his brother James wrote, Joule's 'views have a slight tendency to unsettle one's mind'. Joule's results became widely known in continental Europe and, by 1850, Helmholtz and Clausius had formulated what is now known as the *law of conservation of energy*, or the *first law of thermodynamics*. Helmholtz in particular was the first to express the conservation laws in mathematical form in such a way as to incorporate mechanical and electrical phenomena, heat and work. In 1854, Thomson invented the word *thermo-dynamics* to describe the new science of heat to which he was to make major contributions.

1.5.2 The second law of thermodynamics: James Watt and the steam engine

The first law of thermodynamics proved to be much more troublesome historically than the second law. After 1850, the first law enabled a logically self-consistent definition of heat to be formulated, and the law of conservation of energy was given a firm conceptual basis. There must, however, be other restrictions upon thermodynamic processes, in particular the rules about the directions in which heat flows and thermodynamic systems evolve. These are derived from the *second law of thermodynamics*, which can be expressed in the following form presented by Rudolph Clausius:

> No process is possible whose sole result is the transfer of heat from a colder to a hotter body.

It is remarkable that such powerful results can be derived from this simple statement. Equally remarkable is how the fundamentals of thermodynamics were derived from the operation of steam engines. As L.J. Henderson remarked (Forbes, 1958),

> Until 1850, the steam engine did more for science than science did for the steam engine.

Let us illustrate the truth behind this perceptive statement.

The invention of the steam engine was one of the most important developments which laid the foundations for the industrial revolution which swept through developed countries from about 1750 to 1850. Before that time, the principal sources of power were waterwheels, which had to be located close to rivers and streams, and wind power, which was unreliable. Horses were used to provide power, but they were expensive. The importance of the steam engine was that power could be generated at the location where it was needed.

The first commercially viable and successful steam engine was built by Thomas Newcomen in 1712 in order to pump water out of coal mines and to raise water to power waterwheels. His steam engine operated at atmospheric pressure. The principle of operation was that the cylinder, fitted with a piston, was filled with steam, which was then condensed by injecting a jet of cold water. The resulting vacuum in the cylinder caused the piston to descend, resulting in a power stroke which was communicated to the pump by a mechanical arrangement. When the power stroke was completed, the steam value was opened and the weight of the pump rod pulled the piston back to the top of the cylinder.

James Watt was trained as a scientific instrument maker in Glasgow. While repairing a model Newcomen steam engine in 1764, he was impressed by the waste of steam and heat in its operation. In May 1765, he made a crucial invention which was to lead to the underpinning of the whole of thermodynamics. He realised that the efficiency of the steam engine would be significantly increased if the steam condensed in a separate chamber from the cylinder. This invention of the *condenser* was perhaps his greatest innovation, for which he took out his famous patent in 1768.

Figure 1.7 shows one of the steam engines built by Watt in 1788. The engine operated at atmospheric pressure. The key innovation was the separation of the condenser F from the cylinder and piston E. The cylinder was insulated and kept at a constant high temperature T_1 while the condenser was kept cool at a lower temperature T_2. The condenser was evacuated by the air pump H. The steam in E condensed when the condenser valve was opened and the power stroke took place. Steam was then allowed to fill the upper part of the cylinder and, once the stroke was finished, was also allowed into the lower section of the cylinder, so that the weight of the pump rod restored the piston to its original position.

Pressure to increase the efficiency of steam engines came from the mine owners, who sought to reduce the costs of pumping operations. In 1782, Watt patented a double-acting steam engine in which steam power was used in both the power and the return strokes of the piston, resulting in twice the amount of work done. This required a new design of the parallel motion bars, so that the piston could push as well as pull. Watt referred to this as 'one of the most ingenious, simple pieces of mechanism I have contrived'. There was also a governor to regulate the supply of steam. In the version of the steam engine shown in Figure 1.8, the reciprocating motion of the piston was converted into rotary motion by

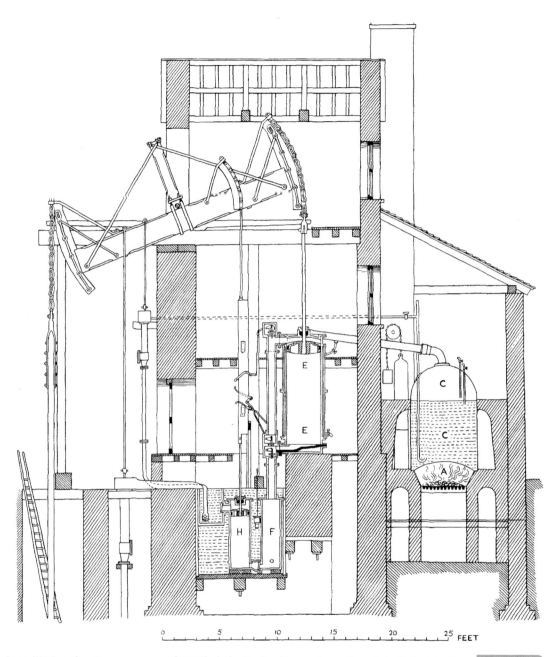

James Watt's single-acting steam engine of 1788. The boiler C is placed in an outhouse. The cylinder E is kept at a high temperature all the time. F is the separate condenser and H an air pump (Dickenson, 1958). **Fig. 1.7**

the sun-and-planet gear. As a result of these innovations, by 1792 the thermal efficiency had increased from 0.5% for Newcomen's steam engine to 4.5% for Watt's double-acting engine. But what was the theoretical maximum efficiency which could be achieved? The answer was provided by Sadi Carnot.

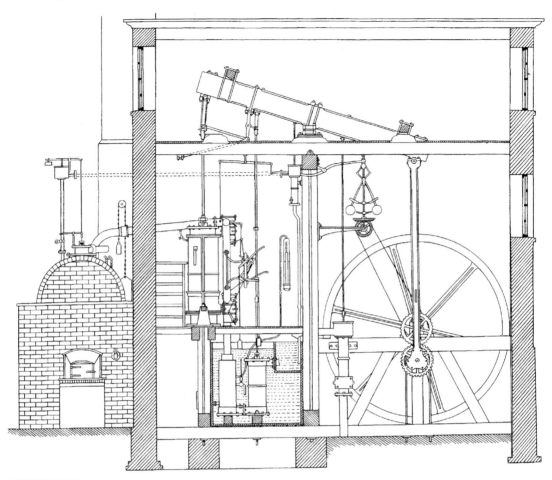

James Watt's double-acting rotative steam engine of 1784. The parallel motion bars are seen on the raised end of the beam. The sun-and-planet gear converts the reciprocating action of the engine into rotary motion (Dickenson, 1958).

1.5.3 Sadi Carnot and the *Réflexions*

The young Sadi Carnot was concerned that France had fallen well behind the United Kingdom in the technology of the steam engine. Although reliable estimates of the efficiencies of heat engines had been presented, those who had developed them were practical inventors and engineers without any deeper appreciation of the fundamental theory of heat engines. His great work *Réflections sur la Puissance Motrice du Feu et sur les Machines Propres à Developper cette Puissance* was published in 1824 when he was 28 years old (Carnot, 1824). The treatise concerned a theoretical analysis of the maximum efficiency of heat engines.

In seeking to derive a general theory of heat engines, he was guided by his father's premise of the impossibility of perpetual motion. In the *Réflexions*, he adopted the caloric theory of heat and assumed that caloric was conserved in the cyclic operation of heat engines. He postulated that the transfer of caloric from a hotter to a colder body is the origin of the work done by a heat engine. The flow of caloric was envisaged as being analogous to the flow of fluid which, as in a waterwheel, can produce work when it falls down a potential gradient. Carnot's basic insights into the operation of heat engines were two-fold. First, he recognised that a heat engine works most efficiently if the transfer of heat takes place as part of a cyclic process, the caloric being transferred by a 'working substance'. The second was that the crucial factor in determining the amount of work which can be extracted from a heat engine is the temperature difference between the source of heat and the sink into which the caloric flows. It turns out that these basic ideas are independent of the particular model of the heat flow process.

By another stroke of imaginative insight, he devised the cycle of operations now known as the *Carnot cycle* as an idealisation of the behaviour of any heat engine. The key feature of the Carnot cycle is that all the processes are carried out *reversibly* so that, by reversing the sequence of operations, work can be done on the system and caloric transferred from a colder to a hotter body. By joining together an arbitrary heat engine and a reversed Carnot heat engine, he demonstrated that no heat engine can ever produce more work than a reversible Carnot heat engine. Otherwise, by joining the two engines together, heat could be transferred from a colder to a hotter body without doing any work, or work could be produced without any net heat transfer, both phenomena being in violation of common experience.

In 1834, Emile Clapeyron reformulated Carnot's arguments analytically and related the ideal Carnot engine to the standard pressure–volume indicator diagram (Figure 1.9). Inspired by Clapeyron's paper, William Thomson went back to Carnot's original analysis in the *Réflexions*. The big problem for Thomson and others was to reconcile Carnot's work, in which caloric was conserved, with Joule's experiments which demonstrated the inter-convertibility of heat and work. The matter was resolved by Clausius and Thomson, who showed that Carnot's theorem concerning the maximum efficiency of heat engines was correct, but that the assumption of no heat loss was wrong – there is a conversion of heat into work in the Carnot cycle. The reformulation by Clausius constitutes the bare bones of the second law of thermodynamics. The law goes far beyond the efficiency of heat engines and serves not only to define the thermodynamic temperature scale through the relation $Q_2/Q_1 = T_2/T_1$, but also resolves the problem of how systems evolve thermodynamically.

Clausius solved the problem by introducing a new function of state, the *entropy*, in 1865. Taking the heats Q_1 and Q_2 to be positive quantities, for any reversible path between the coordinates (p_1, V_1) and (p_2, V_2),

$$\frac{Q_1}{T_1} - \frac{Q_2}{T_2} = 0. \tag{1.21}$$

Whichever reversible path is taken, the same result is always found. Mathematically, for any two points A and B,

$$\sum_A^B \frac{dQ}{T} = \text{constant.} \tag{1.22}$$

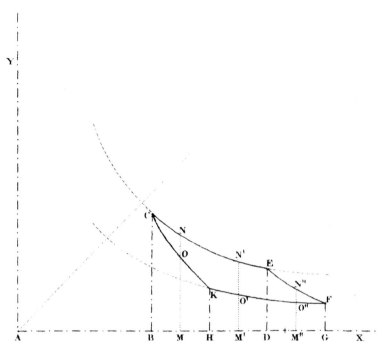

Fig. 1.9   Clapeyron's indicator diagram illustrating the operation of an ideal heat engine according to the principles enunciated by Carnot. The abscissa is volume and the ordinate is pressure. The segment C to E is an isothermal expansion at constant temperature T_2 and that from F to K an isothermal compression at constant temperature T_1. The heat Q_2 enters the working substance at T_2 and is removed from it at T_1. The segment E to F is an adiabatic expansion and K to C an adiabatic compression (Clapeyron, 1834).

Writing this in integral form,

$$\int_A^B \frac{dQ}{T} = \text{constant} = S_B - S_A. \tag{1.23}$$

S is the *entropy* of the system. The generalisation to any complete circuit in the indicator diagram is $\oint dQ/T \leq 0$ and is known as *Clausius's theorem*. From this relation, it follows that, when irreversible processes are present in the cycle, the entropy of the Universe as a whole increases. It is always important to consider the surroundings as well as the system itself in this type of argument.[11]

These examples illustrate the origin of the popular expression of the two laws of thermodynamics due to Clausius:

1. The energy of the Universe is constant.
2. The entropy of the Universe tends to a maximum.

The first and second laws of thermodynamics could be combined into a single relation,

$$T\,dS = dU + \sum_i X_i\,dx_i, \tag{1.24}$$

where the terms in $X_i\,dx_i$ represent all types of work done in generalised coordinates. The remarkable thing about this formula is that the first and second laws are written entirely in terms of functions of state, and hence the relation must be true for all changes. In particular, for any gas, we can write

$$\left(\frac{\partial S}{\partial U}\right)_V = \frac{1}{T},\tag{1.25}$$

that is, the partial derivative of the entropy with respect to the internal energy U at constant volume defines the thermodynamic temperature T. This result would prove to be of particular significance in the development of statistical mechanics.

1.6 Atoms and molecules

Statistical mechanics grew out of the need to develop techniques for constructing a theory which would enable the macroscopic behaviour of matter to be understood in terms of the average properties of vast numbers of interacting atoms and molecules. In the last decades of the eighteenth century, Antoine-Laurent de Lavoisier, the 'father of modern chemistry', had put the science of chemical processes on a firm quantitative experimental basis, including the law of conservation of mass in chemical reactions. In 1808, John Dalton published his treatise on *A New System of Chemical Philosophy* (Dalton, 1808), in which he asserted that the ultimate particles, or atoms, of a chemically homogeneous substance all had the same weight and shape, and drew up a table of the relative weights of the atoms of a number of simple substances (Figure 1.10). In 1811, Amadeo Avogadro realised that the physical unit was not a single atom but a cluster of a small number of atoms, which he defined as *molecules*, namely the smallest particles of the gas which move about as a whole. *Avogadro's hypothesis* stated that:

> Equal volumes of all gases under the same conditions of temperature and pressure contain the same numbers of molecules.

Avogadro's hypothesis was disregarded by chemists for almost 50 years until 1858 when Stanislao Cannizzaro convinced the leading chemists of the validity of the hypothesis.

A major preoccupation was putting some order into the properties of the chemical elements. In 1789, Lavoisier published his list of 33 chemical elements and grouped them into gases, metals, non-metals and earths. It was known that certain groups of elements had similar chemical properties – for example, the alkali metals sodium, potassium and rubidium and the halogens chlorine, bromine and iodine. Dmitri Mendeleyev in 1869 and Julius Meyer in 1870 independently put order into the chemical properties of the elements and published what became known as the *periodic table* of the elements (Figure 1.11).

A major issue for nineteenth-century physics concerned the reality of atoms. Did they really exist or were they simply a convenient calculating tool for describing the quantities of the different substances involved in chemical processes? The issue was not of particular concern for the chemists, but for the physicists, it was closely related to the understanding of the nature of heat. The caloric theory of Lavoisier supposed that heat was a substance which could combine with the chemical elements, while according to the kinetic theory heat was associated with the motions of atoms and molecules. John Waterston, Clausius and Maxwell had no hesitation in developing a kinetic theory of gases, considering them to

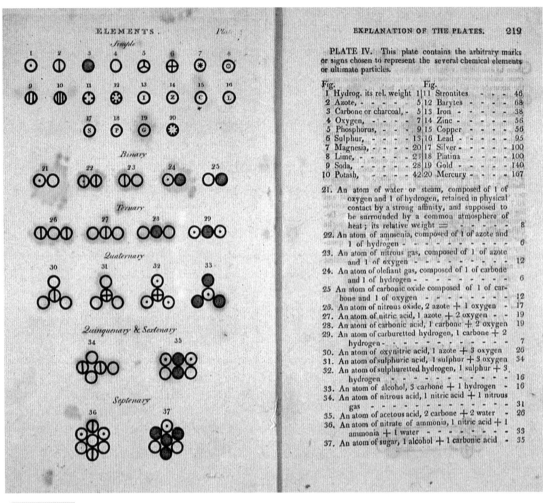

Fig. 1.10 Dalton's symbols for the atoms of various elements and their compounds (Dalton, 1808).

consist of vast numbers of particles making continuous elastic collisions with one another and with the walls of the containing vessel. Clausius provided the first systematic account of the theory in 1857 in his great paper entitled *On the Nature of the Motion, Which We Call Heat* (Clausius, 1857). He succeeded in deriving the equation of state of a monatomic gas by working in terms of the mean velocities of the particles. Whilst accounting for the perfect gas law admirably, it did not give good agreement with the known values of the ratios of the specific heat capacities, $\gamma = C_p/C_V$, of molecular gases, where C_p and C_V are the specific heat capacities at constant pressure and constant volume, respectively. From experiment, γ was found to be 1.4 for molecular gases, whereas the kinetic theory predicted $\gamma = 1.67$. In the last sentence of his paper, Clausius recognised the important point that there must therefore exist some other means of storing kinetic energy within molecular gases which can increase the internal energy per molecule.

Maxwell ran up against exactly the same problem in 1860 in a characteristically novel and profound series of papers entitled *Illustrations of the Dynamical Theory of Gases*

ОПЫТЪ СИСТЕМЫ ЭЛЕМЕНТОВЪ.

ОСНОВАННОЙ НА ИХЪ АТОМНОМЪ ВѢСѢ И ХИМИЧЕСКОМЪ СХОДСТВѢ.

```
                    Ti = 50    Zr =  90    ? = 180.
                    V  = 51    Nb =  94    Ta = 182.
                    Cr = 52    Mo =  96    W  = 186.
                    Mn = 55    Rh = 104,4  Pt = 197,4.
                    Fe = 56    Rn = 104,4  Ir = 198.
                  Ni=Co = 59   Pl = 106,6  O- = 199.
  H = 1             Cu = 63,4  Ag = 108    Hg = 200.
       Be =  9,4 Mg = 24  Zn = 65,2  Cd = 112
        B = 11   Al = 27,4  ? = 68   Ur = 116  Au = 197?
        C = 12   Si = 28    ? = 70   Sn = 118
        N = 14    P = 31   As = 75   Sb = 122  Bi = 210?
        O = 16    S = 32   Se = 79,4 Te = 128?
        F = 19   Cl = 35,6 Br = 80    I = 127
  Li = 7 Na = 23   K = 39  Rb = 85,4 Cs = 133  Tl = 204.
                  Ca = 40  Sr = 87,6 Ba = 137  Pb = 207.
                   ? = 45  Ce = 92
                  ?Er = 56 La = 94
                  ?Yt = 60 Di = 95
                  ?In = 75,6 Th = 118?
```

Д. Менделѣевъ

Mendeleyev's original version of the periodic table of 1869 (Mendeleyev, 1869). The question marks indicate unknown elements inserted so that similar elements would lie along the same row.

Fig. 1.11

(Maxwell, 1860a,b,c). In a few brief paragraphs he derived the formula for the velocity distribution of the particles of the gas and introduced statistical concepts into the kinetic theory and thermodynamics. If only the translational degrees of freedom are taken into account, the value of γ should be 1.67. Maxwell also considered the case in which the rotational degrees of freedom of non-spherical molecules were taken into account as well as their translational motions, but this calculation resulted in a ratio of specific heat capacities $\gamma = 1.33$, again inconsistent with the value 1.4 observed in the common molecular gases. In the last sentence of his great paper, he made the discouraging remark:

> Finally, by establishing a necessary relation between the motions of translation and rotation of all particles not spherical, we proved that a system of such particles could not possibly satisfy the known relation between the two specific heats of all gases.

But matters soon became progressively worse and more complex. The discovery of emission and absorption lines in the spectra of atoms and molecules suggested that there were resonances within atoms and molecules which were characteristic of each species.

Presumably these resonances could store energy within the atoms and molecules of the substance and, if this were the case, the numbers of degrees of freedom available for storing energy within atoms and molecules would increase dramatically, with the consequence that the ratio of specific heat capacities γ would tend to unity. Although not fatal for the kinetic theory, this was undoubtedly a setback and cast doubt upon the validity of the kinetic theory of gases, although most theorists favoured that view. It would not deter Ludwig Boltzmann and Josiah Willard Gibbs from developing highly sophisticated treatments of the theory of statistical mechanics. It would, however, require the concepts of the early quantum theory of Einstein of 1905–06 to account for these discrepancies.

1.7 Reflections

The broad-brush picture of physics in the nineteenth century presented in the last six sections takes the story up to about 1870 when the foundation of a physics laboratory in Cambridge, which was to become the Cavendish Laboratory, was firmly on the agenda of Cambridge University. The project had, however, been put in abeyance because of the lack of funds to provide the building and a Professorship of Experimental Physics.

But what is particularly striking about the history related in this chapter is the singular lack of significant contributions from Cambridge after the epochal contributions of Newton in the seventeenth century. This is vividly demonstrated by the contents of Table 1.1. There were indeed significant contributions by British physicists, but none from Cambridge. The situation is summarised by Whittaker in his monumental *History of the Theories of the Aether and Electricity* (Whittaker, 1951):

> The century which elapsed between the death of Newton and the scientific activity of Green was the darkest in the history of [Cambridge] University. It is true that [Henry] Cavendish and [Thomas] Young were educated at Cambridge; but they, after taking their undergraduate courses, removed to London. In the entire period the only natural philosopher of distinction was [John] Michell; and for some reason which at this distance of time it is difficult to understand fully, Michell's researches seem to have attracted little or no attention among his collegiate contemporaries and successors, who silently acquiesced when his discoveries were attributed to others, and allowed his name to perish entirely from the Cambridge tradition.[12]

How this came about and the forces which led to the remarkable resurgence of Cambridge physics are the subjects of the next chapter.

Mathematics and physics in Cambridge in the nineteenth century

2.1 Pure and mixed mathematics at Cambridge

The Galilean revolution culminated in Newton's great *Philosophiæ Naturalis Principia Mathematica* of 1687. The words of the title are significant: Newton had enunciated the 'mathematical principles of natural philosophy'. In the course of his elucidation of the basic laws of mechanics, dynamics and gravity, he invented his own version of the differential calculus, the method of fluxions. In keeping with the conventions of the time, however, Newton wrote the *Principia* entirely in geometrical terms.[1] Newton's book on the *Method of Fluxions* was completed in 1671 but only published posthumously in 1736. In the meantime, Gottfried Leibniz had independently developed his own version of the calculus about 1673, his book being published in 1684, 50 years before Newton's text. An unpleasant priority dispute developed between Newton and Liebniz. From our present perspective, the significant aspect of the different, but equivalent, approaches was that Newton's fluxions remained the preferred technique in England, particularly in Cambridge, while in continental Europe Leibniz's notation and techniques were adopted and are those commonly used today.

To understand the development of physics and mathematical physics in Cambridge, an important distinction has to be made between two different approaches to the discipline of mathematics. Francis Bacon and Jean d'Alembert made a clear distinction between *pure mathematics* and what was called *mixed mathematics* (Brown, 1991). According to d'Alembert's tree of knowledge, pure mathematics consisted of arithmetic and geometry, the former being divided into elementary arithmetic and algebra, which in turn included elementary algebra and differential algebra, the latter including the differential and integral calculus. Geometry included elementary geometry as applied in the fields of the 'Military, Architecture and Tactics', and transcendental geometry, which concerned the theory of curves. In contrast, mixed mathematics had as its domain

- mechanics, including statics and dynamics
- geometric astronomy, which included cosmography, chronology and gnomonics, meaning the science of sundials
- optics, acoustics and pneumatics
- the art of conjecturing, which meant the analysis of games of chance.

In modern terms, the distinction between pure and mixed mathematics corresponds more or less to that between pure and applied mathematics.

During the second half of the eighteenth century and the nineteenth century, continental mathematicians made very significant advances in the field of analysis, meaning the branches of pure mathematics which include the theories of differentiation, integration and measure, limits, infinite series and analytic functions. Many of these achievements involved providing a secure, purely mathematical foundation for all fields of analysis and so belonged to the realm of pure mathematics. As described by Forfar (1995), the list of French and German mathematicians who pioneered these studies through the period 1753–1909 is formidable: in Germany the names of Gauss, Bessel, Jacobi, Dirichlet, Kummer, Riemann, Dedekind, Kronecker, Weierstrass, Cantor, Klein, Hilbert, Landau and Weyl, and in France, of d'Alembert, Lagrange, Laplace, Legendre, Fourier, Poisson, Cauchy, Liouville, Galois, Hermite, Bertrand, Jordan, Poincaré, Hadamard, Cartan, Borel and Lebesgue, are standard fixtures in mathematical textbooks. The advances in the techniques applicable to natural philosophy by Lagrange, Fourier and Poisson were of particular significance for the study of mechanical and dynamical problems in the early nineteenth century, and also offered new routes for the study of heat, Fourier's *Théorie Analytique de la Chaleur* of 1822 being but one example.

In England, the field of mathematics was dominated by Cambridge and Newton's formidable legacy. Within the University, mathematics was held in particular esteem, and, once the system of written examinations was introduced in the mid eighteenth century, mathematics provided an ideal subject for these purposes. The aim was to train undergraduates in the discipline of logical thinking, preparing their minds for whatever discipline they would ultimately follow. Well into the nineteenth century, students were required to obtain an honours degree in mathematics before they could proceed to studies in the classics, the humanities or training for the Church, which was one of the principal career options open to graduates of the University. The examinations had to cater for students of a wide range of abilities, the serious students, known as 'reading-students', aiming for the highest distinction in the demanding Senate House Examinations. Those students in the top two of eight classes were termed *wranglers*, and the questions and problems that were set became of increasing difficulty over the years. In fact, it was not possible to do well in the final Tripos examinations without intensive training in mathematical techniques, and this was provided by specialist coaches and tutors. The examinations were so difficult that even the best students never achieved more that 50% in the final examinations. But being a high wrangler, particularly a first or second wrangler, was the route to success in whatever profession the student chose to pursue after graduation. The highest-placed students had a second chance to display their abilities by sitting the examinations for the two Smith's Prizes, which had a stronger emphasis upon problems close to the frontiers of research.

In 1827, the mathematics courses and examinations were formally governed by what was termed the Mathematical Tripos. The Classical Tripos was introduced at the same date, but students still had first to qualify in mathematics before proceeding to that course of study.[2]

2.2 Attempts to reform mathematics teaching in Cambridge

While the training in mathematics provided by the Cambridge system was of high quality, it was firmly based upon the traditional programme of pure and mixed mathematics to the exclusion of continental innovations in pure mathematics, specifically in the foundations of analysis. Furthermore, it was strongly geometrical in character and laid considerable stress upon the applicability of the mathematics to the solution of physical problems. These features reflected closely the approach taken by Newton in his *Principia Mathematica*. As Becher (1980) has written,[3]

> In 1800 the mathematics curriculum, the key to a Cambridge degree, included arithmetic, algebra, trigonometry, geometry, fluxions, mechanics, hydrostatics, optics and astronomy. The focus and ultimate end of both pure and applied or 'mixed mathematics' was Newton's *Principia*. This programme was self-perpetuating, for tutors, textbook authors and the Examiners and Moderators had graduated (as) wranglers and they determined the contents of the Tripos. This inbreeding, along with the institutionalized process leading to the degree, made Cambridge the dominant school of mathematics in Britain.
>
> Inbreeding also helped isolate Cambridge from foreign developments. Along with Newton's physics and astronomy, Cambridge retained the geometrical approach of the *Principia* as the foundation of all mathematics and the fundamental means of problem solving.

In the period 1800 to 1820, strenuous efforts were made by a number of more enlightened mathematicians to reform the teaching of mathematics by introducing the new and powerful methods of analysis into the Mathematical Tripos. Robert Woodhouse's attempt failed because, again quoting Becher,

> Cambridge dons saw little or no need for abstraction and generalisation. They were not concerned with the education of career mathematicians. The Tripos was an integral part of a liberal education, and specialised education lay outside its curricular goals.

The torch was taken up by three remarkable undergraduates, George Peacock, Charles Babbage and John Herschel, who jointly created the *Analytical Society* to bring continental analysis to Cambridge. Herschel was senior wrangler in 1813 and Peacock second wrangler in the same year, testifying to their mathematical credentials; Babbage decided not to compete with them. In addition to promoting continental methods of analysis, they made a translation of Lacroix's *Traité Élémentaire du Calcul Différential et du Calcul Intégral* in 1816 and four years later a companion book of examples. Their approach was based upon the procedures of Lagrange, which was not without its technical difficulties. The significant achievement of the members of the Analytical Society was that these books replaced the fluxional texts as standard readings for the Mathematical Tripos. Only Peacock was to remain in Cambridge, as a tutor and subsequently as Lowndean Professor of Astronomy and Dean of Ely.

Over the following two decades, more features of the continental approach to analysis were introduced into the Mathematical Tripos. The tension remained, however, between the rigour of continental analysis and the more intuitive and practical approach of traditional Cambridge mixed mathematics. Nonetheless, the curriculum expanded to include more purely analytic subjects such as analytic geometry, advanced algebra, definite integrals, elliptic integrals and Laplace coefficients. At the same time William Whewell encouraged the inclusion of more physical subjects into the curriculum, including an analytic treatment of Newton's *Principia*, the wave theory of light, electricity, magnetism and heat. A consequence was that the length of the Tripos examinations increased to six days by 1838. Cambridge succeeded in producing a number of outstanding mathematicians, including George Green (1837), George Stokes (1841), Arthur Cayley (1842), John Couch Adams (1843) and William Thomson (1845).

Whittaker emphasises the central importance of George Green, one of the outstanding proponents of the continental methods of analysis, as well as a mathematician of genius. Whittaker writes (Whittaker, 1951):

> Green undoubtedly received his own early inspiration from . . . [the great French analysts], chiefly from Poisson; but in clearness of physical insight and conciseness of exposition he far excelled his masters; and the slight volume of his collected papers has to this day a charm which is wanting in their voluminous writings.
>
> The neglect of his work during his lifetime [he died in 1841] has often been remarked upon. It is, however, to be remembered that in 1840, the Lucasian professor was a man who never wrote anything, that the Plumian professor was Challis, whose attention was engaged in his hydrodynamical researches, that the Lowndean professor was Dean of Ely and lived there, that Airy had removed to Greenwich and was almost altogether occupied with astronomy . . .

But at least the sparks of renewal were there, as revealed by a final quote from Whittaker:

> It is impossible to avoid noticing throughout all Kelvin's work evidences of the deep impression which was made upon him by the writings of Green. The same may be said of his friend and contemporary Stokes; and, indeed, it is no exaggeration to describe Green as the real founder of the 'Cambridge school' of natural philosophers, of which Kelvin, Stokes, Rayleigh, Clerk Maxwell, Lamb, J.J. Thomson, Larmor and Love were the most illustrious in the latter half of the nineteenth century.

The expansion of the curriculum created problems for the role of the Mathematical Tripos in providing a liberal education for students who were not to become mathematicians or mathematical physicists. Whewell, who became Master of Trinity College in 1841, continued to oppose the introduction of pure analysis into the Tripos but reaffirmed his belief that the preferred topics for a liberal education should be firmly based upon Euclid, Newton, geometry and mixed mathematics. Whewell's counter-revolution began to succeed in 1846 with a compromise solution in which from 1848 the Tripos examinations were extended to eight days, the first three concerning elementary mathematics based upon the traditional geometric-Newtonian approach. Then, in 1848, the Mathematical Syndicate, which Whewell had set up during his period as Vice-Chancellor of the University in 1842, further diluted the more advanced parts of the Mathematical Tripos. The excluded subjects included elliptical integrals, Laplace coefficients, the theories of electricity, magnetism, heat and capillary action and the figure of the Earth treated heterogeneously. The Mathematical Tripos was plainly no longer fit for purpose, nor was it a matter of concern for Cambridge alone. The national interest was being adversely affected.

2.3 The Royal Commission report of 1852 and its aftermath

The sorry state of affairs in Cambridge was reflected in the quotation from Whittaker in Section 1.7. But much more than academic distinction and prestige were at stake. The agricultural and industrial revolutions were in full swing by the beginning of the nineteenth century. The development of the steam engine and the new possibilities opened up by discoveries in electricity, magnetism and current electricity could not be ignored. The developing understanding of physics is illustrated by the entries in Table 1.1. The discoveries of the continental scientists and the contributions of Watt, Dalton, Joule and Faraday were clearly of industrial relevance. The British scientists' contributions to experimental physics were carried out largely in private laboratories or workshops. An exception was the Royal Institution of Great Britain, where Faraday carried out his pioneering experiments and which was the closest to a publicly supported laboratory, the funds being provided by a group of wealthy, generally aristocratic, patrons. At the same time, British scientists had made a number of major contributions to theoretical physics, but these had not entered the university curriculum as subjects appropriate for academic study. Conspicuously, none of the contributions in the areas of heat, electricity and magnetism had been made by scientists working in Cambridge, although Thomson and Maxwell were Cambridge alumni.

The United Kingdom, as a seafaring nation, had a strong interest in precise instrumentation for the purposes of navigation. There was a need for sextants, astronomical instruments for use at sea, magnetic compasses, precise rules and accurate clocks, and so there was no lack of skilled craftsmanship available for the new industries. It was in the clear national interest that the underpinning of these disciplines should be fostered in the universities, particularly in the light of the remarkable advances taking place in the industrial and academic centres in Europe – the achievements of the French scientists in electricity, magnetism and the fundamentals of thermodynamics could not be ignored. There was evidently a need to catch up with the remarkable advances made by Britain's industrial and political competitors in continental Europe – the new disciplines of experiment, interpretation and theory in physics should be part of the University curriculum.

Albert, the Prince Consort, was debarred from taking any role in UK politics and so, on the advice of Sir Robert Peel, the Prime Minister from 1841 to 1846, took upon himself in 1841 the role of promoting industry and science in Great Britain. His most spectacular achievement was undoubtedly the Great Exhibition of 1851, held in the magnificent iron and glass structure of the Crystal Palace in Hyde Park, London. The Exhibition was intended to demonstrate that Great Britain was an industrial leader on the global scale. There were exhibitions of scientific instruments including electric telegraphs, microscopes, air pumps and barometers, as well as musical, horological and surgical instruments

Fig. 2.1 Queen Victoria and Prince Albert visit the machinery department of the Great Exhibition. The exhibits include the vertical printing machine invented by Applegarth for the *Times* newspaper, Black's patent folding machine, and a carding machine used in woollen manufacture (courtesy of the Special Collections Department, Library, University of Glasgow).

(Figure 2.1). The Exhibition far exceeded everyone's most optimistic predictions and resulted in a huge profit, which was used to found the Victoria and Albert Museum, the Science Museum and the Natural History Museum, as well as what became known as the 1851 Fellowships to support bright young scientists.[4]

Well before this time, Albert had been convinced of the need for the reform of Cambridge University.[5] He was persuaded to take on the role of Chancellor of Cambridge University by Robert Peel and in that role he proposed a number of reforms in 1848; the general consensus was that the University had fallen behind in the fields of the physical sciences.[6] The proposals were generally welcomed by members of the University, but were opposed by the more conservative element, among whom Whewell played a leading role in his position as Master of Trinity College, the wealthiest of the Cambridge colleges. The Cambridge system of colleges, which were and still are independent legal entities, and the University, which was responsible for the provision of lectures and examinations, had always had a somewhat fraught relationship. In the mid 1800s, the balance was strongly tipped in favour of the colleges, which had profited greatly from the increased revenue from their extensive land ownership thanks to innovations in agriculture – the colleges were rich and the

University poor. Some measure of the challenge facing Prince Albert can be gauged from the response of Whewell to the Prince Consort's proposed reforms:

> [M]athematical knowledge is entitled to *paramount* consideration, because it is conversant with indisputable truths . . . that such departments of science as Chemistry are not proper subject of academic instruction . . .

But the Prince had allies, including Robert Peel, who was fully aware of the significance of chemistry for the economy. The then Prime Minister, Lord John Russell, suspected that the conservative element would block the Prince's proposals for reform and so a Royal Commission was appointed in 1848 to investigate 'the State, Discipline, Studies and Revenues of the University of Cambridge'. Notably, the Commission included John Herschel and George Peacock. The Commission's report was published in 1852 and confirmed the Prince's recommendations of 1848, remarking that

> the operation of social causes little within her control [had caused the University of be] left out of her true position, and become imperfectly adapted to the present wants of the country, so as to stand in need of external help to bring about some useful reforms.

The Commission stressed the need for the teaching of engineering and modern languages. Furthermore, the lectures should be illustrated by experimental demonstrations 'which are daily in use in the hands of working men of science'. Whewell and the conservatives were strongly opposed to the proposed reforms. Whewell wrote:

> . . . that certain Mathematico-physical theories, which had obtained a temporary and questionable footing in the Examination, and which were felt to be in a considerable state of obscurity, involving great mathematical difficulties, and rather marking the frontier of science, than coming as yet fully within its ascertained range (those, namely of Electricity, Magnetism and Heat), should not be admitted as subjects of examination.

The Commission stuck to its guns, recommending that there should be a 'complete and thoroughly equipped laboratory' for chemistry. It also recommended that the University should teach the use of scientific instruments such as 'the handling of meteorological and magnetic instruments and cameras'. If the recommendations were accepted, there did not appear to be any reason

> why Cambridge should not become as great a School of physical and experimental as it is already of mathematical and classical instruction.

An immediate result was the establishment of the Natural Sciences Tripos and the Moral Sciences Tripos in 1851. The Natural Sciences Tripos was to include chemistry, mineralogy, geology, comparative anatomy, physiology and botany. Neither physics nor natural philosophy was included. Physics was not treated as a distinct discipline until 1861 and was not listed as a separate subject in the schedule of lectures until 1873, two years after Maxwell's appointment as the first Cavendish Professor of Experimental Physics.

There was a delay before the money could be found for these developments, but a start was made in 1860. By 1863, lecture theatres and accommodation were completed for zoology, comparative anatomy, chemistry, mineralogy and botany on what is now called the

Fig. 2.2   (*a*) William Cavendish in 1856 (courtesy of Getty Images). (*b*) William Cavendish, 7th Duke of Devonshire as Chancellor of Cambridge University.

New Museums site (see Appendix, Figure A.4), but there was no provision for experimental physics.

2.4 William Cavendish and the founding of the Laboratory

The untimely death of the Prince Consort in 1861 at the age of 42 had the potential to delay the reforms proposed by the Royal Commission, but he was succeeded as Chancellor of the University by William Cavendish (Figure 2.2), who became the 7th Duke of Devonshire on the death of his cousin in 1858. This was an inspired choice. Cavendish, a distant relative of Henry Cavendish of the Cavendish experiment, was a first-rate scientist, certainly not what might be expected of a senior member of the nobility.[7] He excelled in the Mathematical Tripos, being second wrangler and first Smith's Prizeman in 1829, as well as coming eighth in the Classical Tripos. He was the member of Parliament for Cambridge University from 1829 to 1831. When he took charge of his estates in the north of England, he adopted the reforms of the agricultural revolution to make them profitable. Even more important, he was a major investor in the development of the steel industry, in particular in the application of scientific methods to improve industrial steel making. This was stimulated by the discovery of rich deposits of high-grade haematite ore on his estates in North Lancashire. In this role, he became one of the most important figures of the industrial age.

The success of the steel works resulted in a vast expansion of the steel making capacity and of the port of Barrow-in-Furness – the population of the town was 700 in 1851 and had reached 47,000 by 1881. By the time of his death in 1891, he had amassed a fortune of £1,790,870.

Cavendish was the founding president of the Iron and Steel Institute in 1869 and emphasised in his first presidential address the importance of those branches of science of interest to the steel maker. He fully appreciated the application of scientific methods to industrial processes. In addition, he was dedicated to the importance of training for science, technology and industry, chairing the Devonshire Commission on Scientific Instruction and the Advancement of Science which sat from 1870 to 1875.

Despite the success of the Great Exhibition of 1851, Great Britain was lagging behind continental Europe in training in the physical sciences and engineering. This was brought home by what Sviedrys (1976) calls the 'grim lessons for British Industry of the 1867 Paris Exhibition'. Kenneth Beauchamp writes in his book *Exhibiting Electricity* (Beauchamp, 1997):

> Although Britain contributed over 3000 exhibits to the Paris 1867 exposition these were not viewed at all well by influential visitors. In a letter to the RSA, contributed by Dr. Lyon Playfair, who . . . was responsible for the classification of material for the 1851 exhibition, he writes,
>
>> . . . I am sorry to say that, with very few exceptions, a singular accordance of opinion prevailed (in Paris and amongst other European visitors) that our country has shown little inventiveness and made little progress in the peaceful arts of industry since 1862
>
> and puts this down to
>
>> . . . the lack of good systems of industrial education for masters and managers of factories and workshops.
>
> Dr. Playfair was one of the first to recognise the vital role of science and technology in education and devoted much effort to bring about the necessary educational reforms in Britain which followed the 1862 and 1867 exhibitions.

Experimental physics was still not incorporated into the Natural Sciences Tripos and so a syndicate was appointed in 1868 to address this issue; it reported on 27 February 1869. Within the Mathematical Tripos, there had been some increased emphasis on heat, electricity and magnetism, but this was not enough – what was needed was a well-appointed laboratory within which experimental physics could be fostered. It was clear that the task of building such a laboratory for the new discipline could not be undertaken by the existing professors. A possibility was that the Jacksonian Professor of Natural Philosophy could take this role, but, as the Syndicate's report stated,

> The Syndicate find that the Rules which regulate the Jacksonian Professor of Natural Philosophy are 'fanciful and obsolete' and that they require revision.

A report by the Plumian, Jacksonian, Lucasian, Sadlerian and Lowndean professors and the Professor of Chemistry had informed the syndicate in December 1868 that

(Willis, the Jacksonian Professor) already lectures on Mechanism; and [it] can hardly be expected that he should now take up a course of lectures on subjects to which he has not paid special attention.

Indeed, Willis was not appropriate for the task. In 1841 he had published his *Principles of Mechanism*, and in 1851 *A System of Apparatus for the Use of Lecturers and Experimenters in Mechanical Philosophy*. He was also deeply interested in medieval architecture and the construction of English cathedrals. At his death in 1875, he was writing the *Architectural History of the University of Cambridge*, subsequently completed by his nephew John Willis Clark. Clearly he had little expertise in areas of the experimental investigations of electricity, magnetism and heat.[8]

The syndicate recommended the setting up of a new Professorship of Experimental Physics which would terminate when the professor's tenure ended, as well as the construction of a specially designed laboratory for experimental physics. The professor would be supported by a demonstrator in experimental physics to give personal instruction in the laboratories, and a museum and lecture room attendant to service the laboratories as well as the instruments and apparatus. The cost of the building was estimated at £5000 and the instruments at £1300, while the annual cost of employing the professor, the demonstrator and the lecture attendant was £660. The syndicate recommended a further syndicate to determine how the project could be funded.

The usual means of raising the funds for such a project would be to levy a capitation tax on the colleges, but this was a large sum and the colleges were not prepared to make the funds available. The deadlock was broken by the Duke of Devonshire on 10 October 1870, when he wrote to the Vice-Chancellor as follows:

> My dear Mr. Vice-Chancellor,
> I have the honour to address you for the purpose of making an offer to the University, which, if you see no objection, I shall be much obliged to you to submit in such a manner as you may think fit for the consideration of the Council and the University.
> I find in the Report dated February (27), 1869, of the Physical Science Syndicate, recommending the establishment of a Professor and Demonstrator of Experimental Physics, that the buildings and apparatus required for this department of Science are estimated to cost £6,300. I am desirous to assist the University in carrying out this recommendation into effect, and shall accordingly be prepared to provide the funds required for the building and apparatus, so soon as the University shall have in other respects completed its arrangements for teaching Experimental Physics, and shall have approved the plan of the building.
> I remain (&c)
> Devonshire.

On 9 February 1871, the Council of the Senate agreed to accept the chancellor's magnificent gift and also to fund the posts needed to staff what was initially called the Devonshire Laboratory. At last, the University was committed to the incorporation of experimental physics as a discipline into the Natural Sciences Tripos. The project was essentially starting from scratch and it was to be a major challenge for the first incumbents of what was now entitled the Cavendish Professorship of Experimental Physics to achieve the ambitious goals set by the chancellor and the syndicate. But, within 30 years, what was to become the Cavendish Laboratory exceeded everyone's expectations in becoming the premier laboratory for experimental physics. Much of the credit for the vision and its implementation can be justly attributed to the first Cavendish Professor, James Clerk Maxwell.

PART II

1874 to 1879

James Clerk Maxwell (1831–1879)
Oil painting by Jemima Blackburn née Wedderburn

The Maxwell era 3

3.1 The appointment of James Clerk Maxwell

The electors to the Cavendish Professorship of Experimental Physics naturally sought the best possible candidate for the new chair. The obvious first choice was Sir William Thomson, later to become 1st Baron Kelvin of Largs or the Lord Kelvin, who was regarded as the most distinguished British physicist of his day, both in theory and in experiment. As described in Section 1.4.4, he was knighted for his efforts in laying the first successful transatlantic submarine cable in 1866. With his new, well-furnished laboratory at the University of Glasgow and his excellent industrial and domestic arrangements in the area, it is not surprising that he was unwilling to leave Glasgow. Another factor in Thomson's decision was the absence in Cambridge of a network of instrument makers and industrialists upon which much of his work relied. It is said that the electors also tried to attract Hermann von Helmholtz from Germany, but he was too well established in Berlin to consider moving to Cambridge. Some measure of the ambitions of the electors to the Cavendish Chair is provided by the assessment of Helmholtz's stature by R. Steven Turner:

> Helmholtz witnessed the final transition of the German universities from purely pedagogical academies to institutes devoted to organised research. The great laboratories built for him at Heidelberg and Berlin opened to him and his students possibilities for research unavailable anywhere in Europe before 1860. In many respects his career epitomised that of German science itself in his era, for during Helmholtz's lifetime German science, like the German empire, gained virtual supremacy on the Continent. (Turner, 1970)

Next, they turned to Maxwell. Maxwell's health was always somewhat fragile and, after a bout of ill-health, he resigned his chair at King's College, London in 1865. He returned to manage the family estate at Glenlair in the Dumfries and Galloway region of southern Scotland and set about writing his monumental *Treatise on Electricity and Magnetism*, which was eventually published in two volumes in 1873 (Maxwell, 1873). Maxwell was well known in Cambridge and no one doubted his originality and brilliance. He had acted as an examiner for the Mathematical Tripos in the period 1866 to 1870 and, in an effort to make some inroads into the reform of the Tripos, had introduced questions on heat, electricity and magnetism into the examinations. His revolutionary researches into the kinetic theory, thermodynamics, electricity and magnetism and his demonstration that light is electromagnetic radiation were all relatively recent. But it was not then established that his equations for the electromagnetic field were indeed the correct description of electromagnetism – Heinrich Hertz's demonstration that electromagnetic waves have all the properties of light waves were only carried out at the University of Karlsruhe in the period 1886–88 (Hertz, 1893). Nor was it clear that the kinetic theory was the correct description of the properties of gases, as Maxwell himself fully accepted. Nonetheless, he had all the qualifications for the position.

In 1871, John William Strutt, who was to become Lord Rayleigh in 1873 on the death of his father, wrote to Maxwell:

> There is no one here in the least fit for the post. What is wanted by most who know anything about it is not so much a lecturer as a mathematician who has actual experience of experimenting, and who might direct the energies of the younger Fellows and bachelors into a proper channel. There must be many who would be willing to work under a competent man, and who, while learning themselves, would materially assist him . . . I hope you may be induced to come; if not, I don't know who it is to be.

Maxwell was ideally matched to the requirements of the new professorship, although that was not necessarily apparent to the Cambridge academic community at the time. His excellence in mathematics was beyond doubt, having been second wrangler and joint Smith's Prize winner with the senior wrangler Edward Routh in 1854. In 1857 he won the Adams Prize for his remarkable essay on the stability of the rings of Saturn. In addition, he had demonstrated his outstanding experimental ability, for example, in his studies of colour vision (see Longair, 2008b) and his demonstrations of the dependence of the viscosity of gases on pressure and temperature (Maxwell, 1867).[1] His letters and papers are full of sketches for experiments in all branches of physics.[2]

Once Maxwell was reassured that Thomson would not stand, he stood for election to the chair, although he made it clear that he had no experience of teaching experimental physics. But then very few had any experience of teaching the discipline at that time.

3.2 The rise of experimental physics in Great Britain in the latter half of the nineteenth century

The systematic teaching of experimental physics in UK universities only began in the later decades of the nineteenth century. Until that time, advances in experimental physics had been largely made by individuals with an insatiable curiosity to understand the workings of nature through precise measurements and observations – this is vividly demonstrated by the entries in Table 1.1. The history of the development of experimental physics and engineering departments in UK universities is splendidly described by Sviedrys (1976). He describes how laboratories to train students in these disciplines were founded to provide professional training in mechanical and electrical engineering, the result of the dramatic growth of the railway network in the 1830s and 1840s and of the many new practical applications of electricity and electric power.

Sviedrys provides two enlightening tables which list the early nineteenth-century 'Private and nonacademic physics laboratories' (Table 3.1) and 'The first formally

Table 3.1 Private and nonacademic physics laboratories[a]

Location of laboratory	Year founded	Founder of laboratory
Royal Institution, London	1830s	Michael Faraday[b]
King's College, London	1835	Charles Wheatstone
Salford, Manchester	1838	James Prescott Joule
Edinburgh	1840	James David Forbes
Cambridge University	1849	George Gabriel Stokes
Queen's College, Belfast	1850	Thomas Andrews
University of Glasgow	1850	William Thomson
Devon	1853	William Froude[c]
London	1857	Augustus Matthiessen
London	1857	John Peter Gassiot
King's College, London	1862	James Clerk Maxwell
London	1865	David Kirkaldy[d]

[a] Sviedrys, 1976.

[b] Sviedrys regards the Royal Institution as a chemistry laboratory until Faraday's electromagnetic investigations of 1831.

[c] Froude's laboratory was designed for the experimental testing of ship's hull designs using scale models. The *Froude number* is a measure of a ship's drag through water.

[d] Kirkaldy built his laboratory for testing the tensile strength and other properties of samples of iron and steel.

recognised physics laboratories at British institutes of higher learning' (Table 3.2). Of the former laboratories, he writes:

> Private physics laboratories in Britain [see Table 3.1] were small and informal: seldom more than two students would be found working in one at the same time. In the laboratories of James Prescott Joule, William Froude, Augustus Matthiessen, and David Kirkaldy, no student worked at all; in George Gabriel Stokes's laboratory, students were admitted, but only for demonstrations. Academic laboratory work was not a degree requirement; it was usually a voluntary arrangement designed to give a professor free assistance and students practical experience with physical apparatus.

The young Maxwell took full advantage of any opportunity to explore physical phenomena experimentally from an early age, with the strong support and encouragement of his father, John Clerk Maxwell. Maxwell created his own private laboratory in a garret at the age of 15. In 1847, he was taken by his uncle John Cay to visit the optical laboratory of William Nicol, the distinguished optician and inventor of the Nicol prism. Maxwell later recalled that:

> I was taken to see [William Nicol], and so, with the help of 'Brewster's Optics' and a glazier's diamond, I worked at polarisation of light, cutting crystals, tempering glass, etc. (Harman, 1990)

Table 3.2 The first formally recognised physics laboratories at British institutes of higher learning[a]		
Institution	Year founded	Founder–director
University of Glasgow	1866	William Thomson
University College, London	1866	George C. Foster
University of Edinburgh	1868	Peter Guthrie Tait
King's College, London	1868	William G. Adams
Owens College, Manchester	1870	Balfour Stewart
University of Oxford	1870	Robert B. Clifton
Royal School of Mines (later Royal School of Science)	1872	Frederick Guthrie
Royal School of Science, Dublin	1873	William Barrett
Queen's College, Belfast	1873	Joseph D. Everett
Cambridge University	1874	James Clerk Maxwell

[a] Sviedrys, 1976.

He had the full run of Forbes' Physical Laboratory in the University. There, he undertook a series of experiments in 1848 on the chromatic effects of polarised light in doubly refracting materials, crystals and mechanically strained glasses, remarking on the 'gorgeous entanglements of colours' in strained glasses. These experiments and their interpretation were presented to the Royal Society of Edinburgh when Maxwell was only 19 (Maxwell, 1853). His initiatives are an example of what would now be called 'curiosity-driven research', without there being necessarily any practical application of the results. William Thomson in Glasgow ran his laboratory along similar lines, but with the major advantage that the fruits of the many researches which the students carried out were of direct application in the telegraphic and electrical industries.

These arrangements were not adequate, however, to cater for the growing needs of industry and the national interest. Industry had an obvious need for individuals trained in what would now be called physics. In addition, formal qualifications in experimental physics were required for civil service examinations and for the Indian telegraphic services. Later in the century, there was an increased demand for school teachers able to provide courses in the experimental sciences. Courses in experimental physics had to be developed for these purposes and the universities listed in Table 3.2 took up the challenge, thanks to the efforts of the individuals identified in the table.

In 1866, the gold standard had been set by the newly knighted Sir William Thomson at Glasgow, with his excellent new facilities for experimental physics. He had the very considerable advantage of having strong connections with the telegraphic and electrical industries and so was able to maintain the resources to support his research activities as well as producing an excellent stream of trained students. As Sviedrys points out, the other new laboratory directors struggled to maintain their research output in the face of heavy teaching loads, largely because they did not receive adequate resources to provide all the teaching and new equipment needed for running an experimental physics laboratory. The exceptions were at the universities of Oxford and Cambridge.

In both cases, major capital funds were made available for the construction of physics laboratories, in the case of Oxford from the Clarendon Trust.[3] In Cambridge, the philanthropy of the Duke of Devonshire made the construction of the Laboratory a practical possibility. In both cases, the gifts included both the construction of custom-built laboratories and the purchase of scientific instruments.

3.3 The changing face of natural philosophy

There were deeper forces at work which were to play out very much in Cambridge's favour. Germany was at the forefront of the development of theoretical physics as an underpinning discipline for natural philosophy. The two-volume work entitled *The Intellectual Mastery of Nature* by Jungnickel and McCormmach (1986a,b) provides a splendid perspective on how the discipline developed. Natural philosophers had to adopt more and more abstract mathematical tools to describe the natural phenomena being uncovered by discoveries in heat, electricity and magnetism. The summary of discoveries in physics in the nineteenth century listed in Table 1.1 illustrates the need to adopt more and more sophisticated mathematical tools to describe the natural phenomena of physics. Symbolic of this change was the appointment of Rudolph Clausius as the first Professor of Mathematical Physics at Zurich in 1855; he was to be a driving force behind the establishment of the discipline of theoretical physics – the welding of the whole of natural philosophy into single discipline, supported by powerful mathematical tools.

This was also the period when physics began to emerge as a professional discipline which became increasingly inaccessible to the educated public. While there were individuals such as Johann Wolfgang von Goethe who could effortlessly span the arts, humanities and natural philosophy in the early part of the nineteenth century, this became increasingly unfeasible in the second half of the century. The middle years of the nineteenth century can be regarded as the time when the arts–sciences divide became increasingly apparent.

This divergence was undoubtedly associated with conceptual changes in the physical sciences themselves. Until the dramatic discoveries in heat, electricity and magnetism in the late eighteenth century and nineteenth century, the universe of natural philosophy was Newtonian. The innovations of Lagrangian and Hamiltonian techniques provided very powerful tools for the analysis of what still remained problems of Newtonian mechanics. For innovators such as Maxwell, the discovery of his equations for the electromagnetic field was based upon a mechanical model of the aether, despite the fact that the essence of his equations is based upon the concept of fields rather than of particles. As Einstein wrote in the 1931 commemorative volume for the centenary of Maxwell's birth:

> We may say that, before Maxwell, Physical Reality, in so far as it was to represent the process of nature, was thought of as consisting in material particles, whose variations consist only in movements governed by partial differential equations. Since Maxwell's

time, Physical Reality has been thought of as represented by continuous fields, governed by partial differential equations, and not capable of any mechanical interpretation. This change in the conception of Reality is the most profound and the most fruitful that physics has experienced since the time of Newton. (Einstein, 1931)

This change of perspective is also emphasised by Freeman Dyson in his paper, *Why is Maxwell's Theory So Hard to Understand?* (Dyson, 1999):

Maxwell's theory had to wait for the next generation of physicists, Hertz, Lorentz and Einstein, to reveal its power and clarify its concepts. The next generation grew up with Maxwell's equations and was at home with a Universe built out of fields. The primacy of fields was as natural to Einstein as the primacy of mechanical structures had been for Maxwell.

Thus, to be successful in the new era of professional experimental physics, the incumbent of the Cavendish Chair had to be expert in experiment and in mathematics and to fully appreciate the needs of industry and society at large. Rayleigh's perspective, quoted in Section 3.1, that what was wanted was 'a mathematician who has actual experience of experimenting, and who might direct the energies of the younger Fellows and bachelors into a proper channel', encompasses all these requirements. Furthermore, by good fortune, the Cambridge Mathematical Tripos, despite the reservations of the mathematicians about the introduction of experimental physics, was ideally placed to provide the right sort of training for 'useful' mathematicians. But this was some considerable challenge. The general view from outside Cambridge was that Cambridge was too conservative to make a success of experimental physics. As remarked by Norman Lockyer, the influential founding editor of *Nature* in an unsigned editorial in 1874,

[I]t may take Cambridge thirty or forty years to reach the level of a second-rate German university in physical research. (Lockyer, 1874a)

Lockyer had reckoned without the almost unique qualities which Maxwell brought to the new Chair of Experimental Physics. As remarked by Sviedrys,

(Maxwell) was one of the very few British physicists who combined the experimental and philosophical Scottish tradition with the mathematical training provided at Cambridge.

Maxwell's agenda was set forth in his inaugural lecture as Cavendish Professor of Experimental Physics in October 1871.

3.4 Maxwell's manifesto: his inaugural lecture, 1871

In characteristic fashion, Maxwell delivered his inaugural lecture, *Introductory lecture on experimental physics* (Maxwell, 1890), not in the Senate House but in an obscure lecture room and without much publicity, on 25 October 1871. There was at best a modest audience, but what he said was deep and far-reaching, and set the tone for what he hoped the Laboratory would achieve. It is a remarkable vision, designed to set out the distinctive features of experimental physics and how the discipline was complementary to traditional

Cambridge mixed mathematics. The whole lecture is worthwhile reading for its graciousness, in both content and expression. He clearly meant every word of it. He begins:

> The University of Cambridge, in accordance with that law of its evolution, by which, while maintaining the strictest continuity between the successive phases of its history, it adapts itself with more or less promptness to the requirements of the times, has lately instituted a course of Experimental Physics. . . . The familiar apparatus of pen, ink and paper will no longer be sufficient for us, and we shall require more room than that afforded by a seat at a desk, and a wider area than that of the black board. We owe it to our Chancellor that, whatever be the character in other respects of the experiments which we hope hereafter to conduct, the material facilities for their full development will be upon a scale which has not hitherto been surpassed.

From the outset, Maxwell's aim was to create a world-leading laboratory for experimental physics. Note the delicious 'more or less promptness' in the first sentence, undoubtedly with tongue firmly in cheek. Next, he set out the goals of the Laboratory, contrasting these with Cambridge's traditional view of mixed mathematics and natural philosophy:

> When we shall be able to employ in scientific education, not only the trained attention of the student, and his familiarity with symbols, but the keenness of his eye, the quickness of his ear, the delicacy of his touch, and the adroitness of his fingers, we shall not only extend our influence over a class of men who are not fond of cold abstractions, but, by opening at once all the gateways of knowledge, we shall ensure the association of the doctrines of science with those elementary sensations which form the obscure background of all our conscious thoughts, and which lend a vividness and relief to ideas, which, when presented as mere abstract terms, are apt to fade entirely from the memory.

After a description of the distinction between 'experiments of illustration and experiments of research' comes the famous quotation, which needs to be placed in its proper context:

> The characteristics of modern experiments – that they consist principally of measurements – is so prominent, that the opinion seems to have got abroad, that in a few years all the great physical constants will have been approximately estimated, and that the only occupation which will be left to men of science will be to carry on these measurements to another place of decimals.

Maxwell immediately rejects this view. In a following paragraph he goes on:

> But the history of science shews that even during that phase of her progress in which she devotes herself to improving the accuracy of the numerical measurement of quantities with which she has long been familiar, she is preparing the materials for the subjugation of new regions, which would have remained unknown if she had contented with the rough guide of the earlier pioneers.

As an example, he discusses the remarkable advances of Gauss and Weber in organising 'experiments in concert' in which an international network of magnetic observatories was set up, the Magnetic Union. These precise measurements opened up new fields of research including the discoveries of magnetic disturbances associated with Sun and Moon, magnetic storms and the secular variation of the Earth's overall magnetic field. As Maxwell remarks, 'the interior of the Earth is subject to the influence of heavenly bodies'. He also refers to Weber as 'the future founder of the science of electro-magnetic measurement' and continues:

> The new methods of measuring forces were successfully applied by Weber to the numerical determination of all the phenomena of electricity, and very soon afterwards the electric telegraph, by conferring a commercial value on exact numerical measurement, contributed largely to the advancement, as well as to the diffusion of scientific knowledge.

Maxwell fully appreciates that the most difficult problem for students of physics is relating the physical phenomena to the appropriate mathematical structures. It is difficult to imagine the problem and its solution being better expressed.

> It is not till we attempt to bring the theoretical part of our training into contact with the practical that we begin to experience the full effect of what Faraday has called 'mental inertia' – not only the difficulty of recognising, among the concrete objects before us, the abstract relation which we have learned from books, but the distracting pain of wrenching the mind away from the symbols to the objects, and from the objects back to the symbols. This however is the price we have to pay for new ideas.
>
> But when we have overcome these difficulties, and successfully bridged over the gulph between the abstract and the concrete, it is not a mere piece of knowledge that we have obtained: we have acquired the rudiment of a permanent mental endowment. When, by repetition of efforts of this kind, we have more fully developed the scientific faculty, the exercise of this faculty in detecting scientific principles in nature, and in directing practice by theory, is no longer irksome, but becomes an unfailing source of enjoyment, to which we return so often, that at last even our careless thoughts begin to run in a scientific channel.

Then he advocates a role for mathematics in the understanding of the physical sciences which is not so remote from the premises of the value of mixed mathematics as well as a mild rebuke concerning the role of the Mathematical Tripos:

> There may be some mathematicians who pursue their studies entirely for their own sake. Most men, however, think that the chief use of mathematics is found in the interpretation of nature. Now a man who studies a piece of mathematics in order to understand some natural phenomenon which he has seen, or to calculate the best arrangement of some experiment which he means to make, is likely to meet with far less distraction of mind than if his sole aim has been to sharpen his mind for the successful practice of the Law, or to obtain a high place in the Mathematical Tripos.

3.5 The building of the Cavendish Laboratory

As soon as his appointment was approved by the University, Maxwell set about the design of the new laboratory with his usual enthusiasm and attention to detail. He visited

Thomson's laboratory in Glasgow and Clifton's Clarendon Laboratory at Oxford, seeking to create a state-of-the-art laboratory for teaching and research in experimental physics. A site in Free School Lane was selected for the Laboratory, a prime consideration being that it had to be sufficiently far from the main thoroughfares to minimise traffic vibrations. Plans for the building were developed by the architect William M. Fawcett, and these were scrutinised by Maxwell, who altered the locations of walls and the disposition of the rooms according to his perceived needs (Figure 3.1(a)). Crowther (1974) gives a detailed description of the contents of each of the rooms (Figure 3.1(b)). The cost of the building increased to £8450, but the Duke of Devonshire was prepared to cover this cost, in addition to providing the initial complement of scientific instruments.

The formal opening of the Laboratory took place on 16 June 1874 and in the following week an extended article appeared in *Nature* by Lockyer describing its layout in detail (Lockyer, 1874b). The splendid building (Figure 3.2(a)) was on three floors, the use of the various rooms on the ground, first and second floors being indicated on the plans reproduced by Fitzpatrick and Whetham (1910) (Figure 3.1(b)). Some of the special features of the design and attention to detail can be appreciated from the following examples:

- On the ground floor, the highest stability was needed in the Magnetism room. A pier was built, isolated from the rest of the floor area and resting on concrete foundations 18 inches thick, the base of which was three feet below the level of the floor. This pier was used to support the apparatus for the absolute determination of the standard of resistance which would then be used to calibrate all the other electrical apparatus in the Laboratory. The Kew magnetometer was placed on another stone slab in this room (see Box 3.1).
- The room designated Pendulums contained a pier for the principal laboratory clock, to which the other clocks in the Laboratory could be synchronised electrically.
- A lift in the Stores room enabled heavy equipment to be transferred to the upper floors.
- The 15-foot-high room to the left of the entrance was the battery room and was located immediately below the Lecture Room on the floor above so that electric power would always be available for demonstrations. This room also contained gas-holders containing oxygen so that oxy-coal-gas limelight would always be available in the Lecture Room above.
- On the first floor, the Large Laboratory was intended for use by the students. There were ten tables in the original laboratory, each containing a standpipe to which four Bunsen burners could be attached. Water was laid on in all the rooms.
- The Professor's Private Room had two hatches which communicated with the students' laboratory, so that he could keep an eye on what they were up to.
- The Professor's Laboratory was a preparation room for the experiments to be carried out in the Lecture Room, now known as the Maxwell Lecture Theatre.
- The Lecture Room is a splendid example of its type, with steeply raking seats for up to 180 students, who could clearly observe the experiments being carried out on the long oak lecture bench (Figure 3.2(b)). The shutters could be completely closed for optical experiments and there were hatches in the ceiling for experiments such as Foucault's pendulum.[4]

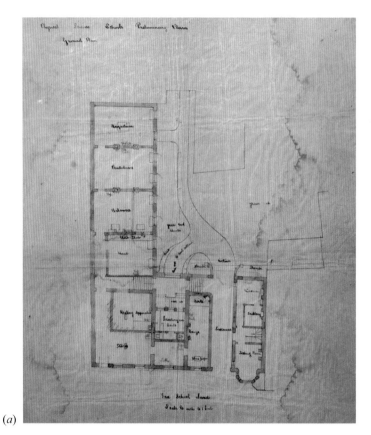

(a)

(b)

Fig. 3.1 (a) Fawcett's plan for the ground floor of the Cavendish Laboratory with Maxwell's changes indicated in his own hand in red pencil. (b) Fawcett's plans for the three-storey Cavendish Laboratory with Maxwell's changes incorporated (Fitzpatrick and Whetham, 1910).

(a)

(b)

(a) The entrance to the Cavendish Laboratory. The photograph, showing the Free School Lane frontage of the Laboratory, was taken between 1895, when the south wing seen on the extreme right of the picture, was added by J.J. Thomson, and before the Rayleigh Wing was built in 1907. (b) The Maxwell Lecture Theatre.

Fig. 3.2

- On the top floor, the Electricity room was kept dry by a special arrangement involving an endless flannel band so that the dry air provided good insulation for electrical instruments.
- The height of the building allowed the construction of a Bunsen's water pump, with

  > a vertical fall of considerably more than 50 ft . . . used to exhaust a large receiver, from which pipes will communicate with the different rooms, so that if it be desired to exhaust the air from any vessel it will only be necessary to connect it with one of these pipes and *turn on a vacuum*. (Falconer, 2014)

- Heating for the building was provided by hot water pipes made of copper so that the magnetic instruments would not be disturbed.
- The floors of the Laboratory had many hatches so that wires and tubes could be easily distributed throughout the building.

Maxwell produced an extensive list of instruments to be located in the various rooms of the new Laboratory. These were acquired through a combination of new purchases funded by the Duke's gift, instruments and demonstrations which Maxwell brought with him on his appointment and new items which Maxwell paid for from his own pocket. Many of these instruments have been preserved and are in the Cavendish Collection of Historic Scientific Instruments. They provide a vivid picture of the state-of-the-art physics laboratory in 1874 (Figure 3.3(*a*) and (*b*)). In his fourth Annual Report of 1877, Maxwell could report to the chancellor that the collection of instruments was complete.

3.6 Maxwell's Cavendish Laboratory: research

The research activities got off to an encouraging start. Maxwell wrote to his uncle Robert Cay on 12 May 1874:

> The Cavendish Laboratory is now open to students for practical work, and several good bits of work are being done already by the men. I expect some of them will have matter for publishing before long. (Maxwell, 1874)

Maxwell was fully aware of the many challenges he faced, but he was in a strong position to make a broad attack on a very wide range of problems in experimental physics. A number of factors were running strongly in his favour.

3.6.1 Maxwell and the determination of fundamental standards and constants

Maxwell had already been deeply involved in issues concerning the determination of fundamental constants and of absolute standards of resistance. The 1861 meeting of the British Association had appointed a committee to oversee the determination of fundamental standards, and Maxwell joined the committee in 1862. At that time, he had just published his first enunciation of his equations for the electromagnetic field and the prediction that light

(a)

(b)

(a) A selection of the scientific instruments and demonstrations which Maxwell brought with him to Cambridge in 1871. (b) A selection of the scientific instruments and demonstrations which Maxwell purchased for the new Laboratory in 1874.

Fig. 3.3

is electromagnetic radiation. He was deeply involved in testing the theory by precise experiment, in particular the determination of the ratio of electrostatic to electromagnetic units of charge (see notes 7 and 8 of Chapter 1). The activities of the committee became much more mathematical and theoretical, playing directly to Maxwell's strengths. He set about supervising the design and construction of apparatus to make a very accurate determination of the ohm with his colleagues Balfour Stewart and Henry Fleeming Jenkin at King's College, London (Maxwell *et al.*, 1863). The success of these experiments convinced the committee that the absolute value of the ohm determined by this and similar techniques was the clearly preferred standard. The committee reported that

> [the system of absolute electrical units was not only] the best yet proposed, [but also that it was] the only one consistent with our present knowledge both of the relations existing between the various electric phenomena and of the connection between these and the fundamental measurements of time, space and mass.

The work on standards fragmented in subsequent years, but Maxwell maintained his strong interest and leading role in the subject and made it one of the central themes of the research programme of the new Laboratory. The result was that the work on determining the absolute standard of resistance was transferred from the Kew Observatory to the Laboratory. This work was to remain one of the central roles of the Laboratory until it was taken over by the National Physical Laboratory on its foundation in 1900.

3.6.2 The *Treatise on Electricity and Magnetism* (1873)

Maxwell began work on his *Treatise on Electricity and Magnetism* as soon as he had settled back at his home at Glenlair in 1865. It is significant that, while he was writing the *Treatise*, he was also an examiner for the Cambridge Mathematical Tripos and realised the dire need for suitable textbooks. Without these, topics such as electricity and magnetism could not be introduced into the Cambridge Tripos structure. The two-volume *Treatise* is unlike many of the other great treatises such as Newton's *Principia* in that it is not a systematic presentation of the subject but a work in progress, reflecting Maxwell's own approach to these researches. In a later conversation, Maxwell remarked that the aim of the *Treatise* was not to expound his theory finally to the world, but to educate himself by presenting a view of the stage he had reached. Disconcertingly, Maxwell's advice was to read the four parts of the *Treatise* in parallel rather than in sequence.

The advantage of this approach was that Maxwell laid out clearly his own perception of the physical content of the theory and also how it could be confronted by precise experiment. These topics were strongly influenced by his work for the British Association committee on fundamental standards. It is intriguing that he devotes a large part of the *Treatise* to basic measurements and electrical apparatus, in the process providing many experimental challenges which could be taken up by the research students in the Laboratory. For example, he devotes 34 pages to the analysis of the Kew magnetometer (Box 3.1), describing in great detail all the precautions and corrections needed to obtain an accurate measurement. Another example is his analysis of five methods of making absolute determinations of the standard of resistance, which occupies the whole of his Chapter XVIII.

Then, on reaching Section 585, halfway through Volume 2, Maxwell states that he is to 'begin again from a new foundation without any assumption except those of the dynamical theory stated in Chapter VII'. As summarised by Peter Harman, Maxwell emphasises the expression of physical quantities free from direct representation by a mechanical model. This needed new mathematical approaches to electromagnetism, including quaternions, integral theorems such as Stokes's theorem, topological concepts and Lagrangian–Hamiltonian methods of analytic dynamics. One of the most important results appears in Section 792 of Volume 2, in which Maxwell works out the pressure which radiation exerts on a conductor. This profound result provides a relation between the pressure p and the energy density ε of radiation derived purely from the properties of electromagnetic fields: $p = \varepsilon/3c^2$. This result was to be used by Boltzmann in his paper of 1884 in which he derived the Stefan–Boltzmann law from classical thermodynamics.

Published in 1873, the *Treatise* had an immediate impact and, together with Thomson and Tait's *Treatise on Natural Philosophy*, provided students with a comprehensive overview of both experimental and theoretical physics. With its publication, there could be no doubt about Maxwell's authority or his fitness to take up the challenges he faced.

3.6.3 Maxwell and his graduate students

Maxwell had another great advantage over the directors of the other newly founded UK physics laboratories which resulted from the particular structure of the Tripos examinations. With the dominance of the Mathematical Tripos, the standard route for students to join the Laboratory to carry out experimental research was first to complete their studies in mathematics and then to proceed to research under Maxwell's direction. Many of these students had done well as wranglers and so they came to the Laboratory with many of the mathematical skills needed to appreciate the most advanced problems in theoretical and experimental physics. Because of their distinction in the Mathematical Tripos, these students would often win college fellowships or college posts which enabled then to carry out substantial long-term research investigations, much like a modern PhD programme. In contrast, except for universities such as Glasgow where Thomson's industrial contacts ensured continuing support, it was difficult to support graduate students for a long period.

Maxwell's interests spanned the whole of experimental and theoretical physics and the apparatus he acquired enabled the students to address essentially any problem at the frontiers of physics. But there was no formal training in experimental research as such. Maxwell firmly believed that experimental expertise should be gained by the students learning for themselves the art of precise experiment. A favourite route to honing these skills was to measure the horizontal component of the Earth's magnetic field using the Kew magnetometer (Box 3.1), which Maxwell had purchased for this very training purpose. As alluded to above, the theory and practice of this experiment is set out in very considerable detail in Sections 449–464 of Maxwell's *Treatise*, which describe all the corrections and precautions the student should take to obtain a good result. Maxwell was undoubtedly inspired by the experiments of Gauss and Weber to map the Earth's magnetic field, about which he had enthused in his inaugural lecture. Participation in this and similar projects was set as a long-term aim of the Laboratory (see Section 3.4).

| Box 3.1 | The Kew magnetometer |

The Kew magnetometer, which was purchased by Maxwell in December 1873 for £60, as an experiment for developing practical experimental skills.

> The Kew Magnetometer was a good way of starting. Using the instrument to measure the Earth's magnetic field provided the student with valuable training in the careful reading of scales and the difficulties of time measurement of the oscillating magnet. Levelling the instrument and isolating it from extraneous magnetic fields already taught important lessons about the special requirements of exact physical measurement. If the student needed help, he sought out the demonstrator William Garnett, who could explain the techniques involved. (Bennett *et al.*, 1993)

> 'The instrument, though formidable in appearance, seemed easy to understand. The method of counting swings was novel and interesting, but I ran up against the "moment" of the magnet, an abstraction to which I had not been introduced. I appealed to Garnett . . . Garnett tried to show me how the moment of variously shaped bodies was calculated. I could not take this in because I had read nothing of differential or integral calculus . . . My ignorance did not prevent the production of fairly good results.' (A.P. Trotter, *Elementary Science at Cambridge, 1876–1879*. Quoted in Bennett *et al.*, 1993)

The agreement to found the Laboratory and the Cavendish Chair also included provision for a Demonstrator, whose role was to supervise and assist the students in their training in experimental physics. The subjects of heat, electricity and magnetism had been reintroduced into the Mathematical Tripos and in 1873 Maxwell was the examiner for these subjects. Among the successful candidates was William Garnett, who impressed Maxwell by his knowledge of physics and who attained the rank of fifth wrangler. Maxwell offered him the position of Demonstrator and Garnett continued in this role until Maxwell's death in 1879. Garnett subsequently became Maxwell's biographer with Lewis Campbell

(Campbell and Garnett, 1882), and Professor of Natural Philosophy at University College, Nottingham. After a brief period at Nottingham, he became Professor of Mathematics and Principal of Durham College of Science at Newcastle, and finally Educational Advisor to the London County Council.

One of Garnett's important innovations was the creation of the Cavendish workshop, which was to assume increasing importance as the need for expert technical skills increased in the subsequent years. Maxwell had generally commissioned his scientific instruments from the best instrument makers while he was at Aberdeen and King's College, London, but that was to change during the Thomson and Rutherford eras. Garnett employed two instrument makers in the workshop, but, when Willis died, they were transferred to the workshops of the Engineering Department under the new Professor of Mechanism, James Stuart. Garnett claimed that 'The little workshop in the Cavendish Laboratory was the starting-point of the University Mechanical Engineering Department.'[5]

Maxwell took an active daily interest in all the researches of the graduate students in the Laboratory, often accompanied by his dog Tobi. He was always available for advice and encouraged the students to develop their own ideas, even if they were unlikely to produce a positive result. This light touch, and the serious involvement of the senior professor with the work of all the graduate students, set the tone for the Laboratory for the coming years and was to be fostered by his successors.

3.6.4 Research during the Maxwell era

The range of research activities in which Maxwell and his students were involved was remarkable.[6] There was the expectation that the Laboratory would make its impact through the publication of the results of research in the prominent academic journals and other academic publications. Tables 3.3 and 3.4, taken from *A History of the Cavendish Laboratory, 1871–1910* (Fitzpatrick and Whetham, 1910), show the record of publications for the years of Maxwell's tenure of the Cavendish Chair and provide a revealing insight into the broad range of activities encouraged by Maxwell. It was in the spirit of Maxwell's extensive approach that not all the experimental endeavours would result in publications in learned journals. It should also be noted that publication often occurred some time after the research had been completed.

Maxwell

The dominant impression of these tables is the extraordinary range of Maxwell's continuing interests, recalling that he was also supervising the construction and development of the fledgling Laboratory. It is evident that he was at the height of his powers across the full range of his interests – pure mathematics, kinetic theory and statistical physics, optics, electricity and magnetism, heat and so on. In addition, he wrote authoritative contributions for the *Encyclopaedia Britannica*, the essays appearing yearly in alphabetical order (his contributions ended with Harmonic Analysis, published in the year after his death). Equally significant was the large effort he devoted to the publication of Henry Cavendish's papers in the authoritative volume entitled *Electrical Researches of the Hon. Henry Cavendish*

Table 3.3 A list of publications by members of the Cavendish Laboratory, 1874–1879[a]	
Author	Publication

1874

Maxwell, J.C.	Plateau on Soap Bubbles, *Nature*, **10**
Maxwell, J.C.	Grove's 'Correlation of Physical Forces', *Nature*, **10**
Maxwell, J.C.	On the Application of Kirchhoff's Rules for Electric Circuits to the Solution of a Geometrical Problem, *Nature*, **10**
Maxwell, J.C.	Van der Waals on the Continuity of the Gaseous and Liquid States, *Nature*, **10**
Maxwell, J.C.	On the Relation of Geometrical Optics to other Parts of Mathematic and Physics, *Proc. Camb. Phil. Soc.*, **2**
Maxwell, J.C.	On Double Refraction in a Viscous Liquid in Motion, *Proc. Roy. Soc.*, **22**

1875

Maxwell, J.C.	On the Dynamical Evidence of the Molecular Constitution of Bodies, *Nature*, **11**
Maxwell, J.C.	On the Application of Hamilton's Characteristic Function to the Theory of an Optical Instrument symmetrical about its Axis, *Proc. Lond. Math. Soc.*, **6**
Maxwell, J.C.	On the Centre of Motion of the Eye, *Proc. Camb. Phil. Soc.*, **2**
Maxwell, J.C.	Articles: Atom; Attraction, *Enc. Brit.*
Saunder, S.A.	On the Variation of the E.M.F of a new form of Leclanché Cell, *Nature*, **12**

1876

Chrystal, G.	On the Effect of Alternating Induction Currents on the Galvanometer, *Proc. Camb. Phil. Soc.*, **3**
Chrystal, G.	On Bi- and Unilateral Galvanometer Deflection, *Phil. Mag. V*, **2**
Chrystal, G. & Saunder, S.A.	Results of a Comparison of the B.A. Units of Resistance *Brit. Assoc. Report*
Clayden, A.W. & Heycock, C.T.	The Spectrum of Indium, *Phil. Mag. V*, **2**
Maxwell, J.C.	Diffusion of Gases through Absorbing Substances, *Nature*, **14**
Maxwell, J.C.	Whewell's Writings and Correspondence, *Nature*, **14**
Maxwell, J.C.	On the Protection of Buildings from Lightning, *Nature*, **14**
Maxwell, J.C.	On Bow's Method of drawing Diagrams in Graphical Statics with Illustrations from Peaucellier's Linkage, *Proc. Camb. Phil. Soc.*, **2**
Maxwell, J.C.	On the Equilibrium of Heterogeneous Substances, *Proc. Camb. Phil. Soc.*, **2**
Maxwell, J.C.	On Ohm's Law, *Brit. Assoc. Report*
Maxwell, J.C.	Article: Capillary Action, *Enc. Brit.*
Maxwell, J.C.	General Considerations concerning Scientific Apparatus, *Kensington Museum Handbook*
Niven, W.D.	On the Calculation of the Trajectories of Shot, *Proc. Roy. Soc.*, **26**

1877

Gordon, J.E.H.	On the Determination of Verdet's Constant in Absolute Units. *Phil. Trans. Roy. Soc*, **167**
Maxwell, J.C.	Hermann Ludwig Ferdinand Helmholtz. *Nature*, **15**
Maxwell, J.C.	On a Paradox in the Theory of Attraction, *Proc. Camb. Phil. Soc.*, **3**
Maxwell, J.C.	On Approximate Multiple Integration between Limits by Summation, *Proc. Camb. Phil. Soc.*, **3**

	Table 3.3 (cont.)
Author	Publication
Maxwell, J.C.	On the Unpublished Electrical Papers of the Hon. Henry Cavendish, *Proc. Camb. Phil. Soc.*, **3**
Maxwell, J.C.	Articles: Constitution of Bodies; Diffusion; Diagrams, *Enc. Brit.*
Schuster, A.	On the Passage of Electricity through Gases, *Proc. Camb. Phil. Soc.*, **3**
Schuster, A.	Spectra of Metalloids, *Nature*, **15**
Schuster, A.	On the Presence of Oxygen in the Sun, *Nature*, **17**
1878	
Glazebrook, R.T.	An Experimental Determination of the Values of the Velocities of Normal Propagation of Plane Waves in different directions in a Biaxial Crystal, and a Comparison of the Results with Theory, *Phil. Trans. Roy. Soc*, **170**
Maxwell, J.C.	Tait's 'Thermodynamics', *Nature*, **17**
Maxwell, J.C.	The Telephone (Rede Lecture), *Nature*, **18**
Maxwell, J.C.	Paradoxical Philosophy, *Nature*, **19**
Maxwell, J.C.	On Boltzmann's Theorem on the Average Distribution of Energy in a System of Material Points, *Camb. Phil. Trans.*, **12**
Maxwell, J.C.	On the Electrical Capacity of a long narrow cylinder and of a Disk of sensible thickness, *Proc. Lond. Math Soc.*, **9**
Maxwell, J.C.	Article: Ether, *Enc. Brit.*
1879	
Gordon, J.E.H.	Measurements of Electrical Constants. No. 11. On the Specific Inductive Capacities of certain Dielectrics. Part I. *Phil. Trans. Roy. Soc*, **170**
MacAlister, D.	Experiments on the Law of Inverse Square. *Cavendish's Electrical Experiments, Note 19*, Camb. Univ. Press.
Maxwell, J.C.	Thomson and Tait's Natural Philosophy, *Nature*, **20**
Maxwell, J.C.	On Stresses in Rarified Gases arising from Inequalities of Temperature, *Phil. Trans. Roy. Soc*, **170**
Maxwell, J.C.	Reports on Special Branches of Science, *Brit. Assoc. Report*
Maxwell, J.C.	Article: Faraday, *Enc. Brit.*
Maxwell, J.C.	Electrical Researches of the Hon. Henry Cavendish, Camb. Univ. Press.
Schuster, A.	On the Spectra of Metalloids. Spectrum of Oxygen. *Phil. Trans. Roy. Soc*, **170**
Schuster, A.	Some Remarks on the Total Eclipse of July 29 1878. *Mon. Not. R. Astron. Soc.*
Schuster, A.	On the Probable Presence of Oxygen in the Solar Chromosphere. *Mon. Not. R. Astron. Soc.*
Schuster, A.	On Harmonic Ratios in the Spectra of Gases. *Nature*, **20**
Schuster, A.	An Easy Method for Adjusting the Collimator of a Spectroscope. *Phil. Mag. V*, **7**
Schuster, A.	On the Spectra of Lightning. *Phil. Mag. V*, **7**
Shaw, W.N.	An Experiment with Mercury Electrodes. *Proc. Camb. Phil. Soc.*, **3**

[a] Fitzpatrick and Whetham, 1910.

(Maxwell, 1879). This proved to be a fruitful source of research projects for the graduate students. An excellent example is the improved determination of the inverse-square law of electrostatics, devised by Maxwell and carried out by Donald MacAlister. The experiment involves the well-known determination of the electric field inside a conducting sphere, if the inverse-square law were to diverge even slightly from $f \propto r^{-2}$. MacAlister's experiment showed that the exponent had to be -2 to within one part in 21,600, an improvement of three orders of magnitude over Cavendish's estimate. MacAlister was later to become Principal of Glasgow University.

Notice also Maxwell's continuing interest in the role of physics in its social context. He writes *On the protection of buildings from lightning* and, for the 1878 Rede Lecture on *The telephone*, he gave a live demonstration of its operation, to general admiration.

Let us highlight just a few of the research achievements of some of the first generation of research students in the new Laboratory, all of them inspired and encouraged by Maxwell.

William Mitchinson Hicks

The first graduate student to work in the Laboratory under Maxwell was William Hicks, who had attained the position of seventh wrangler in the Mathematical Tripos. After his introduction to experimental techniques through the Kew magnetometer, he developed an electrometer which exemplified Maxwell's dictum of learning by doing rather than being taught. Most significantly, he made among the first efforts to test Maxwell's prediction that electromagentic disturbances should propagate at the speed of light. It was not surprising that this experiment was not a success, but it indicates Maxwell's willingness to allow the students to follow their own initiatives. Hicks went on to become Professor of Physics at Sheffield and its first Vice-Chancellor once Firth College was converted to the University of Sheffield.

James Edward Henry Gordon

J.E.H. Gordon was the second student to work under Maxwell. He was fortunate in having private means and so was not encumbered by having to devote large amounts of time to tutoring. He obtained the rank of Junior Optime[7] in the Mathematical Tripos in 1875. His research was wide-ranging and included the determination of electrical constants, the Faraday rotation of light in water and the more optically active material bisulphide of carbon, and the effects of light upon the electrical conductivity of selenium. He also made measurements of the specific inductive capacities (or permittivities) of various transparent substances, with the aim of testing Maxwell's electromagnetic theory of light, according to which the square of the refractive index of the material should be proportional to the product of its permittivity and its permeability. This experiment was somewhat compromised by the dispersive properties of the materials. He was a pioneer of electric lighting and published a treatise on this subject.

George Chrystal

George Chrystal had the distinction of being second wrangler in 1875 and took up the challenge, posed by Maxwell, Everett and Schuster at the meeting of the British Association in Belfast in 1874, of determining the accuracy with which Ohm's law, $\mathcal{E} = RI$, was a correct description of the relation between current and electromotive force. As recounted in Section 1.4.2, Ohm's law is an empirical relation between voltage and current and is not built into the basic structure of electromagnetism. The problem facing the experimenters was that, when a current is passed through a resistor, the temperature changes and special care has to be taken to ensure that the measurements are made at a single temperature. Chrystal's experiment was designed by Maxwell, and Ohm's law was confirmed with unprecedented accuracy. Maxwell wrote about Chrystal's results as follows:

> It is seldom, if ever, that so searching a test has been applied to a law which was originally established by experiment, and which must still be considered a purely empirical law, as it has not hitherto been deduced from the fundamental principles of dynamics. But the mode in which it has been borne this test not only warrants our entire reliance on its necessary accuracy with the limit of ordinary experimental work, but encourages us to believe that the simplicity of an experimental law may be an argument for its exactness, even when we are not able to show that the law is a consequence of elementary dynamical principles. (Harman, 2002)

Maxwell reported these results as part of the 1876 *British Association Report*, which also included Chrystal and Saunder's careful comparison of the British Association units of electrical resistance, as defined by the British Association's Committee on Electrical Standards. This work was subsequently to be refined by Fleming and Glazebrook during Rayleigh's tenure of the Cavendish Professorship. Chrystal subsequently became Professor of Mathematics at the University of Edinburgh.

Arthur Schuster

Unlike the other graduate students in Maxwell's Laboratory, Arthur Schuster was not a Cambridge graduate, but had had experience of experimental physics at Heidelberg, Göttingen and Berlin. He had private means and was able to support himself during his years at the Laboratory; thanks to the good offices of William Garnett, Schuster was given the title of Fellow Commoner of St John's College. He had a reputation for finding puzzling results which sometimes reflected insufficiently precise attention to detail in his experiments. Maxwell coined the jocular phrase *Schusterismus* for his less well-found discoveries. He devoted considerable attention to the developing field of spectroscopy and, in particular, its application to astronomy. This was to act as the touchstone for the emerging field of astrophysics, by which was meant the application of spectroscopy to the understanding of astronomical phenomena. Astrophysical research was thus present in the Laboratory from the very beginning. Schuster was to become Professor of Physics at Owens College of the University of Manchester in the 1880s.

Author	Publication
Table 3.4 A list of publications by members of the Cavendish Laboratory, 1880[a]	
1880	
Fleming, J.A.	On a New Form of Resistance Balance adapted for comparing Standard Coils, *Phil. Mag. V*, **9**
Glazebrook, R.T.	Notes on Nicol's Prism, *Phil. Mag. V*, **10**
Glazebrook, R.T.	Double Refraction and Dispersion in Iceland Spar: An Experimental Investigation, with Comparison with Huygens' Construction for the Extraordinary Wave, *Phil. Trans. Roy. Soc*, **171**
Glazebrook, R.T.	Note on the Reflection and Refraction of Light, *Proc. Camb. Phil. Soc.*, **3**
Maxwell, J.C.	Article: Harmonic Analysis, *Enc. Brit.*
Maxwell, J.C.	On a Possible Mode of Detecting a Motion of the Solar System through the Luminiferous Ether, *Proc. Roy. Soc.*, **30**
Poynting, J.H.	On the Graduation of the Sonometer, *Phil. Mag. V*, **9**
Poynting, J.H.	On a simple Form of Saccharimeter, *Phil. Mag. V*, **10**
Lord Rayleigh	On the Resultant of a large number of Vibrations of the same Pitch and of Arbitrary Phase, *Phil. Mag. V*, **10**
Lord Rayleigh	On the Resolving Power of Telescopes, *Phil. Mag. V*, **10**
Lord Rayleigh	On the Minimum Aberration of a Single Lens for Parallel Rays, *Proc. Camb. Phil. Soc.*, **3**
Lord Rayleigh	On a New Arrangement for Sensitive Flames, *Proc. Camb. Phil. Soc.*, **4**
Lord Rayleigh	On the Effect of Vibrations upon a Suspended Disc, *Proc. Camb. Phil. Soc.*, **4**
Schuster, A.	Some Results of the last two Solar Eclipses. (Siam, 1875, and Colorado, 1878) *Proc. Camb. Phil. Soc.*, **3**
Schuster, A.	On the Polarisation of the Solar Corona. *Mon. Not. R. Astron. Soc.*
Schuster, A.	Spectra of Metalloids, *Brit. Assoc. Report*
Schuster, A.	On the Influence of Temperature and Pressure on the Spectra of Gases, *Brit. Assoc. Report*
Schuster, A.	Note on the Identity of the Spectra obtained from the Different Allotropic Forms of Carbon, *Man. Lit. Phil. Soc. Proc*, **19**
Thomson, J.J.	On Maxwell's Theory of Light, *Phil. Mag. V*, **9**

[a] Fitzpatrick and Whetham, 1910.

Hugh Frank Newall

An interesting case is that of Hugh Frank Newall, who came up to Trinity College in 1876 as an undergraduate intent upon undertaking experimental research, having been inspired by his school training in physics at Rugby School. At that time there was no provision for undergraduates to carry out experimental physics and so he was given the same opportunities as the graduate students who had completed the Mathematical Tripos. He found this a hard assignment but he had access to all the facilities of Maxwell's Laboratory. He realised that the experimental facilities were targeted at research rather than undergraduate teaching. The need to make provision for the systematic teaching of experimental physics at the undergraduate level was to begin under Rayleigh. Newall returned to the Laboratory as a graduate student and then, after a break of four years, became an assistant demonstrator in 1885. He left to pursue his interests in astronomical spectroscopy, where his expertise in optics enabled him to construct high-resolution spectrographs with which to measure high-precision optical spectra. He became one of the pioneers in the field of astrophysics. In 1909, he was appointed Professor of Astrophysics at the Cambridge Observatories.

1879–1880

Glazebrook, Shaw, Wilberforce and Poynting were all to make very significant contributions to the development of the research and teaching programmes, but the most significant of these were to be made under the directorship of Rayleigh, where they are described in detail.

Maxwell died of stomach cancer on 5 November 1879. Although his health declined rapidly from the spring of 1879 onwards, he remained fully engaged with the work of the Laboratory and the graduate students. The list of Cavendish publications for the year 1880 shows the developing vitality of the programme of experimental researches (Table 3.4). Maxwell's last paper, published posthumously in the *Proceedings of the Royal Society*, concerned a means of measuring the speed of the Earth through the hypothetical aether, and was the inspiration for Michelson and Morley's famous experiments (Maxwell, 1880; Michelson and Morley, 1887).[8] The null result of that experiment was to lead to the Special Theory of Relativity, which was already embedded in Maxwell's equations, although this was not appreciated in 1880.

Table 3.4 shows that Fleming, Glazebrook, Poynting and Schuster all made important contributions which exemplify Maxwell's goal of creating a world-ranking research laboratory for experimental physics. There are also the first fruits of Rayleigh's occupancy of the Cavendish Chair, the beginning of his most productive research phase, as well as the first paper by the young Joseph John Thomson, to become Rayleigh's successor.

3.7 Undergraduate teaching

As soon as Maxwell was appointed to the Cavendish Chair in 1871, he began delivering courses of undergraduate lectures in physics as part of the Natural Sciences Tripos. Only in 1873 was physics recognised as a separate subject in the Natural Sciences Tripos, and then it was constrained not to include mathematical formulations involving the use of calculus and analysis, which were the province of the Mathematical Tripos (Wilson, 1982).[9] Thus, the teaching of heat, electricity and magnetism had to be carried out without the use of differential calculus. Fortunately, Maxwell had already published his book on the *Theory of Heat* (Maxwell, 1870) at more or less the appropriate level, using a minimum of mathematics and with a strong emphasis upon experiment. Maxwell began delivering this course in late 1871 and it was followed in subsequent years by courses on electricity and magnetism and on practical physics.

Instruction in experimental physics had already been introduced into the Natural Sciences Tripos by Coutts Trotter, who had been appointed a college lecturer in physical science by Trinity College in 1869. Trotter had studied experimental physics under

Table 3.5 Physics courses, 1877–1878, as advertised in the schedule for the Natural Sciences Tripos[a]

GARNETT (for first year men)	GARNETT	TROTTER	GARNETT or MAXWELL	MAXWELL	TROTTER
(for first year men)		(for second year men)		(for third year men)	
MICHAELMAS					
Mechanical physics	Electromagnetism, electrical measurements			Thermodynamics	
LENT					
Heat		Sound, light	Practical physics	Electricity, magnetism	Physiological optics, acoustics
EASTER					
Electricity, magnetism		Sound, light	Practical physics	Electromagnetism	

[a] Wilson, 1982.

Helmholtz and Kirchhoff in Germany. His lectures were open to anyone in the University on payment of the appropriate fee. Trotter was to continue to deliver these lectures as a college lecturer through the period of Maxwell's tenure of the Cavendish Chair. Wilson provides an illuminating extract from the Cambridge University *Reporter* of January 1877 of the Physics courses to be delivered in the academic year 1877–78 (Table 3.5).[10]

Maxwell took great pains over the preparation of his lectures, but they were relatively poorly attended. John Ambrose Fleming attended the courses of lectures shown in Table 3.5 and wrote:

> Maxwell generally gave a couple of courses of University lectures per year, which, however, were shockingly neglected, and I remember my surprise at finding a teacher whom I regarded as in the very forefront of knowledge lecturing to three or four students as his only audience. In fact, during one term Professor Maxwell gave a course of splendid lectures on Electro-Dynamics, the only audience being myself and another gentleman whose name I think was Middleton, at that time a fellow commoner of St. Johns. (Schuster, 1910)

There were a number of reasons for the slow development of physics in the Natural Sciences Tripos, among the most significant being the fact that attaining a high rank in the Mathematical Tripos was still regarded as the gold standard for a Cambridge degree. It was permitted to attend the physics courses in the Natural Sciences Tripos, but Maxwell was well aware of the fact that this 'distraction' from preparation for the Mathematical Tripos could result in a lower place in the class list. When Maxwell took up the Cavendish Chair, he remarked to Rayleigh that

> if we succeed too well, and corrupt the minds of youth till they observe vibrations and deflexions, and become senior Op[time]s instead of wranglers, we may bring the whole university and all the parents about our ears. (Harman, 1998)

Although the numbers were not large, Maxwell could report of the 1873 experimental physics papers of the Natural Sciences Tripos that several students had 'sent up answers which showed that Experimental Physics, treated without the higher mathematics, may be learned in a sound and scientific manner'. Wilson regards this cohort as the first who could legitimately be called 'NST physicists'. The numbers of students attending the experimental physics courses increased significantly in 1877 when the courses were judged to be appropriate for medical students; these lectures were given by Garnett.

Throughout the Maxwell era, Garnett provided strong support for Maxwell's vision of how the agenda for experimental physics should be carried out. Garnett described his contributions as follows:

> My own work was mainly to lecture to the more elementary students, and to render mechanical assistance to the Laboratory students and to those engaged in their own researches. I was occupied a good deal in devising experiments for lecture illustration and making suitable apparatus. Whenever I came upon a new piece of work, I showed it to the students, but did not publish it; and while I assisted in a good many of the investigations which Maxwell suggested, I never embarked on a continuous research of my own. (Quoted in Schuster, 1910)

3.8 What had been achieved

By the time of Maxwell's untimely death at the age of only 48, the Laboratory was well established and a good flow of scientific papers in the areas of experimental physics under the Laboratory's name began to appear in the literature (see Tables 3.3 and 3.4). Another piece of evidence for the impact of the training offered by the Laboratory is provided by Falconer's analysis of the career destinations of those students who attended Maxwell's lectures or carried out research in the Laboratory, either as undergraduates or postgraduates (Falconer, 2014).[11] Those whom she designates as postgraduates are shown in Table 3.6, which means that they went to lectures or carried out experimental physics after their first degree in the Mathematical or Natural Sciences Triposes. It is apparent that the vast majority came from the Mathematical Tripos and that they were generally wranglers. This list can be compared with the 'List of those who have carried out researches at the Cavendish Laboratory', an appendix to *A History of the Cavendish Laboratory, 1871–1910* (Fitzpatrick *et al.*, 1910). For the Maxwell era, the workers include only Hicks, Gordon, Saunder, Chrystal, Niven, MacAlister, Fleming, Poynting, Glazebrook and Shaw. It is no surprise that there is a strong overlap with the list of those who published papers in the scientific literature (see Tables 3.3 and 3.4).

Table 3.6 shows that, although many of those attending what was on offer by the Cavendish Laboratory did not proceed to significant research contributions in experimental physics, there was real interest in developments in the subject. The final destinations of the students show that they attained distinguished positions, often in the academic sphere, but also in broader areas of importance to the UK economy. This is even more striking in

Table 3.6 Maxwell's graduate students, 1871–1879[a]			
Graduate Student	Undergraduate degree	Research Topic	Career destination
Pendelbury, R.	MT, senior wrangler 1870	Not known	Cambridge University Lecturer in Mathematics 1888–1901
Tennant, J.	MT, 12th wrangler, 1870	Not known	Called to the Bar, 1875
Lamb, H.	MT 2nd wrangler 1872	Not known	FRS, Prof. Mathematics, Owens College Manchester University, 1885–1920
Webb, R.R.	MT, senior wrangler, 1872	Not known	St John's College lecturer 1877–1911
Hicks, W.M.	MT, 7th wrangler, 1873	See main text	Professor of Maths, Sheffield Vice-Chancellor, Sheffield University
Hart, H.	MT, 4th wrangler 1871	Not known	Professor of Mathematics, Royal Military Academy, Woolwich
Hayden, C.J.	MT, 10th wrangler 1869	Not known	Not known
Taylor, H.M.	MT 3rd wrangler 1865	Not known	FRS, Trinity College Tutor
Garnett, W.	MT, 5th wrangler, 1873	See main text	Professor of Mathematics, Durham College of Science, Newcastle 1884–93
Gordon, J.E.H.	MT, 12th Junior Optime, 1875	See main text	Electrical inventor and engineer
Darwin, G.H.	MT, 2nd wrangler, 1868	Not known	FRS, Plumian Professor of Astronomy, Cambridge 1883
Tottie, O.J.	Classics 3d class, 1873	Not known	Died 1878
Marshall, J.W.	MT, 8th wrangler, 1875	Lippman's capillary electrometer and viscosity liquid films	Assistant Master, Charterhouse
Saunder, S.A.	MT, 13th wrangler, 1875	BA units of resistance	Assistant Master Wellington College Gresham Professor of Astronomy
Chrystal, G.	MT, 3rd wrangler, 1875	See main text	Prof. Mathematics, Edinburgh Univ.
Lord, J.W.	MT, senior wrangler 1875	Voltaic cells	Fellow of Trinity College
Whitmell, C.T.	NST 1st class	Highly refractive liquids	H.M. Inspector of Schools
Niven, W.D.	MT, 3rd wrangler, 1866	Trajectory of shot	FRS, Director of Studies, Royal Naval College Greenwich
Pirie, G.	MT, 5th wrangler, 1866	Resistance measurements	Prof. Mathematics, Aberdeen
Schuster, A.	Fellow Commoner St John's	See main text	Prof. Physics, Manchester University
Sutherland, A.W.	MT, 8th wrangler, 1876	Anomalous dispersion of fuchsin	Actuary, National Life Insurance

Table 3.6 (cont.)			
Graduate Student	Undergraduate degree	Research Topic	Career destination
MacAlister, D.	MT, senior wrangler, 1877	See main text	Vice-Chancellor, Univ. Glasgow
Fleming, J.A.	NST, 1st class, 1880	Resistance balance	FRS, Prof. Elec. Eng. Univ. College, London
Poynting, J.H.	MT, 3rd wrangler, 1876	Sonometer, saccharimeter	FRS, Prof. Physics, Birmingham Univ.
Glazebrook, R.T.	MT, 5th wrangler, 1876	Waves in crystals, reflection and refraction in crystals	FRS, First Director, National Physical Laboratory
Middleton, H.	Fellow Commoner St John's	Not known	Engineer and inventor
Shaw, W.N.	MT, 16th wrangler, 1876	Mercury electrodes	FRS, Director, Meteorological Office
Taylor, S.	MT, 16th wrangler, 1859	Soap films	Fellow of Trinity College

[a] Falconer, 2014.

Falconer's analysis of the destinations of those designated undergraduates where their career destinations were in the universities, often as professors of mathematics, in the law, and in school teaching, often at distinguished public schools. These all contributed to the growing recognition of the central role of experimental physics for the understanding of physical phenomena and, ultimately, for the benefit of society.

These figures are supported by the detailed analysis by Wilson (1982) of the career destinations of those students who came through the Natural Sciences Tripos for the period 1871 to 1900. An abbreviated version of his table is included in the notes.[12]

But there was still a long way to go before the potential of the discipline was to make its full impact. Fortunately, Maxwell's successor, John William Strutt, the 3rd Baron Rayleigh, was the ideal person to continue the implementation of Maxwell's vision.

PART III

1879 to 1884

John William Strutt (1842–1919),
3rd Baron Rayleigh
Chancellor of Cambridge University (1908–1919)
Oil painting by Hubert von Herkomer

Rayleigh's quinquennium 4

4.1 Rayleigh's appointment

The original intention of the University was that the Cavendish Professorship should be a single-term appointment, but the wording of the regulations offered the possibility of extension of the position. It stated that the post was to

> terminate with the tenure of office of the Professor first elected unless the University by Grace of the Senate shall decide that the Professorship shall be continued.

Again, Sir William Thomson was an obvious choice, but by now he saw his future in Glasgow. The next obvious candidate was John William Strutt, who had succeeded to the Barony as the 3rd Lord Rayleigh on the death of his father in 1873. It was not a common occurrence for a senior member of the aristocracy and major landowner to become a professional academic, but Rayleigh had already demonstrated outstanding ability in theoretical and experimental physics. He had been senior wrangler in 1865 and first Smith's Prize winner. He had sought to improve his understanding of experimental physics, but there were limited opportunities. Fortunately, Stokes allowed Rayleigh to observe his experimental researches in his private laboratory and so he gained some familiarity with the challenges involved. By the time his name came forward as a candidate for the Cavendish Chair, he was already known for his explanation of the colour of the sky through the process of *Rayleigh scattering*, and he had already written profusely on a very wide range of topics in the physical sciences, including experimental researches carried out at the family home at Terling Place in Essex.

But there was more to it than simply academic prestige. The Rayleigh estates had made most of their earnings through the sale of wheat. With the opening up of the midwest American prairies for wheat production, the price of wheat had plummeted and made the Rayleigh estates unprofitable. The agricultural depression was to last for a number of years and so Rayleigh decided to change to dairy farming. In the late 1880s, 'Lord Rayleigh's Dairies' supplied milk for the capital (Rayleigh, 1924). Although not decisive in encouraging Rayleigh to put his name forward as a candidate for the Cavendish Chair, the appointment would certainly help the family weather the downturn in their fortunes.

There was a clear consensus among those involved in the physical sciences in Cambridge that Rayleigh was the right man for the job. William Cavendish, the Duke of Devonshire and the Chancellor of the University, wrote in characteristic style:

Chatsworth, Chesterfield, Nov. 15th, 1879.

My Dear Lord, –

I understand that there is a strong wish among the Cambridge residents originating with those who take a special interest in the experimental along with the mathematical treatment of Physics that your Lordship would consent to accept the chair of Experimental Physics, which has become vacant by the death of the much lamented Professor Clerk Maxwell.

Though it is perhaps somewhat unreasonable to ask you to undertake duties the discharge of which would involve heavy demands on your time, and might very probably be attended with no small personal inconvenience, I feel so strongly the advantage the University would derive from your acceptance of the office, that I hope you will allow me as Chancellor of the University, and also as taking a special interest in this Professorship, to support the appeal which I am told is about to be made to you, and to express a hope that you will consent to take the proposal into your favorable consideration.

I remain,

My dear Lord,

Yours faithfully,

Devonshire.

The memorial urging Rayleigh to put his name forward for the chair came from the most distinguished physical scientists in Cambridge, including Adams, Cayley, George Darwin, Dewar, Glazebrook, Niven, Routh, Shaw, Stokes and many others. Although the family were not keen on this move, Rayleigh took a perfectly pragmatic view of the situation. He wrote to his mother on 15 November 1879:

> Did you see in the paper that they are getting up a memorial to me at Cambridge to offer myself for the Professorship of Experimental Physics vacant by poor Maxwell's death? Neither of us much likes the idea of living at Cambridge but perhaps I ought to take it for 3 or 4 years, if they can get no one else fit for the post. It would fit pretty well with the agricultural depression.

On 12 December 1879, he was elected to the chair, both sides understanding that Rayleigh intended holding the post only until other suitable candidates appeared on the horizon.[1]

Rayleigh's scientific activities were characterised by their enormous breadth, spanning essentially the whole of classical physics. Some measure of this is the large number of phenomena in classical physics with which his name is associated, some examples being listed in Table 4.1. His mathematical prowess had been fostered by Edward Routh, the most famous of all the coaches for the Mathematical Tripos, and by Stokes. He was elected a fellow of Trinity College in 1866. His correspondence and friendship with Maxwell began in 1870. In 1871 Rayleigh married Evelyn Balfour, the sister of Arthur James Balfour, who was a friend of Rayleigh's at Trinity College and who was to become Prime Minister of Great Britain in 1902. In the next year, 1872, Rayleigh, whose health was always somewhat weak, suffered a severe bout of rheumatic fever. For convalescence, he and his wife spent the following winter in Egypt, accompanied by Eleanor (Nora) Balfour, Evelyn's sister. In 1876 Nora married the moral philosopher Henry Sidgwick, who with Anne Clough had founded Newnham College for women in the previous year; Nora had been one of the very first students at Newnham. She was an excellent mathematician and was to act as

Table 4.1 Examples of phenomena associated with Rayleigh's name

Topic	Definition	Date
Rayleigh scattering	Elastic scattering of light by particles much smaller than the wavelength of the light	1871
Plateau–Rayleigh instability	Break-up of falling stream of fluid into smaller packets	1879
Rayleigh criterion	Angular resolution of a telescope of aperture d, $\theta = 1.22\lambda/d$	1880
Rayleigh distribution	Continuous probability distribution for positive-valued random variables	1880
Rayleigh disc	Measurement of the absolute intensity of sound waves	1882
Rayleigh–Taylor instability	Instability of an interface between two fluids of different densities under gravity	1883
Rayleigh waves	Surface acoustic waves that travel on solids, now known as solitons	1885
Rayleigh law	The behaviour of ferromagnetic materials at low magnetic flux densities	1887
Rayleigh distance	The axial distance from a radiating aperture to a point at which the path difference between the axial ray and an edge ray is $\lambda/4$	1891
Rayleigh–Jeans law	The spectrum of electromagnetic radiation of a black body in the classical limit	1900
Rayleigh–Bénard convection	Natural convection, occurring in a plane horizontal layer of fluid heated from below	1916
Rayleigh number	Dimensionless number associated with the onset of Rayleigh–Bénard convection	1916
Rayleigh–Plesset equation	Ordinary differential equation which governs the dynamics of a spherical bubble in an infinite body of liquid	1917
Rayleigh flow	Diabatic flow through a constant-area duct where the effects of heat addition or rejection are considered	
Rayleigh–Ritz method	Approximate evaluation of normal modes of complex oscillating systems	
Rayleigh quotient	In mathematics, $R(M, x) = x^*Mx/x^*x$, where M is a complex Hermitian matrix and x a non-zero vector	
Rayleigh fading	A statistical model for the effect of multi-scattering propagation on a radio signal	

Rayleigh's collaborator throughout his period as Cavendish Professor – she was co-author of his papers on the determination of electrical standards. In the meantime, during the period in Egypt, Rayleigh began volume 1 of his great and influential *Theory of Sound*, which was published in 1877; volume 2 appeared in the following year (Rayleigh, 1877, 1878).

Although not trained in experimental physics, from his earliest scientific papers of 1869 onwards the topics studied involved a combination of theoretical analysis and comparison with experiments which generally he had carried out himself at his own laboratory. He continued to expand the scope of his experimental researches as a 'gentleman physicist' over a very wide range of topics in classical physics; the quality of these contributions was of the highest order.

Rayleigh was to occupy the Cavendish Chair for only five years, but this was a crucial period in consolidating the research and teaching programmes of the still-fledgling Laboratory. It was a matter of great fortune that Rayleigh was the ideal person to advance

Maxwell's vision of the role of experimental physics, and immediately he put all his considerable energies into making that vision a sustainable long-term proposition. Two conspicuous areas of Rayleigh's contributions were in the reorganisation of the teaching of experimental physics and the continuation of Maxwell's programme to establish a set of absolute standards of resistance, current and voltage, but now with significantly improved precision and accuracy. But he was phenomenally active in pursuing his own research interests, publishing more papers per year during his quinquennium as Cavendish Professor than at any other period of his career.

4.2 Teaching in Rayleigh's Cavendish: Glazebrook and Shaw

There were two important organisational changes during Rayleigh's tenure. The first resulted from the decision taken during Maxwell's tenure of the Cavendish Chair to split the Natural Sciences Tripos into two parts. From 1876, the examinations of the two parts were separated by six months; in 1881 the examinations of the two parts of the Tripos were separated by one year. Part I, spanning the first two years, covered the core elementary aspects of experimental physics and was taken with other experimental subjects. The third year, Part II, contained more advanced material and was appropriate for those students who wished to study for professional employment as physicists or as experts in cognate fields. In Part II, the students only studied a single subject. These changes consolidated the structure shown in Table N.1 in the notes, and correspond broadly to the structure of the Tripos which has existed ever since.

The second change was the formal opening of all physics classes to women on equal terms with the men. Maxwell had not been sympathetic to the presence of women in the Laboratory, their attendance being restricted to the summer term when Maxwell was at home at Glenlair in Scotland. Undoubtedly, Rayleigh's decision to admit women to the experimental physics courses was influenced by the views of Henry and Nora Sidgwick, who were pioneers in promoting the cause of women in the University.[2] In addition to being Rayleigh's collaborator and an excellent mathematician, Nora Sidgwick was appointed Vice-Principal of Newnham College in 1880 and then Principal in 1892.

The period immediately following Rayleigh's appointment was a time of change. A new laboratory assistant, George Gordon, was appointed who would accompany Rayleigh to Terling Place when he resigned from the Cavendish Chair in 1884. Most significantly, in 1880 Rayleigh appointed two demonstrators, Richard Glazebrook and William Napier Shaw, to replace Garnett. Glazebrook had graduated as fifth wrangler in 1876 and had carried out experiments on the determination of the unit of resistance under Maxwell. Napier Shaw had graduated 16th wrangler with first-class honours in the Natural Sciences Tripos in 1876 and had then worked under Warburg at Frieburg and under Helmholtz in Berlin.

Maxwell's intention was to keep a very light touch on the teaching of experimental physics, allowing the students to learn for themselves the skills of the discipline. While appropriate for the best students, Rayleigh was determined from the beginning to set up a more formal structure for learning the broad range of experimental techniques encountered in contemporary physics. There were only two textbooks on experimental physics, Kohlrausch's *Leitfaden der praktischen Physik (Guidelines to Practical Physics)*, published in 1870, and Pickering's *Elements of Physical Manipulation* (Kohlrausch, 1870; Pickering, 1873, 1876). Edward Pickering's achievement was particularly impressive. He was only 22 when he was appointed Thayer Professor of Physics at the Massachusetts Institute of Technology in 1868 and set about developing the first systematic course of instruction in experimental physics in the United States. These endeavours culminated in the publication of his impressive two-volume work on the *Elements of Physical Manipulation*.

The system of instruction created by Pickering was the model for innovations in the teaching of experimental physics which were to be introduced by Glazebrook and Shaw. In his survey of the Rayleigh years, Glazebrook (1910) quotes from Pickering's text:

> The method of conducting a Physical Laboratory for which this book is especially designed and which has been in daily use with entire success at the Institute is as follows. Each experiment is assigned to a table on which the necessary apparatus is kept and where it is always used. A board called an indicator is hung on the wall of the room, and carries two sets of cards opposite each other, one bearing the names of the experiments, the other those of the students. When the class enters the laboratory each member goes to the indicator, sees what experiment is assigned to him, then goes to the proper table where he finds the instruments required, and with the aid of the book performs the experiment....
>
> By following this plan an instructor can readily superintend classes of about twenty at a time and is free to pass from one to another answering questions and seeing that no mistakes are made.

Pickering's two volumes contain a vast range of physics experiments, including the use of astronomical instruments. Ten years after his appointment at the Massachusetts Institute of Technology, Pickering was appointed Director of the Harvard Observatory, where he applied all his experimental and managerial skills to the construction of telescopes and the reduction of the vast numbers of photographic plates for the Harvard Catalogues of Stars and Stellar Spectra. This enormous contribution to the advance of the new field of astrophysics is the work for which Pickering is best remembered.[3]

Rayleigh, Glazebrook and Shaw adopted the Pickering model and adapted it to what they expected of Cambridge undergraduates. Glazebrook and Shaw wrote a series of manuscript notebooks for each experiment and these formed the core of their textbook, *Practical Physics*, which was published in many subsequent editions (Glazebrook and Shaw, 1885). A flavour of the more organised approach to the teaching of experimental physics can be gained from Glazebrook's own words about how the teaching programme was organised (Glazebrook, 1910):

> The notice for the Michaelmas Term 1880 is to the effect that
>
> > The Professor of Experimental Physics will lecture on Galvanic electricity and Electromagnetism. Dr. Schuster will give weekly courses on Radiation. Mr. Glazebrook will give an elementary course of demonstrations on Electricity and Magnetism. Mr. Shaw will give a course of demonstrations on the Principles of Measurement and the Physical Properties of Bodies.

> The Laboratory will be open daily from 10 to 4 for more advanced practical work under the supervision of the Professor and Demonstrators for those who have the necessary training. Courses of demonstrations will be given during the Lent Term on Heat and Advanced Electricity and Magnetism, during the Easter Term on Light, Elasticity, and Sound.

> The various courses of demonstrations were each for two hours, three days a week.... In the Lent Term 1880, sixteen attended Lord Rayleigh's lectures, eighteen the elementary demonstrations on Heat, and fourteen the advanced demonstrations on Electricity and Magnetism. The numbers, however, soon increased and in the Lent term 1884 the elementary courses were duplicated.
> At the same time J.H. Randell of Pembroke and J.C. McConnell of Clare were appointed Assistant Demonstrators....
> In the Lecture list for the Lent Term 1880 ... the total number of courses for the term was five, while on the list for the Lent Term of 1884, the last term in which Lord Rayleigh lectured, the number of courses was ten.

For Glazebrook and Shaw, the understanding was that each of the two demonstrators was on duty three days a week, which occupied most of their time on these days. The other three days were available for research work, college duties and private teaching. The employment of two assistant demonstrators became possible as the students were required to pay fees for the courses they attended. In addition, the relatively long break during the summer term was an ideal time to make significant progress with their research projects. But there were still no lecturers – the core staff were the Cavendish Professor and his demonstrators, the contributions of the college lecturers being an essential and welcome addition to the growing programme of lectures and demonstrations.

The pattern of lectures and what are now called practical classes was set and has remained more or less the same as that set down by Rayleigh, Glazebrook and Shaw. Photographs of the practical classes in the early 1900s and in 1933 show remarkably little difference in approach, although the apparatus had moved on significantly (Figure 4.1).

As Falconer (2014) has pointed out, it is not trivial to work out how many students actually attended the lecture courses in physics. Figure 4.2 shows the total number of students in all subjects matriculating in the University from the year of the formation of the Natural Sciences Tripos until the end of the Rayleigh era (solid line). The total number in all subjects who graduated with honours after the three- or four-year course was much smaller (dotted line), while there were typically about 100 students completing the Mathematical Tripos (dashed lines) and a much smaller number completing the Natural Sciences Tripos (dot-dashed line). Evidently, many more students attended the Part I courses, which lasted two years, but they did not carry on to graduate with honours. Among these students would have been the medical students taught by Garnett in the late 1870s, until Rayleigh removed that course from the syllabus during his tenure. The curves show that the total numbers of natural science students remained at about 20–30 from 1874 onwards, but only a few of them obtained first-class honours. Even in the late 1880s, the number of natural scientists was only in the high twenties, and never more than one or two obtained first-class honours in physics alone.[4]

(a)

(b)

(a) The Part I practical class in the 1900s. (b) The Part I practical classes in 1933.

Fig. 4.1

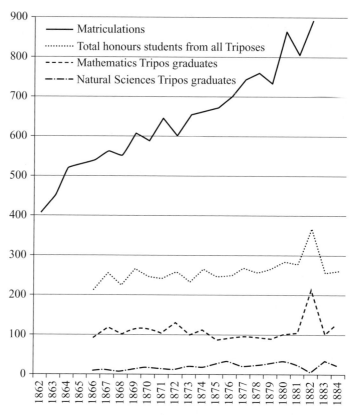

Fig. 4.2   The numbers of students enrolling (matriculating) in Cambridge and those graduating with honours in all subjects from 1862 to 1884. The lower curves show the numbers who graduated with honours in mathematics and in natural sciences. The 'blip' in 1882–83 is associated with the change of interval between Parts I and II examinations from six months to one year (Falconer, 2014).

4.3 Research in Rayleigh's Cavendish

Rayleigh was immediately faced with the problem of the funding of the research programme of the Laboratory. In his Report to the University of 1877, Maxwell had stated that
the chancellor's generous gift for the funding and equipping the Laboratory was complete.
In fact, the apparatus was state-of-the-art at that date, but the demands of experimental
research and teaching were continually evolving and more resources were needed to sustain the programme. Maxwell was reluctant to ask the chancellor for additional funding
for equipment and the University had contributed little to the effort. Maxwell had donated
all the equipment in his possession to the Laboratory and had also arranged for the donation of all the apparatus required for the work of the Electrical Standards Committee,
although it remained the property of the British Association. During the last few years of
his tenure of the Cavendish Professorship, Maxwell had also provided many hundreds of

pounds from his own pocket for the purchase of scientific equipment. More resources were clearly needed to maintain the programme of research and teaching.

The University was not in a position to contribute what was needed to maintain the programme and so Rayleigh sought to remedy the situation by requesting contributions from his friends. He himself contributed £500, a figure matched by the chancellor. Lord Powis contributed £100 and in the end about £1500 was raised. This alleviated the immediate problem, but the funding of an increasingly large programme of experimental research and teaching was always to be a costly business.

4.3.1 The determination of electrical standards

Rayleigh continued his broad range of research interests throughout his tenure of the Cavendish Chair but took the decision to continue Maxwell's programme of the determination of electrical standards, while increasing their precision by an order of magnitude or more, very much in the spirit of Maxwell's dictum that new science would come from improving the precision with which the laws of physics and the fundamental constants were known. The sequence of papers involved collaborations with Schuster and Sidgwick, with the assistance of Glazebrook and J.J. Thomson.

As discussed in Sections 1.4.4 and 3.6.1, the absolute determinations of the quantities of resistance, current and electromotive force were necessary in order to standardise the units of these quantities internationally. Rayleigh took up the challenge of improving these absolute measurements. Glazebrook's contribution to *A History of the Cavendish Laboratory, 1871–1910* gives a very clear account of the issues involved (Glazebrook, 1910). The units of resistance, current and electromotive force could all be expressed in terms of the basic cgs system of centimetres, grams and seconds as follows:

- The electromagnetic unit of current I flowing in a wire of unit length bent into the arc of a circle of unit radius produces unit force on a unit magnetic pole placed at the centre of the circle.
- The unit electromotive force $\mathcal{E}$ does unit amount of work per unit of time when causing unit current to flow round a circuit.
- The unit of resistance R is the resistance of a circuit in which unit electromotive force produces unit current.

It turned out that these values were not very convenient in practice and so a set of practical units was defined to which the names of ampere, volt and ohm were given. Thus:

- the ampere = 0.1 of the cgs unit of current
- the volt = 10^8 cgs units of electromotive force
- the ohm = 10^9 cgs units of resistance.

Since these quantities are related by $\mathcal{E} = RI$, only two of the units need to be determined absolutely. At the time Rayleigh began these experiments, the unit of resistance was only known to about 4% and the unit of current, as measured from the electrochemical equivalent of silver, was only known to 2%.

Box 4.1   **The absolute determination of resistance**

The spinning-coil method of determining the absolute measure of resistance involves using the electromotive force, the emf $\mathcal{E}$, induced in the rotating coil by the horizontal component of the Earth's magnetic field, $B_{\parallel}$. A simple outline of the calculation is as follows. The induced emf is

$$\mathcal{E} = -\frac{\partial \Phi}{\partial t} = \omega m \pi r^2 B_{\parallel}, \quad I = \frac{\omega \pi r^2 B_{\parallel}}{R}, \quad (4.1)$$

where ω is the angular frequency of rotation, m the number of turns of the coil, r the radius of the coils and R their resistance. The current I creates a magnetic field at the centre of the coil which, from the Biot–Savart law (1.9), is

$$B = \frac{\mu_0}{4\pi} \frac{Il}{r^2}, \quad (4.2)$$

where $l = 2\pi r m$ is the total length of the wires of the coils. The effect of this field is to deflect a magnetic dipole at the centre of the coil, the angular deflection being

$$\theta = \frac{B}{B_{\parallel}} = \frac{\mu_0}{4\pi} \frac{Il}{r^2 B_{\parallel}} = \frac{\mu_0}{4\pi} \frac{\omega \pi l}{R}, \quad (4.3)$$

thus determining absolutely the value of R in terms of the units of length, mass and time. In the cgs system, $\mu_0/4\pi$ is unity. Note that the magnetic flux density of the Earth's magnetic field has cancelled out.

Rayleigh and his colleagues made three assaults on the determination of the ohm. First, Rayleigh and Schuster refurbished the apparatus used by Maxwell and the British Association group and made a set of experiments using the spinning-coil method of determining the absolute unit of resistance (Rayleigh and Schuster, 1881). The method, outlined in Box 4.1, was implemented with meticulous care, with particular emphasis upon the need to determine precisely all the dimensions of the apparatus. They found, for example, that the correction needed for the self-inductance of the spinning coil had been underestimated by a factor of ten, and this resulted in a significant error in the B.A. unit of resistance. The coil was rotated at frequencies of 1.06, 2.12, 3.98 and 5.31 Hz using the water engine with the 50-foot drop in the tower section of the Laboratory. The frequency of rotation was measured precisely by a stroboscopic disc which was illuminated by a strong light modulated by a tuning fork driven at a known frequency.

The apparatus was not sufficiently rigid, and so Rayleigh had a larger second version built by the London instrument makers Messrs Elliott Brothers which was much stronger; this is the rotating-coil apparatus shown in Box 4.1. Again, the analysis of all the corrections and sources of error was meticulously carried out, setting the standard for precision

The large pair of precisely wound coils built by Chrystal. These coils were used in two separate determinations of the Fig. 4.3 absolute value of the standard of resistance by Rayleigh and Sidgwick and by Glazebrook.

in the measurement of the basic units of electromagnetism (Rayleigh, 1882b). The result was that the B.A. unit of resistance was found to be

$$1 \text{ B.A. unit} = 0.98651 \times 10^9 \text{ cgs units}, \tag{4.4}$$

only three parts in a thousand different from the value found by Rayleigh and Schuster (1881).

But there were other methods. In Hendrik Lorentz's method, as described by Glazebrook,

> A circular disc rotates about an axis perpendicular to its plane in a magnetic field due to a concentric coil. By balancing the fall in potential between the centre and the edge of the disc against that due to the passage through a resistance of the current producing the field, the value of the resistance is given by the formula $R = nM$, where R is the resistance, M the coefficient of mutual inductance between the coil and the disc and n is the number of revolutions of the disc per second. (Glazebrook, 1910)

Rayleigh realised that his new rotating-coil apparatus could be simply modified in order to carry out this experiment. Fortunately, he had at hand a pair of coils which had been very carefully wound by Chrystal so that the dimensions of the coils were very precisely known. Rayleigh and Sidgwick (1883) describe delightedly how they made use of the existing apparatus:

> [T]he diameter of the disc was chosen so as to be somewhat more than half that of the coils ... The disc was of brass and turned upon a solid brass rod as axle. This axle was mounted vertically in the same frame that carried the revolving coil in the experiments described in a former communication to the Society [Rayleigh, 1882b], an arrangement both economical and convenient, as it allowed the apparatus then employed for driving the disc and for observing the speed to remain almost undisturbed. The coils were supported

horizontally upon wooden pieces screwed on the inner side of the three uprights of the frame.

The experiment involved working out the theoretical value of M, the coefficient of mutual inductance, and great attention had to be paid to the many corrections to the simple relation $R = nM$. The final result of their experiments was

$$1 \text{ B.A. unit} = 0.98677 \times 10^9 \text{ cgs units.}$$

A third method had been developed by Glazebrook using Chrystal's coils, but these had never been used by Chrystal to carry out the estimate of the standard of resistance. The coils were very carefully constructed with very precisely known dimensions. A current was passed through the primary coil and, when this was broken, a current was induced in the secondary winding which depended upon exact knowledge of the coefficient of mutual inductance M between the two coils – this was determined by calculation for the configuration shown in Figure 4.3. The ratio of the currents in the two coils was measured with the same ballistic galvanometer and hence the value of the resistance of the secondary coil could be measured absolutely (Glazebrook *et al.*, 1883). Glazebrook's result was

$$1 \text{ B.A. unit} = 0.98665 \times 10^9 \text{ cgs units.}$$

Rayleigh reviewed the merits of each of these methods of determining the unit of resistance and the reliability of the results in a subsequent paper (Rayleigh, 1882a).

These procedures could be used to calibrate other resistances, which generally consisted of long sections of wire wound on bobbins, using a Wheatstone bridge or the more recent slide-wire version of the bridge developed in 1880 by Fleming, which gave excellent results. There was an alternative way of providing a convenient laboratory standard. Siemens had suggested that a column of mercury 1 mm^2 in cross-section and 1 m long at the temperature of melting ice would provide a convenient standard. As the precision of the determination of the ohm improved, the height of the column was fixed at 106 cm and this was ratified by the 1881 Paris Congress as the legal ohm. Rayleigh recalibrated this value using both his rotating-coil apparatus and Lorentz's technique and found that the length of the mercury column should be 106.21 cm (Rayleigh and Sidgwick, 1882).

Two further contributions to the establishment of electrical standards were undertaken with the same meticulous care for exact measurement. The first was the determination of the absolute value of the unit of current in terms of the amount of silver deposited by electrochemical action. Again, Glazebrook provides the most succinct description of the method.

> A coil was suspended from one arm of a balance with its plane horizontal and midway between two fixed coaxial coils of larger radius. The electrodynamic attraction between the suspended and fixed coils when carrying the same current can be balanced by weights in the opposite pan; it can also be calculated in terms of the current and the dimensions of the two coils; the current is thus measured absolutely in terms of the weight and the dimensions of the coils.
>
> If the same current also traverse a solution of nitrate of silver in a platinum bowl suitably arranged, it can also be measured in terms of the weight of silver it deposits,

and thus we can express a current whose value is known in absolute units in terms of the silver deposited. (Glazebrook, 1910)

The paper by Rayleigh and Sidgwick (1884) is a very impressive achievement, considering in detail every aspect of the experiment and the problems of obtaining a precise result. The final answer was that:

the number of grams of silver deposited per second by a current of 1 ampere

$$= 0.00111794.$$

As part of the same experiment, Rayleigh and Sidgwick determined the emf of the standard cell at the time, the Clark cell. The Clark cell was invented by Josiah Latimer Clark in 1873 and is a wet-chemical cell. In their paper, Rayleigh and Sidgwick describe the H-form of the cell in which an H-shaped glass vessel with zinc amalgam in one leg and pure mercury, surmounted by a layer of mercurous sulphate paste, in the other. The vessel was filled with zinc sulphate solution. The feature of the cell was that it produces a highly stable voltage, which they found to be 1.4345 volts at 15°C. This was essentially the international standard adopted in 1894.

4.3.2 Rayleigh's other researches

In addition to the experimental determination of absolute electrical units and standards, Rayleigh continued to pursue his many other interests in all aspects of classical physics during his five years as Cavendish Professor. A list of the titles of some of his papers published during these years makes impressive reading:

- On the resolving power of telescopes (1880)
- On a new arrangement for sensitive flames (1880)
- On copying diffraction gratings and some phenomena connected therewith (1881)
- On the electromagnetic theory of light (1881)
- On the dark plane which is formed over a heated wire in dusty air (1882)
- On a new form of gas battery (1882)
- On an instrument capable of measuring the intensity of aerial vibrations (1882)
- On the invisibility of small objects in a bad light (1883)
- The soaring of birds (1883)
- On the cuspations of fluid resting upon a vibrating support (1883)
- Investigation of the character of the equilibrium of an incompressible fluid of variable density (1883)
- On Laplace's theory of capillarity (1883)
- Distribution of energy in the spectrum (1883)
- On porous bodies in relation to sound (1883)
- On the circulation of air observed in Kundt's tubes, and on some allied acoustical phenomena (1884)

- Preliminary note on the constant of electromagnetic rotation of light in bisulphide of carbon (1884)
- On telephoning through a cable (1884).

According to J.J. Thomson, the paper which Rayleigh liked best was that on the soaring of birds. He addressed the problem of how large sea-birds such as albatrosses and gulls can soar effortlessly without moving their wings. The argument is simple and elegant and is one of the classics of aerodynamics (Crowther, 1974).

It is intriguing that the same skill in experimental physics and attention to essential detail were at the heart of his subsequent experimental work at Terling Place. This is no more apparent that in the experiments which led to the discovery of argon (Rayleigh and Ramsay, 1895). This relied upon precise measurements of the weights of volumes of atmospheric and laboratory gases and then the subsequent separation of argon from air. This notable achievement, carried out in collaboration with William Ramsay, was the discovery for which he was awarded the Nobel Prize in Physics in 1904.[5]

4.4 Rayleigh's colleagues, graduate students and their future employment

The activities and future careers of the demonstrators and some of the more prominent graduate students may be summarised as follows.

4.4.1 Richard Glazebrook

In addition to his work on the determination of electrical standards, Glazebrook worked on the refraction of light at the surface of uniaxial crystals (Glazebrook, 1882), as well as on numerous studies in birefringence. There are also papers on various theoretical topics on Maxwell's equations and the theory of molecular vortices, as well as topics in optics. As his memoirs of the Rayleigh era make clear, he was a major contributor to developing the role of precise measurement in the determination of fundamental constants. His endeavours were to culminate in his appointment in 1900 as first Director of the National Physical Laboratory, a project advocated by Rayleigh, particularly in the light of the founding of the German Physikalisch-Technische Reichsanstalt in Berlin in 1887.

4.4.2 William Napier Shaw

Shaw became deeply involved in the means of measuring the properties of the atmosphere, his publications including assessments of instruments such as evaporimeters and hygrometers. As part of the programme for establishing absolute measurements of current, he made precise measurements of the atomic weights of silver and copper. In 1905, he was appointed Director of the Meteorological Office, where he is credited with reorganising its operations and introducing physicists into the organisation for the first time, leading to the key discipline of atmospheric physics.

4.4.3 George and Horace Darwin

An example of the breadth of interests in the Laboratory is the abortive attempt by George and Horace Darwin to measure the effects of lunar tides. The idea, prompted by a suggestion by Sir William Thomson, was to investigate experimentally the disturbance of the gravitational acceleration due to the Moon and the effect of these lunar tidal forces upon the solid Earth. The experiment was of extreme delicacy using a bipolar pendulum, ultimately the experimenters being able to measure a disturbance of only 0.01 arcsecond (Darwin and Darwin, 1881). The attempt to measure tidal effects was unsuccessful, but the fact that it was attempted at all indicated the level of sophistication which was becoming expected of the Laboratory.

In the same year, 1881, the Cambridge Scientific Instrument Company was founded by Horace Darwin, who was son of Charles Darwin, and Albert Dew-Smith. The company produced advanced scientific apparatus for many departments in the University, with the Cavendish Laboratory as a major customer.[6]

4.4.4 Joseph John Thomson

Perhaps most significant for the future development of the Laboratory was the series of papers by J.J. Thomson which marked the beginning of his many key contributions to modern physics. Thomson began research at the Cavendish Laboratory in 1880 under Rayleigh. Although normally regarded as a theorist, he published a number of experimental papers and assisted in the experimental determination of the standard unit of resistance. In 1881 he published his paper *On the Electric and Magnetic Effects Produced by the Motion of Electrified Bodies* (Thomson, 1881), in which he demonstrated that, according to Maxwell's theory of the electromagnetic field, the mass of an electrically charged particle must increase with its speed, anticipating the more general result of the special theory of relativity. The reason this effect was found is because Maxwell's equations are implicitly Lorentz-invariant and so contain the increase in mass with velocity. Following the concept of Helmholtz, William Thomson and Maxwell that atoms could be considered as stable vortex rings, he published *The Vibrations of a Vortex Ring, and the Action upon each other of two vortices in a Perfect Fluid* (Thomson, 1882). Then, prophetically, he published the first of a series of papers *On the Theory of Electric Discharge in Gases*, which was to lead to the discoveries with which his name is associated (Thomson, 1883a). By the time of Rayleigh's resignation from the Cavendish Chair, Thomson had built up an impressive portfolio of original experimental and theoretical researches.

4.5 Rayleigh's legacy

Rayleigh's researches completely overshadowed the activities of all the other researchers in the Laboratory during his short period as Cavendish Professor. But his approach and insight were to set the standard and methodology of research in the Laboratory for generations. Indeed, as the story unfolds, Rayleigh's approach became the gold standard for experimental physics until the demands of experiment could no longer be accommodated by the hands-on expertise of a single investigator working with simple but clever apparatus.

Glazebrook, writing in 1910 as Director of the National Physical Laboratory, provides a moving tribute to what Rayleigh had achieved in his brief tenure of the Cavendish Chair. Noting that 50 scientific papers originated in Rayleigh's period at the Laboratory, he writes:

> These writings are all marked by the same characteristics: perfect clearness and lucidity, a firm grasp on the essentials of the problem, and a neglect of the unimportant. The apparatus throughout was rough and ready, except where the nicety of workmanship or skill in construction was needed to obtain the result; but the methods of the experiments, the possible sources of error, and the conditions necessary to success were thought out in advance and every precaution taken to secure a high accuracy and a definite result.
>
> Those of us who were privileged to work with him in those days, to learn of him how truth was to be sought and the difficulties of physical science unravelled, owe him a very real debt. His example and his work have inspired his successors in the Laboratory, and the great harvest of discoveries which has in recent years been reaped in Cambridge springs from seed sown by him.

Rayleigh's role in overseeing the implementation of what is now the standard approach to the teaching of experimental techniques and the appreciation of exact measurement were exactly what was needed for the training of the next generation of leaders of experimental physics (see Section 4.2). In many ways, Rayleigh's five years at the helm were crucial in converting Maxwell's vision into an internationally competitive physics laboratory which was to attract some of the greatest physicists and rapidly assure the ascendancy of experimental physics at Cambridge.

After his five years in Cambridge, he felt his contribution was complete. The family fortunes had revived and he wanted to concentrate on his personal research in his rapidly expanding laboratory at Terling Place, as well as to take on other endeavours in the national interest. He was President of the Royal Society from 1905 to 1908 and Chancellor of the University of Cambridge from 1908 until his death in 1919.

PART IV

1884 to 1919

Joseph John (J.J.) Thomson (1856–1940)
Oil painting by Arthur Hacker

The challenges facing J.J. Thomson   5

5.1 Thomson's election to the Cavendish Chair

William Thomson could still not be enticed back to Cambridge from Glasgow, despite a memorial, spearheaded by J.J. Thomson and sent to him with the signatures of a number of distinguished Cambridge scientists, urging him to stand. The Cavendish Chair was duly advertised and for the first time there was a competitive election. There were five candidates – Richard Glazebrook, Joseph Larmor, Osborne Reynolds, Arthur Schuster and Joseph John (J.J.) Thomson. Larmor was now Professor of Natural Philosophy at Queen's College, Galway, while Reynolds and Schuster both held professorships, in Engineering and Applied Mathematics respectively, at Owens College, Manchester. Somewhat to his own and the University's surprise, the electors offered the chair to the 28-year old Thomson. Glazebrook was Rayleigh's choice, but Davis and Falconer (1997) argue that he was too wedded to the programme of the precise establishment of physical standards to appeal to the electors. Reynolds was thought to be more an engineer than an experimental physicist. The electors took a bold gamble on Thomson, but he undoubtedly had the potential to become a distinguished physicist.

Thomson had entered Owens College, Manchester at the age of 14 and was fortunate to be instructed by inspiring scientists – Thomas Barker lectured on mathematics, Balfour Stewart on natural philosophy and Osborne Reynolds on engineering physics. According to Arthur Schuster, Stewart made extensive use of argument by analogy, very much in the Maxwell tradition, and Reynolds, the pioneer of turbulent flow in fluid dynamics, lectured on the role of vortices in fluid motion. In Thomson's final years at Owens College, Schuster lectured on Maxwell's *Treatise on Electricity and Magnetism* and Poynting, who was to become a lifelong friend, was developing his insights into the interpretation of Maxwell's equations. This training stood him in good stead when he was successful at his second attempt in obtaining funding in the form of an exhibition to Trinity College; in 1876 he matriculated studying for the Mathematical Tripos. Thomson was coached by Edward Routh and in 1880 he graduated second wrangler behind Joseph Larmor and joint first Smith's Prize winner.

In 1881, Thomson applied successfully for a fellowship at Trinity College and then became an Assistant Lecturer in Mathematics at the college. In 1882, the subject of the Adams Prize was 'A general investigation of the action upon each other of two closed vortices in a perfectly incompressible fluid', a topic ideally matched to his areas of interest and mathematical expertise. He duly won the Prize in 1882. Vortices were to play an important part in his subsequent theoretical ideas about the constitution of atoms and molecules, stimulated by the earlier researches of William Thomson and Maxwell, who demonstrated the stability of frictionless vortex rings.[1] In addition to his skills in mathematics, he had a deep interest in experiment, although he was certainly not a natural green-fingered experimentalist. During Rayleigh's period of office, he attempted to make accurate measurements of the ratio of the electromagnetic and electrostatic systems of units, but this was not a successful endeavour. Nonetheless, he developed a deep appreciation of experiment and its underpinning of the edifice of theory to a quite remarkable degree. In addition, he developed another remarkable facility. As Davis and Falconer (1997) write:

> Over the years he developed an extraordinary ability to see what was happening in the discharge tube and interpret it theoretically, as well as diagnose what was wrong when the experiment did not work. E.V. Appleton, one of his later research students, recalled that
>
>> 'J.J., although quite innocent of manipulative skill... was... unique in his ability to conceive some new experimental method or some way of overcoming practical difficulties.'

He was well known in Trinity as a good college man and safe pair of hands, who would not rock the boat. But there turned out to be much more to Thomson than that – his was an inspired choice, a classic example of the right person being in the right place at the right time to take the Laboratory forward to a position of world prominence.

5.2 Pure and applied physics in the 1880s

Lurking in the background was the continuing concern about Great Britain's role as an industrial nation. Lyon Playfair, speaking to the British Association for the Advancement of Science in 1885, stated:

> How is it in our great commercial centres, foreigners – German, Swiss, Dutch and even Greeks – push aside our English youth and take the places of profit? How is it in our colonies German enterprise is pushing aside English capacity? How is it that whole branches of manufactures, when they depend on scientific knowledge, are passing away from this country? The answer to these questions is that our systems of education are still too narrow for the increasing struggle for life. Even Oxford and Cambridge are still far behind a second-class German University. (Bennett *et al.*, 1993)

Werner von Siemens, the great German industrialist, recognised that the international competition in physics-based industries was intensifying. In his report to the Prussian government in 1893, he stated:

> Recently England, France and America, those countries which are our most dangerous enemies in the struggle for survival, have recognised the great meaning of scientific

superiority for their material interests and have zealously striven to improve natural scientific education through improvements in teaching and to create institutions that promote scientific progress. (Bennett *et al.*, 1993)

Siemens recognised early the importance of precise measurement and standards for industry, and had made a fortune through his pioneering work in telegraphy and electric machinery. In 1887, the Physikalisch-Technische Reichsanstalt, the Physical and Technical Institute of the German Reich, was founded, the goal being the supervision and direction of the calibration and establishment of metrological standards. The institute was the joint brainchild of Siemens and his lifelong friend Hermann von Helmholtz, who served as its first president until his death in 1892. Research areas included spectroscopy, photometry, electrical engineering and cryogenics. Siemens was instrumental in its establishment, providing a gift of 20,000 m^2 of land valued at over 500,000 reichmarks – the total cost was about 3.7 million reichmarks, the remainder being provided by the state.

Helmholtz was the key figure in the advance of German physics.[2] He had accepted the position of Professor of Physics at Berlin University in 1870 on condition that the University construct a new Institute of Physics, which was duly opened in 1871. In the succeeding years, a galaxy of great physicists was to work in the institute. His own graduate students included Max Planck, Albert Abraham Michelson, Wilhelm Wien, William James, Heinrich Hertz, Friedrich Schottky, Otto Lummer, Henry Augustus Rowland and Arthur König. Over the succeeding years, others who worked in the Institute included Albert Einstein, Gustav Robert Kirchhoff, Friedrich Kohlrausch, Emil Warburg, Walther Nernst, Max von Laue, James Franck, Gustav Hertz , Erwin Schrödinger, Peter Debye and many others. The combination of the Institute of Physics and the Physikalisch-Technische Reichsanstalt resulted in a formidable centre for experimental and theoretical physics, which was to play a key role in the quantum revolution of the early decades of the twentieth century. While the primary role of the Reichsanstalt was the maintenance of standards and precise measurement, research was also encouraged and the emphasis upon exact measurement was to play a central role in, for example, the determination of the spectrum of black-body radiation.

New experimental techniques were advancing internationally thanks to the development of new technologies. To mention only a few of the more important of these:

- Better vacuums were available to physicists in about 1855 through the invention of the Geissler pump by Johann Heinrich Wilhelm Geissler, the brilliant inventor and glass-blower, who worked at various German universities. The vacuum was produced by trapping air in a mercury column and forcing it down the column by the force of gravity. Typically pressures of 0.1 mm of mercury could be obtained within what became known as Geissler tubes.
- Higher voltages could be placed across discharge tubes with the invention of the Rühmkorff coil, which consisted of a transformer with a small number of windings on the primary and a very large number on the secondary, some of the secondary coils involving 10 km of wire.

- High-precision spectroscopy was revolutionised by the superb diffraction gratings produced by Henry Rowland, who pioneered the construction of high-precision ruling engines at the Johns Hopkins University in Baltimore.
- Also in the United States, Albert Michelson perfected the techniques of optical interferometry, which was to lead to the Michelson–Morley experiment of 1887 and then to the application of interferometry to ultra-high-resolution spectroscopy and the measurement of the angular diameters of red giant stars.
- Samuel Pierpoint Langley perfected the techniques of infrared photometry at the Allegheny Observatory in Pittsburgh through the invention of the platinum bolometer, with which temperature changes as small as 10^{-4} K could be measured. Although Langley's interests were primarily astronomical, the techniques were used to great advantage in the determination of the spectrum of black-body radiation in the final decade of the nineteenth century.
- In France, Hippolyte Fizeau and Léon Foucault's brilliant optical experiments had provided among the best estimates of the speed of light, while in the 1890s Charles Fabry and Alfred Pérot developed the Fabry–Pérot interferometer for high-resolution spectroscopy.

Siemens identification of England, France and America as Germany's main industrial and scientific competitors corresponded closely to those countries where the most advanced research in physics and physics-related subjects was being carried out.

5.3 The developing research and teaching programme

5.3.1 The overall research programme

The style of leadership of the Laboratory was highly dependent upon the character of the Cavendish Professor, and it was the Laboratory's good fortune that Thomson had exactly the necessary intellectual and personal qualities to maintain an enthusiastic and dedicated group of aspiring physicists. The research programme during Thomson's tenure was very much dominated by his own pioneering researches, which inspired all those about him. He did not impose his research priorities on the research students, but rather encouraged them to find their own projects. Newall provides a helpful summary of the various research activities which were carried out in the period 1885 to 1894 (Newall, 1910). The total number of communications amounted to 220 from about 36 contributors. Newall states: 'This number includes short notes, abstracts, memoirs, reports and summaries, but does not include many papers of a mathematical and theoretical nature that did not lead immediately to experimental work'. Further, he analysed the distribution by subject area among the 'more important papers', with the results shown in Table 5.1.

Some impression of the range of topics studied is given in Table 5.2, which shows in broad outline the principal research topics pursued by the assistant directors, University lecturers, demonstrators and assistant demonstrators who were employed in the Laboratory

Table 5.1 Distribution by discipline of the more important papers by subject area, 1884–1894[a]

General physics and properties of matter	13%
Heat	9%
Optics	9%
Electricity and magnetism	20%
Conductivity of gases	20%
Non-experimental	10%
Meteorological	2%
Reports and summaries	7%
Miscellaneous	10%

[a] See Fitzpatrick *et al.*, 1910.

in the period 1884 to 1900; the table also shows their subsequent career destinations.[3] The permanent staff are shown by the first five entries, with the names of the post holders shown in bold font. It is clear that a broad range of topics in experimental physics was being supported and the quality of the those listed can be judged by their career destinations.[4] There were limited opportunities within Cambridge for permanent employment in physics, but the demonstrators attained positions of distinction in other universities, in industry and in teaching. This accounts for the rapid turnover among the assistant demonstrators. As Thomson lamented to Richard Threlfall,

> There is such a demand for physicists that I am at my wits' end to get demonstrators for the next term as my old ones keep going off to new parts. (Wilson, 1982)

In 1893, Thomson introduced the fortnightly meetings of the Cavendish Physical Society, at which all the staff and students, as well as the increasing numbers of visitors, met to discuss their research, recent developments in other laboratories and all other items of topical scientific interest. Thomson was the natural leader of these discussions, with his broad interests in physics, his extensive imagination and good contacts with laboratories in the UK and abroad. Often, staff and students would present their recent research work, which would be discussed in a critical but friendly way before publication. These occasions were enhanced in 1895 when Mrs Thomson provided tea for all the participants.[5]

5.3.2 The growth of student numbers

During Thomson's tenure of the Cavendish Chair, the scale and scope of the activity within the Laboratory increased dramatically. David Wilson has provided an in-depth analysis of how physics developed in Cambridge from the opening of the Laboratory in 1874 to 1900, which includes the first 16 years of Thomson's occupancy of the Cavendish Chair (Wilson, 1982). Figure 5.1 shows the evolution of the numbers of research workers in the Laboratory over this period, the Cavendish Professor being excluded from the numbers. The diagram

Table 5.2 Research by staff, excluding J.J. Thomson, and their career destinations, 1884–1900[a]

Name[b]	Position	Dates	Research	Career destination
R.T. Glazebrook	Assistant Director Demonstrator	1891–98 1880–91	Standards, optical birefringence, optical instrumentation	Director, National Physical Laboratory
W.N. Shaw	Assistant Director University Lecturer Demonstrator	1899–1900 1887–90 1880–87	Hygrometry, meteorology, standards	Secretary and Director, Meteorological Office
L.R. Wilberforce	University Lecturer Demonstrator Asst Demonstrator	1900 1891–1900 1887–90	Optical interference, tests of electromagnetic theory	Professor of Physics, Liverpool University
G.F.C. Searle	University Lecturer Demonstrator	1900–1947 1891–1900	Properties of matter, practical physics	University Lecturer, Cambridge University
C.T.R. Wilson	University Lecturer Demonstrator	1901–1925 1900	Cloud physics, meteorological physics	Jacksonian Professor, Cambridge University
H.F. Newall	Demonstrator Asst Demonstrator	1887–90 1886–87	Conduction through poor conductors, tests of electromagnetic theory	Professor of Astrophysics, Cambridge University
S. Skinner	Demonstrator	1891–1904	Compressibility of liquids, emf standards	Principal, Southwest Polytechnic, Chelsea
J.H. Randell	Asst Demonstrator	1884–87	Practical physics	Died 1888
J.C. McConnell	Asst Demonstrator	1884–85	Optical atmospheric effects in ice crystals	Died 1887
R. Threlfall	Asst Demonstrator	1885–86	Conduction of electricity through gases	Professor of Physics, Sydney University Chemical manufacturer
H.L. Callendar	Asst Demonstrator	1887–88	Electrical determination of temperature	Professor Physics, Imperial College, London
T.C. Fitzpatrick	Asst Demonstrator	1888–1906	Electrolytic conduction, standards	President, Queen's College, Cambridge
R.S. Cole	Asst Demonstrator	1890–92	Pneumatics	Chief Engineer
C.E. Ashford	Asst Demonstrator	1892		Headmaster, Royal Naval College, Dartmouth
W.C.D. Whetham	Asst Demonstrator	1892–94	Velocities of ions in electrolytes	Fellow/Tutor, Trinity College, Cambridge
J.W. Capstick	Asst Demonstrator	1894–96	Specific heats of gases	Junior Bursar, Trinity College, Cambridge
P.E. Bateman	Asst Demonstrator	1896–99		

Table 5.2 (cont.)				
Name[b]	Position	Dates	Research	Career destination
R.G.K. Lempfert	Asst Demonstrator	1899–1900	Meteorology	Superintendent, Statistical Branch, Meteorological Office
J.S. Townsend	Asst Demonstrator	1900–01	Electrolysis, diffusion and collisions of charged ions	Wyckham Professor Physics, Oxford University

[a] Newall, 1910.
[b] Names in boldface are permanent staff members.

illustrates vividly the various changes in the composition of the cadre of research workers. The particularly striking features are:

- In the early years up to 1881, most students came through the route of the Mathematical Tripos (MT). A small number came through the Natural Sciences Tripos (NST).
- From about 1881 to 1895, the majority of students came through the NST or a combined MT + NST route to research. The numbers grew more or less steadily through these years.
- From 1895 to 1900, there was a marked increase in activity, particularly associated with the influx of students from outside Cambridge which was allowed by the change of regulations of 1895. Note that most of the Cambridge researchers came through the route of NST or combined MT + NST, while the number of those who had only passed through the MT remained at a relatively low level.

This last point is further illustrated by the statistics for the recruitment of graduate students by year, compiled by Falconer (1989). While the number of Cambridge graduates remained more or less constant through the most creative of Thomson's years as Cavendish Professor, the numbers from outside Cambridge increased dramatically. Furthermore, as she convincingly demonstrates, the latter students were attracted to the Laboratory to work with Thomson in his ground-breaking researches in atomic physics. The more traditional approach to research, involving increasing the precision of experimental measurement, tended to be the areas to which Cambridge graduates were attracted, probably reflecting the courses in experimental physics they had attended as undergraduates.

The problems of determining undergraduate student numbers who attended the physics courses were discussed in Section 4.2. The relatively slow and steady growth in the numbers is described by Wilson (1982) and by Falconer (1989) and illustrated in Figure 5.1(b). As emphasised in that section, the figures refer to the numbers of students completing the natural sciences course, but the number of physicists was only a small fraction of these. The total number taught was, however, much larger since less than half of the students who matriculated remained to complete the full course. In addition, these figures exclude

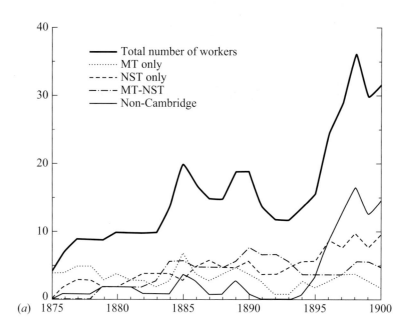

Recruitment of graduate students to the Cavendish Laboratory from
Cambridge and Elsewhere

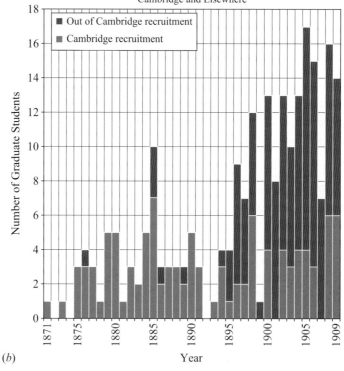

Fig. 5.1 (*a*) The growth in the number of research workers in the Laboratory from 1874 to 1900, from David Wilson's analysis (Wilson, 1982). Wilson excludes from the graph the Cavendish Professors and the six women researchers because 'women's educational and career expectations and patterns differed from men's. Of the six, only two had taken a Tripos (one NST chemistry, the other the MT) and apparently none came to the Cavendish after taking a degree elsewhere or being employed as a physicist in another university.' MT means Mathematical Tripos, NST means Natural Sciences Tripos. (*b*) The recruitment of graduate students per academic year to the Laboratory from 1871 to 1909, distinguishing those who were Cambridge graduates and those from other universities (Falconer, 1989).

			GLAZEBROOK			PRACTICAL
STOKES	THOMSON	ATKINSON	and SHAW	SHAW	HART	PHYSICS

Table 5.3 Physics courses, 1886–1887[a]

STOKES	THOMSON	ATKINSON	GLAZEBROOK and SHAW	SHAW	HART	PRACTICAL PHYSICS
			MICHAELMAS			
Hydrostatics, hydro-dynamics	E&M, Applications of dynamical principles		Physics (adv)	Physics (elem)	Physics for NST, Pt I	E&M (adv), mechanics and heat (elem), heat (adv)
			LENT			
Physical optics	Properties of matter	Heat (elem) (experimental)	Physics (adv)	Physics (elem)	Light for NST, Pt I	E&M (adv), optics and electricity (elem), mechanics and heat (elem)
			EASTER			
	Papers for candidates for NST, Pt II		Physics (adv)	Physics (elem)		Light (adv), sound (adv), optics and electricity (elem)

[a] Wilson, 1982.

the Part I medical students who were admitted following the reintroduction of classes in experimental physics for medical students in 1888. These developments put pressure on the use of space and equipment, as well as on the support of the research and teaching programme.

The lecture courses on offer in 1886–87 (Table 5.3) appeared to be more extensive than those offered in 1877–78 (see Table 3.5) but, as Falconer has pointed out, this is also subject to interpretation. If anything, the comparison of the tables shows the growing influence and scope of the Board of the Natural Sciences Tripos. For example, although Stokes delivered lectures throughout the period as part of his duties as Lucasian Professor, they were not included in the NST syllabus in 1878, but were included in the 1886–87 NST list. Conversely, in 1878 Trotter was included in the NST schedule as a College Lecturer, but he was only one of a number of such lecturers who lectured on physics subjects. By 1886–87 the NST lecture list had expanded to include others, such as Atkinson, who had nothing to do with the Laboratory, and Hart, who had minimal involvement with the Laboratory.

Thomson faced a number of challenges in developing the undergraduate lecture courses. The first change took place in the year before he took up his appointment, with the allowance that students from the Mathematical Tripos could proceed directly to the advanced part (Part II) of the Natural Sciences Tripos. But this produced as many problems as benefits. The central problem was that the Natural Sciences Tripos was deemed to be an integrated course in the natural sciences, and to introduce any mathematics beyond elementary manipulations would have made the course inaccessible to students from other areas of the Tripos, spoiling its integrated nature. Furthermore, the integration of the Mathematical

Tripos students into the course posed problems because they lacked the training in the basic skills required for experimental physics. This was partly remedied by the introduction of a special summer course for students who came from the Mathematical Tripos. But the problems of inadequate preparation of natural sciences students for the more mathematical aspects of the physics course posed a problem. Even as late as 1910, Thomson was writing:

> I feel very strongly that most of the students in the Laboratory would benefit if they studied mathematics more thoroughly than they do at present, while those who are studying applied mathematics would be much better equipped for research in that science if they came into touch with the actual phenomena in the Laboratory. (Thomson, 1910)

After various abortive attempts at reform, the problem was partially solved by the introduction of a mathematics course, given by Thomson and later by Wilberforce, which included the essential elements of mathematics, such as differential and integral calculus, which were necessary for the course in experimental physics. This changed the nature of the physics course, making it less accessible to the other experimental sciences. There remained tensions between the rigours of the mathematics taught in the Mathematical Tripos and the more pragmatic mathematical skills needed for the physics course.

5.3.3 The teaching staff

The staff increased gradually in response to these increased pressures. Despite his disappointment in not being appointed to the Cavendish Chair, Glazebrook continued in his role as demonstrator until he was promoted to Assistant Director of the Laboratory in 1890. He continued in this role until 1898 when he became the first Director of the National Physical Laboratory.

Napier Shaw had also been appointed by Rayleigh in 1879 to the post of Demonstrator at the same time as Glazebrook, and he held this post until 1887 when he was promoted to University Lecturer in Physics. With Glazebrook's departure in 1898, Shaw became Assistant Director of the Laboratory until he too left, to become Secretary of the Meteorological Council in 1900 and then Director of the Meteorological Office in 1905.

With the increasing teaching load, Thomson appointed a succession of assistant demonstrators from the time of his appointment in 1884. Very often these were chosen from among the most successful of the graduate students. Typically, those appointed to these posts stayed for a few years before being attracted to professorships or other more remunerative employment, reflecting the increased needs of the universities, schools and industry for qualified physicists. A typical example is Lionel Robert Wilberforce, who proceeded through the roles of research student (1879), Assistant Demonstrator (1887), Demonstrator (1890) and Lecturer (1900) before succeeding Oliver Lodge as Professor of Physics at the University of Liverpool in the same year.

George Searle was a research student of Thomson's from 1888 and then proceeded to a demonstratorship in 1890, followed by promotion to a University lecturership in 1900. He was a brilliant inventor of new and ingenious experiments, many of them being designed to illustrate points of theory. He was meticulous in the preparation of the material, both the equipment and the theory behind the experiments. The result was an outstanding training in experimental physics and precise measurement. He was to remain the backbone of the teaching of experimental physics until his retirement in 1935, but returned to the Laboratory during the Second World War and only finally retired in 1947 at the age of 83.

Following his pioneering researches as a research student, Charles (C.T.R.) Wilson was appointed a Demonstrator in 1899 and a University lecturer in 1900.

The pattern of the staff of the Laboratory consisting of the Cavendish Professor, two lecturers, two or three demonstrators and two assistant demonstrators was established by 1900 and was to remain roughly the same throughout Thomson's tenure of the Cavendish Chair.

5.3.4 The assistant staff

The University had made provision for one assistant to help the professor with his demonstrations and research, but this was inadequate to cope with the increasing requirements of teaching and research. Rayleigh's assistant George Gordon had joined him at Terling Place after Rayleigh resigned from the Cavendish Chair in 1884. D.S. Sinclair was appointed and held the post until 1887 when he went to an appointment in India. He was succeeded by A.T. Bartlett, who left in 1892 to become involved in electrical engineering, subsequently becoming head of the research department of the English General Electric Company. Next, W.G. Pye was appointed laboratory superintendent until he left in 1899 to found the scientific instrument company W.G. Pye and Co. Ltd. The original aim of this company was to provide reasonably priced apparatus for teaching in schools and universities, but it also evolved into Pye Radio Ltd, which became a successful business supplying a wide range of radio equipment. Pye was succeeded as laboratory superintendent by Frederick (Fred) Lincoln, who was to serve the Laboratory for more than 50 years and who managed the workshops with a rule of iron.

In addition, Thomson had inherited James Rolph, who had been hired by Rayleigh, and also appointed W.H. Hayles, an expert photographer who was responsible for perfecting Thomson's lecture demonstrations and illustrations. Ebeneezer Everett was appointed as Thomson's personal assistant, to construct and carry out the experiments designed by Thomson. Figure 5.2 shows the laboratory assistants in 1900. While it was expected that the research students would largely build their own equipment, expert technical assistance was on hand at all times.

5.4 Accommodation

With the great increases in the student population, accommodation within Maxwell's Laboratory became very crammed. The problem was exacerbated by the influx of medical students in 1888, bringing the total number of graduate and undergraduate students to over 150. As a temporary measure, a former anatomy room was taken over for the physics lectures, but this was not a satisfactory solution. The site immediately to the south of the

Fig. 5.2 The assistant staff in 1900. Front row from left: F.J. Welch, W.G. Pye, E. Everett. Back row from left: W.H. Hayles, J. Rolph, F. Lincoln, G.A. Bennett.

Laboratory became available in 1893 and this was assigned to the Laboratory by the University (Figure 5.3). The proposed three-storey building would have cost £10,000, which exceeded what the University was able to support. In view of the financial stringency, it was recommended that only the ground floor should be built to cope with the desperate need for experimental laboratories.

The extension was built to a design by William Fawcett, the designer of the original Laboratory, at a cost of £4000, of which £2000 was provided by Thomson from the undergraduate tuition fees which he was accumulating to provide apparatus for teaching and research. The extension was opened in 1896, just in time for the arrival of the influx of research students from outside Cambridge. But this was to provide only temporary relief from the chronic lack of accommodation for the Laboratory's programme.

The space problem was soon as bad as ever, some of the students having to find room in other departments. Some measure of the problem can be gauged from the increase in the number of graduate students to be housed in the Laboratory. From about 12 students in 1900, the numbers increased to about 25 by 1908 and to 30 immediately before the outbreak of the First World War.

Relief came thanks to the generosity of Lord Rayleigh, who donated £5000 of his Nobel Prize money, awarded in 1904 'for his investigations of the densities of the most important

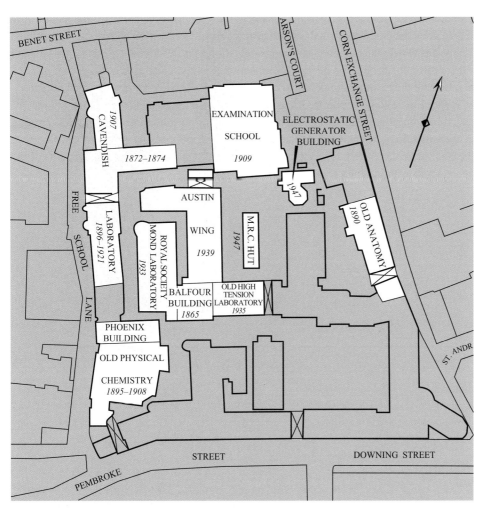

A map showing some of the major building developments at the New Museums site in the centre of Cambridge. Details of the evolution of the site from the earliest times is illustrated and described in the Appendix. There, Figures A.5–A.8 illustrate developments during the Thomson era. The map above is taken from the cover of the booklet *Cavendish Laboratory: The Need for a New Building*, written by Nevill Mott in June 1966, and which led to the move of the Laboratory to the West Cambridge site.

Fig. 5.3

gases and for his discovery of argon in connection with these studies'. Thomson immediately proposed applying this gift to the construction of an extension of the Laboratory. Part of the need for the new building arose from the fact that by 1900 the original buildings were contaminated by radioactive materials, which played havoc with the many electroscopes needed in the most sensitive experiments.

Plans were again drawn up by Fawcett, for a building costing between £7000 and £8000. Thomson supplemented Rayleigh's gift with a further £2000, which he had saved from student fees. Sindall of Cambridge came in with an offer for a building costing £7135 which was accepted by the University, and the building was completed in 1908. Lord Rayleigh succeeded William Cavendish as the Chancellor of Cambridge University and his first duty was to open the extension which was to bear his name. The new accommodation provided a large basement and a number of small rooms for research, a library, a chemical room and a demonstrator's room. The new accommodation was immediately put into use and for the first time all the lectures and practical teaching in physics could be given in the Laboratory. In earlier times, some of the lecture courses were given in the colleges, reflecting the latter's reluctance to give up the control of teaching. The consolidation of all physics teaching in the expanded Laboratory was also a reflection of the growing power of the Board of the Natural Sciences Tripos and the centralisation of responsibility for the curricula of the courses and examinations.

The Thomson era, 1884–1900: the electron   6

6.1 Thomson's agenda

Prior to his appointment as Cavendish Professor, Thomson was uncertain about his ability to sustain a career in experimental physics and had the option of concentrating on theoretical studies. Once his appointment was announced, however, he put all his energies into fostering the experimental activities of all his colleagues in the Laboratory and developing his own experimental agenda through the appointment of an expert laboratory assistant, whose salary he paid from his own pocket. Ebeneezer Everett was appointed to this post in 1888 and was to remain until 1930 when he retired due to ill health. In his touching obituary of Everett, published in *Nature* in 1933, Thomson wrote:

> Everett took a very active and important part in the researches carried on in the Laboratory, by students as well as by the professor. The great majority of these involved difficult glass blowing, which was nearly all done by Everett, as it was beyond the powers of most of the students. In addition to this, he made all the apparatus used in my experiments for the more than 40 years in which he acted as my assistant. I owe more than I can express to his skill and the zeal which he threw into his work. (Thomson, 1933)

For many reasons, 1895 was a turning point in the development of the Laboratory and so it is illuminating to list the titles of Thomson's papers up until that year (Table 6.1). It is apparent that his interests were wide-ranging, but increasingly there was a strong emphasis upon the conduction of electricity in gases, which dates back to his paper of 1883, the year before he was appointed to the Cavendish Chair. Before looking in more detail into these researches, let us review the role of analogy and model-building as tools for the understanding of physical phenomena.

6.1.1 Analogy and model-building

A distinctive feature of Maxwell's approach to the application of mathematics to natural philosophy was his ability to work by *analogy*. As early as 1856, when he was in his mid-20s, he described his approach in an essay entitled *Analogies in Nature*, written for the Apostles Club at Cambridge (Maxwell, 1856a). The essence of the technique can be caught in the following passage.

Table 6.1 A list of publications by J.J. Thomson, 1880–1895[a]

1880

On Maxwell's Theory of Light, *Phil. Mag. V*, **9**

1881

On some Electromagnetic Experiments with Open Circuits, *Phil. Mag. V*, **12**

On the Electric and Magnetic Effects produced by the Motion of Electrified Bodies, *Phil. Mag. V*, **12**

1882

On the Vibrations of a Vortex Ring, and its Action upon each other of two Vortices in a Perfect Fluid,
    Phil. Trans. Roy. Soc., **173**

On the Dimensions of a Magnetic Pole in the Electrostatic System of Units, *Phil. Mag. V*, **13**

On the Dimensions of a Magnetic Pole in the Electrostatic System of Units, *Phil. Mag. V*, **14**

1883

On the Determination of the Number of Electrostatic Units in the Electromagnetic Unit of Electricity,
    Phil. Trans. Roy. Soc., **174**

On the Theory of Electric Discharges in Gases, *Phil. Mag. V*, **15**

1884

On the Chemical Combination of Gases, *Phil. Mag. V*, **18**

On Electrical Oscillations and the Effects produced by the Motion of an Electrified Sphere,
    Proc. Lond. Math. Soc., **15**

1885

On some Applications of Dynamical Principles to Physical Phenomena, *Phil. Trans. Roy. Soc.*, **176**

Note on the Rotation of the Plane of Polarisation of Light by a Moving Medium, *Proc. Camb. Phil. Soc.*, **5**

The Vortex Ring Theory of Gases. On the Law of the Distribution of Energy among the Molecules,
    Proc. Roy. Soc., **39**

Report on Electrical Theories, *Brit. Assoc. Report.*

On the Formation of Vortex Rings by Drops falling into Liquids and some Allied Phenomena (with H.F. Newall),
    Proc. Roy. Soc., **39**

1886

Some Experiments on the Electric Discharge in a Uniform Electric Field, with some Theoretical Considerations
    about the Passage of Electricity through Gases, *Proc. Camb. Phil. Soc.*, **5**

Electrical Oscillations on Cylindrical Conductors, *Proc. Lond. Math. Soc.*, **17**

On an Effect produced by the Passage of an Electric Discharge through Pure Nitrogen (with R. Threlfall),
    Proc. Roy. Soc., **40**

1887

Some Applications of Dynamical Principles to Physical Phenomena, *Phil. Trans. Roy. Soc.*, **A178**

On the Dissociation of some Gases by the Electric Discharge, *Proc. Roy. Soc.*, **42**

Experiments on the Magnetisation of Iron Rods (with H.F. Newall), *Proc. Camb. Phil. Soc.*, **6**

On the Rate at which Electricity leaks through Liquids which are Bad Conductors of Electricity (with
    H.F. Newall), *Proc. Roy. Soc.*, **42**

1888

Electrical Oscillations on Cylindrical Conductors, *Proc. Lond. Math. Soc.*, **19**

The Effect of Surface Tension on Chemical Action (with J. Monckman), *Proc. Camb. Phil. Soc.*, **6**

Table 6.1 (cont.)

1889

The Resistance of Electrolytes to the Passage of very rapidly alternating Currents, with some Investigations on the Times of Vibration of Electrical Systems , *Proc. Roy. Soc.*, **45**

Note on the Effect produced by Conductors in the neighbourhood of a wire on the Rate of Propagation of Electrical Disturbances along it, with a Determination of this Rate, *Proc. Roy. Soc.*, **46**

Specific Inductive Capacity of Dielectrics when acted on by very rapidly alternating Electric Forces, *Proc. Roy. Soc.*, **46**

The Application of the Theory of Transmission of Alternating Currents along a Wire to the Telephone, *Proc. Camb. Phil. Soc.*, **6**

On the Effect of Pressure and Temperature on the Electric Strength of Gases, *Proc. Camb. Phil. Soc.*, **6**

On the Magnetic Effects produced by Motion in the Electric Field, *Phil. Mag. V*, **28**

1890

On the Passage of Electricity through Hot Gases (two papers), *Phil. Mag. V*, **29**

Some Experiments on the Velocity of Transmission of Electric Disturbances in Gases and their application to the striated discharge through Gases, *Phil. Mag. V*, **30**

Determination of 'v', the Ratio of the Electromagnetic Unit of Electricity to the Electrostatic Unit (with G.F.C. Searle), *Phil. Trans. Roy. Soc.*, **A181**

1891

On the Rate of Propagation of the Luminous Discharge of Electricity through a Rarified Gas, *Proc. Roy. Soc.*, **49**

On the Illustration of the Properties of the Electric Field by means of Tubes of Electrostatic Induction, *Phil. Mag. V*, **31**

Note on the Conductivity of Hot Gases, *Phil. Mag. V*, **31**

On the Discharge of Electricity through Exhausted Tubes without Electrodes (two papers), *Phil. Mag. V*, **31**

On the Electric Discharge through Rarified Gases without Electrodes, *Proc. Camb. Phil. Soc.*, **7**

On the Absorption of Energy by the Secondary of a Transformer, *Proc. Camb. Phil. Soc.*, **7**

1893

The Electrolysis of Steam, *Proc. Roy. Soc.*, **53**

On the Effect of Electrification and Chemical Action on a Steam Jet and of Water Vapour on the Discharge of Electricity through Gases, *Phil. Mag. V*, **36**

1894

On the Electricity of Drops, *Phil. Mag. V*, **37**

On the Velocity of Cathode Rays, *Phil. Mag. V*, **38**

Electric Discharge through Gases, *Jour. Roy. Inst.*, **14**

1895

On the Relation between the Atom and the Charge of Electricity carried by it, *Phil. Mag. V*, **40**

On the Electrolysis of Gases, *Proc. Roy. Soc.*, **58**

A Method of Comparing the Conductivities of badly conducting Substances for rapidly Alternating Currents, *Proc. Camb. Phil. Soc.*, **8**

[a] Fitzpatrick and Whetham, 1910.

Whenever [men] see a relation between two things they know well, and think they see there must be a similar relation between things less known, they reason from one to the other. This supposes that, although pairs of things may differ widely from each other, the *relation* in the one pair may be the same as that in the other. Now, as in a scientific point of view the *relation* is the most important thing to know, a knowledge of the one thing leads us a long way towards knowledge of the other.

In the context of mathematical physics, the approach consists of recognising mathematical similarities between quite distinct physical problems and seeing how the successes of one theory can be applied to different circumstances. In relation to electromagnetism, he found formal analogies between the mathematics of mechanical and hydrodynamical systems and the phenomena of electrodynamics. Throughout this work he acknowledged his debt to William Thomson, who had made substantial steps in mathematising electric and magnetic phenomena. Indeed, as illustrated in Section 1.4.4, Thomson had already made use of the analogy between the diffusion of heat and the transmission of signals down a transmission line in his analysis of the propagation of signals in a transatlantic cable (Thomson, 1855). Maxwell's great contribution was not only to take this process very much further, but also to give it real physical content.

As early as 1856, Maxwell published the first of his great papers on electromagnetism, *On Faraday's lines of force* (Maxwell, 1856b). In the first part of the paper, he enlarged upon the technique of analogy and drew particular attention to its application to the stream-lines of incompressible fluid flow and magnetic lines of force. In deriving the first complete description of electromagnetism in 1861–62, Maxwell adopted a physical model for electromagnetic phenomena. In his remarkable series of papers entitled *On physical lines of force*, magnetism was considered to be essentially rotational in nature. His aim was to devise a mechanical model for the medium filling all space which could account for the stresses that Faraday had associated with magnetic lines of force – in other words, a mechanical model for the *aether*, which was assumed to be the medium through which light was propagated. It was through the analysis of this mechanical model of the aether (or vacuum) that he first showed that electromagnetic waves propagate at the speed of light.[1] In his intriguing book *Innovation in Maxwell's Electrodynamics*, David Siegel has carried out a detailed analysis of these papers, vividly demonstrating the richness of Maxwell's insights in drawing physical analogies between mechanical and electromagnetic phenomena (Siegel, 1991). This is no more apparent than in Maxwell's remarkable mechanical model of 1865 to illustrate the process of electromagnetic induction between two current loops (Box 6.1). This mechanical arrangement could reproduce all the known phenomena of electromagnetic induction.

These examples make the important point that, whilst Maxwell's mathematical analysis of varying electric and magnetic fields pointed the way to the future, his thinking was strongly based upon mechanical thinking and the construction of more or less ad hoc models which were intended to cast light on the relations between physical quantities. It is striking that this approach of model-building to describe physical phenomena was in vivid contrast to those of the French and German physicists, who adopted a more abstract and mathematical approach. Heilbron (1977) provides an illuminating and entertaining exposition of this dichotomy. He writes:

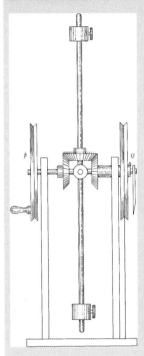

Maxwell was a master in the use of analogy to guide his thinking and analysis of physical problems. This example is described in a footnote on page 228 of the second edition of volume 2 of his *Treatise on Electricity and Magnetism* (Maxwell, 1873).

NOTE.– There is a model in the Cavendish Laboratory designed by Maxwell which illustrates very clearly the laws of the induction of currents. It is represented in [the figure]. P and Q are two discs, the rotation of P represents the primary current, that of Q the secondary. The disks are connected together by a differential gearing. The intermediate wheel carries a fly-wheel the moment of inertia of which can be altered by moving weights inwards or outwards. The resistance of the secondary circuit is represented by the friction of a string passing over Q and kept tight by an elastic band. If the disk P is set in rotation (a current starting in the primary) the disk Q will turn in the opposite direction (inverse current when the primary is started). When the velocity of rotation of P becomes uniform, Q is at rest (no current in the secondary when the primary current is constant); if the disk P is stopped, Q commences to rotate in the direction in which P was previously moving (direct current in the secondary on breaking the circuit). The effect of the iron core in increasing the induction can be illustrated by increasing the moment of inertia of the fly-wheel.

[FitzGerald had] in mind mechanical models, that is, detailed representations of physical phenomena, especially light and electromagnetism, in terms of the motions and interactions of hypothetical particles or media . . . The representations were not meant to be taken literally. To quote Fitzgerald again:

> To suppose that (electromagnetic) aether is at all like the model I am about to describe (which is made from tennis balls and rubber bands) would be almost as bad a mistake as to suppose a sphere at all like $x^2 + y^2 + z^2 = r^2$ and to think that it must, in consequence, be made of paper and ink.

In the same vein, (J.J.) Thomson, in his *Recent Researches in Electricity*, urges his readers to shop about for a model of electrodynamic interactions with which they feel comfortable: 'The question as to which particular model (of illustration) the student should adopt is for many purposes of secondary importance, provided that he does adopt one.' The same physicist might on different occasions use different and even conflicting pictures of the same phenomena, a practice much approved of by Lord Kelvin who, of course, repeatedly insisted that his models were not 'true of nature', but a disjointed series of tableaux which appeal to the imagination.

This approach was not to the taste of the continental physicists. Again quoting Heilbron (1977):

The leading continental theoreticians – the Kirchhoffs, Helmholtz', the Poincarés – did not hold English model making in high esteem. Kirchhoff, according to Hertz, found it 'painful . . . to see atoms and their vibrations willfully stuck in the middle of a theoretical discussion.' Helmholtz, who much admired Kelvin, freely confessed his want of taste or talent for doing physics in the English manner. As for Poincaré, in his experience all Frenchmen were oppressed by 'a feeling of discomfort, even of distress', at their first encounter with the writings of Maxwell.

As we will see, Thomson belonged firmly to the English school of model-building but, through his success in the Mathematical Tripos and his winning of the Adams Prize, had demonstrated his outstanding ability in mathematical physics. These abilities, combined with his dedication to his personal support of experimental physics, made him the ideal person to tackle the challenges of late nineteenth-century physics. In one way or another it turned out that his researches revolved around the elucidation and understanding of atomic structure.

6.1.2 Thomson and vortex models of atoms and molecules

Nowhere was the English approach to model-building more apparent than in Thomson's espousal of vortex models of atoms and molecules. The attraction of this approach to understanding atoms was clearly stated by Thomson in the introduction to his book *A Treatise on the Motion of Vortex Rings* (Thomson, 1883b), the published version of his Adams Prize submission:

> This [vortex] theory, which was first started by Sir William Thomson, . . . has *à priori* very strong recommendations in its favour. For the vortex ring obviously possesses many of the qualities essential to a molecule that has to be the basis of a dynamical theory of gases. It is indestructible and indivisible; the strength of the vortex ring and the volume of liquid composing it remain forever unaltered; and if any vortex ring be knotted, or if two vortex rings be linked together in any way, they will retain for ever the same kind of be-knottedness or linking. These properties seem to furnish us with good materials for explaining the permanent properties of the molecule.

The influence of Maxwell in endowing the vacuum with physical properties and of Reynolds in his lectures on vortices is apparent. In Thomson's model, the atoms were no more than frictionless vortex rings in the aether which preserved their form, like smoke rings in air. Furthermore, linked vortices would form stable structures which he identified with the combination of atoms into molecules (Box 6.2). The appeal of the model for Thomson was that atoms were abolished and were replaced by stable mechanical structures in the aether. The mathematical elaboration of these concepts was demanding, and indeed most of his book of 1883 is a mathematical treatment of vortices, systems of vortices and their stability when linked together.

Thomson was strongly influenced by another analogy suggested by the American physicist Alfred Mayer's experiments with floating magnets under the influence of an external

Maxwell, J.J. Thomson and continuum mechanics **Box 6.2**

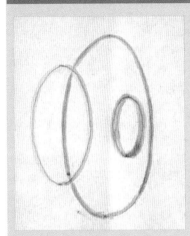

Navarro (Navarro, 2005) and Falconer[2] have provided insight into Maxwell's and Thomson's adherence to vortex models. The basic physics of their models of atoms involved the application of continuum mechanics to the aether, which was presumed to be present throughout all matter and space. The hypothesis of vortices provided a means of creating discrete objects in the continuum of the aether. For this reason it is preferable to use Thomson's expression 'corpuscle' rather than the term 'particle', which conjures up the image of a billiard-ball. In the vortex model, the corpuscle acquires charge as a boundary effect at the end of a Faraday tube of force. The aether was regarded as primary, and matter a secondary phenomenon. In contrast, Rutherford had no hesitation in regarding the electron and the nucleus as charged 'billiard-balls' which were subject to the laws of Newtonian mechanics for discrete particles.

(left) A frame from Maxwell's zoetrope strip showing three interacting vortices, a stable configuration representing a triatomic molecule.

magnetic field (Mayer, 1878). The magnets were allowed to float with their north poles all pointing upwards and so they would naturally repel each other. Then he placed the south pole of a strong magnet above the floating magnets, counteracting their tendency to disperse. He found that the magnets took up the remarkable ring-like structures shown in Figure 6.1. The intriguing feature was that only a certain number of magnets could form the stable structures shown in the diagrams. These images were to have a long-lasting impact upon Thomson's thinking, long after he had dispensed with vortex models of atoms. In 1883, Thomson argued that, by analogy with Mayer's magnets, the interacting vortices would adopt similar configurations – the legacy of Maxwell's analogy between fluid dynamical vortices and magnetic fields reappears in a different guise. In his heroic mathematical analysis of vortex atoms and molecules in his Adams Prize essay, Thomson showed that only certain configurations of vortices would be stable, similar to those described by Mayer, and he identified these with the structures of molecules (Thomson, 1883b). He was to maintain this vortex picture of atoms and molecules into the 1890s, until a quite different picture of the structure of the atom was forced upon him by experiment, but reminiscences of the floating magnets would reappear in his 'plum-pudding' model of the atom in the 1900s.[2]

Thomson was to continue his theoretical work throughout his career, but the work which was to lead to the discovery of the electron was focussed upon experiments which he led and inspired in an effort to understand the phenomena associated with the conduction of electricity through gases.

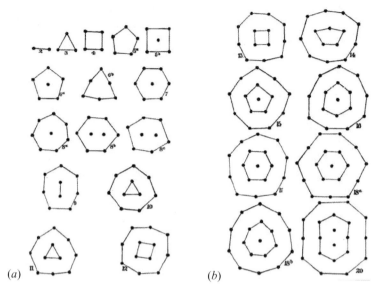

Fig. 6.1 (*a*) Geometric arrangements found by Mayer (1878) for 1 to 12 floating magnetic needles with north poles pointing upwards in the field of the south pole of a strong magnet. (*b*) The same as (*a*) but for 13 to 20 floating magnets.

6.2 The conduction of electricity through gases

The vacuum tube was to play a central role in the discoveries of the 1890s. The basic apparatus consisted of a thin-walled glass vacuum tube with positive and negative electrodes between which a high voltage could be maintained. It was realised in the early nineteenth century that gases are poor conductors of electricity since they are electrically neutral. At very low pressures and high voltages, however, electrical discharges were observed in vacuum tubes. Geissler tubes became popular scientific toys and revealed a remarkable range of coloured discharges. William Crookes (Figure 6.2(*a*)) began his systematic studies of the conduction of electricity through gases in 1879. The appearance of the discharge in the vacuum tube changed as the pressure of the gas decreased (Figure 6.2(*b*)). He discovered that as the pressure was lowered, the negative electrode, the cathode, appeared to emit rays. At low enough pressures, the walls of the tube glowed green due to phosphorescence. It was found that objects placed in the tube cast a shadow on the walls of the tube opposite the cathode. Crookes investigated the properties of these *cathode rays*, inferring that they travel in straight lines and cause fluorescence in the glass at the end of the tube. Furthermore, the impact of the cathode rays on the glass walls resulted in the generation of a great deal of heat. It was inferred that a stream of cathode rays was the cause of the fluorescence observed on the walls of the tube.

Maxwell considered that the understanding of the phenomena observed in Crookes tubes was a suitable subject for research, but it was clearly a very complex problem. In addition, the subject was regarded as suspect by serious physicists. Davis and Falconer (1997) write:

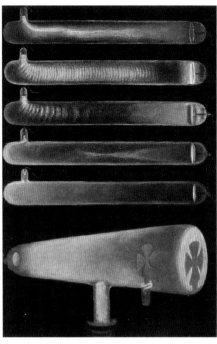

(*a*) (*b*)

(*a*) William Crookes in a cartoon by Spy (courtesy of the Science Photo Library). (*b*) The top five sketches show the phenomena observed as a Crookes tube is evacuated. In the fifth sketch, at the lowest pressure, the inner surface of the tube shines with a greenish glow. The last sketch shows the shadow cast by a Maltese cross placed in the tube when it is operated under a vacuum (courtesy of Jean-Loup Charmet/Science Photo Library).

Fig. 6.2

> At this time experimental work on discharge was very poorly regarded by the scientific establishment. No serious academic would touch it, and it was left in the hands of those for whom science was a sideline, such as Warren de la Rue and Hugo Muller, William Spottiswoode, and Henry Moulton, or those outside conventional academic circles whose livelihood relied on more spectacular science, such as William Crookes.

None of this discouraged Thomson, who threw all his energies into the experimental and theoretical study of the conduction of electricity through gases. Unlike previous experimental work in the Laboratory, which had involved precise and controlled experiment, the discharge experiments were exploratory and generally more qualitative than quantitative. Thomson's approach was quite unlike that of this predecessors but proved to be ideal for unravelling of the complex physics of gaseous discharges.

Thomson regarded the source of the discharge as the disruption of vortex molecules by the action of an electric field. He drew an analogy with the process of electrolysis, the breaking of chemical bonds in atoms, or more properly, ions. Because of this analogy, some insight might be gained from the dielectric breakdown in poorly conducting liquids or the decomposition of polyatomic gases by spark discharges. Thomson and his students explored both of these topics experimentally in the papers entitled *On the Rate at which Electricity leaks through Liquids which are Bad Conductors of Electricity*

(Thomson and Newall, 1887) and *On the Dissociation of some Gases by the Electric Discharge* (Thomson, 1887). The latter studies carried on well into the 1890s (Thomson, 1893b, 1895), but these endeavours resulted in little advance in the understanding of the nature of the electrical breakdown. The paper of 1893, entitled *On the Effect of Electrification and Chemical Action on a Steam Jet and of Water Vapour on the Discharge of Electricity through Gases*, was, however, to have future significance in that it described how the electrification of water droplets could reduce their tendency to evaporate. C.T.R. Wilson was about to join the Laboratory and begin his painstaking and far-reaching studies of the conditions for the formation of water droplets.

In the meantime, two major discoveries strongly impacted the course of research in the Laboratory. The first was Hertz's brilliant series of experiments, first reported in 1887, which demonstrated beyond any doubt that electromagnetic waves have all the properties of light waves. Examples of the types of emitter and detector he used are shown in Box 6.3. Electromagnetic radiation was emitted when sparks were produced between the large spheres by the application of a high voltage from an induction coil. The experiments demonstrated convincingly that there exist electromagnetic waves of frequency about 1 GHz and wavelength 30 cm which behaved in all respects like light. These great experiments were conclusive proof of the validity of Maxwell's equations. These demonstrations had eluded the research workers in the Cavendish Laboratory, but they responded enthusiastically to Hertz's triumph. The impact of this work was even greater in continental Europe where Maxwell's theory of electromagnetism had been received with some caution, but now the theory moved to centre stage and had ramifications for the whole of physics.

The second discovery was much more directly related to Thomson's experimental programme. In 1892, Hertz discovered that cathode rays could pass through thin metal foil, a process which was not possible for rays of typical atomic dimensions. The alternative view, aggressively promoted by Philipp Lenard, was that the rays were 'aether disturbances', similar to ultraviolet radiation, and these caused the fluorescence observed in discharge tubes. Because of the ability of the cathode rays to penetrate metal foils, he was able to create a thin foil 'window' in the vacuum tube which enabled the vacuum to be maintained while at the same time the rays could be studied outside the tube. Those rays which appeared to have escaped from the discharge tube through the metallic window were named *Lenard rays*. In late 1895 Jean Baptiste Perrin showed, by collecting the cathode rays in a Faraday cylinder inside a discharge tube, that they were negatively charged particles (Perrin, 1896). In the same experiment he confirmed the bending of the trajectories of the cathode rays by a magnetic field.

By the early 1890s, many physicists had lost faith in vortex models of atoms, arguing that the atoms, or ions, in molecules had to be discrete particles and were responsible for the phenomena of electrolysis. Thomson wrote up the fruits of his experiments and theoretical studies in his book *Notes on Recent Researches in Electricity and Magnetism* (Thomson, 1893a), intended as a supplement to Maxwell's great *Treatise on Electricity and Magnetism*. Thomson's remarkable volume describes all the experimental and theoretical advances in electricity and magnetism since the publication of Maxwell's *Treatise*. It includes a comprehensive survey of Hertz's experiments, as well as the theoretical apparatus needed to interpret these experiments in terms of Maxwell's theory of the electromagnetic

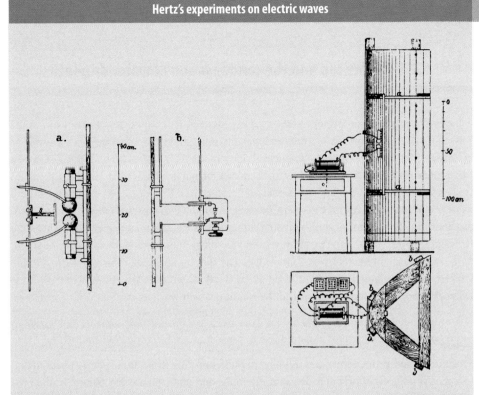

The figure shows Hertz's apparatus for the generation and detection of electromagnetic radiation (Hertz, 1893). The emitter a produced electromagnetic radiation in discharges between the spherical conductors. The detector b consisted of a similar device with the jaws of the detector placed as close together as possible to achieve maximum sensitivity. The emitter was placed at the focus of a cylindrical paraboloid reflector to produce a directed beam of radiation.

After a great deal of trial and error, Hertz was able to generate waves of relatively short wavelength. The frequency of the waves at resonance could be found from the resonant frequency of the emitter and detector, which he took to be $\omega = (LC)^{-1/2}$, where L and C are the inductance and capacitance of the dipole. He measured the wavelength of the radiation by placing a reflecting sheet at some distance from the spark gap emitter so that standing waves were set up along the line between the emitter and the sheet. By measuring the positions at which a minimum signal was observed, he was able to determine the wavelength of the resonant waves. The speed of the waves could be found from the relation $c = \nu\lambda$ and turned out to be almost exactly the speed of light in free space. The chapters of his great book have the headings: rectilinear propagation, polarisation, reflection, refraction.

field. The second chapter, which summarises everything he knew about electrical discharges in gases at that time, is particularly impressive. His interpretation of the phenomena observed in these experiments involved Faraday tubes which linked atoms or ions together to form molecules – the application of an electric field disrupted the molecules.

Meanwhile, events took a somewhat unexpected turn and these were associated with the name of C.T.R. Wilson.

6.3 C.T.R. Wilson and the condensation of water droplets

Charles Thomson Rees Wilson, always known as 'C.T.R.', won a scholarship to Sidney Sussex College, Cambridge in 1888 and was the only student to graduate with physics as his main subject in the Natural Sciences Tripos in 1892. In 1894, he was appointed as a 'temporary observer' for two weeks at the meteorological observatory on the summit of Ben Nevis in Scotland, the highest mountain in the United Kingdom. His experiences there were to determine the rest of his scientific career. Wilson was struck by the beauty of coronas and 'glories', coloured rings surrounding shadows cast on mist and cloud, such as the Brocken spectre. He decided to attempt to imitate these natural phenomena experimentally in the laboratory on his return to Cambridge. The following spring, Wilson returned to the Ben Nevis region and was caught in an electric storm while on the summit of Carn Mor Dearg, the mountain to the east of Ben Nevis. In his own words,

> Suddenly I felt my hair stand up. I did not await any further developments, but started to run down the long scree slope leading to the bottom of the corrie. (Wilson, 1954)

Much of the rest of his career was devoted to understanding these atmospheric phenomena, specifically the origin of water droplets in clouds and atmospheric electricity. In the process, he was to invent the cloud chamber, which was eventually to reveal the trajectories of high-energy charged particles for the first time. In 1894, he must have been well aware of Thomson's interest in the electrification of water droplets and of his paper of 1893 on this topic.

In early 1895, Wilson began a series of experiments using the method of expansion of moist air developed by Paul-Jean Coulier and John Aitken to create artificial clouds. Aitken had concluded that

- when water vapor condenses in the atmosphere, it always does so on some solid nucleus
- the dust particles in the air form the nuclei on which water vapour droplets condense
- if there were no dust in the air, there would be no fogs, no clouds, no mists and probably no rain.

Interestingly, John Aitken of Falkirk in Scotland had carried out a continuous series of measurements of the dust content of air with his dust counter at the summit of Ben Nevis from 1890 to 1894 (Roy, 2004). Thus, it is likely that Wilson would have been familiar with his experiments and conclusions. Wilson greatly improved the performance of the condensation apparatus, in particular so that the expansion of the saturated water vapour could occur rapidly and the ratio v_1/v_2 of the initial to final volumes of the water vapour could be controlled very accurately. He established that, contrary to Aitken's conclusions, dust particles were not necessary to cause the condensation of water droplets. But no condensation occurs in dust-free air unless the volume expansion ratio is greater than 1.252. For

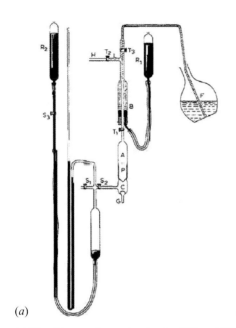

(a)   (b)

(a) Wilson's improved expansion chamber (Wilson, 1897). The expansion chamber is labelled A and the piston P. (b) The expansion chamber in the Cavendish Collection of Historic Scientific Instruments.

Fig. 6.3

larger expansions, a few drops are always produced, however many times the expansions are repeated. Therefore, some kind of condensation nuclei must always be present in small numbers.

Wilson now set about constructing a more accurate version of the expansion chamber in which the expansion from v_1 to v_2 could take place in one hundredth of a second (Figure 6.3). It was a tricky piece of apparatus to build. In Wilson's words (Wilson, 1960),

> The making involved a rather difficult piece of glass blowing, followed by many hours of accurate glass grinding which had to be done in the knowledge that the whole apparatus was almost certain to fly to pieces when the final glass blowing was done. I had about three successes out of a very large number of attempts . . .

The technical difficulty was associated with the expansion of the volume of moist air contained in the cylindrical chamber A in Figure 6.3(a). The piston P had to be ground precisely in order to provide a perfect seal lubricated by the water vapour, but the piston also had to drop rapidly to the bottom of the cylinder to produce an essentially perfect adiabatic expansion without shattering the apparatus. With great skill and patience, Wilson perfected the expansion chamber and completed the initial objectives of his research programme. In his words,

> With this apparatus, owing to the very rapid expansions, a series of wonderful colour phenomena appeared as the expansion increased; the tube looking as if it were filled with

a beautifully coloured liquid, with the colours changing in a perfectly regular way as the increase in the number of droplets caused a decrease in their size.

This research was completed in 1895 and provided an explanation for the appearance of the Brocken spectre and similar atmospheric phenomena.[4] In the meantime, dramatic discoveries were changing the direction of research in the Cavendish Laboratory.

6.4 The revolutions of 1895 and 1896

The great turning points which were to usher in the new physics of the twentieth century took place in 1895 and 1896 with the discoveries of X-rays and radioactivity. At the same time, changes took place in Cambridge University which were to be of profound importance for the future of the Laboratory.

6.4.1 The discovery of X-rays

In 1895, Wilhelm Conrad Röntgen at the University of Würzburg discovered what he termed X-rays (Röntgen, 1895). The biographical sketch on the Nobel Prize website admirably summarises what he did:

> On the evening of November 8, 1895, [Röntgen] found that, if the discharge tube is enclosed in a sealed, thick black carton to exclude all light, and if he worked in a dark room, a paper plate covered on one side with barium platinocyanide placed in the path of the rays became fluorescent even when it was as far as two metres from the discharge tube. During subsequent experiments he found that objects of different thicknesses interposed in the path of the rays showed variable transparency to them when recorded on a photographic plate. When he immobilised for some moments the hand of his wife in the path of the rays over a photographic plate, he observed after development of the plate an image of his wife's hand which showed the shadows thrown by the bones of her hand and that of a ring she was wearing, surrounded by the penumbra of the flesh, which was more permeable to the rays and therefore threw a fainter shadow. . . . In further experiments, Röntgen showed that the new rays are produced by the impact of cathode rays on a material object. Because their nature was then unknown, he gave them the name X-rays. (Röntgen, 1901)

His startling X-ray image of his wife's hand, showing the bones of the hand and her massive ring, had an immediate impact (Figure 6.4(*a*)). Overnight, X-rays became a matter of the greatest public interest and were very rapidly incorporated into the armoury of the doctor's surgery. In 1896 there were already more than 1000 articles on X-rays. The X-rays were more penetrating than cathode rays, since they could blacken photographic plates at a considerable distance from the hot spot on the discharge tube. Their identification with 'ultra-ultraviolet' radiation was only convincingly demonstrated in 1906, when Charles Barkla found that the X-radiation was polarised (Barkla, 1906), and in 1912, when Max von Laue had the inspiration to look for their diffraction by crystals

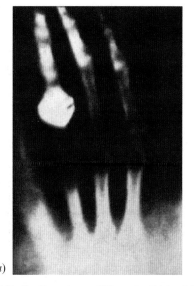
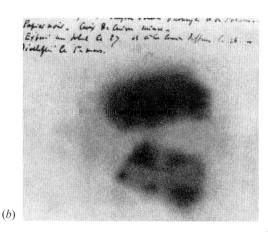

(a)   (b)

(a) The first X-ray picture of Röntgen's wife's hand, illustrating the ability of X-rays to penetrate through soft tissue and reveal the bone structure. (b) Becquerel's developed plate showing a strong image, despite the fact that the radioactive salt had not been exposed to sunlight (courtesy of Creative Commons).

Fig. 6.4

(Friedrich *et al.*, 1912; Laue, 1912), in the process opening up the new field of X-ray crystallography. Measurements of the interaction of X-rays with matter was to prove to be a central tool in disentangling the structure of atoms. For this momentous discovery, Röntgen was awarded the first Nobel Prize in Physics in 1901.

6.4.2 The discovery of radioactivity

The association of X-rays with fluorescent materials led to the search for other sources of X-radiation. In 1896, Henri Becquerel, who came from a distinguished family of French physicists, tested several known fluorescent substances before investigating samples of potassium uranyl disulphate. The photographic plates were wrapped in several sheets of black paper, the phosphorescent material was exposed to sunlight and then the plate developed to find if it had been darkened by X-rays. Becquerel's remarkable discovery was that the plates were darkened even when the phosphorescent material was not exposed to light (Figure 6.4(b)). This was the discovery of natural radioactivity (Becquerel, 1896). In further experiments carried out in the same year, Becquerel showed that the amount of radioactivity was proportional to the amount of uranium in the substance and that the radioactive flux of radiation was constant in time. Another important discovery was that the radiation from the uranium compounds discharged electroscopes. Pierre Curie and Marie Skłodowska-Curie repeated Becquerel's experiments in 1897 and showed that the intensity of radioactivity was proportional to the amount of uranium in different samples. Other radioactive substances were soon identified. Thorium was discovered in 1898

Fig. 6.5 J.J. Thomson with his graduate students in 1897. Front row from left: E.B.H. Wade, G.B. Bryan, W. Craig-Henderson, J.J. Thomson, J.S. Townsend, E. Rutherford. Back row from left: S.W. Richardson, C.T.R. Wilson, J. Henry, J. McClelland, L. Blaikie.

(Schmidt, 1898). By concentrating uranium residues, the Curies discovered the new element polonium, named after Marie's native land. The new element decreased in radioactivity exponentially – each radioactive substance has a particular half-life. In September 1898, very strong radioactivity was found in the barium group of residues. Enough of the radioactive substance was isolated to show that a new element had been discovered, radium (Curie and Skłodowska-Curie, 1898; Curie *et al.*, 1898).

6.4.3 Graduate students from abroad join the Laboratory

In 1895, Cambridge University made a crucial change to its regulations which allowed graduates from other universities to be admitted as 'research students'. After two years of residence, they could submit a dissertation on their research work and, subject to the work being of a satisfactory standard, they were awarded the degree of Bachelor of Arts by Research; this research degree would become a Doctor of Philosophy (PhD) in 1921. The first physics graduate students from abroad included Ernest Rutherford from New Zealand and the Irish physicists John Townsend from Dublin and John McClelland from Galway (Figure 6.5). Rutherford was 'awarded an Exhibition of 1851 scholarship to go anywhere in the world to carry out research of importance to New Zealand's industries' (Campbell, 1999). Townsend was supported by a Clerk Maxwell Studentship, which had been endowed by Maxwell's widow on her death in 1887, while McClelland obtained a fellowship from the Royal University of Ireland. Over the period 1895 to 1898, these graduate students were joined by John Henry from Ireland, Paul Langevin from France, John Zeleny from the United States, Vladimir Novak from Austria and John McLennan from Canada. The result was that, during this period, more than half the active research workers in the Laboratory were from overseas. These students added enormously to the research strength of the Laboratory since they came with secure funds to support their studies in Cambridge and had to be outstanding students to win their fellowships. C.T.R. Wilson remarked on the sudden change of atmosphere within the Laboratory as a result of this influx of bright young research students. At the same time, numerous professors from other universities worldwide began to take their sabbatical leave years in Cambridge, adding further to the burgeoning research activity and exposing the students to new ways of thinking about and doing physics.

6.5 The discovery of the electron and its universality

Röntgen's discovery of X-rays had an immediate impact, not only on medical physics but also on the study of the conduction of electricity through gases. As soon as the announcement of Röntgen's discovery reached Cambridge, Thomson and his students reproduced Röntgens' experiments, as well as using the X-ray images to assist medical practitioners. Thomson had everything at hand to make a major assault on understanding the nature of the X-rays. The culmination of this burst of activity was the publication of his book *Conduction of Electricity Through Gases* (Thomson, 1903a), which involved numerous contributions from research students from abroad: McClelland, Rutherford, Townsend and Zeleny. Through a series of technically brilliant but demanding experiments, all built by Everett, Thomson can rightly be regarded as the discoverer of the electron and of its universality in atomic physics.[5]

Thomson quickly demonstrated that X-rays greatly increased the electrical conductivity of gases within the vacuum tube and in the surrounding air. Furthermore, the voltages necessary to make the gas in the vacuum tube conducting were very much lower (Thomson, 1896a,b). In the spring of 1896, he and Rutherford turned to the systematic investigation of the nature of the conductivity of gases induced by X-rays. This was a major departure for Rutherford, who until then had been perfecting his magnetic detector for radio waves (Box 6.4).

Thomson had to confront Lenard's challenge that the cathode rays were waves and not particles. Hertz had not been able to detect the deflection of the trajectories of cathode rays by a magnetic field, but with the better vacuums which Thomson and Everett obtained in their experiments, deflections were observed when either electric or magnetic fields were applied to the discharge tubes, confirming that particle nature of the cathode rays. Furthermore, the deflections of the trajectories in the magnetic field were found to be independent of the nature of the gas in the evacuated tube; in a uniform field, the trajectories were roughly arcs of a circle.

Box 6.4 **Rutherford's magnetic radio wave detector**

Ernest Rutherford began his physics career in New Zealand at the end of 1893 with an investigation into the magnetisation of iron by high-frequency discharges. He continued this work when he came to Cambridge in 1895, and developed a magnetic detector for electromagnetic waves. He demonstrated the detector in the Cavendish Laboratory at Free School Lane in December 1895 at a distance of 200 yards from a Hertzian spark transmitter. His first outdoor use of the detector occurred on 22 February 1896 when he set up a spark transmitter on Jesus Green and detected the radiated pulses at a distance of 350 yards in a house on Park Parade. On the following day, he continued this work with a successful transmission over a distance of nearly three-quarters of a mile.

Rutherford held the world record for the reception distance at the time, but in the same year Marconi came to England and started to develop a system for the transmission of Morse code signals by means of electromagnetic waves, which was the starting point of wireless telegraphy. Rutherford turned from his work on electromagnetic waves, first to the ionising properties of X-rays in gases and then to the study of radioactivity.

Rutherford's radio detector consisted of several strands of fine steel wire tightly wound with many turns of insulated copper wire, the ends of which were connected to the two arms of a dipole antenna. The steel wire was first magnetised in a separate solenoid and then brought close to a sensitive magnetometer. The magnetometer deflection was then reset to zero with a compensating magnet. When a pulse of electromagnetic radiation arrived at the antenna from a Hertzian spark transmitter, the steel wire was partly demagnetised and the magnetometer needle deflected. The whole procedure was repeated for the next transmission.

6.5.1 The m/e experiments of 1897

Thomson had all the technology at hand to learn more about the nature of the cathode rays from magnetic deflection experiments. Balancing the magnetic Lorentz force against the centripetal acceleration,

$$evB = \frac{mv^2}{r}, \qquad \frac{e}{mv} = Br, \qquad (6.1)$$

where it is assumed that the magnetic field B is perpendicular to the cathode ray beam and r is the radius of curvature of the circular path of the cathode rays in that field. The aim was to estimate the value of m/e, which could then be compared with the values found for the hydrogen atom. The speed v of the particles had to be estimated. In early 1897, Thomson estimated v using a modified version of Perrin's discharge tube to measure not only the total charge Q of the flux of cathode rays but also the total kinetic energy W deposited by the rays which could be estimated by precise calorimetry of the temperature increase of the charge collector (Figure 6.6). Then v could be found since the charge deposited was $Q = Ne$ and the heat generated was $W = \frac{1}{2}Nmv^2$. The speed of the rays, in combination

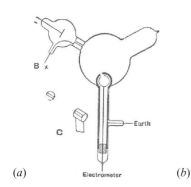

(a)   Electrometer   (b)

(a) Diagram illustrating Thomson's version of the Perrin discharge tube (Thomson, 1897a). The cathode rays are produced at A and deflected by a magnetic field perpendicular to the plane of the diagram. The cathode rays pass through the hole in the anode collector and the temperature increase is measured in the calorimeter at the end of the tube. (b) An example of a Perrin tube built by Everett for Thomson.

Fig. 6.6

with e/m, could then be found:

$$\frac{Q}{W} = \frac{e}{2mv^2} \quad \text{and so} \quad \frac{e}{m} = \frac{B^2 r^2}{2} \frac{Q}{W}. \tag{6.2}$$

Inserting the results of his experiments into these formulae, Thomson found $m/e = 1.6 \times 10^{-7}$ g emu^{-1}. He presented these results at a Friday Evening Discourse at the Royal Institution of Great Britain on 30 April 1897, and his paper was published in the *Proceedings of the Royal Institution* (Thomson, 1897a) as well as in *The Electrician* (Thomson, 1897b). In the concluding paragraphs of the paper, he wrote:

> This [value of m/e] is very small compared with the value 10^{-4} g emu^{-1} for the ratio of the mass of an atom of hydrogen to the charge carried by it. If the result stood by itself we might think it was probable that e was greater than the atomic charge of [the] atom rather than that m was less than the mass of the hydrogen atom. Taken, however, in conjunction with Lenard's results for the absorption of the cathode rays, these numbers seem to favour the hypothesis that the carriers of the charges are smaller than those of atoms of hydrogen.
>
> It is interesting to notice that the value of m/e, which we have found from the cathode rays, is of the same order as the value 10^{-7} g emu^{-1} deduced by Zeeman from his experiments on the effect of a magnetic field on the period of the sodium light.

The remark in the last sentence was of particular significance. Pieter Zeeman had discovered the broadening of the D lines of sodium when a sodium flame was placed between the poles of a strong electromagnet (Zeeman, 1896a). Zeeman used a high-quality Rowland grating with a radius of 10 feet and 14,938 lines per inch, but the 10 kG produced by the magnet was insufficient to resolve the broadened lines. By the end of October 1896, Zeeman was convinced that the broadening of the spectral line was a real effect and found it to be proportional to the applied magnetic flux density. Within days of Zeeman presenting his results to the Science Section of the Dutch Academy of Sciences, Hendrik Lorentz had interpreted them in terms of the splitting of spectral lines by the magnetic field, due to the motion of the 'ions' in the atoms.

Lorentz showed that the splitting of the sodium D lines from the line centre is given by

$$\Delta\omega_0 = \pm\frac{eB}{2m} \quad \text{or} \quad \Delta\nu_0 = \pm\frac{eB}{4\pi m}. \tag{6.3}$$

This is now referred to as the *normal Zeeman effect*.[6] Thus, the broadening of the lines depends upon m/e. The analysis also made predictions about the polarisation of the broadened lines. Zeeman continued his careful measurements over the next two months and discovered that the polarisation properties of the broadened lines agreed with these expectations (Zeeman, 1896b). The significance of Thomson's remark is that the cathode rays, which he termed 'corpuscles', seemed to be the same types of 'ions' which are involved in the production and broadening of spectral lines. In other words, Thomson's corpuscles had to be part of the structure of the atom and involved in the physics of spectral lines.

In his paper of April 1897, Thomson thought that the Lenard rays were secondary phenomena caused by the collision of the cathode rays with the material of the foil windows. He also recognised that, whatever they were, their mean free paths in air were very much greater than would be expected for air molecules. This suggested to him that, combined with their ability to pass through thin metallic films, the rays had to be very much smaller than atoms. It took some time before he concluded that the Lenard rays were in fact cathode rays themselves escaping from the discharge tube.

Thomson next devised an improved means of determining the speeds of the cathode rays by passing the beam through crossed electric and magnetic fields (Figure 6.7). The electric force eE on the particles was balanced by the Lorentz force evB and so $v = E/B$. This famous version of the discharge tube experiment had the advantage that the speed of the particles was measured directly. Thomson repeated the experiments with different gases in the vacuum tube and different electrode materials. The same results were found in all cases, the values of m/e of the corpuscles ranging from 1.1×10^{-7} to 1.5×10^{-7} (Thomson, 1897c). There follows Thomson's famous statement about the nature of the corpuscles, which he interprets as subatomic particles:

> Thus on this view we have in the cathode rays matter in a new state, a state in which the subdivision of matter is carried very much further than in the ordinary gaseous state: a state in which all matter – that is, matter derived from different sources such as hydrogen, oxygen, etc – is of one and the same kind; this matter being the substance from which all the chemical elements are built up. (Thomson, 1897c)

His great paper concludes with a re-discussion of the possible configurations of the corpuscles in the atom on the basis of Mayer's experiments with floating magnets in a magnetic field (see Figure 6.1). Thomson continued to use the term 'corpuscle' for a number of years, but the community of physics soon adopted the word *electron*, the name coined by Johnstone Stoney in 1891 for the 'ions' which conduct electricity (Stoney, 1891).

About the same time, other physicists were performing similar experiments.

- In January 1897, Emil Wiechert used the magnetic deflection technique to obtain a measurement of m/e for cathode rays and concluded that these particles had mass between

Fig. 2.

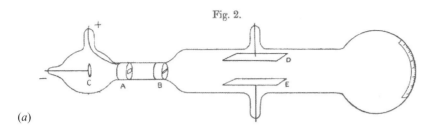

(a)

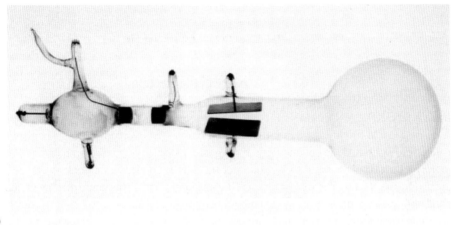

(b)

(a) The diagram from Thomson's paper of October 1897 showing the configuration of the discharge tube with crossed electric and magnetic fields (Thomson, 1897c). The electric field is maintained between the plates D and E. The magnetic field is oriented perpendicular to it so that the $e\boldsymbol{v} \times \boldsymbol{B}$ force opposes the force of the electric field $e E$. (b) A photograph of the vacuum tube with which J.J. Thomson measured the mass-to-charge ratio of cathode rays.

Fig. 6.7

2000 and 4000 times smaller than that of hydrogen, assuming their electric charge was the same as that of hydrogen ions. He obtained only an upper limit to the speed of the particles since it was assumed that the kinetic energy of the cathode rays was $E_{kin} = eV$, where V is the accelerating voltage of the discharge tube.

- Walter Kaufmann's experiment was similar to Thomson's. He found the same values of m/e, no matter which gas filled the discharge tube, a result which puzzled him. He found a value of e/m 1000 times greater than that of hydrogen ions and concluded

> that the hypothesis of cathode rays as emitted particles is by itself inadequate for a satisfactory explanation of the regularities I have observed.

Thomson was well aware of these activities, but he carried on with his systematic experiments and explored every aspect of their properties. In his papers of 1897, he had assumed that the charge of the corpuscles was of similar order to that of the hydrogen ion, but now he devised a means of measuring the charge experimentally. By good fortune, C.T.R. Wilson's cloud condensation experiments provided the means of measuring the charge of the corpuscles directly.

Table 6.2 The dependence of saturated vapour pressure upon temperature

Temperature (°C)	Vapour pressure (kPa)	Vapour pressure (mm Hg)
0	0.6	4.5
5	0.9	6.8
10	1.2	9.0
15	1.7	11.2
20	2.3	17.3
25	3.2	24.0

6.5.2 The charge of the electron

In February 1896, Wilson used an early X-ray tube to illuminate his expansion chamber. A dense fog was produced when the expansion ratio v_1/v_2 exceeded 1.252. Thomson and Rutherford attributed the conductivity of the gas in a discharge tube illuminated by X-rays to the production of ions and so Wilson inferred that the condensation nuclei were the ions created by the X-rays. The role of ions as condensation centres was confirmed by including condenser plates within the cloud chamber, the type of apparatus used being illustrated in Figure 6.3(*b*) – the condenser plates are enclosed within the topmost horizontal part of the expansion chamber. When an electric field was applied to the plates, the ions were swept onto the plates before they could act as condensation centres.

Wilson continued his long and meticulous experiments to determine exactly the degree of supersaturation necessary for condensation to take place – supersaturation being defined as the ratio of the density in the vapour to the density of the saturated vapour pressure at the end of the expansion. Since the expansion took place within a hundredth of a second, it was a pure adiabatic expansion and so the initial and final densities were determined by the adiabatic law,

$$\frac{\rho_1}{\rho_2} = \frac{p_1}{p_2} \left(\frac{v_1}{v_2}\right)^{\gamma}. \tag{6.4}$$

As shown in Table 6.2, the saturated vapour pressure depends very strongly upon the temperature *in degrees Celsius*. A small adiabatic expansion results in a relatively small change in temperature according to the adiabatic law $T/T_0 = (v/v_0)^{-(\gamma-1)}$, where the temperature is measured in kelvins. This small temperature change corresponds, however, to a large decrease in the saturated vapour pressure, as can be seen from Table 6.3, and consequently to a large increase in the supersaturation. Wilson's results were described in three major papers of 1897 (Wilson, 1897) and 1899 (Wilson, 1899a,b). He established that different types of condensation process occur at different expansion ratios and supersaturations, as summarised in Table 6.3.

When the supersaturation reached the value 1.252, 'rain-like' condensation occurred and the condensation centres of the droplets were negative ions. If the expansion was repeated, water droplets continued to be created at about the same rate, indicating that some steady

Table 6.3 Critical supersaturations in air and the corresponding types of condensation

v_1/v_2	Super-saturation	Observed condensation	Condensation nuclei
1.252	4.2	Rain-like condensation	negative ions
1.31	6	Enhanced rain-like condensation	positive ions
1.38	7.9	Fog-like condensation	air molecules

source of ionisation was entering the expansion chamber. At the slightly greater super-saturation of 1.31, there was 'enhanced rain-like' condensation, and this was associated with positive ions. Finally, at an expansion ratio of 1.38, the condensation changed from 'rain-like' to 'fog-like' which Wilson associated with the air molecules themselves acting as condensation nuclei. Wilson attributed the difference between the supersaturations associated with negative and positive ions to the fact that the positive ions moved more slowly than the negative ions, an effect which had been discovered in the clever experiments carried out by Zeleny and Townsend (Zeleny, 1898, 1900; Townsend, 1900). As Townsend wrote,

> Professor Zeleny has shown that negative ions travel faster under an electromotive force than positive ions, the ratio of velocities being 1.24 for air and oxygen, 1.15 for hydrogen, and 1.0 for carbonic acid.

He found similar results in his diffusion experiments (Townsend, 1900).

In his paper of 1897, Wilson estimated the radius r of the water droplets using William Thomson's expression for the equilibrium vapour pressure over a curved surface, $r = 2T/(R\theta \ln S)$, where T is the surface tension, θ the temperature, R the gas constant and S the supersaturation. Then he used J.J. Thomson's expression for the electric charge e which would balance the surface energy and so prevent the drop evaporating, $e^2 = 16Tr^3$. He found a value for the charge of the ion of $e = 5 \times 10^{-19}$ C, about three times the present standard value for the charge of the electron (Wilson, 1897).

Building on Wilson's innovations, in 1898 J.J. Thomson devised an improved means of measuring the charge of the ions produced by exposure to a flux of X-rays (Figure 6.8). Each ion acted as a condensation centre and the current carried by the charged drops was measured. The radius r of the water drops was found from Stokes's formula for the terminal velocity v of the drops under gravity, $r = (9v\eta/2\rho g)$, where η is the kinematic viscosity, ρ the density of the water droplet and g the acceleration due to gravity. From the total mass of condensing gas and the known radii of the drops, the number of ions could be found. From these data, Thomson estimated the charge of the electron to be $e = 2.2 \times 10^{-19}$ C, compared with the present standard value of $e = 1.602 \times 10^{-19}$ C (Thomson, 1898). This remarkable experiment confirmed his view that the mass of the corpuscles was about 1000 to 2000 times less than those of atoms of hydrogen.

The experiment was refined in subsequent experiments, Thomson (1903b) finding the value $e = 1.13 \times 10^{-19}$ C while his colleague Harold A. Wilson (1903) found

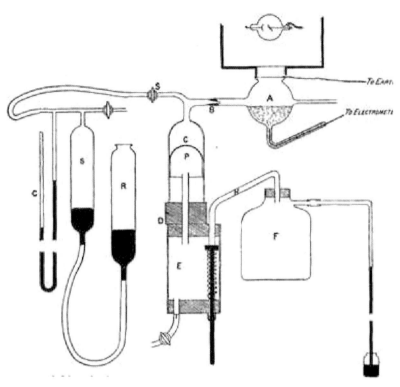

Fig. 6.8 J.J. Thomson's experiment to estimate the charge of the electron, incorporating many of Wilson's innovations (Thomson, 1898).

$e = 1.03 \times 10^{-19}$ C, both on the low side compared with the present standard value. In Wilson's experiment, the charged droplets were accelerated by an electric field, which improved the accuracy of the estimates and also demonstrated that the droplets could pick up 1, 2, 3, . . . units of electric charge. These experiments were the precursors of the famous oil drop experiment carried out by Robert Millikan at the University of Chicago. Millikan replaced the water-vapour droplets by fine drops of a heavy oil, which did not evaporate during the course of the experiment (Millikan, 1913). He also used an electric field to balance the fall of the oil drops under gravity and was able to observe individual drops for very long periods. His estimate of the charge of the electron was 1.592×10^{-19} C, with an accuracy of about 1%.

6.5.3 The particles ejected in the photoelectric effect

Continuing his inspired burst of experimental innovation, Thomson now turned to the particles ejected in the photoelectric effect. The idea of the experiment is illustrated schematically in Figure 6.9. The ultraviolet radiation is incident on the zinc plate by passing through the gauze sheet. The ejected particles then feel the joint effect of an electric field perpendicular to the plate and an orthogonal magnetic field with the result that, as the particles

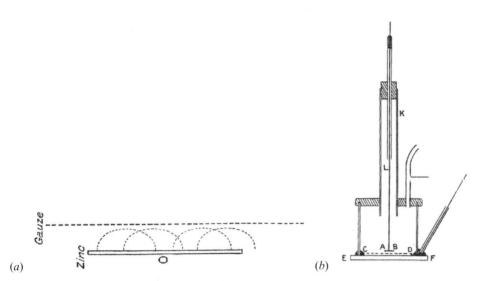

Fig. 6.9

J.J. Thomson's experiment to estimate the mass-to-charge ratio of the particles ejected by ultraviolet radiation in the photoelectric effect (Thomson, 1899; Davis and Falconer, 1997).

are accelerated in the vertical direction, the Lorentz $v \times B$ force causes them to follow the cycloidal paths shown in Figure 6.9(a). It is straightforward to show that the paths of the particles are given by the equations of motion

$$m\frac{d^2x}{dt^2} = eE - eB\frac{dy}{dt}, \quad m\frac{d^2y}{dt^2} = eB\frac{dx}{dt}, \tag{6.5}$$

where the electric field is in the x-direction perpendicular to the zinc sheet and the magnetic field is perpendicular to the electric field in the z-direction. The motion of the particle is cycloidal:

$$x = \frac{m}{e}\frac{E}{B^2}\left[1 - \cos\left(\frac{e}{m}Bt\right)\right], \quad y = \frac{m}{e}\frac{E}{B^2}\left[\frac{e}{m}Bt - \sin\left(\frac{e}{m}Bt\right)\right]. \tag{6.6}$$

Thus, the maximum distance travelled in the x-direction depends upon m/e. The experimental arrangement is shown in Figure 6.9(b), with the zinc plate at EF and the gauze screen at CD. By determining the distance x_{max} at which the current ceased to flow between the zinc plate and the gauze, the value of m/e could be found. From this experiment, Thomson found a value of $m/e = 1.37 \times 10^{-7}$ g emu^{-1}, in reasonable agreement with his estimate of this ratio for the ions released by X-rays and the cathode rays. By repeating his measurement of the charge of the particles released by the same cloud-chamber technique, the charge of the particles was found to be 6.8×10^{-10} esu, in good agreement with his previous estimate of 6.5×10^{-10} esu from the X-ray experiments. Thomson concluded that the particles liberated in the photoelectric effect were the same as the cathode rays and the particles released in the X-ray experiments.

6.5.4 Rutherford, radioactivity and β-particles

As soon as the discovery of radioactivity was announced, Ernest Rutherford took up the study of the rays released in radioactive decays. This work was to dominate his research following the completion of his degree at Cambridge and his subsequent outstanding work on the elucidation of radioactive decay chains at McGill University in Canada. For these studies, he was awarded the Nobel Prize in Chemistry in 1908 'for his investigations into the disintegration of the elements, and the chemistry of radioactive substances'. In his first publication on the subject, Rutherford established that there are at least two separate types of radiation emitted by radioactive substances (Rutherford, 1899). He called the component which is most easily absorbed α-radiation (or α-rays) and the much more penetrating component β-radiation (or β-rays). It was to take Rutherford another ten years before he demonstrated conclusively that α-radiation consisted of the nuclei of helium atoms (Rutherford and Royds, 1909). To complete this story, γ-radiation was discovered in 1900 by Paul Villard as an extremely penetrating form of radiation emitted in radioactive decays; the γ-rays were undeflected by a magnetic field (Villard, 1900a,b). The γ-rays were conclusively identified as electromagnetic waves 14 years later when Rutherford and Edward Andrade observed the reflection of γ-rays from crystal surfaces (Rutherford and Andrade, 1913).

The α-, β- and γ-rays were the only known radiations which could cause the ionisation of air. The characteristic property which distinguished them was their penetrating power. In quantitative terms, these were:

- The α-particles ejected in radioactive decays produce a dense stream of ions and are stopped in air within about 0.05 m. This distance is called the *range* of the particles.
- The β-particles have greater ranges, but there is not a well-defined value for any particular radioactive decay.
- The γ-rays were found to have by far the longest ranges, a few centimetres of lead being necessary to reduce their intensity by a factor of ten.

The β-rays were convincingly shown by Walter Kaufmann to have the same mass-to-charge ratio as the recently discovered electron (Kaufmann, 1902).

As a footnote to these developments, we have already noted that, in his experiments of 1895, C.T.R. Wilson had shown that some kind of nuclei must always be present in small numbers to act as condensation centres. In 1900, he carried out experiments on behalf of the Meteorological Council while on vacation at his brother's home at Peebles with highly insulated electroscopes which had negligible leakage of electric charge. The results of his careful experiments were as follows:

> The mean rate of leak in an ordinary room amounted to 6.6 divisions of the micrometer scale per hour. An experiment was made in the Caledonian Railway Tunnel near Peebles (at night after the traffic had ceased) and gave a leakage of 7.0 divisions per hour. There is thus no evidence of any falling off of the rate of production of ions ... although there were many feet of solid rock overhead. (Wilson, 1901)

Earlier in the same paper of 1901, he made the prophetic remark:

Experiments were now carried out to test whether the production of ions in dust-free air could be explained as being due to radiation from sources outside our atmosphere, possibly radiation like Röntgen rays or like cathode rays, but of enormously greater penetrating power. (Wilson, 1901)

Because of the experiments in the Peebles railway tunnel, he concluded that they were not, but 'a property of air itself'. But the idea of *cosmic radiation* was now in the literature. It would be another 11 years before his intuitive guess was proved to be correct as a result of Victor Hess's experiments (Hess, 1913).[7]

6.5.5 The universality of the electron

The summation of all these activities was that, by about 1900, Thomson's conviction that the electron is a fundamental constituent of matter was generally accepted by the physics community. The same subatomic particles were identified with the cathode rays, with the particles ejected in the photoelectric effect, with the β-rays ejected in the decay of the nucleus and with the particles involved in the formation of the lines in atomic spectra. Thomson's energetic and continued experimental and theoretical involvement in all aspects of these studies make a convincing case for his designation as the 'discoverer of the electron'. He held the view that the electrons had to be part of the fundamental structure of atoms and continued his imaginative construction of models for the atoms of the elements, with a view to understanding chemical and physical properties. This is the subject of the next chapter.

There was an atmosphere of real excitement in the Laboratory, very much the result of Thomson's dedication and enthusiasm for the research programme, which in turn inspired the research students. This atmosphere is delightfully caught in this quotation from Davis and Falconer:

> John Zeleny . . . was working in Berlin when Thomson announced his corpuscle hypothesis. Finding that no one else in Berlin believed in corpuscles, he packed his bags and came to Cambridge. Once there, Thomson inspired the researchers with his enthusiasm by 'his vital personality, his obvious conviction that what he and we were all doing was something important and his camaraderie.' They believed they were making history, for Thomson 'realised that what he had made was a revolution.' Their confidence and mutual enjoyment found expression in the songs written for the Cavendish dinner, instituted in 1897.[8]

6.6 Physics in 1900

Despite the somewhat pessimistic predictions of the 1870s, the Laboratory had gained an international reputation for excellence in experimental physics by 1900. This can largely be attributed to the outstanding scientific leadership provided by Maxwell, Rayleigh and Thomson and their personal contributions. Their approaches were quite different and complementary. Maxwell was an intuitive visionary with a reputation in theory and experiment

Table 6.4 Invitees to the 1911 Solvay Conference, Brussels					
Germany	France	UK	Netherlands	Austria	Denmark
Einstein	Brillouin	Jeans	Lorentz	Hasenöhrl	Knudsen
Nernst	Curie	Rayleigh[a]	Kamerlingh Onnes		
Planck	Langevin	Rutherford			
Rubens	Perrin				
Warburg	Poincaré				
Wien					

[a] Did not attend.

which was only fully recognised long after his death. Rayleigh followed his lead and broadened considerably the scope of the research activities. He also secured for the Laboratory a strong reputation for precision experiment and the establishment of international standards. Thomson was quite different from his predecessors in adopting a much more pragmatic approach to the understanding of nature, basing his insights on clever but demanding experiments which commanded international attention, and guided by the imaginative use of analogy.

Meanwhile, physics in continental Europe was developing rapidly. A simple measure of the diversity of the work being carried out can be appreciated from the invitees to the 1911 Solvay Conference, who are listed by country in Table 6.4. That meeting took place a decade after the point we have reached in our story, but the physicists involved were active in all their respective research areas by 1900. This inaugural Solvay Conference, entitled *La Théorie du rayonnement et les quanta*, was devoted to quantum phenomena, and this was to be the focus of the most dramatic developments in physics during the first three decades of the twentieth century (Langevin and de Broglie, 1912). The experimental physicists present worked in the areas of radioactivity (Curie, Rutherford), black-body radiation (Rubens, Warburg, Wien), low-temperature physics (Kamerlingh Onnes, Nernst), kinetic theory (Knudsen, Perrin) and solid state physics (Brillouin, Langevin). Those who were primarily theorists included Einstein, Hasenöhrl, Jeans, Lorentz, Planck and Poincaré. This summary does scant justice to the broad interests of the participants, but it serves to illustrate the very broad front over which progress in physics was developing and the major problems which were emerging. The challenge for Thomson at the beginning of the new millennium was to understand the structure of atoms. His discoveries were to make major contributions to the changing the face of basic physics.

During the second half of Thomson's long tenure as Cavendish Professor until he stood down in 1919, research centred largely on the problems of atomic structure from the experimental perspective. The names which were to feature most prominently included Charles Barkla, Charles (C.T.R.) Wilson, W. Lawrence Bragg, Geoffrey Taylor and Francis Aston. A long shadow was cast by the remarkable researches of Thomson's former student Ernest Rutherford and his colleagues, first at McGill University in Montreal, Canada from 1998 to 1907, and then from 1907 until 1919 at Manchester where Rutherford was Langworthy Professor of Physics. In 1919 he succeeded Thomson as Cavendish Professor.

It was during the period 1900 to 1919 that the hallmarks of twentieth-century physics were indelibly etched into its infrastructure with the discoveries of the special and general theories of relativity by Albert Einstein and the recognition of the key role of quantisation at the atomic level by Planck, Einstein, Bohr and many others. It was a period when it gradually became apparent that classical physics was no longer adequate to account for physical phenomena at the atomic level, but there was no coherent quantum theory to replace classical physics. Intriguingly, the ferment of continental physics and theoretical physics had relatively little impact upon the experimental programme of the Laboratory, which continued the tradition of carrying out ingenious experiments and model-building, which informed the deliberations of the theorists.

7.1 The problems of building models of atoms

Thomson summarised the researches which led to his discovery of the electron and its universality in his influential book *Conduction of Electricity through Gases* (Thomson, 1903a). As he wrote in the introduction:

> The study of the electrical properties of gases seems to offer the most promising field for investigating the Nature of Electricity and the Constitution of Matter, for thanks to the Kinetic Theory of Gases our conceptions of the processes other than electrical which occur in gases are much more vivid and definite than they are for liquids or solids; in consequence of this the subject has advanced very rapidly and I think it may now fairly be claimed that our knowledge of and insight into the processes going on when electricity passes through a gas is greater than it is in the case either of solids or liquids. The possession of a charge by the ions increases so much the ease with which they can be traced and their properties studied that, as the reader will see, we know far more about the ion than we do about the uncharged molecule.

> With the discovery and study of Cathode rays, Röntgen rays and Radio-activity a new era has begun in Physics, in which the electrical properties of gases have played and will play a most important part; the bearing of these discoveries on the problems of the Constitution of Matter and the Nature of Electricity is in most intimate connection with the view we take of the processes which go on when electricity passes through a gas. I have endeavoured to show that the view taken in this volume is supported by a large amount of direct evidence and that it affords a direct and simple explanation of the electrical properties of gases.

This was followed in 1906 by the award of the Nobel Prize in Physics 'in recognition of the great merits of his theoretical and experimental investigations on the conduction of electricity by gases'.

The problems of formulating models of atoms were, however, formidable. As Planck remarked in 1902,

> If the question concerning the nature of white light may thus be regarded as being solved, the answer to the closely related but no less important question – the question concerning the nature of light of the spectral lines – seems to belong among the most difficult and complicated problems, which have ever been posed in optics or electrodynamics. (Planck, 1902)

Heilbron (1977) conveniently lists six basic questions which the model-builders faced at the turn of the twentieth century:

- The *nature of the positive charge* necessary to create electrically neutral atoms. Was charge neutrality provided by an equal number of positively charged electrons or by some other distribution of positive charge?
- The *number of electrons* in the atom was uncertain. From their charge-to-mass ratio, there could be thousands of electrons in the atom and that was perhaps not unreasonable in view of the large numbers of spectral lines observed in atomic spectra. Even the lightest elements have large numbers of spectral lines, while several thousands of lines are observed in the spectrum of, for example, iron.
- There is nothing in classical physics which could establish a *natural length scale for atoms*. A clue was at hand with the introduction of Planck's constant h in his epochal paper of 1900, but it was to be over a decade before Bohr showed how the concept of quantisation could be applied to determine the size of the hydrogen atom.
- The problem of the *collapse of the atom* due to the radiation of electromagnetic radiation by the orbiting electrons was a major stumbling block for atomic theorists. There were ways of minimising the problem, but these were to prove inadequate.
- Even if these problems could be overcome, there was still the problem of understanding the *origin of the various formulae for the spectral lines* discovered by Balmer and Rydberg.
- Finally, there remained the fundamental problem of understanding the *nature of the oscillators* responsible for the emission of spectral lines.

These issues were to be addressed by a combination of experiment and theory over the following decades, the pieces of the jigsaw gradually falling into place. Two of the problems were to find definitive answers through the experiments of Thomson and Rutherford.

7.2 Thomson and the numbers of electrons in atoms

Thomson addressed the key question: how are the electrons and the positive charge distributed inside atoms? However they are distributed, they cannot be stationary because of *Earnshaw's theorem*, which states that any static distribution of electric charges is mechanically unstable, in that they either collapse or disperse to infinity under the action of electrostatic forces. The alternative is to place the electrons in orbits, in what were often called 'Saturnian' models of the atom, as advocated by Perrin (1901) and Nagaoka (1904a,b). Nagaoka was inspired by Maxwell's model of Saturn's rings and attempted to associate the spectral lines of atoms with the small stable vibrational perturbations of the electrons about their equilibrium orbits discovered by Maxwell.

7.2.1 The radiative and mechanical instability of atoms

A major problem with the Saturnian picture was the radiative instability of the electron. Suppose the electron has a circular orbit of radius a. Then, equating the centripetal force to the electrostatic force of attraction between the electron and the nucleus of charge Ze,

$$\frac{Ze^2}{4\pi\epsilon_0 a^2} = \frac{m_e v^2}{a} = m_e|\ddot{r}|, \tag{7.1}$$

where $|\ddot{r}|$ is the centripetal acceleration. The rate at which the electron loses energy by radiation had been derived by Joseph Larmor, but Thomson gave his own derivation, outlined in Box 7.1, in 1906 (Thomson, 1906a). The calculation illustrates beautifully his remarkable physical insight into the origin of the radiation loss of an accelerated charged particle. Notice his use of Poynting's theorem, named after his lifelong friend John Henry Poynting, who had become Professor of Physics at Birmingham.[1]

The kinetic energy of the electron is $E = \frac{1}{2}m_e v^2 = \frac{1}{2}m_e a|\ddot{r}|$. Therefore, the time it takes the electron to lose all its kinetic energy by radiation is

$$T = \frac{E}{|dE/dt|} = \frac{2\pi a^3}{\sigma_T c}, \tag{7.2}$$

where $\sigma_T = e^2/6\pi\epsilon_0^2 m_e^2 c^4$ is the Thomson cross-section. Taking the radius of the atom to be $a = 10^{-10}$ m, the time it takes the electron to lose all its energy is about 3×10^{-10} s. Something is profoundly wrong. As the electron loses energy, it moves into an orbit of smaller radius, loses energy more rapidly and spirals into the centre.

The pioneer atom model-builders were well aware of this problem. Fortunately, the wavelength of light λ is very much greater than the size of atoms a and so the solution was to

Thomson's derivation of the loss rate of an accelerated electron

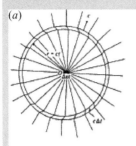

(a)

Figure (a) shows schematically the configuration of electric field lines at time t due to a charge accelerated to a velocity Δv in time Δt at $t = 0$. Figure (b) shows an expanded version of part of (a) used to evaluate the strength of the azimuthal component E_θ of the electric field due to the acceleration of the electron. It is a simple calculation to show that E_θ is

$$E_\theta = \frac{q|a|\sin\theta}{4\pi\varepsilon_0 c^2 r} = \frac{|\ddot{p}|\sin\theta}{4\pi\varepsilon_0 c^2 r},$$

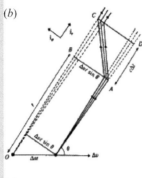

(b)

where $p = er$ is the electric dipole moment of the charge with respect to some origin. This electric field component represents a pulse of electromagnetic radiation, and hence the rate of energy flow per unit area per second at distance r is given by the magnitude of the Poynting vector $S = |E \times H| = E^2/Z_0$, where $Z_0 = (\mu_0/\varepsilon_0)^{1/2}$ is the impedance of free space. The rate of energy flow through the area $r^2\,\mathrm{d}\Omega$ subtended by solid angle $\mathrm{d}\Omega$ at angle θ and at distance r from the charge is therefore

$$Sr^2\,\mathrm{d}\Omega = -\left(\frac{\mathrm{d}E}{\mathrm{d}t}\right)\mathrm{d}\Omega = \frac{|\ddot{p}|^2\sin^2\theta}{16\pi^2\varepsilon_0 c^3}\,\mathrm{d}\Omega.$$

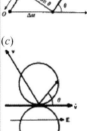

(c)

To find the total radiation rate $-\mathrm{d}E/\mathrm{d}t$, we integrate over solid angle:

$$-\left(\frac{\mathrm{d}E}{\mathrm{d}t}\right) = \frac{|\ddot{p}|^2}{6\pi\varepsilon_0 c^3} = \frac{q^2|a|^2}{6\pi\varepsilon_0 c^3}.$$

Precisely the same result, sometimes referred to as *Larmor's formula*, comes out of the full theory.

Figure (c) shows the *polar diagram* of the radiation field E_θ emitted by an accelerated electron. The magnitude of the electric field strength as a function of polar angle θ with respect to the acceleration vector a is $E_\theta \propto \sin\theta$ (Longair, 2003). The electric polarisation of the radiation is parallel to the acceleration vector.

place the electrons in orbits such that there would be no net acceleration when the acceleration vectors of all the electrons in the atom are added together. This requires, however, that the electrons are well ordered in their orbits about the nucleus. If, for example, there are two electrons in the atom, they can be placed in the same circular orbit on opposite sides of the nucleus and so, to first order, there is no net dipole moment as observed at infinity, and hence no dipole radiation. There is, however, a finite electric quadrupole moment, and hence radiation at the level $(\lambda/a)^2$ relative to the intensity of dipole radiation is expected. Since $\lambda/a \sim 10^{-3}$, the radiation problem can be significantly relieved. By adding sufficient electrons to each orbit, the radiation problem can be reduced to manageable proportions. Thus, before Thomson's paper of 1906, the radiative instability could be overcome by assuming that a huge number of electrons were so disposed as to result in no net multipole moments of the electron distribution. Each orbit had to be densely populated with a well-ordered system of large numbers of electrons. This was the basis of Thomson's *'plum-pudding' model*, in which the orbits were embedded in a sphere of positive charge.

The other problem with the Saturnian picture was its mechanical instability. Nagaoka had been inspired by Maxwell's model of Saturn's rings, in which stable oscillations were found when rings of particles were perturbed. In the Maxwellian case the perturbations were stable under the attractive force of gravity between particles of the ring, but in the case of the repulsive electrostatic forces between electrons, the perturbations were unstable. This mechanical instability was inevitable, even if the radiative instability could be eliminated.

7.2.2 Thomson and Barkla's experiments

In 1906, Thomson published his analysis of three different ways of estimating the number of electrons in atoms (Thomson, 1906a). The first approach involved working out the dispersion of light of different frequencies when it passes through a gas. He applied the method successfully to hydrogen and found that the number of electrons had to be approximately equal to the atomic mass number $A = 1$.

In the second approach, X-rays were scattered by the electrons in atoms by *Thomson scattering*, the theory of which was worked out by Thomson using the classical expression for the radiation of an accelerated electron (Box 7.1) (Thomson, 1906b). It is straightforward to show that the cross-section for the scattering of a beam of incident radiation by an electron is

$$\sigma_T = \frac{e^4}{6\pi \epsilon_0^2 m_e^2 c^4} = \frac{8\pi r_e^2}{3} = 6.653 \times 10^{-29} \text{ m}^2, \tag{7.3}$$

the *Thomson cross-section*, where $r_e = e^2/4\pi \epsilon_0 m_e c^2$ is the classical electron radius.[2]

Charles Barkla completed his physics degree at the University of Liverpool in 1899 and then won a Royal Commission of 1851 Fellowship which he held at Trinity College, his intention being to work with Thomson. His initial studies involved the measurement of the speed of propagation of electromagnetic waves along wires, using Rutherford's magnetic detector. Then the direction of his research changed to the study of X-rays, which became the main topic of his researches for the rest of his career. At this stage, the nature of X-rays was far from resolved. Thomson's view was that X-rays were electromagnetic pulses generated when cathode rays were decelerated by collisions with the walls of the discharge tube. If this were the case, the theory of the emission process outlined in Box 7.1 shows that the dipole pattern of the radiation would mean that the scattered radiation, referred to as the *secondary radiation*, would be polarised and that the intensities of the scattered radiation would be greatest in the direction perpendicular to the X-ray beam. Furthermore, if it was assumed that the electrons in the atom behaved like free electrons, the number of scatterers could be determined from the intensity of the scattered radiation.

Barkla began these studies in Cambridge before he moved to the University of Liverpool to take up the Oliver Lodge Fellowship. Already he had made significant contributions to the understanding of the nature of the X-rays, in particular on the issue of the polarisation of the X-rays (Barkla, 1903, 1904). As Thomson states in his paper,

> Barkla has shown that in the case of gases the energy in the scattered radiation always bears, for the same gas, a constant ratio to the energy in the primary whatever be the nature of the rays, that is, whether they are hard or soft; and secondly, that the scattered energy is proportional to the mass of the gas. The first of these results is a confirmation of the theory, as the ratio of the energy scattered to that in the primary rays . . . is independent of the nature of the rays; the second result shows that the number of corpuscles per cubic centimetre is proportional to the mass of the gas: from this it follows that the number of corpuscles in an atom is proportional to the mass of the atom, that is, to the atomic weight.

For a beam of X-rays passing through air, Barkla had measured the scattered fraction of the X-ray intensity to be to 2.4×10^{-4} cm^{-3} and consequently there should be about 25 corpuscles per air molecule, roughly equal to the atomic mass number A of the air molecules. He followed up these experiments by demonstrating that the secondary radiation indeed followed the expectations of the dipole theory of Thomson scattering of the X-rays. These X-ray experiments were continued by Barkla after his move to Liverpool, where he showed that in fact, except for hydrogen, the number of electrons is roughly half the atomic weight: $n \approx A/2$ (Barkla, 1911a); the correlation is with the atomic number rather than with the atomic weight.

At Liverpool, Barkla went on to make major contributions to the understanding of the physics of X-rays. In particular, by 1906 he had convincingly demonstrated that, by studying the tertiary radiation, that is, the scattering of the secondary radiation by a further scatterer, that the secondary radiation was almost 100% polarised (Barkla, 1906). He also discovered that the scattered radiation consisted of two types, the anisotropic radiation due to Thomson scattering and a second isotropic component which had a quite different dependence of absorption length upon the energy of the incident X-rays, recalling that the Thomson scattering process is independent of the energy of the X-rays. These studies led to the discoveries of the K and L absorption series of the elements and of the *characteristic X-ray signature* which was correlated with the atomic weight of the material (Barkla and Sadler, 1908). Barkla was awarded the 1917 Nobel Prize in Physics 'for his discovery of the characteristic Röntgen radiation of the elements'.

The third approach involved the scattering of β-rays by matter. Thomson derived the formula for what was called 'multiple-scattering theory', the energy loss due to multiple electrostatic interactions between the fast electron and the electrons in atoms; this process is also known as *ionisation losses* in the context of the interactions of electrons with atoms.[3] Using estimates by Rutherford for the mean free path of β-rays, Thomson again found the result $n \sim A$.

As Falconer (1988) has written,

> Thomson himself was largely responsible for undermining his theory. He suggested three ways of determining the number of corpuscles in the atom, and performed the

calculations himself, using other people's data. They were based on the dispersion of light in gases (using Rayleigh's and Ketteler's data), the scattering of x-rays by gases (using Barkla's data), and the absorption of β-rays by gases (using Becquerel's and Rutherford's data).

The conclusion of Thomson's paper was dramatic – most of the mass of atoms could not be due to the negatively charged electrons but must reside in the positive charge which held the atoms together. These experiments demolished Thomson's plum-pudding model of the atom and from about 1906 onwards he directed his energies to the study of the positive rays which accompanied the generation of cathode rays in discharge tubes. Before we continue that story, however, mention should be made of another key aspect of the conduction of electricity in gases, the phenomenon of thermionic emission.

7.3 Richardson and the law of thermionic emission

Owen Richardson obtained outstanding results in the Natural Sciences Tripos in 1900. Under the guidance of Thomson, he carried out his postgraduate research on the phenomenon which he named *thermionic emission*, the emission of electrons from hot bodies. His early researches concerned thermionic emission by platinum, which has a high melting temperature of 1755°C. As early as 1901, he found that the intensity of thermionic emission increased very rapidly with temperature and established what became known as *Richardson's law* for the number of electrons emitted per unit time per unit surface area from the surface of platinum,

$$j = AT^{1/2} \exp(-w/kT), \tag{7.4}$$

where w is the work function of the material, meaning the minimum energy necessary to eject an electron from the material, and A is a constant (Richardson, 1901). Richardson regarded the process of thermionic emission as being similar to the process of evaporation of molecules from the surface of a liquid, for which the above expression can be derived according to classical statistical mechanics (Box 7.2). Later, it was shown that a more satisfactory relation for the current associated with thermionic emission is

$$j = AT^2 \exp(-w/kT), \tag{7.5}$$

but the current j is still dominated by the exponential factor $\exp(-w/kT)$, and either (7.4) or (7.5) was a satisfactory fit to the experimental data.

This discovery led to a number of important developments in the application of thermionic emission. Richardson took up an appointment as Professor of Physics at Princeton University in 1906, where he continued his studies of the thermionic effect. A particularly impressive experiment was the first demonstration of the Maxwell distribution. By holding the anode of the thermionic valve at a negative potential and changing the temperature of the cathode, he and Fay Cluff Brown demonstrated that the current follows

Box 7.2	The theory of thermionic emission

Initially, Richardson modelled the process of thermionic emission by analogy with the kinetic theory of evaporation. In random close-packing within the body of a fluid, the coordination number $\mathcal{N}$, meaning the numbers of nearest neighbours, is 10. If the interaction energy between pairs of molecules is Φ_0, then the *binding energy* of atoms or molecules in the liquid would be $5N_0\Phi_0$ per mole, where N_0 is Avogadro's number. In the vapour phase, ignoring the thermal motion of the particles, the binding energy is zero and so the *latent heat* needed to convert the liquid into vapour would be $L = 5N_0\Phi_0$ per mole.

If the molecule were within the body of the fluid, it would have ten nearest neighbours, but at the surface it has only five bonds, all of them pulling it back into the fluid. Therefore, we would expect that the molecule would have to have energy $5\Phi_0$ to be able to escape from the surface, that is, just the latent heat per molecule. This is an example of a *thermally activated process* in which the probability of the molecule escaping from the surface is proportional to the Boltzmann factor, $\exp(-L/kT)$.

In equilibrium the loss of particles from the surface is balanced by the return of particles from the vapour phase to the liquid state. The rate of arrival of molecules at the surface is given by the usual formula $\frac{1}{4}n\bar{v}$ and so, in equilibrium,

$$A \exp(-L/kT) = \tfrac{1}{4}n\bar{v},$$

where A is some constant. Therefore the pressure of the gas is

$$p = nkT = \tfrac{1}{3}nm\overline{v^2} \propto \exp(-L/kT)(\overline{v^2})^{1/2}, \qquad p \propto T^{1/2}\exp(-L/kT),$$

where we have used the fact that $(\overline{v^2})^{1/2} \propto \bar{v}$. This is the form of the expression for the thermionic effect initially adopted by Richardson. Note that the power-law dependence on temperature is very weak compared with the very strong dependence upon temperature within the exponential term.

A better calculation requires the use of the Fermi–Dirac distribution of electron momenta in the heated material; this distribution was discovered independently by Fermi and Dirac only in 1926.[4] The distribution is integrated over all electrons with energy sufficient to escape from the surface of the material. The classical result of this calculation is that the electric current density escaping per unit time is

$$J = AT^2\,\mathrm{e}^{-W/kT}, \text{ where } A = \frac{4\pi emk^2}{h^3} = 1.2 \times 10^6\,\mathrm{A\,m^{-2}\,K^{-2}}.$$

$W = (U - E_F)$ is the work function of the material, U the potential barrier and E_F the Fermi level of the material. Notice that, in this approximation, the quantity A is a universal constant.

Maxwell's velocity distribution (Richardson and Brown, 1908). A further example is the invention by Lee De Forest (1906) of the thermionic triode valve, in which a negative potential is introduced between the cathode and anode and, by varying the negative potential, the current passing through the triode amplifies the variations in the negative grid potential. This marked the beginning of the electronic revolution. Richardson won the 1928 Nobel Prize in Physics 'for his work on the thermionic phenomenon and especially for the discovery of the law named after him'.

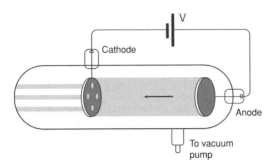

A schematic diagram showing the experimental arrangement for the production of canal rays. Positively charged
particles are accelerated towards the cathode and pass through the holes in it. The result is a set of 'canal rays', as
illustrated in the diagram, which strike the walls of the discharge tube behind the cathode.

Fig. 7.1

7.4 Thomson, Aston and positive rays

With the collapse of the 'plum-pudding' model, Thomson's attention turned to the nature
of the positive charges. Positive rays had been discovered by Eugen Goldstein in 1886. He
carried out cathode ray tube experiments with perforated cathodes and found a glow at
the cathode end of the tube (Figure 7.1). He concluded that there must be another type
of ray which moved in the opposite direction to the cathode rays. Since the new rays
passed through the holes in the cathode, Goldstein named them *Kanalstrahlen*, or *canal
rays* (Goldstein, 1886).

By 1906 when Thomson began his major assault on understanding their nature, relatively little was known about them except that they had to be positively charged and, unlike
the electrons, their mass-to-charge ratio was similar to that of hydrogen. As Falconer has
emphasised, Thomson's approach to the study of canal rays was strongly influenced by his
theoretical speculations (Falconer, 1988). He inferred that they might correspond to the
positively charged component of atoms. In a similar approach to that which he had adopted
in his measurement of the mass-to-charge ratio of the electrons, he studied the motion of
the canal rays in combined electric and magnetic fields, with the difference that in the new
experiments the electric and magnetic fields were parallel to each other. The result was
that particles of the same mass-to-charge ratio should lie along separate parabolae. Thomson's clear and elegant exposition of how this comes about is given in Box 7.3, which is an
extract from this paper on this subject of 1913.

In 1906, Thomson traced the forms of the parabolae from the locations of scintillations
on a willemite fluorescent screen at the end of the tube. The maximum value of e/m of 10^4
was always found for the hydrogen ion H^+. Although Thomson made several attempts to
find well-defined evidence of other elements, no clear results were obtained except for the
hydrogen ion H^+. He did not attempt a systematic improvement of any one of his numerous
experiments until Francis Aston arrived as his research assistant in 1910. Falconer (1988)
provides a detailed account of the various false starts and theoretical speculations which
occupied Thomson in the period 1906 to 1910 and which did not lead to any significant

Box 7.3   **Thomson's analysis of the deflection of canal rays**

It is simplest to quote Thomson's clear exposition of the theory behind the experiment (Thomson, 1913):

> The composition of these positive rays is much more complex than that of the cathode rays, for whereas the particles in the cathode rays are all of the same kind, there are in the positive rays many different kinds of particles. We can, however, by the following method sort these particles out, determine what kind of particles are present, and the velocities with which they are moving. Suppose that a pencil of these rays is moving parallel to the axis of x, striking a plane at right angles to their path at the point 0; if before they reach the plane they are acted on by an electric force parallel to the axis of y, the spot where a particle strikes the plane will be deflected parallel to y through a distance y given by the equation $y = (e/mv^2)A$, where e, m, v, are respectively the charge, mass, and velocity of the particle, and A a constant depending on the strength of the electric field and the length of path of the particle, but quite independent of e, m, or v.
>
> If the particle is acted upon by a magnetic force parallel to the axis of y, it will be deflected parallel to the axis of z, and the deflection in this direction of the spot where the particle strikes the plane will be given by the equation $z = (e/mv)B$, where B is a quantity depending on the magnetic field and length of path of the particle, but independent of e, m, v. If the particle is acted on simultaneously by the electric and magnetic forces, the spot where it strikes the plane will, if the undeflected position be taken as the origin, have for coordinates

$$(1) \qquad x = 0, \qquad y = \frac{e}{mv^2} A, \qquad z = \frac{e}{mv} B.$$

Thus no two particles will strike the plane in the same place, unless they have the same value of v and also the same value of e/m; we see, too, that if we know the value of y and z, we can, from equation (1), calculate the values of v and e/m, and thus find the velocities and character of the particles composing the positive rays. From equation (1) we see that

$$(2) \qquad z^2 = \frac{e}{m} y \frac{B^2}{A}, \qquad z = yv \frac{B}{A}.$$

Thus all the particles which have a given value of e/m strike the plane on a parabola, which can be photographed by allowing the particles to fall on a photographic plate.

advance in the understanding of the nature of the positive rays. The situation changed dramatically with the arrival of Aston, who was a highly skilled experimenter and an expert in high-vacuum techniques. In 1909 he was appointed to a lectureship at the University of Birmingham, but on the recommendation of Thomson's friend Poynting, he joined Thomson in Cambridge the following year as his assistant. He was to be responsible for the many technical advances which were to lead to the understanding of the nature of the canal rays and to the invention of the mass spectrograph.

Through the period 1910 to 1913, Aston was responsible for the following improvements in the experiments (Figure 7.2(a)).

- Thomson had not appreciated that the willemite screen was only sensitive to hydrogen ions and not to heavier ions. When the willemite screen was replaced by a photographic plate, the *characteristic curves*, meaning the parabolae associated with different ions, became apparent.

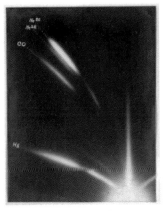

(a)   (b)

Fig. 7.2

(a) Thomson and Aston's photographic positive-ray tube. As described by Davis and Falconer (1997), the apparatus now incorporates Aston's large discharge bulb, liquid-air cooled charcoal to improve the vacuum and a very large electromagnet to provide a strong magnetic field. The photographic plate is contained in the tall vertical tube to the left of the picture. (b) Thomson and Aston's photograph of the isotopes of neon, showing clearly the separation of the isotopes ^{20}Ne and ^{22}Ne (Thomson, 1912a).

- There was a dichotomy between the need for reasonable pressures to create the fluxes of ions in the discharge tube and the need for a very much higher vacuum in which the deflection by the electric and magnetic fields took place. This was achieved by making the volume of the discharge tube very much larger and separating it from the higher-vacuum region where the deflections took place.
- A much improved vacuum was achieved by exhausting the deflection part of the apparatus with liquid-air-cooled charcoal.

With these improvements, Thomson soon observed the characteristic curves of numerous atomic and molecular ions, including CH, CH_2, CH_3 and the hydroxyl ion OH^+. The most important of these discoveries was the discovery of the isotopes of neon, ^{20}Ne and ^{22}Ne (Figure 7.2(b)) (Thomson, 1912a). These were the first non-radioactive isotopes to be discovered. Thomson was somewhat reluctant to believe in the reality of these isotopes and was more excited by the discovery of the molecule which he labelled X_3 and which in the end turned out to be the H_3 ion.

These studies were interrupted by the outbreak of the First World War. After 1914, Thomson turned his attention to other matters but Aston realised the importance of the new lines of enquiry opened up by the positive-ray tube and, after hostilities had ceased, he developed a series of three mass spectrometers of increasing mass resolution and sensitivity. The first mass spectrograph of 1919 was a great success, establishing beyond doubt the existence of isotopes of neon, chlorine, bromine and krypton. Within the accuracy of his measurements, all the elements had masses which were integral multiples of the unit of mass on the oxygen mass scale. This gave rise to Aston's *whole-number rule*, that the masses of all isotopes are integral multiples of the standard unit mass. By 1925, he had used this mass spectrograph to measure the masses of the isotopes of over 50 elements.

A key exception to the whole-number rule was the mass of the hydrogen nucleus, which was about 1% greater than the standard mass unit; this was a result of the highest importance and opened up the possibility of nuclear energy generation. This was not lost on Arthur Eddington, the Plumian Professor of Astronomy at the Cambridge Observatories. In a remarkably prescient paragraph of his presidential address to the Mathematical and Physics Section of the British Association for the Advancement of Science at its annual meeting, held in Cardiff, he stated (Eddington, 1920):

> Certain physical investigations in the past year . . . make it probable to my mind that some portion of this sub-atomic energy is actually being set free in the stars. F.W. Aston's experiments seem to leave no room for doubt that all the elements are constituted out of hydrogen atoms bound together with negative electrons. The nucleus of the helium atom, for example, consists of 4 hydrogen atoms bound with two electrons. But Aston has further shown conclusively that the mass of the helium atom is less than the sum of the masses of the 4 hydrogen atoms which enter into it; and in this at any rate the chemists agree with him. There is a loss of mass in the synthesis amounting to about 1 part in 120, the atomic weight of hydrogen being 1.008 and that of helium 4 . . . Now mass cannot be annihilated, and the deficit can only represent the mass of the electrical energy set free in the transmutation. We can therefore at once calculate the quantity of energy liberated when helium is made out of hydrogen. If 5 per cent of the star's mass consists initially of hydrogen atoms, which are gradually being combined to form more complex elements, the total heat liberated will more than suffice for our demands, and we need look no further for the source of a star's energy.

This prediction was made before there was any understanding of the mechanism by which the energy could be released – quantum mechanics had not then been discovered.

Aston was elected to Fellowship of Trinity College in 1920 and awarded the Nobel Prize in Chemistry in 1922 'for his discovery, by means of his mass spectrograph, of isotopes, in a large number of non-radioactive elements, and for his enunciation of the whole-number rule.' Realising the significance of the precise measurement of the masses of isotopes, Aston completed his second mass spectrograph in 1925, which had a mass resolution of about 10,000. This was followed by a third mass spectrograph in which the mass resolution approached one part in 100,000 (Figure 7.3). It is estimated that he identified 212 isotopes of the chemical elements. This was a triumph of experimental genius. Aston's list of publications from the 1920s and 1930s shows the remarkable catalogue of isotopes he identified for elements throughout the periodic table.[5]

7.5 Towards the old quantum theory

Returning to the first decade of the twentieth century, the beginnings of the revolution which was to result in the establishment of the theory of relativity and the creation of quantum mechanics was gathering momentum. As indicated in Section 6.6, the preoccupations of many of the leading physicists in Europe concerned the role of quanta in fundamental

Aston's third mass spectrograph. Aston devoted the whole of his research career at the Cavendish Laboratory to
perfecting the mass spectrograph, eventually reaching a mass resolution of almost one part in 100,000. It is interesting
to compare Aston's final version of the spectrograph with Thomson and Aston's positive-ray tube shown in
Figure 7.2(*a*).

Fig. 7.3

physics. Planck had introduced his constant h in his epochal paper of 1900 (Planck, 1900)
and Einstein had proposed his revolutionary concept that, under certain circumstances,
light could be considered to consist of discrete quanta, each with energy $E = h\nu$, rather
than waves (Einstein, 1905). It was not until the Solvay Conference of 1911 that these concepts began to be taken really seriously by the majority of the leading European physicists.
These endeavours were to result in the formulation of what can be called the *old quantum
theory*, a patchwork of quantum concepts which lacked overall systematic coherence until
it was superseded by the new theory of *quantum mechanics* from 1925 onwards.[6]

The Cavendish Laboratory's contributions to these developments continued to be in the
establishment of the experimental evidence, rather than in the theoretical debate. An example of this, which attracted little attention, was the early work of Geoffrey (G.I.) Taylor
on the interference of light rays of very low intensity.

7.5.1 G.I. Taylor and the interference of light waves

Taylor had a distinguished undergraduate record in the Natural Sciences Tripos which
resulted in his winning a major research scholarship at Trinity College in 1908. He
asked Thomson about possible areas of research and Thomson suggested investigating the

interference of light waves of very low intensity. Thomson was well aware of Einstein's paper on the quantum nature of radiation and speculated that the quanta might be associated with irregularities in the wavefront of the radiation. Then, he reasoned, there might well be observable changes in the interference of light if the packets of light were to arrive one at a time at the detector.

Taylor carried out the experiment at his home and it involved observing the interference fringes associated with the diffraction of light by a needle. The intensity of light from a gas jet could be reduced by introducing four smoked glass screens between the gas jet and a photographic plate. The lengths of the exposures were calibrated so that the same image intensity was produced on the plate as more glass screens were introduced. In the longest exposure, which was for three months, the resulting diffraction fringes were just as sharp as in the case of a short exposure without any smoked glass screens.[7] The flux of radiation incident upon the photographic plate in the three-month exposure was 5×10^{-6} erg cm^{-2} s^{-1}, corresponding to a flux of optical quanta of about 10^6 photons cm^{-2} s^{-1}. In other words, despite the fact that the quanta were arriving one at a time at any point on the photographic plate, sharp interference fringes were still observed. The full significance of this result for the concept of the wave–particle duality was only appreciated much later. Taylor never returned to his work, but carried out very distinguished research in fluid mechanics and dynamics.

7.5.2 Rutherford, the nature of α-particles and the nuclear structure of atoms

The nature of β-rays as electrons was quickly assimilated into the armoury of the physicist, but what about the nature of the α-particles? In 1902, while Rutherford was at McGill University in Canada, he showed that the α-particles were deflected by electric and magnetic fields and that their value of e/m was roughly that of hydrogen ions (Rutherford, 1903). Rutherford took up the Chair of Physics at Manchester University in 1907 and, in the following year, demonstrated convincingly that α-particles are helium nuclei (Box 7.4) (Rutherford and Royds, 1909).

The discovery of the nuclear structure of atoms resulted from a brilliant series of experiments carried out by Rutherford and his colleagues, Hans Geiger and Ernest Marsden, in the period 1909–12 at Manchester. Rutherford had been impressed by the fact that α-particles could pass through thin films rather easily, suggesting that much of the volume of atoms is empty space, although there was clear evidence of small-angle scattering. Rutherford persuaded Marsden, who was still an undergraduate, to investigate whether or not α-particles were deflected through large angles on being fired at a thin gold foil target. To Rutherford's astonishment, a few particles were deflected by more than 90°, and a very small number almost returned along the direction of incidence. In Rutherford's words:

> It was quite the most incredible event that has ever happened to me in my life. It was almost as incredible as if you fired a 15-inch shell at a piece of tissue paper and it came back and hit you. (Andrade, 1964)

In 1911 Rutherford hit upon the idea that, if all the positive charge were concentrated in a compact nucleus, the strong scattering could be attributed to the repulsive electrostatic

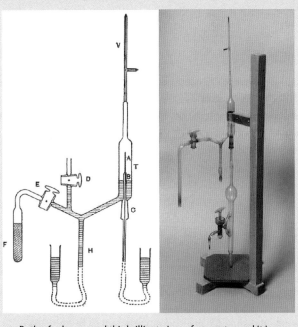

A source of α-particles was inserted into a fine glass tube (A) which could be inserted into an evacuated discharge tube. Before the 'needle' was inserted, a high voltage was maintained across the discharge tube and no evidence for the emission lines of helium was observed. Once the needle was inserted into the tube, the α-particles passed through the thin walls of the glass tube, which were only 0.01 mm in thickness, and the characteristic lines of helium were observed in the discharge tube. This was convincing evidence that α-particles are the nuclei of helium atoms.

Rutherford preserved this brilliant piece of apparatus and it is now on display in the Cavendish Collection.

force between the incoming α-particle and the positive nucleus. He derived his famous *Rutherford scattering* formula for the probability that the α-particle is scattered through an angle of deflection ϕ,

$$p(\phi) \propto \frac{1}{v_0^4} \, \mathrm{cosec}^4 \, \frac{\phi}{2}, \qquad (7.6)$$

where v_0 is the initial velocity of the α-particle. This $\mathrm{cosec}^4(\phi/2)$ law was found to explain precisely the observed distribution of scattering angles of the α-particles (Geiger and Marsden, 1913).

The fact that the scattering law was obeyed so precisely, even for large angles of scattering, meant that the inverse-square law of electrostatic repulsion held good to very small distances indeed. They found that the nucleus had to have size less than about 10^{-14} m, very much less than the sizes of atoms, which are typically about 10^{-10} m. Rutherford attended the first Solvay Conference in 1911, but made no mention of his remarkable experiments, which led directly to the nuclear model of the atom.

7.5.3 Nicholson's and Bohr's models of atoms

The first physicist to attempt to introduce quantum concepts into the construction of atomic models was a Viennese doctoral student Arthur Erich Haas, who realised that, if Thomson's sphere of positive charge were uniform, an electron would perform simple harmonic motion through the centre of the sphere. Haas argued that the energy of oscillation of the electron, $E = e^2/4\pi\epsilon_0 a$, should be quantised and set equal to $h\nu$. Therefore,

$$h^2 = \frac{\pi m_e e^2 a}{\epsilon_0}. \qquad (7.7)$$

Haas used this expression to show how Planck's constant could be related to the properties of atoms, taking for the frequency ν the short-wavelength limit of the Balmer series (Haas, 1910a,b,c). Haas's efforts were discussed by Lorentz at the 1911 Solvay Conference, but they did not attract much attention. According to Haas, Planck's constant h was simply a property of atoms, whereas those already converted to quanta preferred to believe that h had a much deeper significance.

The next clue was provided by the work of the mathematician and physicist John William Nicholson. He studied first at Manchester University, where he met his lifelong friend Arthur Stanley Eddington. Together, they proceeded to Trinity College, Cambridge. Nicholson had an outstanding academic career, being a Smith's Prize winner and twice winning the Adams Prize. For some time, he was a lecturer in the Cavendish Laboratory.

Nicholson's major contribution to the understanding of the structure of atoms was his discovery of the rule for the quantisation of angular momentum. He showed that, although the Saturnian model of the atom is unstable for perturbations in the plane of the orbit, perturbations perpendicular to the plane are stable for orbits containing up to five electrons; he assumed that the unstable modes in the plane of the orbit were suppressed by some unspecified mechanism (Nicholson, 1911, 1912). The frequencies of the stable oscillations were multiples of the orbital frequency and he compared these with the frequencies of the lines observed in the spectra of bright nebulae, particularly with the unidentified 'nebulium' and 'coronium' lines.[8] Performing the same exercise for ionised atoms with one less orbiting electron, further matches to the astronomical spectra were obtained. The frequency of the orbiting electrons remained a free parameter, but when he worked out the angular momentum associated with them, Nicolson found that they turned out to be multiples of the quantum unit of angular momentum $h/2\pi$.

Niels Bohr completed his doctorate on the electron theory of metals in 1911. Even at that stage, he had convinced himself that this theory was seriously incomplete and required further mechanical constraints on the motion of electrons at the microscopic level. He spent the following year in England, working for seven months with Thomson at the Cavendish Laboratory, and four months with Rutherford in Manchester. On his arrival in Cambridge in the autumn of 1911, Thomson showed little interest in Bohr's dissertation on the theory of metals, but Bohr's interest was strongly aroused by Thomson's attempts to model the electronic structure of atoms. In 1912 Bohr moved to Manchester, where he found Rutherford much more encouraging and sympathetic to his ideas on the structure of atoms.

Bohr was immediately struck by the significance of Rutherford's model of the nuclear structure of the atom and began to devote all his energies to understanding atomic structure on that basis. He quickly appreciated the distinction between the chemical properties of atoms, which are associated with the orbiting electrons, and radioactive processes which are associated with activity in the nucleus. On this basis, he could understand the nature of the isotopes of a particular chemical species. Bohr also realised from the outset that the structure of atoms could not be understood on the basis of classical physics. The obvious way forward was to incorporate the quantum concepts of Planck and Einstein into the models of atoms. Einstein's statement,

> for ions which can vibrate with a definite frequency, . . . the manifold of possible states must be narrower than it is for bodies in our direct experience. (Einstein, 1906)

was precisely the type of constraint which Bohr was seeking to understand how atoms could survive the inevitable instabilities according to classical physics.

In the summer of 1912, Bohr wrote an unpublished memorandum for Rutherford in which he made his first attempt at quantising the energy levels of the electrons in atoms (Bohr, 1912). His memorandum was principally about issues such as the number of electrons in atoms, atomic volumes, radioactivity, the structure and binding of diatomic molecules and so on. There is no mention of spectroscopy, which he and Thomson considered too complex to provide useful information. When Bohr returned to Copenhagen later in 1912, he was perplexed by the success of Nicholson's model, which seemed to provide a successful, quantitative model for the structure of atoms and which could account for the spectral lines observed in astronomical spectra.

The breakthrough came in early 1913, when Hans Marius Hansen told Bohr about the Balmer formula for the wavelengths, or frequencies, of the spectral lines in the spectrum of hydrogen,

$$\frac{1}{\lambda} = \frac{\nu}{c} = R_\infty \left(\frac{1}{2^2} - \frac{1}{n^2} \right), \tag{7.8}$$

where $R_\infty = 1.097 \times 10^7 \, \mathrm{m}^{-1}$ is the *Rydberg constant* and $n = 3, 4, 5, \ldots$. As Bohr recalled much later,

> As soon as I saw Balmer's formula, the whole thing was clear to me.[9]

In the first paper of his famous trilogy (Bohr, 1913a,b,c), Bohr acknowledged that Nicholson had discovered the quantisation of angular momentum in his papers of 1912. These results were the inspiration for what became known as the *Bohr model of the atom*.[10] It is convenient to regard Bohr's application of quantum concepts to atoms and molecules in his papers of 1913 as marking the beginning of the development of the old quantum theory, which was to dominate theoretical and experimental physics in continental Europe until it was superseded by quantum mechanics in the period 1925 to 1930. Meanwhile, the research activities in the Laboratory remained firmly in the experimental domain. Two major contributions were made in the period immediately before the First World War which provided new techniques for studying fundamental processes in atomic and nuclear physics: C.T.R. Wilson's invention of the cloud chamber and the Bragg's discovery of the law of X-ray diffraction.

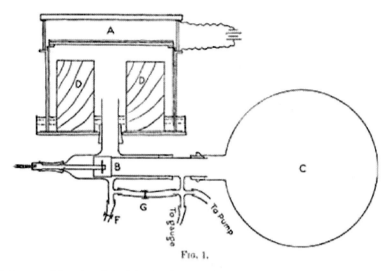

FIG. 1.

Fig. 7.4   Wilson's 1910 version of the enlarged cloud chamber (Wilson, 1912).

7.6 Wilson's cloud chamber

We left C.T.R. Wilson in Chapter 5 at the point where he had refined his water vapour condensation techniques to such a degree that estimates could be made of the charge of the electron (Section 6.5.2), and made his abortive attempts to understand the origin of atmospheric radioactivity at sea level (Section 6.5.4). For the first decade of the twentieth century, Wilson devoted a considerable effort to understanding the origins of atmospheric electricity, but in 1910 he returned to refine what became known as the *Wilson cloud chamber*. By that time, the nature of the rays emitted by radioactive substances was better understood; Rutherford and Royds (1909) had shown that α-particles were the nuclei of helium atoms. Wilson realised that he might be able to image the tracks of the particles by photographing the streams of water droplets condensing onto the ions created by the ionisation losses of the high-energy particles passing through the active chamber. The 1910 version of the cloud chamber (Figure 7.4) was similar to the well-known final version (Figure 7.5). The configuration was a very much enlarged version of his earlier experiment (see Figure 6.3), with the cylindrical expansion chamber much increased in diameter.

An intriguing feature of the apparatus was the means by which the rapid expansion of the cloud chamber was synchronised with the firing of the 'flash-bulb' to take a photograph of the particle tracks (Box 7.5). Wilson obtained the first images of the tracks of α- and β-particles with this cloud chamber in 1911. The final perfected cloud chamber is shown in Figure 7.5. The quality of the images taken by Wilson was superb. As Blackett remarked,

> (The many exquisite photographs) ... still remain among the technically best photographs ever made. (Blackett, 1960)

The final version of the Wilson cloud chamber.

Fig. 7.5

Synchronisation of the cloud chamber expansion and the flash to illuminate the particle tracks (Wilson, 1912)

Box 7.5

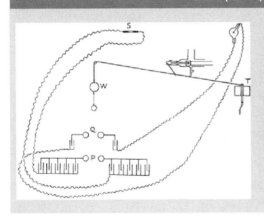

A capillary mercury vapour spark (S), actuated by Leyden jars, was used to provide illumination. When the trigger T is pulled, the string is released, the weight begins to fall and opens the valve of the cloud chamber, causing the expansion. The string breaks and the steel sphere W falls under gravity, short-circuiting both the X-ray tube (Q), if that is required in the experiment, and the poles of the Leyden jar to make the spark (P, S) illuminate the chamber. (Wilson, 1912)

Figure 7.6(a) shows examples of the images of α- and β-particles while Figure 7.6(b) shows ions released by a flux of X-rays. This was the only version of the final cloud chamber Wilson ever built. According to J.G. Crowther in his history, *The Cavendish Laboratory, 1874–1974*,

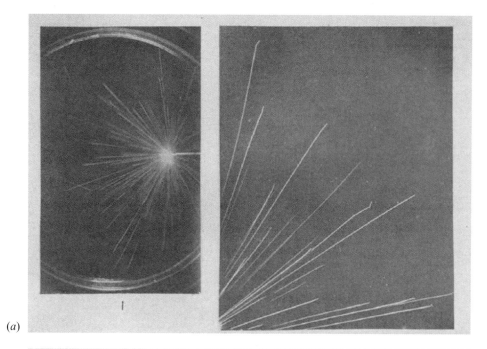

(a)

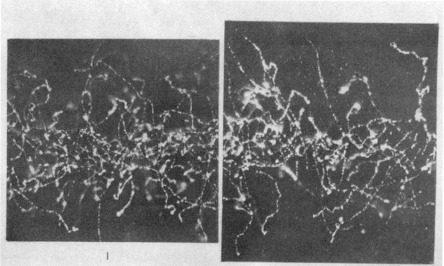

(b)

Fig. 7.6   (a) Cloud chamber images of the tracks of α-particles from the radioactive decay of radium. (b) The ionisation tracks caused by the irradiation of the cloud chamber by X-rays (Wilson, 1912).

When W.L. Bragg was assembling historic apparatus for the museum at the Cavendish, he asked Wilson whether the cloud chamber for photographing atomic tracks, which he presented, was the original. Wilson replied in his strong Scottish accent,

'Therr was neverr but the one'.

Its construction had cost only about £5. (Crowther, 1974)

In 1913, Wilson was appointed Observer in Meteorological Physics at the Solar Physics Observatory of the Cambridge Observatories, now the Institute of Astronomy. After the First World War, he moved the apparatus to the Solar Physics Observatory, where he conducted his experiments from then on. In 1921–22, he took 500 superb stereoscopic images of particle tracks. Blackett's assessment of what Wilson had achieved was as follows:

> There are many decisive experiments in the history of physics which, if they had not been made when they were made, would surely have been made not much later by someone else. This might not have been true of Wilson's discovery of the cloud method. In spite of its essential simplicity, the road to its final achievement was long and arduous: without C.T.R. Wilson's vision and superb experimental skill, mankind might have had to wait many years before someone else found the way. (Blackett, 1960)

Blackett enumerates some of the major discoveries in particle physics which were made with the Wilson cloud chamber: the positive electron, pair production and cosmic ray shower phenomena, the μ-meson and its spontaneous decay, the charged and neutral V-particles and the negative cascade hyperon. J.J. Thomson was equally effusive about Wilson's achievement:

> This work of C.T.R. Wilson, proceeding without haste and without rest since 1895, has rarely been equalled as an example of ingenuity, insight, skill in manipulation, unfailing patience and dogged determination. . . . For many years he did all the glass-blowing himself . . . how often, when the apparatus is all but finished, it breaks and the work has to begin again. [Wilson] would take up a fresh piece of glass, perhaps say 'Dear, dear' but never anything stronger, and begin again. (Thomson, 1936)

Although couched in the language of an obituary, Wilson was very much alive and thriving when Thomson wrote these words in his reminiscences. It is striking that the invention of the cloud chamber resulted from the experimental study of the pure science problem of understanding the formation of water droplets in clouds but ended up being applied in an area of the greatest scientific importance for the understanding of nuclear interactions.

Wilson was awarded the 1927 Nobel Prize in Physics, jointly with Arthur Holly Compton. The citation for Wilson reads:

> [F]or his method of making the paths of electrically charged particles visible by condensation of vapour.

7.7 Bragg's law and the X-ray spectra of the chemical elements

Throughout the early years of the twentieth century, the issue of whether X-rays were particles or waves remained contentious. The identification of X-rays with 'ultra-ultraviolet' radiation was strongly suggested by Charles Barkla's demonstration that the scattered X-radiation was polarised (Barkla, 1906) (Section 7.2.2). Then, in 1912, Max von Laue realised that, if the X-rays were indeed electromagnetic radiation, they should be

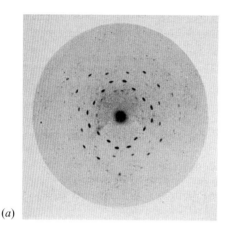
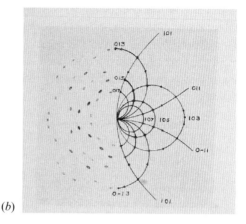

(a)   (b)

Fig. 7.7   (a) Von Laue's X-ray diffraction pattern for a crystal of cubical zinc blende. (b) Lawrence Bragg's match of the spots on Laue's diffraction pattern to his theory of diffraction by planes in a face-centred crystal (Bragg, 1912a).

diffracted by crystals. He persuaded his colleagues Paul Knipping and Walter Friedrich to attempt an experiment in which the crystal materials were used as transmission gratings – the diffracted X-rays passed through the crystal and the diffraction pattern was recorded on a photographic plate. They succeeded in observing diffraction spots in crystals of copper sulphate and later of ZnS (zinc blende) and diamond (Friedrich *et al.*, 1912; Laue, 1912) (Figure 7.7(a)). This was compelling evidence for the wave nature of X-rays.

In 1908, Lawrence Bragg came up to Trinity College, Cambridge as an undergraduate, having recently arrived from Australia with his father, William Bragg, who had been appointed Cavendish Professor of Physics at the University of Leeds. The younger Bragg first studied mathematics and then completed his degree in physics with high distinction. In his recollections, he recalls the inspiration of the lectures of C.T.R. Wilson:

> C.T.R. Wilson lectured on optics. The lectures were the best, and the delivery was the worst, of any lectures I have ever been to. He mumbled facing the board, he was very hesitant and jerky in his delivery, and yet the way he presented the subject was quite brilliant. I think his lectures on optics set the standard for similar lectures all over the country when later his pupils got chairs. . . .
>
> On one occasion, a suggestion from him was wonderfully fruitful. I had worked out theoretically that one could consider X-ray diffraction as due to reflection of the X-rays in the crystal plane. It was Wilson who suggested I should try specular reflection from cleavage sheets of mica. The experiment came off and I published it in *Nature* and this caused quite a furore of similar experiments. . . . It was C.T.R.'s treatment of a grating diffracting white light, which he had given us in his lectures, which set me thinking on the right lines when I gave a simpler treatment in November 1912 of Laue's diffraction experiments. (Bragg, 1912b)

Lawrence Bragg and his father William turned their attention to the discoveries of Knipping, Friedrich and von Laue, but initially they adopted quite opposite views about the origin of the spots. William considered the spots to be associated with X-ray particles travelling down channels within the crystal structure, whereas Lawrence was more convinced by the wave picture. In his first year as a graduate student under Thomson, he came up with the profound insight that the spots could be associated with the diffraction patterns of the various planes defined by the atoms in the crystal. Von Laue had been able to account for some of the spots in Figure 7.7(*a*) by the scattering of the X-rays from individual atoms, but he could not explain them all, nor the absence of other spots.

Although the spectrum of the incident X-rays contained a wide range of frequencies, Lawrence Bragg realised that only those associated with certain wavelengths would result in a coherent reflected signal. As he wrote in his paper published in 1912,

> Thus, it is to be expected that the intensity of the spot produced by a train of waves from a set of planes in the crystal will depend upon the value of the wavelength, viz. $2d \cos \theta$ [where d is the shortest distance between successive identical planes in the crystal]. When $2d \cos \theta$ is too small the successive pulses in the train are so close that they begin to neutralise each other and when $2d \cos \theta$ is large the pulses follow each other at large intervals and the train contains little energy. Thus the intensity of the spot depends on the energy in the spectrum of the incident radiation characteristic of the corresponding wavelength. (Bragg, 1912a)

This was the first appearance of what became known as *Bragg's law*,

$$n\lambda = 2d \sin \theta, \tag{7.9}$$

where λ is the wavelength of the radiation, d is the lattice spacing and θ is the angle between the crystal planes and the direction of incidence of the X-ray beam; n is an integer. In a brilliant analysis, Lawrence went on to show how *all* the spots in the X-ray image of zinc blende (ZnS) could be explained in terms of the constructive interference of waves scattered coherently from the different planes of the crystal structure (Figure 7.7(*b*)). In particular, he showed that the structure of zinc blende had to be that of a face-centred rather than a body-centred cube.

William Bragg was soon converted to his son's point of view and from then on they collaborated on the use of what was to become the new discipline of X-ray crystallography. Lawrence Bragg appreciated the importance of the new generation of X-ray spectrometers which his father was constructing at Leeds. Pure crystal samples, such as rock salt, could be used as a diffraction grating and the spectrum of the X-rays found in reflection according to Bragg's law. The invention of the X-ray spectrometer enabled X-ray spectra to be recorded on photographic plates with high spectral resolution (Figure 7.8).

The elder Bragg was happy to leave most of the further determinations of crystal structures to his son. The next paper concerned the structure of rock salt (NaCl), which Ewald described as 'the great break-through to actual crystal structure determination and to the absolute measurement of X-ray wavelengths' (Bragg, 1913b; Ewald, 1962). As Phillips remarked (Phillips, 1979), Bragg's conclusion that

> in sodium chloride the sodium atom has six neighbouring chlorine atoms equally close with which it might pair off to form a molecule of NaCl

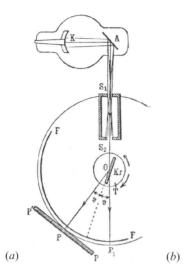
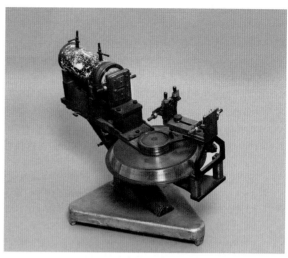

(a)   (b)

Fig. 7.8 (a) A diagram illustrating the operation of William and Lawrence Bragg's rotating crystal X-ray spectrometer. This apparatus was used by Moseley in his experiments in which the nature of the K and L lines of the elements was elucidated (Sommerfeld, 1919). (b) W.H. Bragg's second spectrometer, built by his instrument maker C.H. Jenkinson and sent to his son Lawrence Bragg in Cambridge in 1914.

was to disturb the chemists for many years. Next, the Braggs tackled the structure of diamond and discovered its tetrahedral structure (Bragg and Bragg, 1913). This was a key result for the full acceptance of the method. Again, quoting Ewald (1962),

> Diamond was the first example of a structure in which the effective scattering centres did not coincide with the points of a single [Bravais] lattice. Whereas in the structures of rock salt, zinc blende and fluorite the absence of molecules in the accepted sense created an element of bewilderment, the beautiful confirmation of the tetravalency of carbon on purely optical grounds made this structure and the method by which it was obtained, immediately acceptable to physicists and chemists alike.

Bragg's next great paper (Bragg, 1913a) described the structures of zinc blende (ZnS), fluorspar (CaF_2), iron pyrites (FeS_2) and calcite (CaC_3).

These studies were interrupted by the outbreak of the First World War in 1914, during which Lawrence worked on acoustic techniques for determining the positions of enemy gun emplacements (see Section 7.8). He was joint winner with his father of the Nobel Prize in Physics in 1915 'for their services in the analysis of crystal structure by means of X-rays'. After the war, Lawrence Bragg was appointed Langworthy Professor of Physics at Manchester, in succession to Rutherford who was appointed Cavendish Professor of Physics in the Cavendish Laboratory in 1919. X-ray crystallography was to develop into a major discipline in its own right.

Barkla at Liverpool University had continued his studies of the scattered X-ray emission of different elements and in 1908 he and Charles Sadler discovered that each element had a *characteristic X-ray signature* which was correlated with the atomic weight of the material

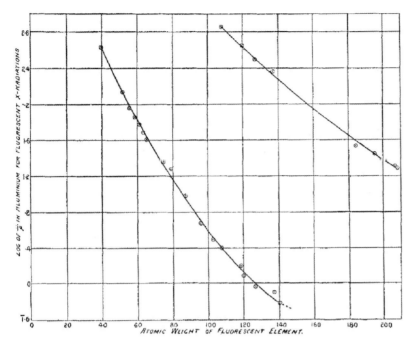

Barkla's summary of his experiments on the K (left curve) and L (right curve) components of fluorescent X-rays from samples of different elements, plotted against atomic weight (Barkla, 1911b). The ordinate is the logarithm of the quantity λ / ρ, where λ is the absorption coefficient defined by $I = I_0 \, e^{\lambda x}$.

Fig. 7.9

(Barkla and Sadler, 1908). In these experiments, the absorption of the X-rays by thin aluminium sheets was used to measure the 'hardness' or 'softness' of the X-ray emission. For a number of elements the fluorescent emission consisted of two components, a 'soft' component which was readily absorbed and a 'hard' component which suffered very much less absorption. In 1911, he summarised the results of his numerous absorption experiments (Figure 7.9), demonstrating that the materials had both hard and soft components, which he labelled K and L (Barkla, 1911b).

The nature of the K and L series was elucidated by Moseley's brilliant experiments carried out at Manchester using the type of X-ray spectrograph perfected by William Bragg. Different pure materials were inserted into an X-ray tube and the reflected X-rays analysed spectroscopically by reflecting them from a carefully prepared sample of rock salt. He discovered that the reflected spectrum consisted of continuum radiation superimposed upon which were strong X-ray lines. The continuum radiation was the *bremsstrahlung*, or braking radiation, of the energetic electrons decelerated in the material. The lines were responsible for the K and L components of the X-ray emission identified by Barkla. The K lines were split into two components which Moseley labelled K_α and K_β. Corresponding splittings were observed in the L lines at somewhat longer wavelengths. The K_α line was about five times stronger than the K_β line but the K_β lines had frequencies about 10% greater than those of the K_α lines.

The most spectacular result of Moseley's experiments was the discovery of the correlation between the frequency of the X-ray lines and the atomic number Z, corresponding to the number of electrons in the neutral atom (Moseley, 1913, 1914). In his papers, he plotted these correlations separately for the K_α, K_β, L_α and L_β lines (Figure 7.10). The remarkable linear correlation between the square root of the frequencies of the lines and the atomic number had several crucial consequences. Moseley wrote the correlation between the square root of the frequencies of the K_α lines and the atomic number in the somewhat provocative form

$$K_\alpha \text{ lines:} \quad \nu_\alpha = R_\infty (Z-1)^2 \left(\frac{1}{1^2} - \frac{1}{2^2} \right), \tag{7.10}$$

where R_∞ is the same constant which appears in Rydberg's formula. For the L series, the correlations were described by

$$L_\alpha \text{ lines:} \quad \nu_\alpha = R_\infty (Z-7.4)^2 \left(\frac{1}{2^2} - \frac{1}{3^2} \right). \tag{7.11}$$

Moseley wrote to Bohr in November 1913 that these results are 'extremely simple and largely what you would expect'. Rewriting Bohr's formula for the frequencies of the spectral lines for a nucleus of charge Ze as

$$\frac{\nu}{c} = R_\infty Z^2 \left(\frac{1}{m^2} - \frac{1}{n^2} \right), \tag{7.12}$$

a perfect fit to Moseley's data was obtained. Rutherford immediately appreciated the significance of Moseley's discovery. In his words,

> The original suggestion of van den Broek that the charge of the nucleus is equal to the atomic number and not to half the atomic weight seems to me very promising. The idea has already been used by Bohr in his theory of the constitution of atoms. The strongest and most convincing evidence in support of this hypothesis will be found in a paper by Moseley in the *Philosophical Magazine* of this month. He there shows that the frequency of the X-radiations from a number of elements can be simply explained if the number of unit charges on the nucleus is equal to the atomic number. It would appear that the charge of the nucleus is the fundamental constant which determines the physical and chemical properties of the atom, while the atomic weight, although it approximately follows the order of the nuclear charge, is probably a complicated function of the latter depending upon the detailed structure of the nucleus. (Rutherford, 1913)

Moseley's fifth conclusion of his paper of 1914 read:

> (5) Known elements correspond with all the numbers between 13 and 79 except three. There are here three possible elements still undiscovered.

This conclusion resulted in the prediction of elements with atomic numbers 43, 61 and 75 (see Figure 7.10), which were only discovered many years later as the elements technetium (Tc), promethium (Pm) and rhenium (Re), respectively. Tragically, Moseley was killed in action during the Gallipoli campaign in Turkey on 10 August 1915.

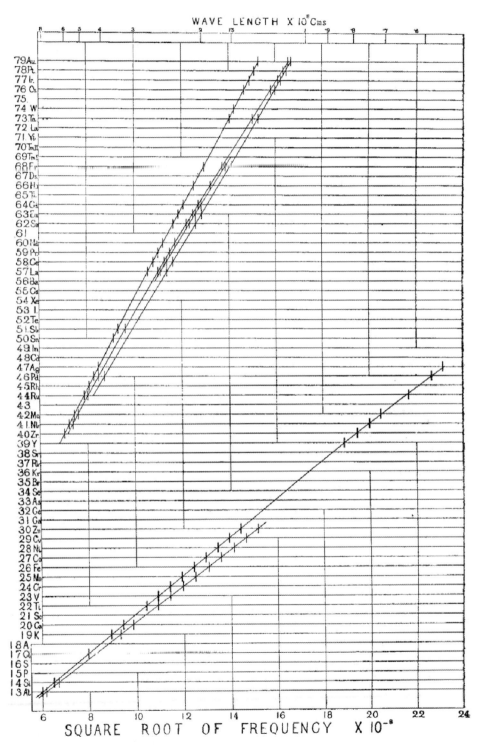

Moseley's correlation diagram between frequency of the various X-ray lines and atomic number. The K_α, K_β lines (bottom half of diagram) and L_α, L_β lines (top half of diagram) are plotted separately and show perfect correlations with atomic number (Moseley, 1913, 1914).

Fig. 7.10

7.8 The war years

Research in the Laboratory effectively stopped during the period of hostilities from 1914 to 1918. Almost all the research students joined the officer's training corps and many saw action at the front. Some of the best physicists joined military scientific establishments but in general there was little concept that basic physics could contribute to military matters. In December 1914, soldiers were billeted in the large experimental room in the Laboratory used for teaching medical students, and the workshops were used for the manufacture of gauges for use in the armament industries.

In 1915 J.J. Thomson became a member of the Board of Invention and Research, which was set up to review all scientific ideas and inventions which might be useful for the war effort. According to Crowther, about one hundred thousand inventions were considered, of which about 30 proved to be of some use; perhaps the most useful part of the board's initiatives was the work of William Bragg on anti-submarine devices.

Among the more successful war endeavours was Lawrence Bragg's development of acoustic techniques for locating the positions of enemy gun emplacements (van der Kloot, 2005). The blast wave from the gun was not useful for this purpose since the wave travels supersonically. What Bragg realised was that the later low-frequency boom, at about 25 Hz, travelled at the local sound speed and so the trick was to measure the arrival times of the low-frequency signal. Ordinary microphones were not sensitive at these low frequencies. The detector invented by Bragg and William Tucker consisted of an empty wooden ammunition case with a small hole and platinum wire attached to a sensitive Wheatstone bridge. When the boom arrived at the detector, the puff of air lowered the temperature of the wire, causing its resistance to change, and resulted in a strong signal in the galvanometer. By having a number of such recording stations distributed well behind the front line, it was only a matter of geometry to use triangulation techniques to locate the source of the explosion. This technique proved to be a great success and it is estimated that using these accurate positions less than 5% of the enemy gun emplacements were not identified. Bragg received the Military Cross (MC) and the Order of the British Empire (OBE), and was mentioned in dispatches three times. By the end of the war, he had risen to the rank of Major.

The human cost of the war was dreadful. Among the most distinguished scientists, the astrophysicist Karl Schwarzschild died from the illness pemphigus on the eastern front, Henry Moseley was killed in the Gallipoli campaign and Friedrich Hasenöhrl, the brilliant theorist and teacher of Erwin Schrödinger, died at the front.

7.9 The end of an era

The researches highlighted in this chapter are simply the tips of the iceberg of the vibrant research activity by the many graduate students who carried out research in the Laboratory in the period 1900 to 1919.[11] Many of these activities contributed directly to the successful studies of the conduction of electricity in gases. For example, John Zeleny showed that, when created by a flux of X-rays, the negative ions always move faster than the positive ions, and that the movement of ions in the electric field set up convection currents in the gas. John A. McClelland studied the conductivity of gases passed over heated wires or flames. John Townsend's experiments elucidated the ionisation of gases by electronic and ionic collisions. Richard S. Willows investigated the origin of striations in discharge tubes and Robert John Strutt, son of William John Strutt and later the 4th Lord Rayleigh, studied the processes of ionisation by α- and β-particles. Strutt also studied the use of radioactivity to estimate the age of the Earth and made other applications of radioactivity in geology and geophysics (see Section 8.2.1). Other notable researches include Searle and Bedford's much improved techniques for measuring hysteresis in magnetic materials, James A. Crowther's experiments on the absorption of X-rays and Alexander Wood and Norman R. Campbell's demonstration that potassium and rubidium are radioactive. All these activities contributed to the intellectual health of the Laboratory.

Two further observations should be made. The first is that, by the end of Thomson's tenure of the Cavendish Chair, the Laboratory had secured its place as a centre of excellence in experimental physics and attracted many of the most accomplished physicists as visitors to the Laboratory. The Laboratory was, however, not greatly involved in the theoretical developments of relativity and quantum theory which were preoccupying the leaders of the experimental and theoretical institutions in continental Europe, particularly in Germany, Denmark and the Netherlands. Some interest was shown by the mathematicians, but nothing like the scale of the endeavours in Göttingen, Munich, Heidelberg and Copenhagen.

The second observation is that, with the appointment of Rutherford to the Langworthy Chair of Physics at Manchester University and Moseley's brilliant experimental achievements, the centre of gravity of frontier research in experimental physics in England was shifting from Cambridge to Manchester. Rutherford's achievement in discovering the nuclear structure of the atom and Moseley's use of high-resolution X-ray spectra to establish the Bohr model of the atom and the ordering of the elements by atomic number were outstanding achievements by the very highest international standards.

In 1918, Montagu Butler, the Master of Trinity College, died and Thomson, by then the most distinguished British scientist, was the obvious choice as Butler's successor. Thomson resigned from the Cavendish Professorship in favour of Rutherford, who was the obvious choice for the Cavendish Chair. Rutherford was duly elected on 2 March 1919.

PART V

1919 to 1937

Ernest Rutherford (1871–1937),
1st Lord Rutherford of Nelson
Oil painting by Oswald Hornby Joseph Birley

Rutherford at McGill and Manchester universities: new challenges in Cambridge

8

8.1 The changing frontiers of physics research

The outbreak of the First World War brought much of the research activity in UK universities to a halt. In particular, as indicated in Section 7.8, a whole generation of young men enlisted with tragic loss of life in the trenches of northern France, Belgium and the Low Countries. Although the attentions of experimental physicists had to turn to military-related topics, the theorists who remained in the universities throughout Europe continued to produce work of the highest quality. Perhaps most spectacular of all were the contributions of Einstein. As expressed by Pais,

> Einstein's productivity was not affected by the deep troubles of the war years, which, in fact, rank among the most productive and creative of his career. During this period, he completed the general theory of relativity, found correct values for the bending of light and the displacement of the perihelion of Mercury, did pioneering work in cosmology and gravitational waves, introduced his A and B coefficients for radiative transitions, found a new derivation of Planck's radiation law – and ran into his first troubles with causality in quantum mechanics. During the war, he produced, in all, one book and about fifty papers, . . . (Pais, 1982)

At the same time, Einstein became an outspoken radical pacifist, which provoked a hostile reaction from the authorities in the midst of a devastating war.

By the end of the war, there had been major advances in the efforts to incorporate quantum concepts into physics at the atomic level, what is conveniently referred to as the *old quantum theory*. This was to continue in the years immediately following the war until the quantum revolution of 1925–27, associated with the names of Heisenberg, Born, Jordan, Dirac, Schrödinger, Pauli and many others. It is revealing to list these advances up to the summer of 1925, most of them taking place in continental Europe:[1]

- The Bohr model of the atom and the concept of *stationary states* which electrons can occupy in orbits about the atomic nucleus (Bohr, 1913a)
- The introduction of additional quantum numbers to account for multiplets in atomic spectra and the associated selection rules for permitted transitions (Sommerfeld, 1915a,b, 1916; Rubinowicz, 1918)
- Millikan's determination of the relation between wavelength and ejected electron energy in the photoelectric effect (Millikan, 1916a,b)

- The explanation of the Stark and Zeeman effects in terms of the effects of quantisation upon atoms in electric and magnetic fields (Epstein, 1916; Schwarzschild, 1916; Sommerfeld, 1916)
- Bohr's correspondence principle, which enabled classical physical concepts to inform the rules of the old quantum theory (Bohr, 1918)
- Landé's vector model of the atom, which could account for the splitting of spectral lines in terms of the addition of angular momenta according to the expression for the Landé g-factor (Landé, 1919)
- The experimental demonstration of space quantisation by the Stern–Gerlach experiment (Gerlach and Stern, 1922)
- The confirmation of the existence of light quanta from Compton's X-ray scattering experiments (Compton, 1922)
- The concept of electron shells in atoms and the understanding of the electronic structure of the periodic table (Bohr, 1922; Stoner, 1924)
- The introduction of de Broglie waves associated with the momentum of the electron, and the extension of the wave–particle duality to particles as well as light quanta (de Broglie, 1923a,b,c)
- The discovery of Bose–Einstein statistics, which differed very significantly from classical Boltzmann's statistics (Bose, 1924; Einstein, 1924, 1925)
- The discovery of Pauli's exclusion principle and the requirement of four quantum numbers to describe the properties of atoms (Pauli, 1925)
- The discovery of electron spin as a distinct quantum property of particles – although inferred from angular momentum arguments, the intrinsic spin of the electron is not 'angular momentum' in the sense of rotational motion (Uhlenbeck and Goudsmit, 1925).

But there was no sense in which this was a self-consistent, coherent quantum theory. Jammer's words encapsulate the unsatisfactory state of the theory:

> In spite of its high-sounding name and its successful solutions of numerous problems in atomic physics, quantum theory, and especially the quantum theory of polyelectron systems, prior to 1925, was, from the methodological point of view, a lamentable hodgepodge of hypotheses, principles, theorems and computational recipes rather than a logical consistent theory. Every single quantum-theoretic problem had to be solved first in terms of classical physics; its classical solution had then to pass through the mysterious sieve of the quantum conditions or, as it happened in the majority of cases, the classical solution had to be translated into the language of quanta in conformance with the correspondence principle. Usually, the process of finding the 'correct solution' was a matter of skillful guessing and intuition, rather than of deductive or systematic reasoning. (Jammer, 1989)

It is scarcely surprising that the experimental physicists in the Cavendish Laboratory distanced themselves somewhat from the mathematically complex tools which had been developed to cope with the old quantum theory. Jammer records the basic dichotomy between the classical and quantum world pictures by quoting William Bragg:

This state of affairs was well characterised by Sir William Bragg when he said that physicists are using on Mondays, Wednesdays and Fridays the classical theory and on Tuesdays, Thursdays and Saturdays the quantum theory of radiation. (Jammer, 1989)

Thomson had, in a sense, attempted to 'quantise' the aether with his identification of vortices with atoms, and he remained a proponent of such continuum aether models. His endeavour was, however, a very different proposition from the old quantum theory as formulated by Planck, Einstein and Bohr, in which Planck's constant h appeared as a new fundamental constant of nature. Rutherford adhered to his charged 'billiard-ball' model of protons, electrons and nuclei and concentrated upon what he was uniquely qualified to do – to unravel the nature of physical processes at the atomic and nuclear level by imaginative experimentation and careful observation. First, we need to review Rutherford's spectacular research activities prior to his return to Cambridge in 1919 as Cavendish Professor.

8.2 Rutherford at McGill and Manchester universities

8.2.1 McGill University, 1898 to 1907

Rutherford accepted the Macdonald Research Professorship of Physics at McGill University in Montreal in 1898. The following years were a period of remarkable experimentation during which the processes of the transmutation of the elements by chains of radioactive decays were established. Rutherford's close attention to detail in these experiments and his careful investigation of any anomalies were the hallmarks of his genius. These qualities, combined with his astonishing energy and enthusiasm for physics research, made him a natural leader and role model for the succeeding generations of experimental physicists.[2]

The known radioactive elements at the turn of the century were uranium, discovered by Becquerel in 1896, polonium and radium by Pierre and Marie Curie in 1898, thorium also in 1898 by Gerhard Schmidt, and actinium by André-Louis Debierne in 1899. Clues to the understanding of the processes of radioactivity were provided by the complex behaviour of the decay products of thorium compounds. Rutherford's preferred means of studying quantitatively the properties of radioactive substances was by sensitive electroscopic measurements of the ionisation of the gas within the sensitive volume of the electroscope caused by the ionising radiation of the radioactive products.

In 1900, Robert Owens, Rutherford's colleague at McGill University and recently appointed Professor of Electrical Engineering, was studying the radioactive decays of thorium, in preparation for a visit to the Cavendish Laboratory. He was perplexed by the remarkable sensitivity of the measured degree of ionisation to draughts and air movements in the vicinity of the radioactive material. Rutherford began a systematic experimental study of these phenomena and inferred that their cause was the formation of radioactive gas, which he called *thorium emanation*. The emanation could be isolated by blowing air over the surface of the thorium sample and collecting it in a flask. The gas, which is now known as radon, was extremely radioactive and could account for much of the radioactivity of thorium.

| Box 8.1 | Rutherford's experiments on thorium X emanation |

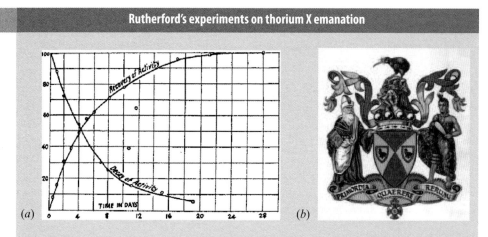

(a) Illustrating the decay of thorium X and the recovery of the original sample of thorium (Rutherford and Soddy, 1902a). This behaviour could be understood from the transfer equation for radioactive decay, $dN/dt = -N/\tau + Q$, where N is the number of atoms of thorium X, τ is the decay constant of the thorium X and Q is the rate of formation of thorium X from the original thorium sample, which, because of its half-life of a few billion years, could be taken to be constant. If there were no production of thorium X, $Q = 0$ and $N = N_0\,e^{-t/\tau}$. If $Q = $ constant and there is no thorium X to begin with, $N = N_0(1 - e^{-t/\tau})$, where $N_0 = Q\tau$. (b) Rutherford's coat of arms, including the decay and recovery of thorium X in the quartering of the heraldic shield (courtesy of John Campbell).

In addition, the thorium emanation had the property that it rendered other substances radioactive, what was referred to as *induced* or *excited activity*. The induced radioactivity also decayed in strength in a matter of minutes. Soon it was found that the other primary radioactive substances also produced emanations. For example, Friedrich Ernst Dorn and Marie Curie discovered similar phenomena associated with radium, in what became known as *radium emanation*.

The chemical nature of the radioactive products needed to be established and, fortunately for Rutherford, a brilliant young chemist from Oxford, Frederick Soddy, had been appointed in 1900 as a demonstrator in chemistry at McGill University. They began systematic studies of the chemical nature of thorium emanations and the various induced radioactive substances. In a series of physics and chemistry experiments, Rutherford and Soddy inferred that thorium emanation was the product of another extremely radioactive substance, which they called thorium X. In the key experiment, they found that when thorium X was chemically separated from the sample of thorium, it decayed radioactively with a half-life of about four days. If the thorium X was chemically removed from the original thorium sample, the radioactivity recovered with the same time constant of about four days (Box 8.1(a)).

These results contained the inspiration for the development of the theory of radioactive decay chains, the way in which Rutherford accounted for these experiments being illustrated in the caption of Box 8.1. Rutherford maintained a lifelong affection for the relations illustrated here, even including the characteristic decay and recovery curves in his coat of arms on being created a life peer as Baron Rutherford of Nelson in 1931 (Box 8.1(b)).[3]

In a remarkable series of combined physics and chemistry experiments, Rutherford and Soddy disentangled the chains of radioactive decays, establishing the transformation, or disintegration, theory of radioactive decay (Rutherford and Soddy, 1902a). They showed, for example, that thorium emanation was an inert gas, radon, which had a half-life of only about 1 minute before decaying into a new radioactive element which they named thorium A – this was the material responsible for the phenomenon of induced radioactivity. In turn, thorium A decayed in about 11 hours into thorium B, which then decayed to thorium C after 55 hours.[4] Note that there are numerous gaps in the chains described by Rutherford in his 1904 Bakerian Lecture to the Royal Society of London (Table 8.1).[5]

In 1902, Rutherford and Soddy published their great papers in which they described these results and propounded the theory of the transformation of the elements by radioactive decay (Rutherford and Soddy, 1902a,b). These were truly revolutionary papers for chemistry in that they overturned the basic concept that atoms are immutable. The process of radioactive decay, Rutherford and Soddy now proposed, was the means by which the elements can be transmuted from one species to another. Despite the initial concerns of the chemists for whom this smacked of alchemy, the evidence presented by Rutherford and Soddy was overwhelmingly compelling.

In the following year two important additions were made to the story. The first was the experimental fact

> that the rate of change of the system at any time is always proportional to the amount remaining unchanged. (Rutherford and Soddy, 1903)

This immediately led to the exponential law of radioactive decay and hence to a characteristic time-scale T, or decay constant $\lambda = T^{-1}$, for each radioactive decay, which uniquely identified the radioactive element. For example, Rutherford had no hesitation in identifying thorium A with polonium on the basis of their identical decay constants. With this result, Rutherford was able to set up the set of transfer equations to describe the radioactive decay chains.

The second addition arose from the need to understand the nature of the particles emitted in radioactive decay chains. Thomson had shown that the β-particles are identical to the recently discovered electron, but the nature of the α-particle was a much more difficult proposition. While the electron was readily deflected by strong electric and magnetic fields, the α-particle, having mass about 8000 times greater than the electron and so roughly of the same order as the hydrogen atom, had not been deflected by the strongest electric or magnetic fields available until Rutherford's experiments of 1903.

The principles Rutherford used were similar to those employed in Thomson's original discharge tube experiments, the electric field deflection determining the quantity Mu^2/E and the magnetic deflection Mu/E, where M and E are the mass and charge of the α-article respectively and u its velocity. He devised an arrangement for producing a collimated flux of α-particles and then estimated the radii of curvature of their paths from the change

Table 8.1 Radioactive decay chains (from Rutherford's 1904 Bakerian Lecture)[a]

Product	Half-life T	Decay constant λ (s^{-1})	Some physical and chemical properties
URANIUM ↓	10^9 years	2.2×10^{-17}	Soluble in excess of ammonium carbonate
Uranium X ↓	22 days	3.6×10^{-7}	Insoluble in excess of ammonium carbonate
Final product	–	–	
THORIUM ↓	3×10^9 years	7×10^{-18}	Insoluble in ammonia
Thorium X ↓	4 days	2×10^{-6}	Soluble in ammonia
Thorium emanation ↓	1 minute	1.15×10^{-2}	Chemically inert gas; condenses about $-120°$C.
Thorium A ↓	11 hours	1.75×10^{-5}	Behaves as solid; insoluble in ammonia; volatilised at a white heat; soluble in strong
thorium B ↓	55 minutes	2.1×10^{-4}	acid; thorium A can be separated from B by electrolysis
Thorium C (final product)	–	–	
ACTINIUM ↓	–	–	
Actinium X ? ↓	–	–	
Actinium emanation ↓	3.7 seconds	1.87×10^{-1}	Behaves as a gas
Actinium A ↓	41 minutes	2.80×10^{-4}	Behaves as solid; soluble in strong acids; A can be partially separated from B by
Actinium B ↓	1.5 minutes	7.7×10^{-3}	electrolysis
Actinium C (final product)	–	–	
RADIUM ↓	800 years	2.8×10^{-11}	
Radium emanation ↓	4 days	2.00×10^{-6}	Chemically inert gas; condenses about $-150°$C; definite spectrum; volume diminishes with time
Radium A ↓	3 minutes	3.80×10^{-3}	Behaves as solid; soluble in strong acids;
Radium B ↓	21 minutes	5.38×10^{-4}	volatilised at white heat; B is more volatile than A or C and can thus be temporarily
Radium C ↓	28 minutes	4.13×10^{-4}	separated from them
Radium D ↓	About 40 years	–	Gives out only β-rays; soluble in strong acids
Radium E ↓	About 1 year	–	Probably active constituent of radio-tellurium; soluble in strong acids; volatilised at a red heat; deposited on bismuth in solution

[a] Rutherford, 1905a.

Rutherford's determination of the value of E/M for α-particles   Box 8.2

The α-particles, emitted from a layer of radium, passed upwards through 25 narrow slits (G) and then through a thin aluminium foil into the hydrogen gas within the sensitive volume of the electroscope. To estimate the radii of curvature of their paths in a magnetic field, a magnetic field was applied perpendicular to the plane of the diagram and so bent the trajectories of the α-particles, which would hit the walls of the slits when the field was strong enough. At that point, the ionisation in the electroscope due to the α-particles would drop to zero. The deflection in the strongest electric field Rutherford could produce was found in the same way, but was even smaller than the magnetic deflection.

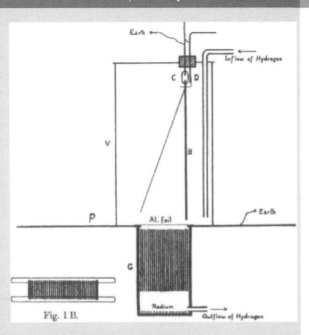

Fig. 1 B.

The flow of hydrogen through the apparatus swept out the radium emanation, which would have caused much greater ionisation within the electroscope (Rutherford, 1903).

in the ionisation in the electroscope (Box. 8.2). He used the high-precision electroscopic techniques developed by C.T.R. Wilson to measure the ionisation of air (see Section 6.5.4). Using the strongest electric and magnetic fields available to him, he determined the values $E/M = 6000$ and $u = 2.5 \times 10^7$ m^{-1} (Rutherford, 1903). It was immediately apparent that radioactive decay was an enormously powerful source of energy, the α-particles travelling at about one tenth the speed of light. Crucially, this was the first demonstration that the α-particles were positively charged particles and that the value of E/M was of the same order as that of hydrogen atoms. Rutherford was duly cautious about these results. As he stated later,

> These results showed that the α particles were of atomic mass comparable with that of the hydrogen atom. The value of e/m for the hydrogen atom is 9643, indicating that, if the α particle carried the same charge, its mass should be about twice that of the hydrogen atom. (Rutherford *et al.*, 1930a)

The issue of the mass and charge of the α-particle was to remain a preoccupation of Rutherford's throughout his time at McGill and Manchester universities.[6] He was personally convinced that the α-particles were helium nuclei, but did not succeed in finding definitive evidence for this until his experiments with Royds when he was at Manchester (Section 8.2.2 and Box 7.4).

The fruits of this remarkable period of sustained collaboration between Rutherford, the physicist, and Soddy, the chemist, were summarised in Rutherford's Bakerian Lecture of 1904 to the Royal Society (Rutherford, 1905a). In this great paper, Rutherford described the results of his researches in some detail and set out in a table the primitive form of the radioactive decay chains (Table 8.1). Rutherford identified four series, the uranium, thorium, actinium and radium series. Later, Rutherford's friend and colleague Bertram Boltwood showed that Rutherford's radium series was part of the uranium series. It is not so surprising that Rutherford was awarded the 1908 Nobel Prize in *Chemistry*. In a sense, Rutherford and Soddy were lucky in that the half-lives of the radio isotopes lay in a very convenient range for laboratory study, as can be seen from the data in Table 8.1. There are several gaps in the chains, some associated with very short-lived isotopes and others with β-decays, but the overall picture was correct and an enormous step forward in understanding of nature of radioactivity.

As early as 1904, Rutherford appreciated the significance of radioactive decay for estimating the age of the Earth and for understanding its internal heating. A battle had been raging for many years between William Thomson, Lord Kelvin, and geologists about the age of the Earth. From the observed temperature gradient close to the Earth's surface, which amounted to an increase of $100°$F by a depth of one mile, it was a straightforward calculation in heat diffusion to estimate the time it would take the Earth to cool from a molten state if there were no internal heat sources. Kelvin found the age of the Earth to be at most 40 million years. This was also of the same order as the Kelvin–Helmholtz time-scale for the Sun, the time it would take the Sun to radiate away its thermal energy if there were no internal heat source. The geologists, on the other hand, found much greater ages from stratigraphic and other analyses, preferring ages of 100 million years or greater.

Rutherford showed that the radioactive decay of radium, with an abundance of only five parts in 10^{14} per unit mass of the material of the Earth, would be adequate to account for the continuous heating of the interior of the Earth (Rutherford, 1905b). In addition, Boltwood had shown that the amount of radium in rocks was proportional to the amount of uranium present and so Rutherford inferred that uranium, with its long half-life of 10^9 years (Table 8.1), was the parent species of radium. Since uranium itself decays by natural radioactivity, the fact that there is any left on Earth at all told him something about the age of the Earth. In his paper of 1905, he suggested an age of 5×10^9 years. In his last publication before he left Canada in 1907, he refined these calculations on the basis of the work of Robert John Strutt, son of Lord Rayleigh, the second Cavendish Professor, who found that the mineral thorianite contained 12% uranium and 7% thorium as well as large quantities of entrapped helium. From the abundance of helium atoms, which Rutherford assumed were the α-particles liberated in the decays of thorium and uranium atoms, he estimated the age of the Earth to be about 700 million years (Rutherford, 1907).

Rutherford announced his first results in 1904 in his Friday Evening Discourse at the Royal Institution, presented the day after his Bakerian Lecture to the Royal Society. In his reminiscences, he states:

I came into the room which was half dark and presently spotted Lord Kelvin in the audience and realised that I was in for trouble with the last part of my speech dealing with the age of the Earth where my views conflicted with his. To my relief, Kelvin fell asleep, but as I came to the important point, I saw the old bird sit up, open an eye and cock a baleful glance at me! Then a sudden inspiration came and I said Lord Kelvin has limited the age of the Earth, provided no new source was discovered. That prophetic utterance refers to what we are now considering tonight, radium! Behold! the old boy beamed upon me. (Eve, 1939)

Rutherford's analyses marked the beginning of the disciple of radioactive dating.[7]

8.2.2 Manchester University, 1907 to 1919

Despite his remarkable successes at McGill, Rutherford felt remote from the active centres of physics research, which were unquestionably in Europe. The opportunity to return to England resulted from the initiative of Arthur Schuster, then the Langworthy Professor of Physics at the Victoria University of Manchester. After his period at the Cavendish Laboratory from 1876 to 1881, Schuster had taken up the Beyer Chair of Applied Mathematics at Owens College at Manchester and then in 1888 succeeded Balfour Stewart as Langworthy Professor. Schuster was instrumental in the construction of new physics laboratories, which were opened in 1900 by Lord Rayleigh. Schuster resigned from the Langworthy Chair in 1907 and played a major role in ensuring that Rutherford was appointed as his successor. Rutherford's tenure of the Langworthy Chair was a period of outstanding achievement. He inherited from Schuster a well-equipped Laboratory to which he attracted many graduate students – soon the Laboratory was overcrowded. The list of distinguished physicists who worked in the Department of Physics under Rutherford included Moseley, Chadwick, Geiger, Bohr, Andrade, Darwin, Marsden, von Hevesy and many others. The leadership of innovation in physics was shifting from Cambridge to Manchester.

Rutherford was fortunate in having inheriting from Schuster his research assistant, Hans Geiger, who had completed his doctorate on the electrical discharge of gases from the University of Erlangen in 1906. The traditional way of counting the number of charged particles liberated in radioactive decays was by observing the flashes on a fluorescence screen, which was a tiring and demanding procedure. Rutherford sought a better way of counting the charged particles. He and Geiger realised that if a discharge tube were maintained at a voltage just below breakdown, the thousands of ions and electrons liberated when an energetic charged particle passed through the sensitive volume would cause an electrical breakdown of the gas, resulting in a strong electric pulse in the circuit maintaining the central wire at a high potential.[8] The classic form of what became known as a *Geiger counter* is shown in Figure 8.1(a), in which the central wire is maintained at a voltage of about 1000 volts (Rutherford and Geiger, 1908b). The electric pulse was detected by a string electrometer and, to record each event, a spot of light was reflected from the electrometer suspension onto a moving strip of photographic paper (Figure 8.1(b)). This device was to be perfected by Geiger over the succeeding years, the ultimate version being the Geiger–Müller detector of 1928 (Geiger and Müller, 1928, 1929). This early form of the Geiger counter was used by Rutherford and Geiger to calibrate the number of α-particles liberated

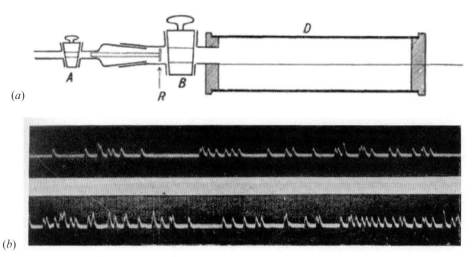

(a)

(b)

Fig. 8.1   (a) Rutherford and Geiger's original α-particle detector of 1908 (Rutherford and Geiger, 1908b). (b) An example of the photographic recording of the times of arrival of α-particles. The upper record corresponds to particles being detected at a rate of 600 per minute and the lower to a rate of 900 per minute (Rutherford *et al.*, 1930a).

Box 8.3   **Rutherford and Geiger's experiment to measure the total charge of a flux of α-particles**

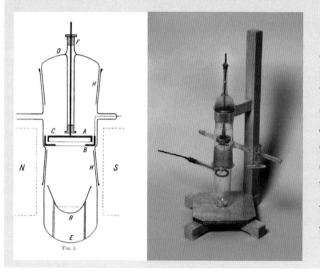

Rutherford and Geiger measured the total charge of a flux of α-particles using the apparatus shown (Rutherford and Geiger, 1908a). The radium sample was contained in the vessel R and the α-particles collected on the plate CA. Precautions were taken to ensure that only the charges of the α-particles were collected.

This apparatus is now on display in the Cavendish Collection.

by radium, which they found to be 3.4×10^{10} per second from one gramme of radium (Rutherford and Geiger, 1908b).

Rutherford and Geiger next measured the total charge liberated by a sample of radium using the apparatus shown in Box 8.3. The radium sample was placed in the shallow vessel

R and the emitted α-particles were collected on the plate CA which was attached to a sensitive electrometer. To ensure that only α-particles reached the collector CA, the apparatus was placed in a strong magnetic field which deflected electrons and secondary products created when the α-particles passed through the aluminium and mica sheets. From the rate of arrival of charge at the collector, they inferred that the α-particles had charge twice that of the electron (Rutherford and Geiger, 1908a).

The final identity of α-particles with the nuclei of helium atoms was convincingly demonstrated by the Rutherford and Royds experiment of 1909, described in Section 7.5.2 and Box 7.4 (Rutherford and Royds, 1909). This apparatus required extraordinarily skilful glass-blowing, and Rutherford was fortunate that from Schuster he had also inherited Otto Baumbach, who constructed the apparatus, including the long, thin needle in which the radioactive radium sample was contained – its walls were only $1/100$ mm thick. The helium in the outer discharge tube built up over a few days.

Undoubtedly, Rutherford's greatest scientific achievement was the discovery of the nuclear structure of the atom, described in Section 7.5.2. Geiger and Marsden (1909) published their discovery of the rare, but significant, large deflections of α-particles in 1909, but it was not until 1911 that Rutherford hit upon the idea that all the positive charge had to be contained in a single compact nucleus to account for the enormous force needed to return the particle back along its incoming trajectory. As Geiger wrote:

> One day (in 1911) Rutherford, obviously in the best of spirits, came into my room and told me that he knew now what the atom looked like and how to explain the large deflections of the α-particle. (Eve and Chadwick, 1938)

Rutherford was notoriously sceptical of theory, but on this occasion he worked out the formulae for what became known as *Rutherford scattering* himself. In particular, he predicted the angular distribution of the α-particles $p(\phi)\,d\phi \propto (1/v_0^4)\,\mathrm{cosec}^4\,(\phi/2)\,d\phi$ (Rutherford, 1911).

With his titanic energy and imagination for profound, but simple, experiments, Rutherford continued to advance the study of α-particles and their properties. While his friend Bertram Boltwood was on sabbatical leave at Manchester, they measured accurately the rate of formation of helium from radium and then, knowing the rate of liberation of α-particles, they were able to make a direct estimate of the value of Avogadro's constant to an accuracy of about 1% (Boltwood and Rutherford, 1911).

Another example is his improved apparatus for the deflection of α-particles by an electric field. This had been a challenging experiment in 1903 but, in 1914 with Harold Robinson, he used stronger electric fields and increased considerably the length of the region of strong electric field (Figure 8.2). The result was the determination of a value of $E/M = 4820$ in cgs units with a probable error of about 1 part in 400. This figure was half the value of E/M for the hydrogen nucleus.

It had been established through his careful experiments that, in α-particle decays, the energies of the liberated particles were of the same energy for a given radioactive decay. It became evident, however, that the energies of the α-particles in the different radioactive decays were not the same, typical energies lying between about 3 and 10 MeV. A key result, which was to play an important part in the development of quantum mechanics, was the

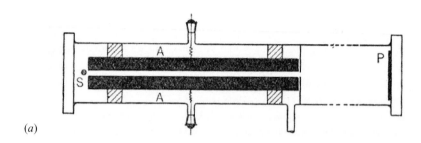

(a)

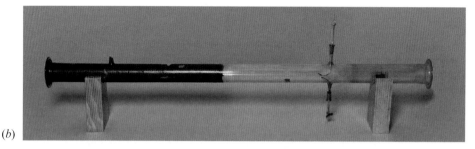

(b)

Fig. 8.2 (a) Rutherford and Robinson's apparatus to measure the deflection of α-particles in an electric field (Rutherford and Robinson, 1914). A potential difference of 2000 V was maintained between the long plates AA, and the deflected trajectories were measured on the photographic plate P. (b) The apparatus in the Cavendish Collection.

relation between the energies E of the α-particles and their radioactive decay constants λ, the *Geiger–Nuttall law* (Geiger and Nuttall, 1911). The law can be written in the form

$$\log R = A \log \lambda + B, \tag{8.1}$$

where R is the range of the particle in centimetres, which is a measure of the energy of the emitted α-particle, A is a constant and B takes different values for the various radioactive series (Figure 8.3). An alternative form of the relation is

$$\ln \lambda = -a_1 \frac{Z}{\sqrt{E}} + a_2, \tag{8.2}$$

where Z is the atomic number and the constants a_1 and a_2 take different values for the various radioactive series. The striking feature of the relation is the rather narrow range of particle energies compared with the enormous range of decay constants (see also Table 8.1).

Rutherford's energy and enthusiasm certainly inspired Henry Moseley in his studies of the X-ray lines of the chemical elements (Section 7.7). These brilliant experiments were of central importance in establishing the Bohr model of the atom and in understanding the structure of atoms.

The war years had a similar effect upon Manchester University as it did in Cambridge. Research essentially ground to a halt and the students joined the military forces. In 1915, Rutherford became a member of the panel of the Admiralty Board of Invention and Research, taking a particular interest in the acoustic detection of submarines. The large ground

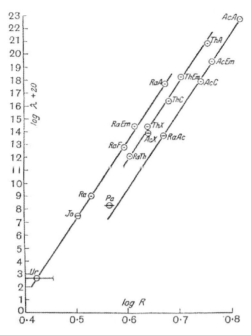

The Geiger–Nuttall law of radioactive α-particle decay (Geiger and Nuttall, 1911). The units on the abscissa are the logarithm of the range R of the particle in centimetres, which is a measure of the energy of the emitted α-particle. The ordinate is the logarithm of the decay constant λ.

Fig. 8.3

floor of his Manchester laboratory was converted into an acoustics laboratory, with a large tank for studying underwater acoustics.

Despite the much depleted staff, Rutherford was able to continue his research in radioactivity, and towards the end of his time in Manchester made another great discovery: the artificial destruction of nitrogen nuclei by fast α-particles. His experiments on the bombardment of different elements by energetic α-particles were the subject of four major papers in 1919, the year in which he accepted the offer of the Cavendish Professorship of Experimental Physics. We will take up that story once we have Rutherford relocated and ensconced in Cambridge.

8.3 The aftermath of war

Rutherford was 47 years old when he returned to Cambridge with a quite outstanding record of achievement in experimental physics. He brought with him a great deal of the apparatus he had used in his earlier experiments, as well as the continuing loan of 250 milligrams of radium from the Vienna Institute of Radium. The war had interrupted his work and that of his students while they were engaged in war work. In 1919, Rutherford wrote to Geiger, who had survived the war in Berlin:

> We are all feeling very rusty scientifically after the War, and it will be some years before we can get going properly, for apparatus is very dear and difficult to get . . . (Crowther, 1974)

The challenges were very considerable. Besides the problems of regenerating the research programme, there was a flood of students wishing to study physics, the result of a backlog of four years of student intake. The result was that, when Rutherford arrived, teaching was required for about 600 undergraduates and 50 naval officers, about twice as many as could be accommodated in the Laboratory. To cope with the increased demand, Rutherford wrote a memorandum to the University explaining the urgent need for investment in further buildings and infrastructure. Rutherford was in no doubt about the social, industrial and economic importance of physics, emphasising that 'the Laboratory was soon recognised as the chief centre of research activity in physics'. He reiterated the importance of training in physics for the benefit of society and industry:

> The rising prestige of this country and the eclipse of Germany will lead to an increase in the number of research students from neutral and allied countries who wish to work in the Laboratory. The international importance of attracting highly-qualified research workers not only from our own Empire but from friendly States, needs no emphasis . . .
>
> The need of a laboratory specially devoted to training in research in Applied Physics is of pressing importance if we are to play our part in the researches required by the State, and in providing well-qualified research men for various branches of industry and for the scientific departments of State.

He summarised his requirements as follows:

- increased laboratory and lecture space for the teaching of physics
- provision of new, well-equipped laboratories for applied physics, optics and properties of matter
- provision of three additional lecturers of high standing, competent to direct advanced study in research in the new departments mentioned above
- endowment of another Chair of Physics in the University.

The cost of the new building would be £75,000 and the endowment £125,000. In 1919, the University was in no position to provide such funds, nor was it convinced of the necessity of investing in large-scale physics projects. This was a setback for Rutherford and he was unwilling to make the effort to raise the funds independently. As he stated, 'if the British want research, they can pay for it'. Rutherford was forced to continue in his traditional mode of devising simple, ingenious experiments, relying on insight and experimental skill.

Despite this setback, there was no lack of remarkable talent wishing to join the Laboratory (Figure 8.4). The most prominent among the new intake were the following:

- *James Chadwick* had been a student of Rutherford's in Manchester. In 1913, he won a studentship of the Royal Commission of 1851 to work under Geiger in Berlin. With the outbreak of the war, he was interned at Ruhleben for the duration of the war. Fortunately, this did not disrupt his research activities since he was allowed to set up a small laboratory in the camp and received support and equipment from prominent German

Rutherford and Thomson with the graduate students in 1920. Back row, Left to right: A.L. McAulay, C.J. Power, G. Shearer, Miss Slater, Miss Craies, P.J. Nolan, F.P. Slater, G.H. Henderson, C.D. Ellis. Middle row: J. Chadwick, G.P. Thomson, G. Stead, Prof. Sir J.J. Thomson, Prof. Sir E. Rutherford, J.A. Crowther, A.H. Compton, E.V. Appleton. Front row: A. Muller, Y. Ishida, A.R. McLeod, P. Burbidge, T. Shimizu, B.F.I. Schonland.

Fig. 8.4

scientists, including Rubens, Nernst and Warburg. After the war, Rutherford brought him to Cambridge and in 1920 he was awarded a Clerk Maxwell studentship. At the end of his fellowship, he was appointed to the post of Assistant Director of Research, funded by the Department of Scientific and Industrial Research to support Rutherford's work. In addition to carrying out a major programme of research, he took responsibility for much of the administration of the Laboratory.

- In 1913 *Charles Ellis* was a cadet in the Royal Military Academy in preparation for a military career. In 1914, he was sent to Germany for military studies and was interned with Chadwick on the outbreak of the war. Although he had no physics training, he learned a lot from Chadwick in his laboratory at Ruhleben and, at the end of the war, returned to Cambridge determined to become a physicist. He became Rutherford's research assistant. He won a research fellowship at Trinity College in 1921 and became an assistant lecturer in the same year, followed by his promotion to University lecturer in 1926.

- *Patrick Blackett* was a naval officer who had commanded a destroyer during the war and embarked upon a six-month course at Cambridge designed for returning naval officers. Within three weeks, he resigned his commission in order to become a physics undergraduate. On completing his degree in 1921, Blackett became Rutherford's research student.

He won the Moseley Fellowship of the Royal Society and was appointed a University lecturer in 1930.

- In 1919, *Arthur Compton* was awarded one of the first two US National Research Council fellowships which enabled students to study abroad. He chose to work in the Cavendish Laboratory, working principally with George Thomson on the scattering and absorption of γ-rays. After a year he returned to the USA to the Wayman Crow Professorship of Physics at Washington University, St Louis. There he made his key measurements of the change of wavelength by scattering of energetic X-rays, which was to become known as *Compton scattering* and which established beyond doubt the particle nature of electromagnetic radiation (Compton, 1922).

- *Pyotr Kapitsa* joined the Laboratory in 1921. He had been part of a delegation from the Soviet Union seeking to renew scientific relations after the Communist revolution and the Russian civil war. Under Joffe's leadership the delegation was also responsible for purchasing scientific equipment. When the party reached Cambridge, Kapitsa asked if he could join the Laboratory and, despite the severe overcrowding, he was admitted by Rutherford.

- *George Thomson*, the son of J.J. Thomson, had studied physics in the Laboratory as an undergraduate, but then was commissioned during the war. After service in the Queen's Regiment, he was assigned to the Royal Aircraft Establishment, where he studied the stability of aircraft. Returning to Cambridge after the war, he worked on positive rays before taking up the Professorship of Physics at Aberdeen in 1922, where he carried out his pioneering experiments on electron diffraction (Thomson, 1928).

- *Edward Appleton* completed his undergraduate degree in physics at Cambridge in 1914. Upon joining the Army, he was assigned to signal duties. This introduced him to the subjects of thermionic valves and radio propagation, both topics which he espoused on his return to Cambridge in 1919 as a graduate student. He was appointed an Assistant Demonstrator in 1920.

These are only a few of the graduate students who were attracted by the reputation of the Laboratory and, in particular, by the opportunity to work under Rutherford. One striking feature of the graduate population was the network of former Cavendish students and former colleagues of Rutherford, such as Joseph Gray from Kingston, Ontario in Canada, who sent their best students from various parts of the British Empire to work with Rutherford. Others included Thomas Laby who went to Melbourne, Australia, Arthur Eve to McGill University in Canada and John McLennan to Toronto. Many of these excellent research students were supported by Research Scholarships and Senior Research Studentships provided by the Royal Commission for the Exhibition of 1851.[9]

A further source of support for the expanding research programme resulted from the formation in 1915 of the UK government's Department of Scientific and Industrial Research, the DSIR. The stimulus for this initiative was the recognition at the outbreak of the First World War that 'Britain found . . . it was dangerously dependent on enemy industries'. The role of the DSIR was

> to finance worthy research proposals, to award research fellowships and studentships [in universities], and to encourage the development of research associations in private industry and research facilities in university science departments.

In 1915, the DSIR's budget was £1 million, and Rutherford and his colleagues took full advantage of these opportunities.

Another significant development was the formal introduction of the degree of Doctor of Philosophy, PhD, created by Royal Patent in May 1920. Among the first research students who took advantage of the new degree were Chadwick (1921), Kapitsa and Ellis (1923) and Stoner (1924).[10] This was to become the formal qualification for employment in the universities, although for a number of distinguished students the winning of a Cambridge college fellowship was adequate recognition of excellence. At the same time, Rutherford and other senior colleagues in the University succeeded in introducing the degree of Master of Science, MSc, as a one-year research degree for those who did not wish to continue to the full three-year PhD.

During Rutherford's tenure of the Cavendish Chair, 1919–37, the nature of physics research was to change dramatically.

- There were now numerous centres of excellence in experimental physics in Europe and the USA, and fierce competition and the potential for controversy increased considerably.
- From 1925 onwards, the discovery of quantum mechanics provided a new lease of life for theoretical physics, which could not be ignored by the experimentalists – quantum mechanics had to be central to the understanding of the nature of atoms, molecules and nuclei.
- The electronic revolution was well underway with the development of electronics as well as radio communications.
- The move towards 'Big Science' was gathering momentum, and this would require investment on scales previously undreamt of. Despite Rutherford's brilliance at simple, profound experiments, the way in which research was carried out had to evolve to keep up with and exploit these new opportunities.

The changing experimental, theoretical and social atmosphere in which research was carried out in the Laboratory is brilliantly caught in Jeffrey Hughes' PhD dissertation (Hughes, 1993). To put some coherence into the story, it will be convenient to describe, in Chapter 9, the developments in radioactivity which were to give birth to the new discipline of nuclear physics. In Chapter 10, other aspects of the physics research programme are described – these developments would lead to an enormous expansion of the scope of physics research, which was to have profound implications for post-Second World War science and for management of the Laboratory.

8.4 The undergraduate teaching programme

An indicator of how the undergraduate courses in physics developed is provided by the undergraduate syllabuses for the period 1927–29. Fortunately, not only were the announcements of the courses published each year in the *University Reporter*, but lecture notes of a number of those who took the physics courses have been preserved.[11] Table 8.2 shows the courses on offer for the academic year 1927–28, recalling that the Heisenberg breakthrough had only taken place in mid 1925 and that Schrödinger's six great papers were published in

Table 8.2 Lectures proposed by the Board of the Faculty of Physics and Chemistry, 1927–1928			
Lecturers	Michaelmas	Lent	Easter
ELEMENTARY COURSES			
Mr Wood & Mr Ratcliffe	Mechanics, Hydrostatics, Magnetism	Electricity & Light	Light (cont), Sound, Heat
Mr Stead (practical)	Mechanics, Hydrostatics, Magnetism	Electricity & Light	Light (cont), Sound, Heat
NAT. SCI. TRIPOS PtI			
Lectures			
Mr Thirkill	Heat & Elementary Thermodynamics		
Mr A. Wood	Mechanics & Properties of Matter	Sound & Light	Light (cont)
Dr Ellis		Electricity & Magnetism	Electricity & Magnetism
Practical work			
Mr Bedford	Mechanics, Properties of Matter & Heat	Light & Magnetism	Electricity & Magnetism
Dr Ellis	Electricity & Magnetism	Electricity & Magnetism	Electricity & Magnetism
Dr Searle	Mechanics, Heat	Light	Mechanics Heat & Light
Advanced courses			
Prof. Sir E. Rutherford	Constitution of Matter	Electrical Oscillations & Radio Telegraphy (with Mr Ratcliffe)	Ionisation & Radioactivity
Prof. C.T.R. Wilson	Condensation on Nuclei	Atmospheric Electricity	(cont)
NAT. SCI. TRIPOS PtII			
Dr Searle	Heat	Electrical & Magnetic Measurements	Electrical & Magnetic Measurements
Mr Blackett	Properties of Matter		
Mr Thirkill		Light	
Mr A. Wood		Wave Motion & Sound	
Mr Stead	Thermionics & X-rays		
Mr Thirkill	Demonstrations	Demonstrations	Demonstrations
Dr Hartree			Light
Special courses			
Prof. Sir J.J. Thomson		Electrical Discharge through Gases	
Prof. Sir J. Larmor	Dynamical & Physical Principles	Electric Waves, Free & Controlled	
Prof. Sir E. Rutherford	Problems of the α-particle		
Prof. Newall	Solar Research		
Prof. Eddington			Stellar Astronomy
Dr Lamb	Tides & Waves		
Dr Bromwich	Electric Waves	Electric Waves (cont)	

	Table 8.2 (*cont.*)		
Lecturers	Michaelmas	Lent	Easter
Mr Cunningham	Electron Theory	Dynamics of Crystal Lattices	
Mr Stratton			Stellar Physics
Mr Fowler	Dynamical Theory of Gases	Statistical Mechanics with Applications to Electrolytes	Statistical Mechanics
Mr Birtwistle	Quantum Theory of Spectra		
Dr Aston	Isotopes		
Dr Kapitsa		Recent Researches in Magnetism	
Dr Carroll	Line & Band Spectra	Stellar Physics	
Mr Southwell	Elasticity. Introductory Course	Elasticity (special problems)	Elastic stability
Dr Hartree		Physics of Quantum Theory	(cont)
Dr Dirac	Modern Quantum Mechanics	(cont)	
Mr Pars	General Dynamics		

1926. Heisenberg's great paper on the uncertainty principle (Heisenberg, 1927) was only submitted to the *Zeitschrift für Physik* at the end of March 1927.

A number of features of the courses are of special interest. What is immediately apparent is that the students were exposed to lectures by many of the leaders of physics research in the Laboratory and in the Mathematics Faculty. The courses were delivered by prominent physicists and theorists, including J.J. Thomson in retirement, rather than by pedagogues, a tradition which has been maintained over the years. It is natural that there is a strong emphasis upon classical physics in Part I of the course, with substantial amounts of practical laboratory work, much of it overseen by the redoubtable George Searle. In Part II, the core courses continue the development of classical physics, but the special courses contain a wide range of topical subjects from current research, many involving quantum physics. In particular, there are courses specifically on quantum topics by Birtwistle, Carroll, Hartree and Dirac, all of them members of the Mathematics Faculty. The course by Birtwistle is of interest since it summarises frontier topics in quantum theory and the different approaches taken by the pioneers. It is also significant that Birtwistle's books *The Quantum Theory of the Atom* and *The New Quantum Mechanics* were essentially the 'books of the courses' which he delivered during these years. Birtwistle surveyed many of the key developments of the newly formulated quantum mechanics, without developing new insights beyond what was in the literature. In contrast, Dirac's lectures were the precursors of his monumental *The Principles of Quantum Mechanics*, published in 1930. Dirac's book was the first complete exposition of the new quantum mechanics. It was to go through four editions, the 1947 third edition introducing Dirac's economic bra and ket notation, which he had invented in 1939 (Dirac, 1939).

Table 8.3 Natural Sciences Tripos. Part II. Physics (4)

Three questions should be attempted.

19. Compare the classical and quantum theory treatments of the normal Zeeman effect. Discuss the rôle played by the correspondence principle in connection with the polarisation of the components and with the selection rules.

20. Describe the methods of obtaining very low temperatures, and discuss the main facts of superconductivity.

21. Describe the main theoretical and experimental investigations of the Brownian movement. Explain the application of the theory to the behaviour of sensitive galvanometers.

22. Discuss the theory of the thermionic effect and its relation to the photo-electric effect.

23. Discuss the theory and significance of both the Michelson–Morley experiment and Fizeau's experiment to determine the velocity of light in a moving fluid.

or

Show that the Lorentz transformations satisfy the postulate of the constancy of the measured velocity of light. Derive from the transformations the variation with velocity of the mass of the electron.

24. What information can be obtained as to the structure of radio-active nuclei from the study of beta-ray spectra?

Further evidence about what was expected of the students is provided by the examination questions which were set for the final Part II examinations in physics for the Natural Sciences Tripos. There were four three-hour papers, each containing six questions, of which the students had to answer three in each paper. As an example of the style and content of the examinations, the questions set in Paper 4 of the May 1929 examination are shown in Table 8.3. The students were expected to write three one-hour essays on three of the topics. It is apparent that the questions addressed topical physics issues. There is not a strong emphasis upon the mathematical elaboration of the topics, although there was scope for doing so if the student was inclined to do so. It is intriguing that in, for example, Question 24, the question was set before Pauli's proposal of the existence of the neutrino. Notice also the strong emphasis upon experiment. It is apparent that any student who did well in these examinations would have been brought to the frontiers of current research by the end of the undergraduate course.

8.5 Accommodation, finance and management

A striking aspect of the story of experimental physics in Cambridge between the wars was the lack of accommodation, funding and administrative support for what was generally regarded as the premier experimental physics laboratory. The difficulties are repeatedly emphasised in the reminiscences of those who carried out research in the Laboratory. The circumstances were not favourable for investment in physics research because of the lack of funds following the First World War and the financial crises of the 1920s and 1930s. These problems were compounded by Rutherford's unwillingness to go out and seek the funds necessary for the support of the Laboratory. Rutherford's genius in devising brilliant simple experiments could only go so far, and physicists such as Chadwick, Cockcroft, Blackett and Oliphant fully appreciated that this approach was not sustainable – they all left the Laboratory in the 1930s for organisations more willing to invest in the future. Nonetheless, individuals such as Kapitsa and Cockcroft could persuade Rutherford of the need for additional funds to support experiment, and he eventually developed strong links with the Metropolitan-Vickers Company, but it ran against his natural instincts.

The pressure for space was an equal problem. The Laboratory was full to bursting in the 1920s and the University was unable to provide the £200,000 Rutherford requested in his 1919 proposal for much expanded accommodation. Various parts of other buildings were requisitioned as their previous occupants moved out – Kapitsa built his dynamo in a shed which had been vacated by the Department of Physical Chemistry and Cockcroft and Walton put together their 1930 accelerator in a corner of a chemistry store room. Buildings which had housed the Engineering Department were quickly commandeered by Cockcroft and Walton for the definitive design of their famous accelerator of 1932. Relief first appeared through the efforts of Kapitsa in persuading Rutherford to accept funds for the construction of the Mond Laboratory in 1930 (Section 10.3).

Then, on 29 April 1936, Sir Herbert Austin, founder of the Austin Motor Company, who was elevated to the peerage in the same year, wrote to the Chancellor of the University, Stanley Baldwin, as follows:[12]

> I have for several years been watching the very valuable work done by Lord Rutherford & his colleagues at Cambridge in the realm of scientific research & knowing that, as Chancellor, you are keenly interested in obtaining sufficient funds with which to build, equip & endow a very much needed addition to the present resources, I shall be very pleased indeed to present securities to the value of approximately £250,000 for this purpose.
> May I request, if it is in order that my name will be associated with the extension.
> I am,
> Yours sincerely,
> H. Austin

This magnificent gift was accepted with gratitude by the chancellor on behalf of the University. Rutherford died unexpectedly on 19 October 1937 and did not live to see the realisation of his 1919 proposal – responsibility for the construction of the Austin Wing was taken over by Cockcroft. The principal architect was Charles Holden and work by the local Cambridge builders Rattee and Kett started in May 1938. The wing was completed in June 1940 (Figure 8.5), but was immediately used to support the war effort by an Army

Fig. 8.5   An early picture of the Austin Wing.

ballistics unit and a Navy signals unit. In the end, the total cost was only £91,500 for both the construction and the equipping of the new wing. Cockcroft was instrumental in persuading Rutherford to agree that the residue of the Austin donation should be used to fund a high-tension laboratory and a cyclotron.

The management of the Laboratory was the responsibility of the Cavendish Professor, but much of the day-to-day administration was carried out by Chadwick in his role as Assistant Director. When he departed for Liverpool in 1935, Cockcroft, a brilliant organiser with invaluable experience in high-power electrical equipment, assumed many of his responsibilities. His management and technical abilities were invaluable both to Kapitsa in his development of extremely high magnetic fields and to the success of the Cockcroft–Walton experiment.[13]

The head of the workshops through the 1920s and 1930s was Fred Lincoln (see Figure 5.2) who was notoriously parsimonious in providing the equipment and materials needed by the experimentalists. Stories abound about his unwillingness to release basic materials needed for experiments.

The other side of the coin was that the cramped and impecunious conditions undoubtedly forced a communal spirit of cooperation and 'make-do'. The general impression was one of good-hearted enthusiasm and dedication under the charismatic and sympathetic leadership of a truly great and inspiring scientist. Rutherford fostered the freedom to carry out speculative experiments and the free exchange of ideas among an outstanding group of physicists. Communal activities included the Cavendish Physical Society, which met every two weeks with tea provided by Lady Rutherford. The Kapitsa Club continued as an informal forum for the robust exchange of ideas and views, even after Kapitsa was unable to leave the Soviet Union after 1934.[14]

The Rutherford era: the radioactivists

The Rutherford era will always be remembered for the extraordinary achievements in radioactivity and the beginnings of the new field of nuclear physics. Jeffrey Hughes (1993) refers to the protagonists in this story as the *radioactivists*; they were to transform the field and open up new, and expensive, areas of basic research.

9.1 Rutherford and nuclear transformations

The α-particles remained the projectiles of choice used by Rutherford and his colleagues to study the nucleus, but the experiments were dependent upon securing sources of radioactive materials, which were generally in short supply and expensive.[1] In 1908 it was discovered that zinc sulphide laced with 0.01% copper was extremely sensitive to α-particles and emitted most of its luminosity in the yellow-green region of the optical spectrum to which the eye is most sensitive – about a quarter of the incident energy was converted into light (Hendry, 1984). This became Rutherford's preferred method of detecting α-particles and was the technique which Geiger and Marsden used in their experimental demonstration of the validity of the Rutherford scattering law (Section 7.5.2) (Geiger and Marsden, 1913). The energies of the α-particles could be estimated from their *ranges R* in air at a standard temperature and pressure, which Rutherford and his colleagues took to be 15°C at normal pressure (Box 9.1). It can be seen from the abscissa of Figure 8.3 that the typical ranges of the particles were between 2 and 8 cm, corresponding to particle energies of roughly 4 to 8 MeV. The ranges of the α-particles could be measured in terms of the mass traversed by the particle per unit cross-section ξ until they were brought to rest by the process of ionisation losses. To measure ξ, the amount of absorbing material between the source of the particles and the screen was increased until there was a sudden drop in the number of scintillations counted. The range ξ could then be translated into a physical distance R in air at the standard reference conditions. The ranges of the α-particles were characteristic of each radioactive decay.

In 1914, Marsden investigated the ranges of hydrogen nuclei produced when α-particles were projected through a volume filled with hydrogen. As predicted by Darwin (1912) and Bohr (1913d), the ranges of the hydrogen nuclei were expected to be four times that of the α-particles, about 28 cm in air, and far greater than for any of the α-particles produced by the known radioactive elements (Marsden, 1914). This experiment was repeated by Rutherford in his last years at Manchester with the apparatus shown in Figure 9.1, using the energetic α-particles produced in the radioactive decays of radium-C (bismuth). In

Ionisation losses and the energies and ranges of α-particles   Box 9.1

The process by which energetic α-particles lose energy is *ionisation losses*.[2] The electrons are removed from the atoms of the medium through which the α-particles pass by the electrostatic attraction between the particles and the electrons. The theory of this process was worked out by Thomson (1912b), Darwin (1912) and Bohr (1913d), whose calculation is similar to the modern version presented here.

The charge of the high-energy particle is ze and its mass m; b, the distance of closest approach of the particle to the electron, is called the *collision parameter*. It is straightforward to show that the kinetic energy transferred to the electron is

$$\frac{p^2}{2m_e} = \frac{z^2 e^4}{8\pi^2 \varepsilon_0^2 b^2 v^2 m_e} = \text{energy loss by high-energy particle.} \qquad (9.1)$$

Integrating over all collision parameters, the total energy loss of the high-energy particle, $-\mathrm{d}E$ in length $\mathrm{d}x$ is

$$-\frac{\mathrm{d}E}{\mathrm{d}x} = \frac{z^2 e^4 N_e}{4\pi \varepsilon_0^2 v^2 m_e} \ln\left(\frac{b_{max}}{b_{min}}\right). \qquad (9.2)$$

b_{max} and b_{min} are the maximum and minimum collision parameters which can contribute to the loss of kinetic energy of the α-particle. Since they appear inside the logarithm they need not be known very precisely. In the simplest non-relativistic approximation, the loss rate can be written

$$-\frac{\mathrm{d}E}{\mathrm{d}x} = \frac{z^2 e^4 ZN}{4\pi \varepsilon_0^2 v^2 m_e} \ln\left(\frac{m_e v^2}{\overline{I}}\right) = z^2 NZ\, f(v), \qquad (9.3)$$

where $2m_e v^2$ is the maximum kinetic energy which can be transferred to the electron, $\overline{I}$ is a mean ionisation potential, Z is the atomic number of the medium and N is the number density of atoms of the medium. The range R of the particle is found by integrating the energy loss rate from the particle's initial energy E_0 until it is brought to rest:

$$R = \int_0^{E_0} \frac{\mathrm{d}E}{(\mathrm{d}E/\mathrm{d}x)}. \qquad (9.4)$$

This calculation breaks down at the very smallest kinetic energies but the particle travels only a very short distance once its kinetic energy falls below that at which our calculation is valid. Then $-\mathrm{d}E/\mathrm{d}x = z^2 NZ f(v), E = \frac{1}{2}mv^2, \mathrm{d}E = mv\, \mathrm{d}v$ and so

$$R = \frac{m}{z^2 NZ} \int_0^{v_0} \frac{v\, \mathrm{d}v}{f(v)}, \qquad (9.5)$$

where m is the mass of the α-particle and the integral is independent of the material through which the particle passes. To a good approximation, $f(v) \propto v^{-2}$ and so $R \propto v_0^4$, where v_0 is the initial velocity of the particle, as found by Bohr (1913d).

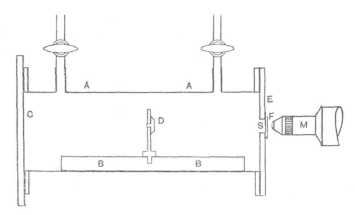

Fig. 9.1 The apparatus with which Rutherford demonstrated the nuclear disintegration of nitrogen atoms (Rutherford, 1919a,b,c,d, 1920). The source of energetic α-particles was radium-C and was located at D. S is the fluorescent zinc sulphide screen.

addition, he studied the collision products when the gas in the cylinder AA was oxygen, nitrogen or carbon dioxide (Rutherford, 1919a,b,c,d).

In his great series of four papers of 1919, Rutherford gave a detailed account of the apparent anomalies observed in experiments in which different gases were bombarded by α-particles. In the first of the papers, he confirmed Marsden's result that energetic protons bombarding hydrogen gas had ranges four times greater than the most energetic α-particles. If a head-on collision takes place between an α-particle and a hydrogen nucleus, the maximum velocity of a hydrogen atom is $v_H = (2M_\alpha/(M_\alpha + M_H))v_\alpha = 8/5v_\alpha$ and the range depends strongly upon the velocity of the particles, roughly as v^4 (Box 9.1). Precautions had to taken to ensure that the result was not due to contamination.

Long-range particles were observed in α-particle collision experiments with air and nitrogen, but not with oxygen and carbon dioxide. In the fourth paper of the series, Rutherford concluded that the long-range particles only appeared in collisions with nitrogen gas. They had to be hydrogen nuclei, what he was to refer to as *protons*, liberated from the nucleus in collisions between the α-particle and the nuclei of nitrogen atoms. As he wrote in the conclusion of his paper,

> From the results so far obtained it is difficult to avoid the conclusion that the long-range atoms arising from the collision of alpha particles with nitrogen are not nitrogen atoms but probably atoms of hydrogen, or atoms of mass 2. If this be the case, we must conclude that the nitrogen atom is disintegrated under the intense forces developed in a close collision with a swift alpha particle, and that the hydrogen atom which is liberated formed a constituent part of the nitrogen nucleus. (Rutherford, 1919d)

The origin of the energetic protons was in fact the nuclear interaction

$$^{14}N + \alpha \rightarrow \ ^{17}O + p, \tag{9.6}$$

in which the α-particle is absorbed by the nitrogen nucleus with the liberation of a proton and a residual ^{17}O nucleus, rather than the disintegration of the whole nucleus (see

Section 9.2). Rutherford estimated that these collisions were very rare; only 'about one α-particle in every million passed through the nitrogen gave rise to a proton' (Rutherford et al., 1930a). The ^{17}O nuclei were far too rare to be detected and had not been detected in Aston's mass spectrograph experiments on oxygen.[3]

Once Rutherford and Chadwick were settled in Cambridge, they began a programme of training in experimental techniques for the graduate students. This included the methods to be adopted in, for example, the construction of electroscopes, the measurement of the ranges of particles, the construction of ionisation chambers, scintillation counting, as well as lectures on vacuum techniques and the design of experiments. The new students carried out this training course in an attic known as the 'Nursery'.

Of particular relevance for this part of the story was the importance of the protocols which were devised for scintillation counting. The problem was that accurate counting required considerable concentration by a trained experimenter who had to be properly dark-adapted. Specifically, the protocol required: (1) suitable dark adaption of the eye, (2) counting for only one minute at a time, with (3) at least one minute's rest between counts, and in addition complete concentration during the one minute of counting. Chadwick adopted a scheme whereby pairs of experimenters carried out the observations in alternating one-minute counting sessions so that the efficiencies of counting of each observer could be determined. The physiological and psychological problems involved in accurate counting are described in Sections 126 and 126a of *Radiations from Radioactive Substances* (Rutherford et al., 1930a).[4]

With these new protocols in place, Rutherford and Chadwick began a more refined set of experiments, the results of which were published in 1921. Their famous apparatus is shown in Figure 9.2. As they wrote:

> The face of the brass tube T was provided with a hole 5 mm in diameter, which was closed by a silver foil of 6 cm air equivalent. The source R of α-particles was a disc coated with radium (B + C). The zinc sulphide screen S was fixed to the face of the tube so as to leave a slot in which absorbing screens could be inserted. (Rutherford et al., 1930a)

With this apparatus, they found that, with dry air in the brass tube, particles with ranges up to 40 cm were observed. When the dry air was replaced by carbon dioxide or hydrogen, no ranges greater than 29 cm were observed. They found that long-range tracks, greater than 29 cm, were found in the disintegrations of nitrogen, aluminium, boron, fluorine, sodium and phosphorus.

There was still the concern that some of the long-range particles might originate from the radioactive source itself, rather than from nuclear disintegration. It was discovered, however, that the long-range particles were emitted at a wide range of angles and so a new experiment was undertaken in which the long-range particles were only detected at 90° to the direction of the beam of fast α-particles. These experiments confirmed the conclusions of the first experiment. In summary, they found evidence for the disintegration of all elements from boron to potassium, with the exceptions of carbon and oxygen. They also found no certain evidence for the disintegration of lithium, beryllium or the elements beyond potassium (Rutherford and Chadwick, 1924a,b).

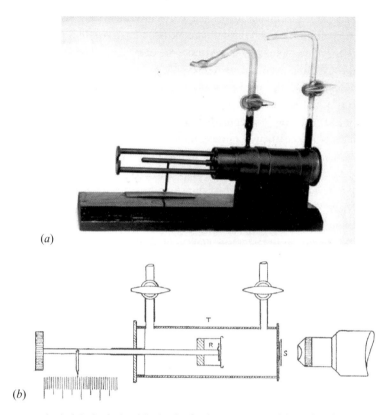

(a)

(b)

Fig. 9.2   (a) The apparatus with which Rutherford and Chadwick refined measurements of the nuclear disintegration of different atomic nuclei. (b) A diagram showing the internal workings of the disintegration experiment (Rutherford and Chadwick, 1921). The source of α-particles is at R and the fixed zinc sulphide screen at S.

These results were not without controversy. Although Rutherford and his colleagues had undoubtedly dominated the field of artificial radioactive decay during the early 1920s, the groups in Paris led by the Curies and Stefan Meyer's Radium Institute at Vienna, which had the benefit of a considerable repository of radioactive material, were hot on the trail. In particular, two young researchers from the Radium Institute, Hans Pettersson and Gerhard Kirsch, began building up their expertise in disintegration experiments and repeating the Cavendish experiments. They found that, contrary to the experiments of Rutherford and his colleagues, most of the light elements, including beryllium, magnesium, lithium and silicon, showed evidence of disintegration with ranges between 10 and 18 cm (Kirsch and Pettersson, 1923). In addition, Kirsch and Pettersson found that nuclear disintegrations were a much more frequent occurrence than in the Cavendish experiments. The disagreements were compounded by the different models which were espoused by the Cavendish and Viennese researchers to explain their results. In 1926, Kirsch and Pettersson published their book *Atomzertrümmerung (Atomic Fragmentation)*, in which they set out their very different perspective on the results of the disintegration experiments (Kirsch and Pettersson, 1926). The dispute became somewhat virulent and cast doubt upon the Cavendish results. Rutherford made every effort to contain the dispute and avoid public disagreements. The gravity of the situation has been described in depth by Hughes (1993).

At the heart of the dispute were the scintillation techniques used to estimate the ranges and frequency of occurrence of nuclear disintegrations. In 1925, Chadwick carried out a very thorough investigation of the techniques and protocols to ensure the reliability of the results, but even then the source of the disagreements could not be unravelled. Eventually, in 1927, Hans Thirring, Professor of Physics at Vienna University and a colleague of Meyer's, initiated exchange visits between Cambridge and Vienna so that their procedures could be compared. Petterrson visited Cambridge in 1927 and these discussions were cordial without resolving the differences. The return visit by Chadwick took place in December 1927. Crucially, Chadwick was able to repeat the Vienna experiments, but now using the Cambridge protocols for observing α-particles and protons. The experiments carried out under Chadwick's direction confirmed the Cambridge results, much to the annoyance of the Viennese physicists. In the end, Meyer offered to retract the Viennese results, but Chadwick refused, preferring the outcome to remain private. He expected that eventually the Vienna results would fade into obscurity without public humiliation of his colleagues.

The difficulties with the scintillation technique were further exemplified by the problems associated with the claims by Bergen Davis and Arthur Barnes to have found resonances in the capture of electrons by α-particles (Davis and Barnes, 1929). The results conflicted seriously with the expectations of theory. The source of the problem was eventually tracked down by Irving Langmuir to errors in the application of the strict protocols for observing the scintillations (Hughes, 1993). Whilst not fatal to the proper use of the scintillation technique, further doubt was cast upon techniques which involved the subjective judgement of individual observers.

The fact that Rutherford and Chadwick preferred not to publicise the resolution of the dispute between the Cambridge and Vienna physicists and the disquiet in the community about the procedural errors of Davis and Barnes meant that uncertainty continued to overshadow the results of disintegration research using the scintillation counting technique. All parties recognised that a less subjective means of detecting the products of nuclear decay was needed, and already alternatives were available. The cloud chamber pictures produced by C.T.R. Wilson, the Geiger counter and the opportunities offered by the development of radio valves were soon to replace the scintillation technique. In fact, the first of these had already produced dramatic results.

9.2 Shimizu, Blackett and the cloud chamber

Following C.T.R. Wilson's triumph in photographing the tracks of charged particles with his perfected cloud chamber (Section 7.6), the Cambridge Scientific Instrument Company marketed the chambers commercially, but they were tricky to use. With Rutherford's discovery of the disintegration of nitrogen nuclei by fast α-particles, he immediately turned his thoughts to how these events could be photographed using the cloud chamber technique.

Fig. 9.3 The commercial version of the Shimizu–Wilson reciprocating cloud chamber, manufactured by the Cambridge Scientific Instrument Company (Cattermole and Wolfe, 1987).

He gave the problem to a new graduate student, Takeo Shimizu, in the full realisation that only about one in 100,000 of the α-particles would undergo a close collision with a nitrogen nucleus.

Shimizu's first achievement was to devise a reciprocating mechanism which would allow 50 to 200 expansions of the chamber per minute. With this apparatus, Rutherford stated that he could make 'visual observations of branching tracks of α-particles'. But a permanent record was needed and so Shimizu recorded the images of the particle tracks using cinematographic film. The photographs were synchronised with the expansion of the chamber and a pair of cameras at right angles to each other were used to obtain three-dimensional images of the particle tracks. With this apparatus, Shimizu was able to take 1000 images or more per hour. On a reel of 200 feet of film, 3000 α-particle tracks were observed. The Cambridge Scientific Instrument Company assisted Shimizu in obtaining a patent for the Shimizu–Wilson reciprocating cloud chamber, which the company manufactured under a royalty agreement (Figure 9.3). In 1926, in response to the needs of schools and colleges, the company produced a simplified version of the Shimizu cloud chamber with a fixed expansion ratio. Within four years over 50 cloud chambers were sold, not only in the UK but also in Paris, Berlin and elsewhere. Shimizu's description of the apparatus and the first photographs of the branching of α-particles in elastic collisions were published in the *Proceedings of the Royal Society* in August 1921 (Shimizu, 1921a,b).

Shimizu returned to Japan in 1921 and Patrick Blackett, having completed his undergraduate studies, was asked by Rutherford to continue the development of the reciprocating cloud chamber. Blackett made a number of important improvements to Shimizu's apparatus, including replacing the reciprocating action by a spring mechanism and refining the synchronisation of the expansion of the cloud chamber with the photographic exposures to optimise the sharpness of the images. Elastic collisions of α-particles with helium

 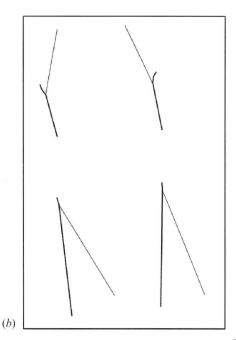

(*a*)　　　　　　　　　　　　　　　　(*b*)

(*a*) The ejection of protons from nitrogen nuclei by α-particles. In the top stereographic pair, the proton is ejected in the forward direction. In the lower image it travels in the backward direction (Blackett, 1925). (*b*) A schematic diagram showing the tracks of the incoming α-particles and the paths of the ^{17}O nuclei in heavy black lines and the paths of the protons in thin lines in the upper and lower panels of part (*a*).

Fig. 9.4

nuclei and hydrogen nuclei were studied to distinguish these events from genuine nuclear transformations. As Blackett described in his Nobel Prize lecture of 1948,

> This preliminary work done, production was started in earnest in 1924 and 23,000 photographs were taken within a few months. With an average of 18 tracks a photograph these gave over 400,000 tracks, each of which had to be scrutinized for anomalous behaviour. On some days when the apparatus worked well, as many as 1,200 photographs were taken. Eight forked tracks were found which had a quite different appearance from those showing normal elastic collisions, and these were readily identified as the sought for transmutation of nitrogen. (Blackett, 1964)

The key features of the eight images were interactions in which a long-range proton was ejected as well as a heavy nucleus, but no α-particle was released. Stereographic images of two of the eight events are shown in Figure 9.4(*a*), with diagrams showing the anomalous inelastic collisions in Figure 9.4(*b*) (Blackett, 1925). These famous images were unambiguous evidence for nuclear transformations – the α-particle was absorbed into the nucleus and then, rather than causing the disintegration of the nucleus as a whole as Rutherford had originally conjectured, the nucleus was transformed into a nucleus of ^{17}O and a fast proton was ejected. The rare ^{17}O nuclei were only discovered several years later in mass spectrograph experiments.

Fig. 9.5 An image of the first photographic record of the arrival of a cosmic ray particle by Skobeltsyn in 1929 (Skobeltsyn, 1929). The track of the particle in the cloud chamber is indicated by the two white arrows and one black arrow (see also Sekido and Elliot, 1985).

9.3 Blackett and Ochiallini: cosmic rays and the discovery of the positron

Following Wilson's prescient suggestion that the ionisation of air might be associated with cosmic radiation (Section 6.5.4), the breakthrough came in 1912 and 1913 when first Victor Hess and then Werner Kolhörster made manned balloon ascents in which they measured the ionisation of the atmosphere with increasing altitude (Hess, 1913; Kolhörster, 1913). By late 1912 Hess had flown to 5 km and by 1914 Kolhörster had made ascents to 9 km, these dangerous experiments being carried out in open balloons. They found the startling result that, above an altitude of about 1.5 km, the average ionisation increased exponentially with respect to the ionisation at sea level. This was clear evidence that a source of ionising radiation must be located above the Earth's atmosphere. The attenuation constant α, defined by $n(l) = n_0 \exp(-\alpha l)$, was found to have values of 10^{-3} m^{-1} or less, meaning that this radiation was much more penetrating than the most energetic γ-rays found in radioactive decays. For example, the absorption coefficient for the γ-rays from radium-C in air has a value of 4.5×10^{-3} m^{-1}, at least five times less penetrating than cosmic radiation.

It was not too much of an extrapolation to assume that the *cosmic radiation* or *cosmic rays*, as they were named in 1925 by Robert Millikan, were γ-rays with very much greater penetrating power than those observed in natural radioactivity. In 1929, Dmitri Skobeltsyn, working in his father's laboratory in Leningrad, constructed a cloud chamber to study the properties of the β-rays emitted in radioactive decays. The experiment involved placing the chamber within the jaws of a strong magnet so that the curvature of their tracks could be measured. Among the tracks, he noted some which were hardly deflected at all and which looked like electrons with energies greater than 15 MeV (Figure 9.5). He identified these with secondary electrons produced by the 'Hess ultra γ-radiation'. These were the first pictures of the tracks of cosmic rays (Skobeltsyn, 1929).

With Geiger and Müller's improvements of what became known as the *Geiger–Müller detector*, individual cosmic rays could be detected and their arrival times determined very precisely (Geiger and Müller, 1928, 1929). In 1929, Bothe and Kolhörster (1929) introduced the technique of *coincidence counting* for studying the cosmic rays. Two counters were placed one above the other, and they found that simultaneous discharges of the two detectors occurred very frequently, even when a strong absorber was placed between the detectors, indicating that charged particles of sufficient penetrating power to pass through both detectors and the attenuating material were very common. In the crucial experiment, they placed slabs of lead and then gold up to 4 cm thick between the counters and measured the decrease in the number of coincidences when the absorber was introduced. The mass absorption coefficient agreed very closely with that of the atmospheric attenuation of the cosmic radiation. The experiment strongly suggested that the cosmic radiation had to consist of charged particles rather than γ-rays. As they put it in their classic paper:

> One can perhaps summarise the whole discussion in a single argument: the mean free path of a γ-ray between two electron ejecting processes would be $1/\mu = 10$ m in water, $1/\mu = 0.9$ m in lead and $1/\mu = 0.52$ m in gold for the high latitude radiation. Hence one can see that a quite exceptional accident must be supposed to happen if two electrons produced by the same γ-ray should display the necessary penetrating power and the correct direction to strike both counters directly. (Bothe and Kolhörster, 1929)

They also showed that the flux of these particles could account for the observed intensity of cosmic rays at sea level and, because of their long ranges in matter, the energies of the particles had to be about 10^9–10^{10} eV, much greater than those produced in radioactive decays.

Blackett continued to make improvements to the performance of the cloud chamber, and at the same time developed a strong interest in the new quantum mechanics, although this subject was not particularly to Rutherford's taste. Interest in the new physics was stimulated by the clubs set up by Kapitsa and Dirac, the Kapitsa Club and the ∇^2 Club, respectively. The ∇^2 Club in particular was the forum for the discussion of quantum mechanics, the meetings including theorists such as Arthur Eddington, Harold Jeffreys, Arthur Milne, Douglas Hartree, Ralph Fowler and Edmund Stoner as well as Chadwick, Kapitsa, Dirac and Blackett. Blackett made an unsuccessful attempt to observe electron diffraction as a prediction of Schrödinger's wave mechanics. Among the fruits of Blackett's interest in quantum mechanics and quantum statistics was his confirmation of Nevill Mott's prediction of the angular distribution of α-particles scattered by helium nuclei (Blackett and Champion, 1931). Mott realised that, because helium nuclei and α-particles are identical bosons, they had to obey Bose–Einstein statistics in their scattering properties. At $45°$ to the direction of incidence, slow α-particles should display twice the probability as compared with the Rutherford scattering formula (Mott, 1928). The experiments of Blackett and Frank Champion were in excellent agreement with Mott's detailed predictions. Despite his antipathy to quantum theory, Rutherford was impressed by Mott's theoretical prediction, remarking, 'If you think of anything like this again, come and tell me' – a considerable encouragement for the 23-year old Mott (Mott, 1984).

Although notoriously reticent, Dirac became good friends with Blackett; as Dirac remarked in his reminiscences, 'I was quite intimate with Blackett at the time and had told him about my relativistic theory of the electron' (Dirac, 1984). Dirac was a regular visitor to the Laboratory through the years when Blackett was developing and perfecting the automatic cloud chamber technique. In one of the great theoretical extensions of quantum mechanics, Dirac succeeded in deriving the relativistic wave equation for the electron. In his two great papers of 1928 he used his equation, now known as the *Dirac equation*, to show that the magnitude of the intrinsic angular momentum, or spin, of the electron is $\hbar s(s + 1)$, where $s = 1/2$, that its magnetic moment is $e\hbar/m$ and that there exist negative-energy solutions as well. There was considerable debate before the physical nature of the negative solutions was understood. At first, they were thought to correspond to protons, but they had to have the same mass as the electron. Only in 1931 did Dirac come down decisively in favour of the interpretation that the negative solutions correspond to positively charged electrons, the positrons (Dirac, 1931).[5] As a result, Blackett and his experimental colleagues who attended the ∇^2 Club were well aware of the concepts of positrons and antimatter.

From the 1930s to the early 1950s, cosmic rays provided a natural source of very high-energy particles, energetic enough to penetrate into the nuclei of atoms. In 1930, Millikan and Anderson used an electromagnet ten times stronger than that used by Skobeltsyn to study the tracks of particles passing through the cloud chamber. Anderson observed curved tracks identical to those of electrons but corresponding to particles with positive electric charge (Anderson, 1932). The problem with this experiment was that Anderson's experiment depended upon the chance passage of a cosmic ray particle through the chamber. Blackett realised that the coincidence technique developed by Bothe and Kolhörster could be combined with the cloud chamber so that the chamber was only triggered when a cosmic ray particle passed through.

At the instigation of Bruno Rossi, who had become an expert in coincidence techniques and the associated electronic circuitry, Guiseppe (Beppe) Occhialini was sent to Cambridge in 1931 to work with Blackett in order to master cloud chamber techniques. As Blackett later remarked, Occhialini 'had came for three months: he stayed for three years'. They quickly mastered the combination of the cloud chamber and the Bothe–Kolhörster coincidence counters, and success was almost immediate (Figure 9.6(*a*)). Occhialini later recalled the excitement of working with Blackett at that time:

> What not everyone had a chance to see was the passionate intensity with which he worked. I can still see him, that Saturday morning when we first ran the chamber, bursting out of the dark room with four dripping photographic plates held high, and shouting for all the Cavendish to hear 'one on each, Beppe, one on each! (Occhialini, 1975, quoted in Lovell, 1975)

By the summer of 1932, they were obtaining cloud chamber images at a rate of one every two minutes with an 80% chance of these being associated with cosmic rays. The only limitation was that the magnetic field had to be operated continuously, limiting the magnetic flux density of the water-cooled magnet to 3000 gauss. They obtained many excellent photographs of the positive electrons, on many occasions observing showers containing

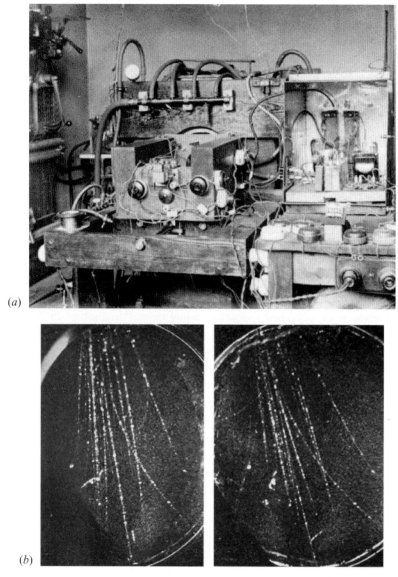

(a) The Blackett–Occhialini automated cloud chamber. (b) A stereographic image of a shower of electrons and positrons created in an interaction between a cosmic ray particle and the material of the cloud chamber (Blackett and Occhialini, 1933).

Fig. 9.6

equal numbers of positive and negative electrons created by cosmic ray interactions with the body of the apparatus (Figure 9.6(b)) (Blackett and Occhialini, 1933). In their paper they discussed the results of 500 photographs which showed the tracks of high-energy particles. The discovery of the *positive electron* or *positron* coincided almost exactly with Paul Dirac's theory of the electron (Dirac, 1928a,b), which Blackett had no hesitation in adopting.

In 1933, Blackett took up the Professorship of Physics at Birkbeck College, where he continued the study of high-energy interactions using cosmic rays with an eleven-ton magnet purchased thanks to a grant of £1500 from the Royal Society's Mond Fund, named after the bequest of the chemist and industrialist Ludwig Mond. This apparatus subsequently followed him to Manchester in 1937. With C.T.R. Wilson's retirement, the cloud chamber studies largely came to an end at the Laboratory, but one of his very few graduate students, Cecil Powell, was to take up the study of particle interactions at high energies by the use of photographic emulsion stacks – but these developments were to take place in Bristol, not in Cambridge.

9.4 Wynn-Williams, thyratrons and the scale-of-two counter

Despite Rutherford's attachment to the scintillation counting technique and its success in many of his great discoveries, more objective and reliable means of counting charged particles were required. The breakthrough came in 1926 when Heinrich Greinacher and his colleagues at the University of Bern constructed high-gain linear amplifiers to detect the ionisation currents of individual α-particles and protons (Greinacher, 1926, 1927). The electronic revolution was underway and a number of the Cavendish graduate students had a particular aptitude and enthusiasm for the burgeoning field of electronic circuitry. Among these graduates, Eryl Wynn-Williams was especially interested in applying these techniques to the counting of the ionising particles released in radioactive decays. Initially, he and his colleagues took the output signal from an ionisation chamber and fed it directly into the grid of the first stage of the amplifier (Figure 9.7). The amplified signal could then be used to drive an Einthoven string galvanometer which recorded the pulses on a cylindrical chart. This was the counting device used by Chadwick in his discovery of the neutron (Section 9.5). Many ingenious improvements were made to the apparatus, which eventually counted up to 100 or more events per second (Rutherford et al., 1930b).

Mechanical counters were already used in telecommunications, but they were inadequate to cope with the large counting rates required. Fortunately, *thyratrons* became commercially available in 1928 – these are triode valves containing a low-pressure gas such as argon or mercury vapour. If a positive pulse is applied to the grid of a thyratron, a self-sustaining anode current is set up that can only be stopped by dropping the voltage of the anode. The device therefore behaves like a two-state system rather than a linear amplifier. The Laboratory could not afford such valves, but the BTH Company generously donated six thyratrons to the Laboratory. In 1930 Wynn-Williams devised an electronic circuit in which several thyratrons were connected in a ring such that only one thyratron at a time could pass a current. Successive electric pulses would activate the thyratrons in sequence. A ring of five thyratrons connected to a mechanical counter could therefore handle five times the pulse rate of the mechanical counter (Wynn-Williams, 1931).

The ring counter was a great success, but Wynn-Williams realised that the circuit could be considerably simplified if the ring were reduced to just a pair of thyratrons, which turned out to result in much improved performance and stability. He then optimised the use of valves for counting by connecting the pairs of thyratrons in series so that each pair counted

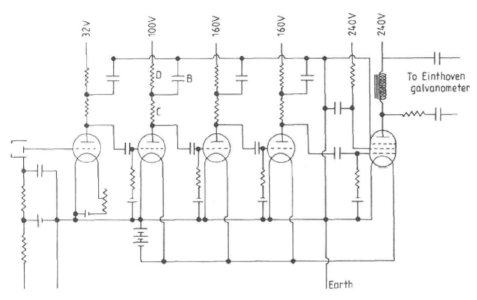

Ward and Wynn-Williams' linear amplifier used to drive an Einthoven galvanometer. The detector on the left of the circuit is connected directly to the grid of the first valve. This was the arrangement used in Chadwick's discovery of the neutron in 1932 (Ward *et al.*, 1929; Chadwick *et al.*, 1931; Chadwick, 1932a,b). **Fig. 9.7**

A 'next vintage' version of the Wynn-Williams scale-of-two counter. In the 1980s, he tried to find the original, but it was long gone. This second-generation counter survived (Wynn-Williams, 1984). **Fig. 9.8**

only every second pulse received by the preceding pair. He termed this invention, which is at the heart of all modern computing, the 'scale-of-two' counter (Figure 9.8) (Wynn-Williams, 1932). The apparatus with its associated batteries could be moved from room to room on a large trolley (Figure 9.9). By this means, particles could be counted at a rate

Fig. 9.9 This iconic photograph of Rutherford and Ratcliffe was taken by Wynn-Williams in 1932 during one of Rutherford's regular visits to the experimental laboratories. Wynn-Williams switched on the illuminated 'Talk Softly Please' sign just before Rutherford's arrival. The sign was built by Bowden and was needed because the lead from the ionisation chamber to the input valve of the linear amplifier was microphonically very sensitive. This laboratory is the recently vacated engineering drawing office which was taken over by the Cavendish Laboratory when the Engineering Department moved to their new building in Trumpington Street. The door open on the left leads to Chadwick's laboratory, where the neutron was discovered. The trolley contains the rebuilt Wynn-Williams and Ward amplifier as well as the three batteries of power supplies. This massive trolley enabled the amplifier and its power supplies to be moved from one laboratory to another as required (Lewis, 1984).

of up to 1250 counts per second.[6] An identical counter was constructed and these were in constant use until 1935 when Wynn-Williams moved to Imperial College, London, where he worked with G.P. Thomson's group studying neutron physics.[7]

This innovation marked the beginning of the use of binary numbers in electronic computation. R.V. Jones, who was the UK government's scientific intelligence advisor during the Second World War, wrote presciently in *Nature* in 1981:

[T]he modern computer is only possible because of an invention made by a physicist, C.E. Wynn-Williams, in 1932 for counting nuclear particles: the scale-of-two counter, which may prove to be one of the most influential of all inventions. (Jones, 1981)

9.5 Chadwick and the discovery of the neutron

Rutherford's Bakerian Lecture to the Royal Society of London in 1920 provides a vivid picture of the state of knowledge of the properties of the atomic nucleus at that time (Rutherford, 1920). The atomic masses of the chemical elements in atomic mass units were known to be about two or more times their atomic numbers. The commonly held explanation was that the nucleus was composed of electrons and protons, these 'inner' electrons neutralising the 'extra' protons. The fact that certain nuclei ejected electrons in radioactive β-decays supported this point of view. Rutherford speculated in his 1920 review that the neutral mass in the nucleus might be associated with a new type of particle, similar to the proton but with no electric charge. In the same spirit, he also proposed the existence of deuterium as an isotope of hydrogen, consisting of one proton and one neutron. As he wrote,

> [I]t seems very likely that one electron can bind two H nuclei and possibly also one H nucleus. In the one case, this entails the possible existence of an atom of mass nearly two carrying one charge, which is to be regarded as an isotope of hydrogen. In the other case, it involves the idea of the possible existence of an atom of mass 1 which has zero nuclear charge. . . . it may be possible for an electron to combine much more closely with the H-nucleus, forming a kind of neutral doublet. Its external field would be practically zero, except very close to the nucleus, and in consequence it should be able to move freely through matter. Its presence would probably be very difficult to detect in a spectrometer, and it may be impossible to contain it in a sealed vessel. . . . The existence of such atoms seems almost necessary to explain the building up of the heavy elements. (Rutherford, 1920)

Chadwick was quickly converted to the concept of what was named the *neutron* in the following year. Little attention was paid to the proposal outside Cambridge. Nonetheless, Chadwick, Rutherford and their colleagues, Joseph Glasson and John Roberts, made a number of unsuccessful attempts to find evidence for these particles. As Chadwick himself wrote much later:

> From time to time in the course of the following years, sometimes together, sometimes myself alone, we made experiments to find evidence of the neutron, both its formation and its emission from atomic nuclei. I shall mention some of the more respectable of these attempts; there were others which were so desperate, so far-fetched as to belong to the days of alchemy. (Chadwick, 1984)

Chadwick appreciated that neutrons might be liberated in the collisions of α-particles with light elements, and the case of beryllium was of particular interest since it did not

emit protons under bombardment. It was conceivable that the beryllium was split into two
α-particles and a neutron. He turned his attention to the possible emission of γ-rays in
α-particle collision experiments, but the detection apparatus was too insensitive and his
polonium source too weak. The first problem was overcome by the use of the recently
developed Geiger–Müller counter in Hugh Webster's experiments.[8] The problem of the
weakness of the polonium source was solved through the generosity of the Kelly Hospital
in Baltimore. Norman Feather was a visitor at the Johns Hopkins University in Baltimore at
the time and persuaded Curtis Burnham and F. West to donate to the Laboratory a number
of old radon tubes which could be recycled to produce a powerful polonium source of
energetic α-particles.

In the meantime, in 1930 Walther Bothe and Herbert Becker published their discovery
of a very penetrating form of radiation emitted when light elements such as beryllium are
bombarded by α-particles (Bothe and Becker, 1930). Because the penetrating particles did
not cause ionisation, they postulated that the neutral particles were high-energy γ-rays. In
1931, Irène Joliot-Curie found that the penetrating particles were much more penetrating
than had been previously thought. She made the assumption that the rays must be similar to
the cosmic rays, which were assumed to be very high-energy γ-rays, probably influenced
by Millikan's recent visit to Paris (Curie, 1931).

Similar experiments to those of Bothe and Becker were carried out at the Laboratory by
Webster, who found that penetrating radiation was also produced when boron was bombarded by fast α-particles. In addition, he discovered the important result that the radiation
emitted in the same direction as the incident α-particle was more penetrating than that
emitted in the opposite direction (Webster, 1932). As Chadwick recounted:

> This fact, clearly established, . . . could only be readily explained if the radiation consisted
> of particles, and, from its penetrating power, of neutral particles. (Chadwick, 1984)

Three weeks after her first paper, Irène Joliot-Curie and her husband Frédéric Joliot
published the results of a further series of experiments in which the penetrating neutral
radiation encountered a block of paraffin wax, a material rich in hydrogen atoms. They
found that energetic protons were emitted with energies up to 4.5 MeV. If the energetic
protons were produced by Compton scattering by high-energy γ-rays, a simple calculation
shows that, in a head-on collision of a high-energy photon of energy $h\nu$ with a stationary
proton of mass m_p, the minimum energy of the γ-ray necessary to impart a kinetic energy
E to the particle is given by the relation[9]

$$E = \frac{2h\nu}{2 + \dfrac{m_p c^2}{h\nu}}. \tag{9.7}$$

In the limit $m_p c^2 / h\nu \gg 2$, which applies in this case, setting $E = 4.5$ MeV, it followed that
the γ-ray had to have energy $h\nu \gtrsim 50$ MeV (Curie and Joliot, 1932).

When news of this result reached Cambridge, Rutherford's uncharacteristic response
was 'I don't believe it'. Chadwick was now convinced that the beryllium radiation was
a flux of the long-sought-for neutrons. He was ideally placed to carry out the necessary

(a)

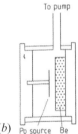

(b)

(a) The 'source vessel' with which Chadwick discovered the neutron (Chadwick, 1932a). (b) The layout of Chadwick's neutron experiment. The source vessel is on the left and the ionisation chamber on the right. This is a redrawn version of the original (Chadwick, 1932a) presented by Hendry (1984).

Fig. 9.10

experiments. His apparatus is shown in Figure 9.10, in which the beryllium radiation is produced in collisions between the α-particle and the beryllium screen in what Chadwick referred to as the 'source vessel' (Figure 9.10(b), left diagram). On the right-hand side of the same diagram, the ionisation chamber is shown, directly connected to the amplifier and oscillograph shown in Figure 9.7. Feather describes succinctly what Chadwick, working at fever pitch, achieved in a virtuoso set of experiments:

> So within ten days Chadwick had measured the range of the protons under various conditions, had detected the recoil of atoms of helium, lithium, beryllium, carbon, nitrogen, oxygen, and argon, and had determined the maximum ionisation produced when these recoil atoms were liberated in the gas in the ionisation chamber. It was obvious at once that the whole picture made sense numerically if the penetrating radiation from beryllium consisted of neutrons of mass roughly equal to the proton mass . . . (Feather, 1984)

One of the advantages of using the electronic method of detecting the particles was that the ionisation associated with the recoil nuclei could be measured directly. In his short paper to *Nature* (Chadwick, 1932b) and his more detailed paper in the *Proceedings of the Royal*

Society of London (Chadwick, 1932a), Chadwick thoroughly demolished the possibility that the beryllium radiation consisted of γ-rays. Among the arguments presented were the fact that the frequency of observation of the emission of the energetic protons was much greater than that predicted according to the Klein–Nishina cross-section. Furthermore, if the reaction chain which produced the beryllium radiation was

$$\alpha + {}^{9}\text{Be} \rightarrow {}^{13}\text{C} \rightarrow {}^{12}\text{C} + \text{X}, \tag{9.8}$$

the masses of ${}^{13}\text{C}$ and ${}^{12}\text{C}$ were sufficiently well known that the product X could at most have energy 14 MeV. On the other hand, associating X with the neutron, everything fell into place if its mass were roughly the mass of the proton. By comparing the ratio of the recoil energies of collisions with hydrogen and nitrogen, the mass of the invisible neutron was found to be 1.15 atomic mass units. Chadwick repeated the experiment using boron rather than beryllium as a target and found again a flux of neutrons, but now better constraints could be placed on the mass of the neutron since the atomic masses of boron and nitrogen had been accurately measured by Aston. The result was a mass of 1.0067 atomic mass units. Chadwick's conclusion was that 'the neutron consists of a proton and electron in close combination'. Earlier in his paper (Chadwick, 1932a), he remarked:

> It is, of course, possible to suppose that the neutron may be an elementary particle. This view has little to recommend it at present, except the possibility of explaining the statistics of such nuclei as ${}^{14}\text{N}$.

Almost immediately, the reality of the neutron as the neutral partner of the proton was established. First of all, Feather filled his cloud chamber with nitrogen and bombarded the gas with neutrons. The cloud chamber pictures revealed the recoil tracks associated with nitrogen nuclei and also double tracks of α-particles created in the disintegration of the nitrogen nucleus (Feather, 1932). As soon as the news of the discovery of the neutron reached Bohr and Heisenberg, they adopted the view that the neutron was indeed a fundamental particle and this stimulated Heisenberg to develop influential papers on the quantum mechanics of nuclei and nuclear forces. In particular, Heisenberg had no hesitation in regarding the neutron as a spin-1/2 particle. Similarly, Robert Bacher and Edward Condon demonstrated that the neutron had to be a spin-1/2 particle and hence a fermion like the proton, and not a boson,[10] which would be the case if it consisted of a proton plus an electron (Bacher and Condon, 1932). The identification of neutrons as fermions solved the problem of the spin of the ${}^{14}\text{N}$ nucleus, as well as many other problems of nuclear spin. By the time of his Bakerian Lecture in 1933, Chadwick had adopted the point of view that the neutron was indeed a new elementary particle (Chadwick, 1933). He was particularly persuaded by the argument that, according to quantum mechanics, the combination of an electron and a proton results in a hydrogen atom. Furthermore, as he wrote, 'If the neutron is a proton and electron why does not the hydrogen atom transform into a neutron with the release of energy?'

The final nail in the coffin of the proton-plus-electron model was provided in 1934 by the experiment suggested to Chadwick by Maurice Goldhaber, who had recently arrived as a graduate student from Germany (Chadwick and Goldhaber, 1934). Goldhaber pointed out that the mass of the neutron could be determined in collisions between γ-rays and deuterium nuclei, what were referred to at the time as *diplons*. The reaction is

$$^2D + \gamma \rightarrow p + n. \tag{9.9}$$

By measuring the kinetic energy of the protons, the mass of the neutron was found to be greater than 1.0077 and less than 1.0086 in atomic mass units. Thus, the neutron was more massive than the hydrogen atom, consisting of a proton plus an electron, and which had an accurately determined mass of 1.0078. By then, even the cautious Chadwick was thoroughly convinced he had discovered a new particle.

9.6 Cockcroft, Gamow and Walton: 'splitting the atom'

Throughout the 1920s, the projectiles used to probe the nucleus were the products of natural radioactivity – α-particles, β-rays and γ-rays. In the case of the α-particles, only a limited range of particle energies was available and so only the light elements could be probed by this means since the heavier elements had greater nuclear potential barriers. A second problem was that the fluxes of the projectiles were low and so only a limited number of experiments could be conducted. The obvious solution was to develop techniques to produce beams of accelerated energetic particles. The fluxes of particles to be accelerated could be provided by positive-ray tubes, which produced fluxes many thousands of times greater than those available from naturally occurring radioactive substances. In his presidential address to the Royal Society of 1927, Rutherford made an impassioned plea:

> It has long been my ambition to have available for study a copious supply of atoms and electrons which have an individual energy far transcending that of the α and β-particles from radioactive bodies. I am hopeful that I may yet have my wish fulfilled, but it is obvious that many experimental difficulties will have to be surmounted before this can be realised, even on a laboratory scale. (Rutherford, 1928)

Already, the Laboratory had considerable experience in the technologies for creating beams of fast particles through Thomson and Aston's pre-war researches with positive-ray tubes (Section 7.4). Further developments of positive-ray tubes and the subsequent series of mass spectrographs were continued by Thomson, Aston and George Thomson after the war. This expertise was of great benefit in the development of particle accelerators in the Laboratory.

The race was on to produce accelerators which would provide the necessary controlled intense beams of high-energy protons and α-particles. Gregory Breit and Merle Tuve at the Department of Terrestrial Magnetism of the Carnegie Institute of Washington exploited the use of Tesla coils, which consisted of a loosely wound transformer to prevent the electrical breakdown which would occur if the coil were tightly wound (Breit and Tuve, 1928). Charles Lauritsen at the California Institute of Technology employed a cascade of transformers to produce a strong current at voltages up to 750 keV (Lauritsen and Bennett, 1928). At the University of California at Berkeley, Ernest Lawrence developed the concept of a linear accelerator, in which the alternating current was tuned so that during each half-cycle the particles experienced an accelerating field. This scheme turned out to result in far too long a chain of accelerating elements, but he struck upon the idea of using the same principle in a circular accelerator in which the particles were accelerated during each half-cycle by the application of a strong magnetic field – this was the invention of the *cyclotron* (Lawrence and Livingston, 1931; Lawrence and Sloan, 1931). But before any of these prototypes could be used for particle physics experiments, the race was won by John Cockcroft and Ernest Walton at the Laboratory – they became famous in the popular press as the physicists who 'split the atom'.[11]

Even before Rutherford's address to the Royal Society, Thomas Allibone had arrived at the Laboratory, sponsored by the engineering firm Metropolitan-Vickers. He was already familiar with the equipment necessary to create voltages of half a million volts and, with Rutherford's encouragement, his project was to build an accelerator which could produce a beam of 0.5 MeV electrons. Rutherford knew that atomic nuclei had potential barriers of 8 MeV or more, but these experiments might be interesting for studies of lower-energy electrons. As part of the cooperation with Metropolitan-Vickers, Allibone constructed 500 kV Tesla coils and succeeded in producing beams of 300 keV electrons.

John Cockcroft was working in the same room in the Laboratory and he too had come to Cambridge with a background in electrical engineering, again with the support of Metropolitan-Vickers. As an undergraduate at Cambridge, he had read mathematics and then embarked upon his PhD in 1924. Besides these attributes, he was an excellent organiser who was trusted by Rutherford.

Ernest Walton arrived in 1928 from Dublin supported by an 1851 studentship, specifically to work on the acceleration of particles to high energies. His initial idea did not work out, but it was an early attempt to use the principles which would eventually lead to the betatron. At Rutherford's suggestion, he changed tack to develop a linear accelerator. Suddenly, the perspective changed with the arrival in Cambridge of a memorandum by the young George Gamow.

When Gamow arrived in Göttingen from the Soviet Union in 1927, he read Rutherford's presidential address of 1927 to the Royal Society on the problem of understanding the α-decay in thorium C', or polonium-212 (^{212}Pu). Geiger's α-scattering experiments had shown that the height of the electrostatic potential barrier within which the nucleons were confined was at least 8.57 MeV, and yet the energies of the α-particles observed in the α-decay of thorium C' were less than half this value, 4.2 MeV. Gamow realised that this was an example of barrier penetration in quantum mechanics. The nuclear potential could be modelled as a deep rectangular potential well, as illustrated in Figure 9.11(*a*). Then, according to the quantum calculation for barrier penetration, although the barrier is impenetrable according to classical physics, there is a finite probability that the α-particles can reach the other side because of the wave properties of the particle. This is illustrated by the diagram from Gamow's paper (Figure 9.11(*a*)), which shows the amplitude of the wave function on both sides of the barrier and within it. In fact, Gamow and independently Ronald Gurney and Edward Condon almost simultaneously solved Schrödinger's equation for the nuclear potential shown in Figure 9.11(*a*) and derived a relationship between the decay constant λ of the nucleus against α-particle decay and the energy of the α-particle (Gamow, 1928; Gurney and Condon, 1928, 1929). This theory of α-decay could account

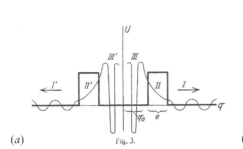

(a) The one-dimensional model used by Gamow in his paper on barrier penetration by α-particles, showing the amplitude of the oscillating wave function within the nucleus, decaying through the potential barrier and then propagating as waves outside the potential barrier (Gamow, 1928). (b) Cockcroft (left) and Gamow (right) discussing barrier penetration.

Fig. 9.11

naturally for the very narrow range of energies of the α-particle and the enormous range of their decay constants, as found by Geiger and Nuttall in 1911 (Figure 8.3).

Gamow explained the theory in a manuscript sent to Rutherford in November 1928. Cockcroft repeated Gamow's calculations and in a memorandum to Rutherford showed that, because of the process of barrier penetration, protons accelerated to only 300 keV could penetrate a boron nucleus with about 0.6% probability. Cockcroft inferred that an accelerating electric potential of only 300 keV would be sufficient to penetrate the boron nucleus and induce nuclear transmutations.

Cockcroft and Walton joined forces to make a determined effort to produce powerful fluxes of high-energy protons, which had a much higher probability of penetrating the nuclei of light elements than the α-particles. To achieve their goals, Cockcroft persuaded Rutherford to obtain £1000 from the University to buy a 300 kV transformer. They also benefitted from the purchase of prototype Apiezon pumps, invented by C.R. Burch at Metropolitan-Vickers, which were made available to them at a remarkably low price. These enabled them to attain high vacua much more effectively than using the existing vacuum pumps. The challenge was the rectification of the alternating output voltage of the transformer to produce a steady accelerating voltage of several hundred thousand volts. With the copious use of the recently introduced plasticine, also produced by Burch, they constructed the apparatus shown in Figure 9.12, the two horizontal large glass bulbs providing the rectified voltage. By May 1930, with the help of Allibone, they achieved a beam energy of 280 keV, but no nuclear interactions were observed. They wrote up a description of their accelerator for the *Proceedings of the Royal Society* (Cockcroft and Walton, 1930).

At this point, Cockcroft and Walton had to vacate their laboratory, which had been on loan from the Department of Physical Chemistry, and move to much larger premises in the former Balfour Library building (Figure 5.3). Rather than move their equipment, they decided to build a new accelerator with the objective of reaching 800 MeV, the negative results of their first accelerator suggesting that higher voltages might be necessary. The

Fig. 9.12 The Cockcroft Walton accelerator with which they achieved voltages of 280 keV by May 1930 (Cockcroft and Walton, 1930).

equipment required a major redesign, including the replacement of the large fragile recti-fier bulbs by robust cylindrical tubes and a clever capacitor circuit designed by Cockcroft which could produce multiples of the rectified voltage of each stage. The voltage achieved was estimated from the size of the sparks between the large prominent spheres seen in Figure 9.13. The protons were directed onto a lithium target and the products of the inter-actions observed on a standard zinc fluoride fluorescent screen. Figure 9.13 is the classic image of the Cockcroft–Walton experiment, with Walton inside the tent observing the flu-orescent screen with Cockcroft on the left.

On the morning of 14 April 1932, Walton succeeded in observing the first artificial nuclear disintegrations by bombarding lithium with high-energy protons (Cockcroft and Walton, 1932). He immediately telephoned Cockcroft, who confirmed the observation, and then Rutherford was invited to observe the scintillations. Squeezing his considerable body into the little tent, he saw the zinc sulphide screen aglow with scintillations. Rutherford exclaimed:

> Those scintillations look mighty like alpha particle ones to me. . . . I should know an alpha particle scintillation when I see one, for I was in at the birth of the alpha particle and have been observing them ever since.

The process involved was

$$^{7}\text{Li} + \text{p} \quad \rightarrow \quad ^{4}\text{He} + ^{4}\text{He}. \tag{9.10}$$

The energies of the accelerated protons were precisely known, as were the rest masses of the lithium and helium atoms. The decrease in mass in the above interaction corresponded

The apparatus with which Cockcroft and Walton artificially disintegrated lithium nuclei (Cockcroft and Walton, 1932).   Fig. 9.13
Walton is sitting inside the little tent, observing the decay products on a luminescent screen. Cockcroft is on the left.

to the liberation of 14.3 ± 2.7 MeV of kinetic energy, shared equally between the two emitted α-particles. From the observed ranges of the α-particles, the total liberated energy was 17.2 MeV. This excellent agreement between theory and experiment provided the first direct experimental test of Einstein's mass–energy relation $E = mc^2$. They carried out co-incidence experiments with zinc sulphide screens on either side of the target and found that 25% of the scintillations on one side coincided with those on the other side.

Disintegrations of lithium nuclei were observed with accelerating potentials as low as 125 keV. Cockcroft and Walton went on to bombard many more light elements with high-energy protons, Be, B, C, O, F, Na, Al, K, Ca, Fe, Co, Ni, Cu, Ag, Pb and U – scintillations were observed from all these elements. The strongest fluxes of α-particles resulted from the disintegrations of lithium $^7_3$Li, boron $^{11}_5$B and fluorine $^{19}_9$F. As they pointed out, these are all elements with atomic mass numbers of the $4n + 3$ type, where n is the number of α-particles, $^4_2$He. The addition of a proton to each of these would result in the formation of a 'new α-particle' inside the nucleus. This experiment marked the beginning of experimental high-energy physics, in which particles are accelerated to high energies and used as probes of the structure of the nucleus and as tools for the discovery of new particles.

Mark Oliphant completed his PhD on the bombardment of metal surfaces with positive ions in December 1929. It was recognised that he had great technical skill and so, with the discovery of the artificial disintegration of light nuclei by Cockcroft and Walton, he was invited by Rutherford to collaborate on further studies in that field. Oliphant quickly designed and built a simplified version of the Cockcroft–Walton machine which operated up to voltages of 200 kV and produced fluxes of protons one hundred times greater than the earlier machine. Oliphant and Rutherford confirmed the earlier results and provided many more details of the products of the disintegrations (Oliphant and Rutherford, 1933).

At the end of 1933, the chemist Gilbert Lewis of the University of California at Berkeley donated to the Laboratory two tiny phials containing a total of 0.5 cm^3 of heavy water, D_2O. Oliphant was given the responsibility of looking after these precious drops and proceeded to carry out a brilliant series of experiments involving deuteron collisions (Oliphant *et al.*, 1933). The apparatus was modified so that the particle collisions could be photographed in a Wilson cloud chamber, and Philip Dee and Walton succeeded in photographing both proton and deuteron collisions with lithium and boron targets (Dee and Walton, 1933).

The most brilliant of Oliphant's discoveries were those of tritium ^3H and helium-3 ^3He. The voltages available in Oliphant's new machine were doubled and beams of high-energy deuterons were collided with compounds containing deuterium. Large numbers of protons and neutrons were liberated in these collisions as well as particles with mass number 3 and atomic numbers 1 and 2. These were identified with the species tritium ^3H and helium-3 ^3He, respectively, which were created by the interactions

$$^2_1H + {}^2_1H \rightarrow {}^3_1H + p \quad \text{and} \quad {}^2_1H + {}^2_1H \rightarrow {}^3_2He + n \tag{9.11}$$

(Oliphant *et al.*, 1934).[12]

9.7 Ellis, Pauli, Fermi and β-decay

Rutherford and Chadwick concentrated their efforts upon the use of α-rays as projectiles for bombarding different materials. The advantage of the α-particles was that, for a given radioactive substance, the emitted α-particle had a well-defined energy. In contrast, the β-decay process resulted in a broad continuum spectrum of electron energies as well as line spectra. Rutherford had carried out important β-ray experiments in Manchester with Harold Robinson and W.F. Rawlinson in 1912–13 following the discovery by Otto Baeyer, Otto Hahn and Lise Meitner that groups of electrons with characteristic speeds are ejected by most β-emitting nuclei (Baeyer and Hahn, 1910; Baeyer *et al.*, 1911a,b, 1912). The magnetic deflection technique shown in Figure 9.14(*a*) was used to focus the ejected electrons onto a photographic plate; the faster the electron, the greater the diameter of its circular path (Rutherford and Robinson, 1913). In the experiments of Rutherford, Robinson and Rawlinson, the γ-rays released in radioactive decays were passed through ordinary matter and similar patterns of emitted electron velocities were observed (Rutherford and Robinson, 1913; Rutherford *et al.*, 1914). Examples of the spectrum of emitted electrons from the radioactive decay of radium-B and of the β-ray spectrum excited by the passage

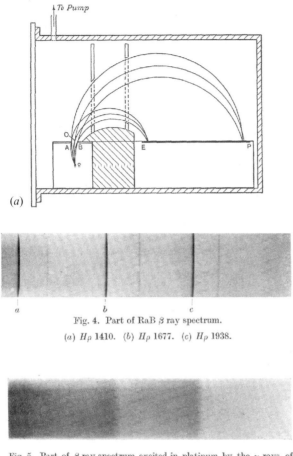

Fig. 4. Part of RaB β ray spectrum.

(a) $H\rho$ 1410.   (b) $H\rho$ 1677.   (c) $H\rho$ 1938.

Fig. 5. Part of β ray spectrum excited in platinum by the γ rays of RaB corresponding to 4.

(a) The magnetic spectrograph used to determine the velocity spectra of β-rays liberated in β-decay. The trajectories of the particles liberated from the source S are focussed by a uniform magnetic field onto the photographic plate EP. The greater the velocity of the particles, the greater the diameter of their orbits in the magnetic field (Rutherford *et al.*, 1930a). (b) Comparison of the β-ray spectra in the radioactive decay of RaB and the same part of the β-ray spectrum excited in platinum by the γ-rays of RaB (Rutherford *et al.*, 1914, 1930a). The maximum energies of the latter β-rays are well aligned with the lines in the former spectrum.

Fig. 9.14

of the γ-rays through platinum are shown in Figure 9.14(*b*). The inference was that the line spectra of the electrons were not associated with the primary β-decay process, but rather with the ejection of electrons from the K, L and M shells of the atoms by nuclear γ-rays. This line of research came to a halt with the outbreak of the First World War.

After the war, at Rutherford's suggestion, Charles Ellis took up these complex challenges in Cambridge and soon became a leading figure in β-ray radioactivity. The challenges facing research into β-radioactivity were much greater than those involving α-particles. As explained by Rutherford, Chadwick and Ellis:

In 1921 further experiments were made on this subject which disclosed not only the origin of these β ray groups, but also gave a method of determining the frequencies of the γ rays. It appears that in a β ray disintegration there are only two main phenomena, the emission of the actual disintegration electron, and the emission of the nuclear γ rays. There are, however, many secondary effects, of which the chief is the relatively frequent conversion of these γ rays in the outer electronic structure and consequent emission of high-speed photo-electrons. These electrons form the characteristic groups already mentioned, and have proved subsequently to be of great importance for the information they yield about the disintegration. This is by no means the only secondary effect, for following this there is a succession of X-ray and electronic emissions from the outer levels. When all these effects are superimposed on the results of the original disintegration, which by itself is sufficiently complicated, the complexity of the whole emission can easily be realised. (Rutherford *et al.*, 1930a)

Ellis showed great experimental skill in establishing the line spectrum of the β-rays emitted in what was called the *inner photoeffect*. The basic equation employed was the standard photoelectric rule, but now applied to the nuclear γ-rays:

$$E_e = h\nu - W_e, \tag{9.12}$$

where E_e is the kinetic energy of the emitted electron, $h\nu$ is the energy of the γ-ray emitted in the radioactive decay and W_e is the binding energy of the orbital electrons in the atom. Since the W_e are quantised, it followed that, since the values of E_e form a line spectrum, the emitted γ-rays must also be quantised. By analogy with the origin of optical spectral lines, there had to be quantised energy levels within the nucleus itself. Ellis devoted a great deal of effort to establishing the nuclear energy levels and also to the determination of *internal conversion coefficients*, meaning the probability that the γ-ray released in the initial radioactive decay suffers internal conversion by ejecting an orbital electron relative to the probability that it escapes from the atom.[13]

During the 1920s, there was an ongoing debate about whether the broad continuum electron energy spectrum could be attributed to what was termed 'ordinary' processes, meaning that the electrons were created with a single energy which was then redistributed by processes such a Compton scattering, or whether the continuous energy spectrum was intrinsic to the radiative decay process itself. An example of the continuous energy spectrum of electrons found in the decay of radium-E (^{210}Bi) is shown in Figure 9.15 (Neary, 1940). In that example, there is an upper limit to the energies of the emitted electrons of just over 1 MeV, but the spread of energies extends to less than 4% of this value, the maximum occurring at just less than 300 keV and the average energy amounting to 390 keV.

After two years of challenging experiments, Charles Ellis and William Alfred Wooster completed calorimetric experiments in which they showed that the average energy deposited in their calorimeter was about 350 keV per disintegration, rather than about 1 MeV as might be expected if all the energy was injected with maximum energy of 1 MeV and then dissipated by 'ordinary' processes (Ellis and Wooster, 1927). In these remarkable experiments, the precision of the calorimetry was such that temperature differences of 10^{-3} K could be reliably measured. The experiment indicated that the observed electron energy spectrum was indeed the intrinsic energy spectrum of the β-decay process.

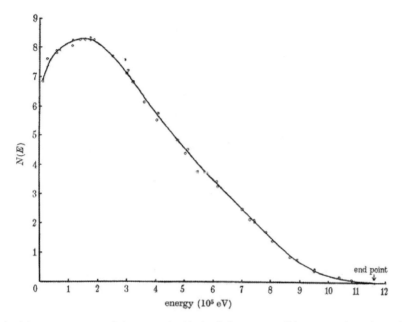

A example of the energy spectrum of electrons emitted in the β-decay process. This spectrum shows the continuous electron energy spectrum found in the radioactive decay of radium-E (^{210}Bi) (Neary, 1940).   Fig. 9.15

The problem with this result was that the process seemed to violate conservation of energy. Furthermore, with the understanding of the quantum mechanical rules for the addition of angular momentum, the hyperfine splitting of atomic lines could be used to determine the magnetic moments and hence the spin of nuclei. Ellis had pursued the study of the hyperfine splitting of nuclear transitions using a powerful electromagnetic designed by Cockcroft at a cost of £250. The design was based on that which Aimé Cotton had developed in Paris for his huge electromagnet, the construction of which began in 1924 – magnetic flux densities as great as 7 Tesla could be achieved with this machine. The process of β-decay seemed to involve a violation of the law of conservation of angular momentum at the nuclear level. Thus, the laws of conservation of both energy and angular momentum were in jeopardy.

For some time, Bohr returned to his earlier concern about the validity of the law of conservation of energy at the atomic level, which he had elaborated in his paper with Kramers and Slater (Bohr *et al.*, 1924). In 1930, in desperation, Pauli suggested that the problem might be solved by invoking the existence of a neutral particle which he called a 'neutron'. At that time, the only known 'subatomic particles' were the proton, the electron and the photon. Pauli's radical proposal was contained in an impassioned letter to his expert colleagues working on radioactivity at their meeting in Tübingen:[14]

Dear Radioactive Ladies and Gentlemen,
I have come to a desperate way out regarding the 'wrong' statistics of the N- and ^{6}Li nuclei, as well as the continuous β-spectrum, in order save the 'alternation law' of statistics and the energy law. To wit, the possibility that there could exist in the nucleus electrically

neutral particles, which I will call neutrons, which have spin 1/2 and satisfy the exclusion principle and which are further distinct from light-quanta in that they do not move with light velocity. The mass of the neutrons should be of the same order of magnitude as the electron mass and in any case not larger than 0.01 times the proton mass . . . The continuous β-spectrum would then become understandable from the assumption that in β-decay a neutron is emitted along with the electron in such a way that the sum of the energies of the neutron and the electron is constant. . . .

For the time being I dare not publish anything about this idea and address myself confidentially to you, dear radioactive ones, with the question how it would be with the experimental proof of such a neutron, if it were to have a penetrating power equal to or about ten times larger than a γ-ray. . . .

Your most humble servant,

W. Pauli

In 1932, Chadwick's discovery of the neutron, meaning the neutral partner of the proton, changed the picture. In the following year, Fermi suggested that Pauli's 'neutron' might be better called a *neutrino*, and that usage was established from then on. In the following year, Fermi published his theory of weak interactions and β-decay (Fermi, 1934). In his famous paper, he treated the process by analogy with the process of the emission of radiation according to the rapidly developing theory of quantum electrodynamics. Neutrinos have a very small cross-section for interaction with matter and it was not until 1956 that they were detected experimentally by Frederick Reines and Clyde Cowan, who used a fission reactor as a source of neutrinos (Reines and Cowan, 1956). The discovery was made only two and a half years before Pauli's death. His response on receiving the news was sent in a congratulatory telegram:

Thanks for message. Everything comes to him who knows how to wait. Pauli.

In fact, Ellis and Mott were very close to discovering the neutrino. Not long after Pauli's proposal, they found that the maximum energy of the emitted electron in β-decay is equal to the differences between the initial and final states of the nuclei concerned, either or both of which might be in excited states. This was an important result since it meant that Pauli's neutrino had to have an extremely small rest mass (Ellis and Mott, 1933). Mott remarked philosophically in his memoir of the period:

We really had in Ellis's work much of the evidence for the existence of the neutrino, and, with hindsight, it is a pity we didn't say so. (Mott, 1984)

9.8 The discovery of nuclear fission

The discovery of the neutron had immediate implications for experimental nuclear physics. Unlike the electron or the α-particle, the neutron is electrically neutral and so could readily penetrate the Coulomb barrier of the nucleus. The discipline of nuclear physics was transformed, since heavy nuclei such as uranium could be bombarded with neutrons, resulting in the formation of new isotopes. Furthermore, the Cockcroft and Walton experiment had demonstrated the validity of Einstein's relation $E = mc^2$ and the production of energy in nuclear fission reactions. Rutherford was not optimistic about the use of these discoveries of nuclear physics for energy generation. In his address to the British Association in 1933, he is reported in *The Times* of 12 September 1933 as stating:

> We might in these processes obtain very much more energy than the proton supplied, but on the average we could not expect to obtain energy in this way. It was a very poor and inefficient way of producing energy, and anyone who looked for a source of power in the transformation of the atoms was talking moonshine. But the subject was scientifically interesting because it gave insight into the atoms.

Leo Szilard was well aware, however, of the significance of these experiments for nuclear energy generation. In 1933, following Chadwick's discovery of the neutron, Szilard realised that a self-sustaining nuclear reaction chain would be possible if the neutrons liberated in the types of interaction involved in the Cockcroft and Walton experiment could be used to initiate further nuclear interactions. He filed patents for this concept and also carried out unsuccessful experiments in which light elements were bombarded with neutrons to demonstrate the effect. Such experiments were initiated by Fermi and his colleagues in Rome in 1934. They believed that they had demonstrated the formation of a new element with atomic number 94 (Fermi *et al.*, 1934), but the result was viewed with some scepticism.

Fermi's experiments were repeated by Otto Hahn, Lise Meitner and Fritz Strassmann. In 1938 following the Anschluss, in which she lost her citizenship, Meitner fled to Sweden and continued her collaboration with Hahn by mail. During this correspondence, Hahn informed Meitner of his discovery of traces of barium when uranium was bombarded with neutrons. This came as a complete surprise since barium has only 40% the atomic weight of uranium. Meitner soon convinced herself and Hahn that the barium resulted from what became known as the *nuclear fission* of the uranium nuclei. The results were published by Hahn and Strassmann (1939). Meitner and her nephew Otto Frisch, who was also working in Sweden, published the results of their calculations that a new type of nuclear reaction had been observed in Hahn and Strassmann's experiments (Meitner and Frisch, 1939).

As soon as the results of Hahn and Strassmann's experiments were published, Szilard and others immediately realised that this provided a route to a nuclear chain reaction, both for the generation of nuclear power and for the creation of nuclear weapons. He urged restraint in the publication of these results because of the impending war, but Joliot and his colleagues in Paris did not hesitate. The theory of nuclear chain reactions was published by both groups in 1939 (von Halban *et al.*, 1939; Szilard and Zinn, 1939). The details of this story and the subsequent development of nuclear weapons is vividly told in Rhodes' classic book *The Making of the Atomic Bomb* (Rhodes, 1986).

Rutherford died suddenly in 1937 and did not live to see the realisation of practical means of generating nuclear energy and nuclear weapons by the application of the great discoveries which he and his colleagues had made during what is often referred to as this 'golden era' of nuclear physics.

N.S.Alexander.   P.Wright.   A.G.Hill.   J.L.Pawsey   G.Occhialini.   H.Miller

W.E.Duncanson.   E.C.Childs.   T.G.P.Tarrant.   J.M.McDougall.   R.C.Evans.   E.S.Shire.   E.L.C.White.   F.H.Nicoll.   R.M.Chaudhri.   B.V.Bowden.   W.B.Lewis.

P.C Ho.   C.B.Mohr.   H.W.S.Massey.   M.L.Oliphant.   E.T.S.Walton.   C.E.Wynn-Williams.   J.K.Roberts.   N.Feather.   Miss.Davies.   Miss.Sparshott.   J.P.Gott.

J.A.Ratcliffe.   P.Kapitza.   J.Chadwick.   R.Ladenberg.   Prof.Sir.J.J.Thomson.   Prof.Lord.Rutherford.   Prof.C.T.R.Wilson.   F.E.Aston.   C.D.Ellis.   P.M.S.Blackett.   J.D.Cockcroft.

Fig. 9.16   Staff and research students in the Cavendish Laboratory in 1932.

9.9 The exodus of the radioactivists

A continuous series of annual photographs of the senior academic staff and graduate students lines the walls of the collection of historic scientific instruments in the Cavendish Laboratory. The photograph of 1932 shows many of the protagonists of this *annus mirabilis* (Figure 9.16) – the photograph includes nine Nobel Prize winners. The key role of Rutherford in leading the research programme through this dramatic period of discovery is unambiguous. His achievements up to 1919 when he took up the Cavendish Chair were extraordinary enough, but to repeat the feat through the subsequent 18 years has scarcely been paralleled in any area of the physical sciences.

The one negative aspect of Rutherford's dominant role in Laboratory for his colleagues and the research programme was his reluctance to seek the resources needed to continue the development of research in nuclear physics, even when they were offered. It is estimated that the annual budget for research and teaching apparatus was about £2000, which was quite inadequate for a laboratory which had to cater for 400–500 students per year. The reason for the success of the research programme was undoubtedly Rutherford's extraordinary ability to invent simple experiments which enabled him to draw profound conclusions by careful attention to detail.

In 1933, Chadwick had recommended the construction of a high-tension laboratory, but at that time Rutherford was against it, and Chadwick formed the view that there was no future for nuclear physics at Cambridge. Cockcroft tried to persuade Rutherford to allow him and his colleagues to go ahead with the construction of a cyclotron, which had been successfully developed by Lawrence and his colleagues at Berkeley. Rutherford was disinclined to go ahead with this quite different approach to the acceleration of high-energy particles, largely because of the success of the 200 keV accelerator. It was to be three years before the construction of a cyclotron was begun at the Laboratory.

This reluctance of Rutherford to invest in larger machines for nuclear physics undoubtedly led to the decision by some of his most distinguished, and more entrepreneurial, colleagues to set up their own laboratories at other UK universities. As Blackett remarked to his student Frank Champion, 'If Physics Laboratories have to be run dictatorially . . . I would rather be my own dictator' (Lovell, 1975). The fame of the Laboratory and the quality of its research workers meant that they had no lack of job offers both in the UK and abroad. The result was a significant exodus of many of the outstanding physicists from Cambridge:

- Blackett took up the Professorship of Physics at Birkbeck College in 1933. Then he moved to the Langworthy Professorship at the Victoria University of Manchester in 1937 and then to Imperial College, London in 1953.
- Chadwick took up the Chair of Experimental Physics at Liverpool University in 1935 but returned to Cambridge to become Master of Gonville and Caius College in 1948.
- Ellis was appointed to the Wheatstone Chair of Physics at King's College, London, succeeding Appleton in 1936.
- At the outbreak of the Second World War, Cockcroft was appointed Assistant Director of Scientific Research in the Ministry of Supply. He then took charge of the Canadian Atomic Energy project in 1944 and became Director of the Montreal Laboratory and Chalk River Laboratories. In 1946, he returned to Britain and set up the Atomic Energy Research Establishment (AERE) at Harwell. He became the founding Master of Churchill College, Cambridge in 1959.
- Walton became a Fellow of Trinity College Dublin in 1934, and in 1946 was appointed Erasmus Smith's Professor of Natural and Experimental Philosophy at the college.
- Wynn-Williams was appointed Assistant Lecturer in Physics at Imperial College, London in 1935.
- Following Chadwick's departure for Birmingham, Oliphant became one of Rutherford's Assistant Directors of Research, but he too left, to become Poynting Professor of Physics at the University of Birmingham in 1937, working with Chadwick, who had secured the funds to build a cyclotron.

Kapitsa was detained in Moscow in 1934 and C.T.R. Wilson retired from the Jacksonian Chair in 1934, to be replaced by Appleton. But a new generation of gifted physicists was beginning to flourish who would take the Laboratory in new directions.

The Rutherford era: the seeds of the new physics

While the ground-breaking discoveries in nuclear physics were taking place, the seeds of many new disciplines were being sown which were to be of central importance for the future development of the Laboratory. The great quantum revolution was in full swing and this required the assimilation of quite new theoretical tools and concepts which were to be essential for the future experimental research programme. We first address how the new theoretical physics was incorporated into the Laboratory's programme before describing the new directions taken in experimental physics.

10.1 Experimental and theoretical physics

Rutherford was notoriously unimpressed by complex mathematical theory, preferring simplicity and model-building as in his 'billiard-ball' model for collisions between charged particles. But he was always prepared to listen and could not ignore the success of Gamow's prediction of barrier penetration. However sceptical he may have been, he was fully supportive of Cockcroft and Walton's experiments which relied upon the correctness of the wave mechanical model of the atomic nucleus (Section 9.6). Likewise, he was impressed by Mott and Blackett's demonstration of the need to use Bose–Einstein statistics in the scattering of helium nuclei by α-particles (Section 9.3).

There were significant and friendly interactions between the applied mathematicians and the experimental physicists in the Cavendish Laboratory, despite the fact that they belonged to different faculties. The Cambridge college system helped bring experimenters and theorists together, but against this was the almost insuperable obstacle that mathematics was taught as part of the Mathematical Tripos and physics as part of the Natural Sciences Tripos. This division placed a significant academic barrier in the formal teaching of the disciplines and to collaboration between physicists and mathematicians.

The development of quantum mechanics from Bohr's pioneering discovery of the structure of the hydrogen atom in 1913, through the old quantum theory from 1913 to 1925 and to the fully fledged theory of quantum mechanics in the period 1925 to 1930 was the product of the brilliant researches of theorists largely centred on Copenhagen, Göttingen and Munich.[1] The schools of Bohr, Born, Sommerfeld and Hilbert produced a galaxy of brilliant young theorists, including Heisenberg, Jordan, Pauli, Kramers and many others – Pauli referred to the new quantum mechanics as *Knabenphysik*, young man's physics. The only quantum theorist in Cambridge on the same level as the great continental theorists was the equally young Paul Dirac. In continental Europe, Bohr, Born, Sommerfeld and Hilbert had well-founded departments or institutes of theoretical physics, which had no counterpart in Cambridge. Of the three Cambridge mathematics professors, Arthur Eddington was the Plumian Professor of Astronomy and Experimental Philosophy, but he was ploughing his own furrow, first in stellar structure and evolution and then, more speculatively, in his fundamental theory. Ralph Fowler was appointed to the newly founded John Humphrey Plummer Professorship of Theoretical Physics in January 1932 and Dirac was elected to the Lucasian Professorship of Mathematics in succession to Larmor later that same year.

There were, however, informal ways in which theory and experiment were brought together. The two key figures were Fowler and Pyotr Kapitsa. Fowler had begun as a pure mathematician and lecturer in the Mathematics Faculty. With Rutherford's arrival in Cambridge, he became deeply involved in theoretical physics – and he married Rutherford's daughter Eileen in 1921. He was allocated an office in the Laboratory next but one to Rutherford's, but he also spent much time at Trinity College, at his home in Trumpington or abroad, in America and also in continental Europe where the instruction and facilities for theoretical physics were far superior to anything in Cambridge.

Fowler was undoubtedly the key figure in fostering theoretical physics in Cambridge and maintaining contacts with the continental theorists. He was Dirac's PhD supervisor and it was he who sent the proofs of Heisenberg's fundamental paper of 1925 to Dirac, who quickly appreciated its significance and converted the whole theory into Hamiltonian form (Dirac, 1925; Heisenberg, 1925). In the Cambridge of the 1920s, keeping up with the great advances in continental Europe and fostering relations between theory and experiment were significant challenges, as described by Alan Wilson:

> The only point of contact was through R H Fowler – because he was Rutherford's son-in-law. But Fowler, like the rest of us, worked in his college rooms – in Trinity – and if you wanted to consult him you had to drop in half a dozen times before you could find him in. He lived in Trumpington and did most of his work there. Also he spent half his time in America. . . . Dirac was unapproachable and also spent a lot of time abroad. . . . There was, in fact a seminar once a week [in the Cavendish] on Wednesday afternoons lasting an hour for research students to present as well as they could some of the enormous number of papers appearing in the *Zeitschrift für Physik* and the *Annalen der Physik*, but with Dirac and Fowler absent the discussion of these was perfunctory. (Wilson, 1984)

According to Wilson, there were only three research students working on quantum mechanics in 1927–28: J.A. Gaunt, William McCrea and Wilson himself. All three were to make important contributions to theoretical physics. Gaunt worked on the quantum mechanical theory of collision and radiation processes, his name being associated with the Gaunt factor in collision processes. McCrea was to become a distinguished astrophysicist and cosmologist and a leader of these disciplines in the UK. Wilson carried out pioneering research in the quantum theory of semiconductors. Nonetheless, Wilson gained most value from his trips to the main centres of theoretical physics in continental Europe. In 1929–30 he gave lecture courses in quantum mechanics while Dirac gave a parallel advanced course which was to lead to his monumental *Principles of Quantum Mechanics* (Dirac, 1930).

The other principal vehicle for promoting collaboration between theoretical and experimental physics was the Kapitsa Club, founded by the redoubtable Kapitsa (see

Section 10.3) soon after he arrived in Cambridge in 1922. This was modelled upon the Russian tradition of seminars sponsored by a senior distinguished physicist at which one or more papers were presented and then opened for discussion. Membership of the club was restricted by invitation and initially for theorists only, with the exception of Kapitsa. By the 1931–32 session, however, there was a good mix of experimenters and theorists, the members being, alphabetically, Bernal, Blackett, Cockcroft, Dee, Dirac, Feather, Fowler, Gray, Kapitsa, Massey, Mott, Oliphant, Powell, Ratcliffe, Roberts, Snow, Tarrant, Walton, Webster, Wilson and Wooster. There the barriers between experimental and theoretical physics were broken down. Of particular importance was the fact that the experimenters became fully conversant with developments in theory and were prepared for the unexpected better than most other experimental physics groups. In addition, visitors to Cambridge would present papers to the club. The galaxy of theorists and experimenters is truly remarkable – in the late 1920s Franck, Bohr, Schrödinger and many others talked to the club, while in the early 1930s the speakers included Eddington, Lemâitre, Millikan, Uhlenbeck, Bohr, Beck, Bethe, Tamm, Morse, Houtermans, Goetz, Weisskopf and many others. The Cambridge researchers in experimental and theoretical physics were kept fully abreast of developments in both areas through the very English institution of 'the club'.

In addition, distinguished theorists such as Mott, Harrie Massey and Rudolf Peierls were members of the Cavendish Laboratory during the 1930s. One of the major outcomes of the collaboration between Mott and Massey was their monumental series, *Theory of Atomic and Molecular Collisions*, which eventually expanded to five volumes (Mott and Massey, 1934).

Another important vehicle for assimilating the most recent advances in theory was the establishment of the Scott Lectures, which became the Laboratory's premier series of invited lectures each year. The lectures were named after A.W. Scott who, although never a member of the Laboratory, was Phillips Professor of Science at St David's College, Lampeter, Wales. Because of his admiration for the work of the Laboratory, he bequeathed an endowment to the Cavendish Laboratory in 1927 'for the furtherance of Physical Sciences'. The roll-call of the lecturers up to the outbreak of the Second World War gives some impression of the level of the lectures: Bohr (1930), Langmuir (1931), Debye (1932) Geiger (1933), Heisenberg (1934), Hevesy (1935), Appleton (1936), de Haas (1937), Siegbahn (1938) and Blackett (1939).

Thus, although there was no theoretical physics department, there were theoretical physicists and they were on good terms with the experimentalists.

10.2 Appleton and the physics of the ionosphere

Edward Appleton began his research career in 1914 under Lawrence Bragg just before the outbreak of the First World War. He helped unravel the structure of metallic crystals, but then was caught up in the fervour for involvement in the war effort. He joined the Royal Engineers and was chosen to work on signal duties. As he wrote,

I was very fortunate in this, because radio was just beginning to influence army com-munications. The thermionic valve, which we now know to be the primary tool in radio, was then a secret device, whose design was little understood, and whose performance had to be discovered experimentally rather than predicted. Then, again, the subject of radio wave propagation was in its infancy, for there were only available the crudest methods of measuring signal strength.

However, during my R.E. career, I was able to put a few problems 'on the shelf' in my mind, to be attacked when the War was over. Two of these occupied my interest when I got back to Cambridge, the theory of the thermionic valve and the theory of long-distance radio propagation. (Quoted in Ratcliffe, 1966)

As good as his word, as soon as the war was over, he returned to Cambridge and in 1919 was appointed a Fellow of St John's College and an Assistant Demonstrator in the Labora-tory. His first research activities were associated with thermionic valves, which he brought back from his wartime service. This work built upon the pioneering research carried out by Richardson (1901) in his discovery of the law of thermionic emission (see Section 7.3). Ap-pleton's particular interest was the non-linear voltage–current characteristic of thermionic valves and, with Balthasar van der Pol, who studied with J.J.Thomson in 1917–19, he inves-tigated how these properties could be exploited in electrical circuits. Among their particular achievements was the demonstration that two coupled non-linear oscillating circuits would lock onto the same frequency (Appleton and van der Pol, 1921, 1922). This pioneering work on non-linear systems was widely applicable and led van der Pol to the development of the concept of the *relaxation oscillator*, what is now called the *limit-cycle behaviour* of a non-linear oscillating system (van der Pol, 1927).[2] Appleton's researches into the physics of thermionic vacuum tubes culminated in his pioneering monograph, written while he oc-cupied the Wheatstone Chair of Physics at King's College, London, which he had taken up in 1924. Entitled *Thermionic Vacuum Tubes*, it provided a coherent and clear account of the various types of vacuum tubes, their operation, non-linear effects and applications (Appleton, 1931).

The work for which Appleton is best remembered, and for which he won the Nobel Prize in Physics, is his elucidation of the properties of radio wave propagation in the ionosphere and the equations which describe these phenomena, the *Appleton–Hartree equations*. This story starts with Heinrich Hertz's demonstration that electromagnetic radiation, produced by his spark gap transmitter, displayed all the properties of light waves – reflection, re-fraction, diffraction, polarisation and so on (Hertz, 1893). Among others, Guglielmo Mar-coni realised that electromagnetic waves might be a replacement for the telegraphic cable which dominated telegraphy. He gradually increased the distance over which Morse code messages could be transmitted by radio waves – for example, crossing the English Channel between Wimereux in France to the South Foreland Lighthouse in England in 1899. But the great challenge was transatlantic radio communications.

In 1901, Marconi reported the transmission of radio signals between Cornwall and New-foundland, which was regarded as a major technological breakthrough. There were, how-ever, some doubts as to whether or not he had actually achieved this remarkable feat. Many physicists were sceptical as to whether radio waves could be propagated around the curved surface of the Earth, their view being that radio waves travelled rectilinearly, as had been demonstrated by Hertz in his experiments. In this case, the signal strength for an isotropic emitter would not have been detectable over transatlantic distances. In 1902, however, Arthur Kennelly and Oliver Heaviside independently proposed that radio waves might be propagated round the curvature of the Earth by reflection from an electrified layer in the upper atmosphere – the signal would propagate as in a waveguide between the surface of the sea and this conducting layer high in the atmosphere (Kennelly, 1902; Heaviside, 1902). This layer became known as the *Kennelly–Heaviside Layer* or the *Heaviside layer*; it is now referred to as the *E-region*. But, at the time, there was no experimental evidence for this layer.

In 1924 Miles Barnett arrived from New Zealand as a graduate student to work with Appleton on wireless propagation. In 1922, the London BBC station had begun its transmissions, and they made recordings of the variations in the signal strength as received in Cambridge. The signal strength was constant during the day, but became variable as night approached. They suggested that the variability was associated with the interference between the direct signal from the London station and a signal reflected from the Kennelly–Heaviside layer which was responsible for long-range radio transmissions. The originality of their suggestion was that the reflections could also take place at small angles to the vertical direction and cause interference over short distances.

Appleton devised two approaches for testing this hypothesis. The first method was called *frequency modulation*. If the distance travelled by the direct wave is h and the distance travelled by the reflected wave is h', the path difference $D = h' - h = N\lambda$, where λ is the wavelength. Hence, $N = (h' - h)/\lambda$. Thus, constructive interference occurs for integral values of N and destructive interference for half-integral values. In the frequency modulation technique, D is fixed but the wavelength is changed until constructive interference occurs again at, say, λ'. Then $D = N\lambda = (N - 1)\lambda'$ and hence

$$\frac{1}{D} = \left(\frac{1}{\lambda} - \frac{1}{\lambda'} \right). \tag{10.1}$$

Appleton and Barnett carried out this experiment by varying the frequency of the transmissions between Bournemouth and Oxford after radio transmissions had finished for the day. They found that the reflections took place from a layer about 90 km in altitude (Appleton and Barnett, 1925a,d). In the second approach, they measured directly the angle of incidence of the reflected waves at the receiver, confirming that the signal indeed came from a reflecting layer high in the atmosphere (Appleton and Barnett, 1925c).

The observations were continued through the period of an eclipse of the Sun on 24 January 1925. Appleton and Barnett demonstrated that the height of the Heaviside layer changed at precisely the time of the total eclipse, showing that the source of ionisation of the upper layers of the atmosphere was associated with emissions from the Sun which travelled at the speed of light (Appleton and Barnett, 1925b). We now know that the ionisation is associated with the ultraviolet radiation from the Sun. In 1927, the observations revealed secondary fringes which corresponded to double reflections between the Earth and the ionosphere. As might be expected, these secondary signals were generally weaker than the primary reflections, but occasionally they were much stronger and sometimes only the 'secondary' reflected signal was observed. Appleton correctly inferred that these signals were being reflected from a yet higher ionised layer in the atmosphere, which he termed the F-layer, in contrast to the Heaviside layer which he referred to as the E-layer (Appleton, 1927). The F-layer, which extends from about 200 to 500 km altitude, is also referred to as the *Appleton layer*. Appleton fully appreciated the significance of these pioneering measurements, which opened up entirely new ways of studying what Robert Watson–Watt, the pioneer of radar, referred to as the *ionosphere*, for understanding the processes involved in radio wave propagation in plasmas. Appleton devoted the rest of his scientific career to their study.

In parallel with these experimental advances, Appleton developed the theory of the propagation, refraction, reflection and absorption of radio waves in the ionosphere. In their initial papers, Appleton and Barnett emphasised the importance of the magnetic field and drew analogies with Lorentz's theory of light propagation in crystals in the presence of a magnetic field (Appleton, 1925; Appleton and Barnett, 1925d). As expressed by Ratcliffe (1966),

> The only difference [from Lorentz's theory] is that the electrons and ions were free from each other in the upper atmosphere but were bound together in atoms in the gas or crystal. Lorentz dealt with two different cases, one where the light travelled along the magnetic field and the other where it travelled at right angles, and he showed that there would be a phenomenon of double refraction in which a single wave was split up into two. Appleton extended this idea to deal with the case in which the electrons and ions were completely free and also where the wave could be travelling in any direction at all, not necessarily just parallel or perpendicular to the magnetic field. He named his theory the *magneto-ionic theory*.

Almost at the same time, similar calculations were published in the USA by H.W. Nichols and J.C. Schelleng, who included the effects of collisions in their analysis (1925a, 1925b). By 1927, Appleton had derived essentially the final version of the theory of electromagnetic wave propagation in ionised media and gave a full account of it in his paper of 1932 (Appleton, 1932). In 1929, Douglas Hartree included the Lorentz term in the equations and gave a treatment of reflection problems according to the theory (Hartree, 1929). The distinctive feature of the Appleton–Hartree approach is that solutions are found for the propagation of the radio waves in terms of a complex refractive index n, some properties of which are outlined in Box 10.1. The imaginary part of the refractive index corresponds to the dissipation or damping of the waves. As remarked by Ratcliffe (1966),

> Since that time, his equation has been of the greatest possible importance, both in ionospheric physics and nowadays also in plasma physics.

These studies elucidated the nature of the reflection of radio waves by the ionosphere. The radio waves propagate through the plasma in which the density of electrons changes with height. The refractive index of the medium consequently changes and the wavefront of a radio wave is deflected. At a certain height, the deflection returns the wave to the Earth (Hartree, 1929). Notice also that the same process means that any radio signals arriving from above the ionosphere are reflected back into space at the very long wavelengths used

The Appleton–Hartree equation

The Appleton–Hartree equation for the complex refractive index n of a magneto-ionic medium can be derived by considering the equation of motion of electrons in an electromagnetic wave in the presence of a magnetic field. In the complete equation, the effects of collisions are included. The expression for n can be written

$$n^2 = 1 - \frac{X}{1 - iZ - \frac{1}{2}Y^2 \sin^2\theta/(1 - X - iZ) \pm \left[\frac{1}{4}Y^4 \sin^4\theta/(1 - X - iZ)^2 + Y^2 \cos^2\theta\right]^{1/2}}.$$

where $X = \omega_0^2/\omega^2$, $Y = \omega_H/\omega$ and $Z = \nu/\omega$. $\omega = 2\pi f$ is the angular frequency of the wave, while f is the wave frequency (in Hz). In the expressions for X, Y and Z, $\omega_0 = 2\pi f_0 = (Ne^2/\epsilon_0 m)^{1/2}$ is the *electron plasma frequency*, $\omega_H = 2\pi f_H = B_0|e|/m$ is the electron gyrofrequency and ν is the electron collision frequency. ϵ_0 is the permittivity of free space in the SI system of units used here. B_0 is the ambient magnetic flux density and θ is the angle between the ambient magnetic field vector and the wave vector.

If the effects of damping due to the collision term Z are neglected, the expression for n becomes

$$n^2 = 1 - \frac{X(1-X)}{1 - X - \frac{1}{2}Y^2 \sin^2\theta \pm \left[\left(\frac{1}{2}Y^2 \sin^2\theta\right)^2 + (1-X)^2 Y^2 \cos^2\theta\right]^{1/2}}.$$

This is the expression for the refractive index in a cold collisionless plasma in the presence of a magnetic field. In the absence of a magnetic field, it becomes even simpler:

$$n^2 = 1 - X = \left(1 - \frac{\omega_0^2}{\omega^2}\right),$$

illustrating the cut-off for radio wave propagation at angular frequencies below the plasma frequency ω_0.

In the first two cases, the plus or minus signs give rise to two solutions of the equations. For propagation perpendicular to the magnetic field, the plus sign represents the 'ordinary mode', and the minus sign the 'extraordinary mode'. For propagation parallel to the magnetic field, the plus sign represents a left-hand circularly polarised mode, and the minus sign a right-hand circularly polarised mode. These can be thought of as the normal modes of electromagnetic waves in the plasma, and they have different phase velocities. In the case of propagation along the magnetic field direction, the different phase velocities of the right- and left-hand circularly polarised modes gives rise to the Faraday rotation of the plane of polarisation of a linearly polarised signal in a plasma.[3]

in these early studies. As the wavelength decreases, however, the ionosphere becomes increasingly transparent to radio waves.

On C.T.R. Wilson's retirement in 1934, Appleton returned to the Laboratory as Jacksonian Professor of Natural Philosophy. There, he set up a field station for ionospheric research, supported by an award of £200. On the death of Rutherford in 1937, he took on the role of Acting Director of the Laboratory until the appointment of Lawrence Bragg to the Cavendish Chair. In 1939, he was appointed Secretary to the Department of Scientific and Industrial Research and soon became deeply involved in the war effort. In 1949, he was appointed Principal and Vice-Chancellor of the University of Edinburgh.

His active involvement in research ceased in 1939, and the leadership of the Radio Group fell to Ratcliffe, whose expertise in radio wave propagation was to prove of immense value to the development of radar during the war (Section 11.6). The distinguished work of the Radio Group was to take a surprising and fruitful turn after the war when Martin Ryle and his colleagues began their attempts to understand the origin of the cosmic radio noise (Section 12.7).

10.3 Kapitsa and the Mond Laboratory

The 27-year-old Pyotr Kapitsa began his protracted stay in Cambridge in July 1921, having lived through the Communist revolution and the Russian civil war in St Petersburg (then Petrograd) and recovered from the deaths of his wife, son and daughter from influenza in 1919. David Shoenberg, who was to become his graduate student in 1932, summarised his character as follows:

> Peter Kapitsa was a legendary figure both in Rutherford's Cambridge of the 1920s and 1930s and subsequently in Moscow to the end of his long life and the legends serve to illustrate his colourful personality. In his scientific work he showed great versatility and brought the skills of an engineer and mathematician to bear on important problems in physics and technology in an entirely original way. (Shoenberg, 1985)

Kapitsa arrived in Cambridge after an excellent training in experimental physics at Joffe's newly founded Physico-Technical Institute in Petrograd. Joffe had succeeded in bringing together a group of outstanding young physicists with a strong emphasis upon the new physics which was developing very rapidly in Europe. Despite the chaos following the Russian revolution and the subsequent civil war and the great shortages of equipment and facilities, many important lines of experimental research were started on what Shoenberg describes as a 'do-it-yourself' basis. The parallels with Rutherford's approach in his much more privileged environment are noteworthy. It was after Joffe had travelled to Europe with sufficient resources to purchase equipment from abroad that Kapitsa persuaded Rutherford to let him join the already overcrowded Cavendish as a graduate student. Chadwick had recognised Kapitsa's originality and technical skill. As required, Kapitsa started the laboratory training course in the attic, but within two weeks he was excused attendance because of his clearly outstanding expertise in experimental physics.

Kapitsa's ebullient character was very different from that of his colleagues – Blackett the naval officer, Chadwick the reserved and cautious physicist, Cockcroft who was famously reticent, and Dirac who was notoriously self-contained. His sense of humour and originality appealed to Rutherford, who recognised a kindred spirit with a strong sense of adventure. Kapitsa made two major contributions during his stay in Cambridge. The first was the institution of the Kapitsa Club, which he set up almost as soon as he was established in Cambridge. Modelled on Joffe's seminar in Petrograd, the club played an essential role in disseminating and discussing the new discoveries in physics, the discussions being masterminded in the Russian style by Kapitsa himself (Section 10.1). The second contribution was in providing technologies which could be exploited by others.

Kapitsa established his experimental credentials by taking up Rutherford's challenge to measure how the energy of an α-particle falls off towards the end of its trajectory. In a tricky experiment using a Boys radiomicrometer, he measured the heating produced in a plate by a beam of α-particles. The work was completed in only nine months (Kapitsa, 1922). He was delighted by Rutherford's commendation of his work. As he wrote to his mother,

> Today the Crocodile summoned me twice about my manuscript . . . It will be published in the Proceedings of the Royal Society, which is the greatest honour a piece of research can receive here . . . Only now have I really entered the Crocodile's school . . . which is certainly the most advanced in the world and Rutherford is the greatest physicist and organiser. (Quoted by Shoenberg, 1985)

This is an early appearance of Kapitsa's nickname 'Crocodile' for Rutherford. As he explained in an interview with Ritchie Calder,

> In Russia the crocodile is the symbol for the father of the family and is also regarded with awe and admiration because it has a stiff neck and cannot turn back. It just goes straight forward with gaping jaws – like science, like Rutherford.[4] (Shoenberg, 1985)

The second paper on Kapitsa's energy-loss experiments never appeared, because he had hit upon a different technique for measuring the energy of the α-particles. The available steady magnetic fields were too weak to allow observation of the bending of the particle tracks towards the end of their ranges, but he had the idea that much larger impulsive fields could be created, which would last only for a very short time. This was a key insight and very significant for the future direction of research in the Laboratory. It is best summarised by Shoenberg:

> This marked a turning point in Kapitsa's career. Once the large impulsive magnetic fields had been achieved he saw whole new vistas opening up for exploiting the technique, and this led him into pioneering work in solid state and low temperature physics. It was also the beginning of the transition of the Cavendish from the string and sealing wax tradition to the age of large machines. (Shoenberg, 1985)

To achieve these much larger fields, Kapitsa designed a large accumulator which could be rapidly discharged. He persuaded Rutherford of the potential of this method, and Rutherford made two important contributions to the project. The first was to allocate £150, and the second to arrange for Kapitsa's technician Emil Yanovich Laurmann to come to Cambridge from Petrograd. Laurmann was very skilled in the techniques of electrical engineering, photography and delicate instrumentation and together he and Kapitsa obtained their first results only three and a half months after the programme began. They measured the tracks of α-particles in magnetic fields which Kapitsa referred to as of 'terrific field strength' (Kapitsa, 1923). The resulting energy loss of the particles towards the ends of their tracks agreed with Kapitsa's previous measurements using the Boys microradiometer. The details of these remarkable experiments were published in the following year (Kapitsa, 1924a,b).

Kapitsa quickly finished his PhD in 1923 and was awarded a Clerk Maxwell Scholarship in the same year. He went on to study the Zeeman effect in magnetic fields up to 130 kG but then, in 1924, he came up with a method for creating impulsive fields up to 1000 kG. The idea was simply to short-circuit a large rapidly rotating dynamo through a suitably designed coil, the field energy originating in the kinetic energy of the dynamo. In this development he was assisted by another Soviet colleague, Michail Kostenko, but the cost was large. Rutherford continued to support Kapitsa and a grant of £8000 was provided by the DSIR with the involvement of the Metropolitan-Vickers Company. The development of the new system was not without its hair-raising moments. As Kapitsa wrote to Rutherford on 12 December 1925,

> I am writing to you in Cairo to tell you that we have already obtained fields of more that 270,000 G ... We couldn't go further because the coil burst with a deafening bang which I am sure would have given you much pleasure.... The accident was the most interesting part of the experiment and finally strengthens our belief in success, since we now know exactly what happens when the coil bursts. We also now know what an arc of 13,000 A looks like. (Shoenberg, 1985)

After numerous further technical difficulties, very strong impulsive fields were obtained in 1927 (Kapitsa, 1927). In this development, Cockcroft was responsible for the detailed design of the coils, which had to be able to withstand currents of up to 30 kA. With his new powerful magnet, Kapitsa made pioneering studies of the magnetoresistance of metals in strong fields, and of magnetostriction in bismuth, the first observation of such an effect in a diamagnetic material. This development in technique led in due course to Shoenberg's work on the de Haas–van Alphen effect.

Kapitsa's career continued to advance with his appointment as Assistant Director of Magnetic Research in the Laboratory in 1925 and a Research Fellowship at Trinity College in the same year. In 1929, he was almost simultaneously elected to the Royal Society of London and as a Corresponding Member of the Soviet Academy of Science, a very rare and remarkable combination of honours.

In 1930, Kapitsa persuaded Rutherford to seek funds for a laboratory dedicated to housing his high magnetic field equipment and cryogenic facilities, with which to extend his researches to very low temperatures. Again, Rutherford was enthusiastically supportive and obtained £15,000 from the Royal Society's Mond Fund. In the same year, Kapitsa was appointed to a Royal Society Messel Professorship. Cockcroft, with his organising skills, was deeply involved in the design of the new Mond Laboratory, which was built on the courtyard site next to the old Cavendish buildings (see Figure 5.3). It was opened by the Chancellor of the University, Stanley Baldwin, in February 1933. Particular features of the building, organised in secret by Kapitsa, were the engraving of a full-size crocodile on the outside wall beside the entrance of the new laboratory and a bust of Rutherford himself, both by the distinguished sculptor Eric Gill (Figure 10.1).

The design of the building involved a long 'magnetic hall', the reason for which was that, despite elaborate damping in its mounting, the high magnetic field dynamo caused a large mechanical disturbance when it was discharged and had to be separated from the delicate recording equipment by a distance of 20 m so that the seismic wave caused by the

Fig. 10.1   The entrance to the Mond Laboratory with the sculpture of the crocodile by Eric Gill.

huge electrical discharge arrived only after the impulsive magnetic field, lasting less than 0.01 s, had died away. To minimise the impact of the shock waves on the instruments, the building was constructed in two halves. The result was a long laboratory with rooms on either side, which later became a natural meeting place for the exchange of ideas.

Kapitsa had also concluded that it was vital to obtain access to much lower temperatures as well as to much larger magnetic fields. James Dewar had in 1898 succeeded in liquefying hydrogen, with a boiling point of 20 K, but in 1908 was beaten to the draw in liquefying helium by Kamerlingh Onnes in Leiden. Helium has a boiling point of 4.2 K at atmospheric pressure, and may be cooled down to about 1 K by pumping on the liquid; and, as we shall see in the next section, the Dutch group had been discovering a remarkable series of new phenomena at these very low temperatures. Before the Mond Laboratory was completed Kapitsa, in collaboration with Cockcroft, had succeeded in developing a low-temperature facility.

Their first endeavour was the construction of a hydrogen liquifier. Extremely pure hydrogen was needed within the liquefier itself to avoid impurities condensing and blocking up the narrow-bore regenerator tubes. They adopted an ingenious approach in which only a small volume of pure hydrogen was used in a closed cycle, which cooled commercially available hydrogen in a heat exchanger (Kapitsa and Cockcroft, 1932). In characteristic Kapitsa style, the roof of the liquefier room was made very light so that it would easily blow off in the event of an explosion – Kapitsa enjoyed explaining this precaution to visitors, although the potential explosion never actually occurred.

The next challenge was the construction of a helium liquifier for which, typically, Kapitsa adopted a method of cooling quite different from the commonly used Joule–Thomson effect. Instead, he perfected an adiabatic expansion technique in which work is done against a piston, using the helium gas itself as lubricant (Figure 10.2). The machine was completed in 1934 and served the Mond Laboratory over the next ten years as the prime source of liquid helium; for a few years Leiden, Toronto and Cambridge were the only laboratories in the world having access to liquid helium temperatures. Kapitsa's

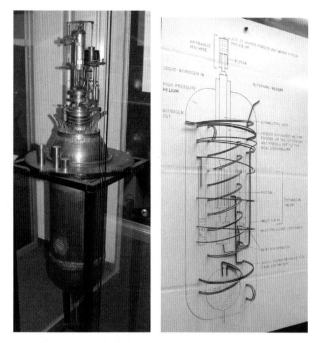

Kapitsa's helium liquifier in the Cavendish Collection, with the accompanying three-dimensional graphic illustrating its mode of operation.

Fig. 10.2

liquefier was commercialised by Samuel C. Collins at the Massachusetts Institute of Technology in 1947, and these machines made liquid helium available to the wider scientific community.

Kapitsa returned to the Soviet Union each summer but, on his visit in 1934, he was detained and not allowed to return to Cambridge, much to his and his colleagues' distress.[5] Although the full story is complex, the period coincided with the beginning of the great purges under Stalin's rule in the Soviet Union. Partly, it must have been associated with the need for the Soviet Union to enhance its generation of electricity. It may well also have been a response to the defection to the West of George Gamow and the desire of the authorities to make a success of the early five-year plans, which also involved a significant expansion of the science budget. Kapitsa was not to return to the West until 1965. The Soviet authorities tempted him to accept his situation by constructing an institute for him, the Institute for the Study of Physical Problems, on an attractive site in the Lenin Hills.

There remained the issue of his Directorship of the Mond Laboratory and the equipment that he had constructed in Cambridge. Eventually an agreement was reached according to which

> [In] return for payment of £30,000, the University would transfer to Moscow the high-field equipment and duplicates of such items (e.g. the liquifiers) as were needed in Cambridge to continue the Mond's activities. (Shoenberg, 1985)

Kapitsa resigned from the Directorship of the Mond Laboratory in 1935 and this role was taken over by Cockcroft. By the end of 1936, Kapitsa was back in business in Moscow in his new Institute, an enlarged replica of the Mond Laboratory, complete with identical Metropolitan-Vickers heavy current switch gear mounted on marble slabs.

10.4 Superconductivity and superfluidity: Kapitsa, Allen, Misener and Jones

10.4.1 A brief history of superconductivity and superfluidity

Until the construction of Kapitsa's helium liquifier, low-temperature physics had played no role in the research programme of the Laboratory. Low-temperature studies had, however, been making a major contribution in persuading the community of physicists to take seriously the concepts of quantum theory. Walther Nernst's measurements of the low-temperature behaviour of the specific heat capacities of solids carried out in Göttingen and Berlin had provided support for Einstein's quantum theory of heat capacities (Einstein, 1906), which was successfully extended by Debye (1912).

In Leiden, Kamerlingh Onnes had succeeded in liquefying helium for the first time in 1908, and very soon his group made two dramatic discoveries. The first was that the electrical resistance of mercury, instead of falling steadily on cooling as expected, vanished abruptly and apparently completely at what became known as the *critical temperature* of 4.15 K, the phenomenon of *superconductivity* (Kamerlingh Onnes, 1911a,b). The same phenomenon was subsequently discovered to occur at various critical temperatures in many other metals, although not in all. Much later it was discovered that the viscosity of liquid helium fell suddenly to zero when it was cooled below 2.17 K, the analogous phenomenon of *superfluidity* (Section 10.4.2).

At first it was thought that superconductivity must be due to a sudden and unexplained removal of electron scattering, the process responsible for the resistivity of the metal and by which it approached equilibrium. But it was subsequently realised that the equilibrium states themselves had also changed in nature. By 1928 it had been established that liquid helium undergoes a higher-order phase transition at 2.17 K at which its specific heat capacity diverges and its density shows an anomaly. Helium I is the higher-temperature phase, and helium II the lower-temperature phase. Because of the shape of the relation between specific heat capacity and temperature, 2.17 K is referred to as the *lambda point* of helium.

Then, in 1933, it was discovered that when a metal becomes superconducting, the magnetic flux is expelled from the interior (Meissner and Ochsenfeld, 1933), a phenomenon quite different from the behaviour of a classical material of infinite conductivity, in which the magnetic flux is frozen into the conductor. This *Meissner effect* indicated that a decrease in the equilibrium free energy must have occurred to provide the energy needed to force the magnetic flux out of the conductor. When it was realised that superconductivity, like superfluidity, involved a thermodynamic phase transition, the Leiden group were encouraged to apply classical thermodynamics for the first time to superconductors. By the mid 1930s the free energies of both the superconducting and the superfluid states had been measured accurately.

It was not at all clear what was going on. The first suggestion was to adopt a *two-fluid model* in which, below the transition temperature, the material was supposed to contain two interpenetrating fluids: the normal fluid, which suffered scattering in the usual way, and the superfluid, which was not scattered and had other remarkable properties. This idea was first put forward for superconductors in 1934 by Cornelius Gorter and Hendrik Casimir (Gorter and Casimir, 1934). They fitted their model to the measured thermodynamic parameters, and concluded that the proportion of superfluid electrons must vary with temperature approximately as $1 - (T/T_c)^4$. A more sophisticated model was developed for helium in 1938 by the Hungarian physicist Laszlo Tisza, then working at the Collège de France in Paris, who suggested that the superfluid carried zero entropy, and used this model to explain the remarkable discovery by Willem Keesom and his daughter Anna Petronella Keesom (Keesom and Keesom, 1936) that helium II had infinite thermal conductivity. This model was further developed by Landau, who analysed the quantum mechanics of a liquid of Bose particles and showed that the low-lying excited states could be described in terms of elementary excitations which carry a definite energy and momentum, corresponding to the normal fluid (Landau, 1941).

In 1935 Fritz and Heinz London (London and London, 1935a,b) proposed that the electrodynamics of superconducting electrons, or superelectrons, should be described by the following pair of equations:

$$\frac{\partial(\Lambda \boldsymbol{J}_s)}{\partial t} = \boldsymbol{E}, \tag{10.2}$$

$$\text{curl}(\Lambda \boldsymbol{J}_s) = -\boldsymbol{B}, \tag{10.3}$$

where $\Lambda = m/n_s e^2$ is the *London parameter* and n_s is the density of superelectrons. The first equation shows how the superelectrons accelerate under the influence of an electric field, while the second in effect explains the Meissner effect: it is straightforward to deduce from (10.3) that the magnetic field $\boldsymbol{B}$ can only penetrate into a superconductor by a small *penetration depth* $\lambda = (m/\mu_0 n_s e^2)^{1/2}$, of order 10^{-6} m in most superconductors.

These two equations proved very successful, and the London brothers pointed out that they could be understood if the electronic many-body wave function possessed what they called 'long-range momentum order'. One way of expressing this is to say that the superfluid behaves as though it has the momentum associated with an effective *superfluid wave function*, which has local amplitude and phase like a single-particle wave function, but is nevertheless guiding the behaviour of many electrons. It was a mystery how this could come about for the wave function of a strongly interacting many body system, especially for fermions, which cannot accommodate more than one particle in a given one-particle state.

In 1938 Fritz London proposed that a similar phenomenon might be occurring in helium II (London, 1938). He suggested that the properties of the superfluid might be associated with the theoretical *Bose–Einstein condensation* of an assembly of non-interacting bosons, in which it was well known that, as the temperature falls, an increasing fraction of the particles become condensed into the one-particle ground state. If this were the case for the superfluid, it would explain why it had zero entropy. But again the picture was of dubious value, because helium atoms are not independent, non-interacting particles. Nonetheless, London's guess that helium II possesses a simple effective superfluid wave function proved to be prophetic.

10.4.2 Low-temperature physics in the Mond Laboratory

Kapitsa had taken an active interest in all these developments. Following his detention in Moscow, Rutherford invited Jack Allen to take charge of low-temperature research in the Mond Laboratory in August 1935. Allen had completed his PhD in superconductivity two years earlier under an old friend of Rutherford's at Toronto, John C. McLennan, who had built up a major group in low-temperature physics. Besides Kamerlingh Onnes's group in Leiden, Toronto was the only physics department which could produce liquid helium until about 1933; already Allen was a seasoned researcher in low-temperature physics and superconductivity. Although Cockcroft was formally Head of the Mond Laboratory, in practice Allen ran the research activities while Cockcroft was deeply involved in numerous large projects – arranging for the transfer of equipment to Kapitsa in Moscow, the building of the Austin Wing, designing the next generation of particle accelerators and being Junior Bursar of St John's College.

The following year Allen was joined by Don Misener as a graduate student, also from Toronto, supported by an 1851 Scholarship. Misener had already participated in an impressive study of the shear viscosity of liquid helium just below the transition temperature using the decay time of torsional oscillations of a rotating cylinder immersed in liquid helium II. The viscosity was found to decrease abruptly below the transition threshold from helium I to helium II. This was evidence that helium II behaved quite differently from a classical fluid (Wilhelm *et al.*, 1935).

In 1937 Allen and his colleagues in the Mond Laboratory made a pioneering investigation of the thermal conductivity of superfluid helium in thin glass capillaries. The rate of heat transfer was found to be very large indeed and quite anomalous – it was not proportional to the applied temperature gradient (Allen *et al.*, 1937). In 1930 Keesom and van den Ende had found the first evidence that helium II had an extremely low entropy (Keesom and van den Ende, 1930). In late 1937, Allen and Misener went on to measure the viscosity of helium II in thin glass capillaries and discovered that it had vanished completely, thus establishing on a firm experimental basis the phenomenon of superfluidity (Allen and Misener, 1938).

Meanwhile, Kapitsa was pursuing a similar line of research in Moscow, stimulated by the papers of Keesom and Keesom (1936) and of Allen, Peierls and Uddin (1937). His original hypothesis was that the heat transport might be associated with convection, in which case the viscosity had to be extremely low. To measure the effect, the liquid flow had to take place along very fine channels, or between two optically flat surfaces, separated by only 1 µm or less. The remarkable result of this experiment was that the viscosity had to be many orders of magnitude less than expected; in fact only an upper limit could be obtained

(a) (b)

The fountain effect is a manifestation of the two-fluid character of liquid helium II, meaning helium below its superfluid transition temperature of 2.19 K. Allen discovered the effect accidentally in 1938 when he shone a pocket torch on his experimental apparatus. A chamber filled with He II was heated by radiation from the torch and a fountain of helium was observed to be forced out of the glass tube.

The fountain effect demonstrates that He II behaves as if it were a mixture of two components, superfluid and non-superfluid He. In Allen and Jones' apparatus a chamber was constructed which was connected to a reservoir of He II by a porous plug through which superfluid helium can leak easily, but through which non-superfluid helium could not pass (a) (Allen and Jones, 1938). If the interior of the chamber is heated, the superfluid helium changes to non-superfluid helium. In order to maintain the equilibrium fraction of superfluid helium, the superfluid component leaks through the porous plug and increases the pressure. This causes a jet of superfluid to be ejected as a fountain out of the top of the tube (b) (Balibar, 2010). (Image taken by J.F. Allen in 1972, courtesy of the University of St Andrews.)

for the coefficient of viscosity, but consistent with Allen's results (Kapitsa, 1938). It was Kapitsa who introduced the term *superfluidity*.

In the following month, February 1938, Allen and Jones published their observation of the *fountain effect*, a thermo-mechanical phenomenon (Allen and Jones, 1938). When superfluid helium is heated on one side of a porous medium or thin capillary, the pressure increases sufficiently to produce a little fountain at the end of the tube (Box 10.2). This was the first experimental demonstration of a two-fluid phenomenon.

All of these effects depended on the fact that the superfluid carries no entropy. As Leggett put it,

> [T]he apparent infinite thermal conductivity is simply due to convective counterflow of the two components. The fact that the superfluid carries zero entropy offers an immediate explanation of the fountain and mechanocaloric effects, since in those experiments it is only the superfluid which flows through the superleak. (Leggett, 1995)

Kapitsa continued to carry out ingenious experiments in which he demonstrated that the heat flow was accompanied by the mechanical flow of fluid. In due course he was awarded the 1978 Nobel Prize in Physics 'for his basic inventions and discoveries in the area of low-temperature physics'.[6]

The Moscow connection was continued thanks to the efforts of David Shoenberg, who had been left to his own devices midway through his PhD as a result of Kapitsa's enforced departure from the Laboratory, and was responsible for two important pre-war contributions.

The first concerned the magnetic penetration depth of superconducting materials. Although this was very small, Shoenberg realised that a measurable magnetisation could be observed if the specimens themselves were of the order of the penetration depth λ in size. He used a compact magnetometer that allowed magnetisation measurements to be made within the confined space of a narrow Dewar flask. Using pharmaceutical techniques, he created a suitably finely divided mixture of mercury and chalk spheres. Then, as the temperature was decreased below the critical temperature of 4.1 K, at which mercury becomes superconducting, the decrease in λ resulted in a steady increase in the magnetic moment (Shoenberg, 1940). According to the London brothers the penetration depth λ should be $(m/\mu_0 n_s e^2)^{1/2}$, as noted above. Combining this with the temperature dependence of n_s obtained from the Gorter–Casimir two-fluid theory, λ should vary as $[1 - (T/T_c)^4]^{-1/2}$; Shoenberg's results fitted this prediction very well.

Shoenberg's other contribution was not related to superfluids but concerned the de Haas–van Alphen effect, a quantum mechanical phenomenon in which the magnetic moment of a pure metal crystal oscillates as the intensity of the applied magnetic field is increased (see Section 12.9.1). The effect had been predicted by Landau (1930) and was discovered in the same year by Wander Johannes de Haas and his student Pieter Marinus van Alphen (de Haas and van Alphen, 1931).

Shoenberg spent a year at Kapitsa's institute in Moscow from 1937 to 1938, during which he exploited the method advocated by Kariamanikkam Krishnan of making magnetic measurements with a torsion balance. He applied this method successfully to the measurement of the de Haas–van Alphen effect in bismuth. These data provided vastly superior precision measurements of the effect, and showed that the electrons obeyed Fermi–Dirac statistics with a tensorial effective mass, meaning that the effective mass attributed to the electron depends upon its orientation relative to the magnetic field direction (Shoenberg, 1939). After the war, he was able to develop even more precise methods of measuring the effect, which eventually provided the best means of determining the detailed shape of Fermi surfaces in metals (see Sections 12.9.1 and 15.1.1).

10.5 Geoffrey Taylor: continuum and fluid mechanics

Geoffrey (G.I.) Taylor has already appeared in this story as the undergraduate who carried out the remarkable experiment in which he showed that the diffraction pattern of a thin needle is observed, even if the intensity of light is so low that the photons only arrive one at a time (Section 7.5.1) (Taylor, 1909). At the same time, he developed an interest in shock waves and won the Smith's Prize on this topic; it was later published in the *Proceedings of the Royal Society* (Taylor, 1910). In 1910 he was elected to a Prize Research Fellowship at Trinity College, providing him with six years of support during which he had considerable freedom to develop his entirely individual approach to research. His particular genius was for simple but profound experiments in what may broadly be termed the continuum mechanics of fluids and solids, which could be confronted with theory. While the focus of most research in the Laboratory was devoted to the burgeoning areas of quanta and the physics of atoms and nuclei, he continued elucidating problems of classical physics, much in the manner of Rayleigh before him. Just like Rayleigh, the number of phenomena associated with his name is impressive:

- Taylor–Couette instability
- Taylor–Proudman theorem
- Taylor–Proudman column
- Taylor dislocations
- Taylor vorticity transfer theorem
- Taylor–Green problem
- Taylor microscale
- Taylor frozen-flow hypothesis
- Rayleigh–Taylor instability
- Taylor dispersion
- Saffman–Taylor fingering.

His brilliant researches are much more rarely cited in histories of the Cavendish Laboratory than they deserve to be, despite the fact that he was a member of the Laboratory for the whole of his research career. He had his office next to Rutherford's, where he and his research assistant worked on a huge range of some of the most challenging problems in experimental continuum and fluid mechanics. By his own admission, his research endeavours were more determined by 'external circumstances' than as part of a well-defined long-term programme of research.[7] It also has to be said that his interests were strongly influenced by his love of sailing and flying as well as his unquenchable sense of adventure.

An immediate example of his opportunistic approach was his appointment as Schuster Reader in Dynamical Meteorology. This three-year position had been created through the generosity of Arthur Schuster, who was independently wealthy, and it could be held at any British university. The first appointee, Ernest Gold, had been considered a success and so Schuster agreed to continue the funding for a further three years. Despite the fact that

Taylor had very little knowledge of the subject at that time, he was appointed to the readership in 1912. Very quickly, he devised experiments to measure the small-scale processes involved in the mixing and vertical transport of momentum and heat in the lowest layers of the atmosphere which resulted from turbulent velocity fluctuations. Using a simple double-jointed wind vane, he established that, close to the ground, the horizontal fluctuating velocity components were much less than the vertical components, but above 10 feet, the turbulent velocities were isotropic (Taylor, 1917b). This research sowed the seeds for his major contributions to the study of turbulence during the inter-war years.

In 1913, he was invited to take the post of meteorologist on board the ship *Scotia*, which was to carry out measurements to assist in the prediction of icebergs in the North Atlantic, which had been the cause of the *Titanic* disaster of 1912. His task was to undertake all types of meteorological observations throughout the trip, including the use of kites and meteorological balloons to measure the properties of the atmosphere as a function of height. Launching the kites successfully was a challenge which he overcame by deploying them from a position at the top of the mizzen mast. He was able to make kite observations up to an altitude of 6000 feet. The scientific aspects of this work involved understanding heat transport from the surface of the sea, and in this work he introduced the key concept of *eddy viscosity* associated with the turbulent flow. Specifically, the heat transfer could be described by a diffusion equation in which the diffusivity was given by $\langle|w|\rangle l$, where $\langle|w|\rangle$ is the root mean square magnitude of the fluctuations in the vertical direction and l is the average vertical distance travelled by an element of the fluid before it acquires a different temperature by mixing with the surrounding fluid (Taylor, 1915). l is nowadays referred to as the *mixing length* and is a key quantity in the description of convective and turbulent processes.

With the outbreak of the First World War, Taylor offered his services to the Army as a meteorologist, but was engaged in an even more appropriate capacity as a research scientist at the Royal Aircraft Factory at Farnborough. This role involved applying his expertise in meteorology, fluid and gas dynamics to various issues in aircraft design. Characteristically, he trained as a pilot and learned the arts of aeronautics at first hand. He pioneered the use of wind tunnels in the understanding of the pressure distribution across airfoils. At that time, there was little confidence in the utility of wind-tunnel measurements in the design of aircraft. To compare them with what actually happened in flight, he arranged for the workshops at Farnborough to place 20 holes across the wing of a BE2C aircraft and connected each hole to a glass manometer inside the cockpit. Once the aircraft was in steady flight, the manometers were photographed over the full range of speeds at which the aircraft could fly. Initially there was poor agreement between the results in flight and the wind-tunnel experiments, but once account was taken of corrections for the effects of the wind-tunnel walls, excellent agreement was found. These were the first full-scale pressure distribution measurements to have been recorded, and opened up the techniques of wind-tunnel scale modelling for aeronautics (Taylor, 1916).

After the war, Taylor returned to Cambridge and in 1919 was appointed to a fellowship and college lectureship in mathematics at Trinity College. This coincided with Rutherford's appointment as Cavendish Professor and they quickly became very good friends. The convivial spirit is epitomised by the golfing foursome of Fowler, Aston, Rutherford and Taylor who played every Sunday morning from 1919 until Rutherford's death in 1937.[8] As already mentioned, Rutherford allocated bench space in the Laboratory next to his own for Taylor's experiments in fluid dynamics.

The period of Rutherford's tenure of the Cavendish Chair coincided with a period of extraordinary creativity on Taylor's part. The immediate post-war years resulted in three major advances.

- *Tidal friction.* In his earlier studies of the frictional force at the surface of the Earth exerted by the turbulent boundary layer, he found that the force per unit area was roughly $0.002\rho V^2$, where V is the mean wind speed near the surface and ρ the density of air. It turned out that a similar formula applied to the frictional force due to turbulent flow through a pipe with rough walls. This formula enabled the frictional tidal force between the sea and the ocean floor to be evaluated (Taylor, 1919). His friend and colleague Harold Jeffreys used these results to show that this force is the cause of the gradual increase in the length of the lunar day (Taylor, 1920). The numerical coefficient 0.002 has still not been explained theoretically.
- *The Taylor column.* Taylor's career was characterised by the carrying out of experiments to test the predictions of theoretical hydrodynamics, a classic example being his experiments on steady flow past a solid sphere or cylinder in a rotating fluid system. The theoretical result had been published slightly earlier by Joseph Proudman, whose theorem stated that if a sphere is moved slowly in any direction through a rotating body of liquid, the sphere is accompanied by the fluid in the circumscribing circular cylinder (Proudman, 1916; Taylor, 1917a). This result is due to the fact that, in rotating fluids in which the Coriolis forces in the fluids are dominant, the flow is necessarily two-dimensional. Taylor demonstrated the effect experimentally, and the cylindrical rotating column is known as a *Taylor–Proudman column* (Taylor, 1923a). Jupiter's red spot has been postulated as an example of a naturally occurring Taylor–Proudman column.
- *Stability of flow between concentric rotating cylinders.* Another brilliant example of Taylor's ability to conceive and carry out experiments which give insight into hydrodynamic flows was his combined theoretical and experimental investigation of the stability of steady flow between two rotating cylinders. The external cylinder was rotated at a fixed speed and the internal cylinder speeded up until the point was reached at which the flow transformed into a sequence of equally spaced rotating tori. Taylor made the changing flow patterns visible by smearing the inner cylinder with dye and making the outer cylinder transparent. The evaluation of the critical speed and wavelength involved lengthy calculations but these agreed essentially perfectly with the experimental results (Figure 10.3) (Taylor, 1923b).

In 1923, Taylor was appointed Royal Society Research Professor, thanks to the benefaction of Sir Alfred Yarrow. This relieved him of all teaching, and he was to hold the post until his retirement in 1951. His research in the period up to the Second World War was dominated by two major themes, the mechanism of plasticity in solid materials and turbulence.

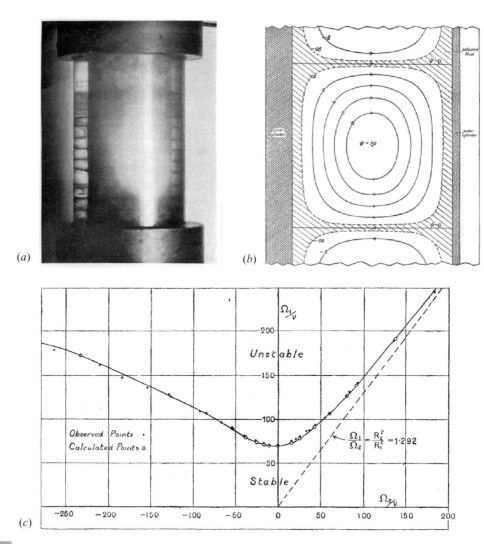

Fig. 10.3 Taylor's experimental demonstration of the instability of flow between concentric rotating cylinders (Taylor, 1923b). (a) Photograph of the formation of vortices in the gap between two rotating cylinders having angular velocities Ω_1 and Ω_2. (b) The predicted streamlines for toroidal motion at a speed just slightly greater than the critical speed. The cylinders are rotating in the same sense. (c) Comparison between the observed and calculated critical speeds for instability in a plot of Ω_1/ν and Ω_2/ν, where ν is the kinematic viscosity. The dashed line shows the stability criterion for an inviscid fluid.

10.5.1 Plasticity of crystalline materials

Taylor's interest in the plasticity of solid materials dated from his studies of the failure of aircraft propeller shafts while working at the Royal Aircraft Factory at Farnborough in 1914. His interest was reawakened in 1922 by the work of Harold Carpenter and Constance Elam (Tipper), who had plastically deformed large crystals of aluminium

(Carpenter and Elam, 1921). In collaboration with Elam, Taylor carried out experiments on the modes of plastic deformation of different metal crystals, frequently finding that the stress–strain relation was of parabolic form, $S \propto s^{1/2}$, where S is the stress and s the resulting strain.

In 1934, *dislocations* were postulated by Taylor, and independently by Orowan and by Polanyi, to explain the plastic behaviour of metals under stress and also the phenomenon of *work hardening*, the strengthening of a metal by plastic deformation (Orowan, 1934; Polanyi, 1934; Taylor, 1934). Dislocations are crystallographic defects, or irregularities, within a crystal structure which explain the remarkable weakness of the structure compared with the theoretical yield strength of a perfect crystal. In a perfect crystal, all the bonds along the slip plane have to break simultaneously. With a dislocation present, only the bonds in its vicinity have to break as the dislocation moves along the slip plane. Furthermore, interactions of dislocations with each other and with other obstacles within the crystal can explain phenomena such as work hardening and alloy strengthening. Taylor considered that work hardening was due to a large number of dislocations becoming tangled up, so that they were unable to move and distortion of the crystal was inhibited.

Taylor presented a quantitative model for the deformation process which could be compared with experiment. As summarised by Mott (1996), Taylor's insights were:

- The strength of a cold worked material is due to the strains around dislocations.
- The flow stress depends on the dislocation density and their arrangement.
- A stress–strain curve could be derived if a dislocation slip distance L could be defined.

Taylor's achievement was to show how these insights lead to a quantitative theory of plasticity in metal crystals, in particular, in accounting for the parabolic stress–strain relation (Box 10.3). Taylor did not return to this area of research after 1937, but it was very influential in the study of the strength and deformation of plastic materials. It was not only the theory, but also his careful thermometric and calorimetric experiments to measure the internal stress energies of the materials, which were of lasting significance. Specifically, he found that $L \sim 10^{-4}$ cm and that the predicted energy associated with the plastic deformation of the material agreed with his calorimetric measurements. These ideas were to influence strongly the thinking of the Metal Physics Group after the war. Mott recalled vividly the day two decades later when his students excitedly brought to him transmission electron microscope images of dislocations in motion obtained by Peter Hirsch and his colleagues (Mott, 1996) (see Section 15.2.1).

10.5.2 Turbulence

Taylor's major contributions during the inter-war years were undoubtedly in the area of the description and understanding of turbulence, the greatest unsolved problem of non-linear hydrodynamics. His important early contributions informed his major advances in the 1930s. As was his practice, he did not carry out the theoretical development of turbulent motions until he had a means of making measurements of the velocity field. This came about through the development of hot-wire anemometry, in which a very fine wire of the order of several microns diameter is electrically heated above the ambient temperature.

Box 10.3　　　　　　　　　　　**Taylor's theory of plasticity**

Plasticity means the permanent deformation of a material, in this case a crystal lattice, when the stress exceeds the elastic limit below which the stress is proportional to the strain.

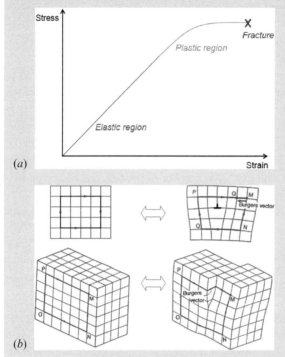

(a)

(b)

(a) The stress–strain relation changes to follow a roughly parabolic relation. The distortions can be described by a *Burgers vector*, named after Dutch physicist Jan Burgers, and is the vector **b** representing the magnitude and direction of the lattice distortion between one side of the dislocation and the other. (b) A dislocation line and its Burgers vector must both lie in the slip plane but can make an arbitrary angle with one another. When they are at right angles to each other we obtain an edge dislocation (top) and when they are parallel to one another we find a screw dislocation (bottom).

According to Mott's order-of-magnitude calculation (see Mott, 1996), Taylor assumed that each dislocation could slip a distance L before being stopped. He adopted a two-dimensional model in which there were n dislocations per unit area. Then the strain produced was $s = nLb$, where b is the magnitude of the Burgers vector. Timpe (1905) and Volterra (1907) had worked out the strain associated with a single dislocation and shown that it falls off inversely with distance. The average distance between dislocations in the two-dimensional model is $n^{-1/2}$ and so the stress needed to move dislocations of opposite sign past each other is of order $Gbn^{1/2}/2\pi$, where G is the shear modulus. Taylor identified this shear stress with the flow stress S and so $S \propto G(bs/L)^{1/2}$, accounting for the parabolic relation between S and s.

Air flow cools the wire and, since the electrical resistance of materials such as tungsten is temperature-dependent, the air flow speed can be found. These anemometers proved to be very sensitive and enabled the rapidly fluctuating velocity field to be measured. As Taylor wrote to Ludwig Prandtl in 1932,

> The same kind of analysis can be applied to hot-wire measurements, and I am hoping to begin some work along these lines. In particular the 'spectrum' of turbulence has not received much attention. By means of a hot wire and oscillograph one can (theoretically) obtain the frequency of occurrence of all turbulent velocities and can construct a frequency diagram. We have made a start on that and find it resembles the pure error curve, but more work must be done before anything definite can be stated. (Batchelor, 1996)

Taylor was well aware that the turbulence was only three-dimensional and isotropic far from the walls of the apparatus, and appreciated that the natural way of describing the fluctuations was through the use of statistics. Now, however, he had to deal with random continuous functions of velocity and pressure. In a sequence of four papers of 1935, he laid out the basic elements of the *statistical theory of turbulence* (Taylor, 1935a,b,c,d). These were revolutionary papers in that the statistical techniques had not been applied to random velocity and pressure distributions.

To quantify the statistical properties, Taylor adopted a Fourier approach and this led naturally to the introduction of power spectra to describe the power per unit wavelength at different wavelengths. Specifically, in his paper of 1938 entitled 'The spectrum of turbulence', he showed that the Fourier transform of the two-point velocity correlation function for the random velocities in the fluid is the power spectrum or spectral energy density function in the fluid (Taylor, 1938). The great advance was that measurements of the velocity field in the fluid could be described quantitatively in terms of the energy density in random motions on different physical scales and so could be compared with the measured energy dissipation in the fluid, specifically by hot-wire anemometry. With the assistance of L.F.G. Simmons at the National Physical Laboratory, hot-wire anemometer measurements were used to evaluate the two-point correlation functions for flow through a grid of bars in an wind tunnel, as well as the frequency spectra of the velocity field at points in the fluid. These experiments confirmed the theoretical relations obtained by Taylor. As expected, the pattern of turbulence changed little as the flow proceeded downstream, leading to Taylor's *frozen-pattern hypothesis*. These studies were to be continued after the war by Alan Townsend experimentally and George Batchelor theoretically (Section 12.10.1).

10.6 The end of an era

Rutherford died unexpectedly in 1937 after a brief illness. His passing marked the end of an extraordinary era of discovery in experimental physics in the Laboratory. He bequeathed to his successor Lawrence Bragg a Laboratory in the process of change. Many of his most distinguished collaborators had left Cambridge for other posts and the Laboratory's role in nuclear physics had declined. Many of the new types of science described in this chapter were to become major themes of the post-war years. But the war was to overshadow everything and led to quite new approaches to physics research. Many other new areas of research began to make an impact during the latter years of the Rutherford era, but it is best to describe these in the context of the reorganisation of the Laboratory and its new directions under Lawrence Bragg.

PART VI

1938 to 1953

William Lawrence Bragg (1890–1971)
Oil and watercolour by Homi J. Bhabha

Bragg and the war years

11.1 Lawrence Bragg at Manchester and the National Physical Laboratory

The Vice-Chancellor of Manchester University, Sir Henry Miers, having been Professor of Crystallography, was keen to develop the discipline. In his words,

> In my opinion, the importance of the study of crystals has now become so great, not only for the identification of substances by crystal measurements but also on account of the new knowledge which modern crystal study is contributing to problems belonging to different sciences, that there is a real need for a department of pure crystallographic research, one in which such studies can be carried out quite independently of elementary teaching or of immediate applications, and without being tied to mineralogy. (Quoted by Phillips, 1979)

Lawrence Bragg, although only 29, was the obvious choice to succeed Rutherford as the Langworthy Professor of Physics at the University in 1919, but it was to prove a real challenge. He had never taught physics and now was in charge of a large department which had become a leader in the dynamic field of nuclear physics under Rutherford. Despite a rocky ride in getting on top of teaching and the management of the growing laboratory during his first few years, he now had his own research laboratory and concentrated upon making the discipline of crystallography an exact quantitative science. This was to be his major and distinguished contribution during his Manchester years. As expressed by David Phillips,

> Bragg's unique contribution here was to see the value of making experimental measurements of the absolute intensities of the X-ray reflexions which showed directly the effective number of electrons contributing to each reflexion. The work on rock-salt also stimulated work on the theoretical derivation of the atomic scattering factors that were needed to calculate the intensities of reflexions corresponding to any model structure for comparison with the observed values. (Phillips, 1979)

His research programme got off to an excellent start with his acquisition of a new X-ray spectrometer and the gift of a state-of-the-art Coolidge X-ray tube from the General Electric Company at Schenectady. From his continuing studies of crystalline structures, he developed the concept of the sizes of the common ions so that interatomic distances in atomic compounds could be determined by an additive law; these dimensions agreed quite well with the measured atomic distances (Bragg, 1921). In 1924, he turned again to aragonite ($CaCO_3$), which he had first studied at Cambridge before the war. The structural analysis involved the determination of nine variable parameters using absolute measurements of the X-ray intensities – the most complex structure yet attempted. It turned out to have an orthorhombic dipyramidal structure (Bragg, 1924b). In this work, he introduced into the analysis the symmetries of the crystals in terms of formal space group theory. Next, he turned briefly to the inverse problem of determining the physical properties of crystals, such as their refractive indices, from their atomic structures (Bragg, 1924a).

A key conceptual breakthrough came with the introduction of Fourier techniques for the determination of the electron density distribution in crystals. The problem facing the investigators was that, in order to reconstruct an image from the Fourier data, both the amplitudes and phases would be required, but the X-ray intensities provided measurements only of the amplitudes, but not the phases.[1] Because of this problem, a crystal structure was postulated and then the predicted positions and intensities of the X-ray spots found by Fourier inversion. By a process of trial and error, the structure of the crystal could be determined. In their pioneering studies of beryl ($Be_3Al_2Si_6O_{18}$), Bragg and J. West (1926) determined its structure by a series of one-dimensional Fourier inversions of the X-ray patterns in the crystal planes. These techniques led to the determination of crystal structures with 20 to 30 parameters, as described in one of Bragg's favourite papers (Bragg and West, 1928).

In the same year, Bragg and Warren carried out an important analysis of diopside ($CaMg(SiO_3)_2$), using crystal samples provided by Arthur Hutchinson, Professor of Mineralogy at Cambridge University, in which the crystals had been cut perpendicular to the principal axes (Bragg and Warren, 1928). Of particular significance was the fact that the tetrahedral SiO_3 units were joined at their corners to form long chains which ran through the whole crystal structure. A year later, Bragg returned to the structure of diopside and, with the two-dimensional data which had been obtained in these experiments combined with phases inferred from the model derived with Warren, he was able to determine the projected electron distribution by Fourier series in two dimensions on the three facets of the unit cell (Bragg, 1929). In those days, the Fourier inversions had to be done by hand and various methods were developed to accomplish this, including the use of Patterson functions and Beevers–Lipson strips.

Bragg encouraged his first Manchester graduate student, Albert Bradley, to investigate the use of powder methods for determining the structures of metals and alloys after he completed his PhD (Box 11.1). After a year learning these techniques with Arne Westgren and Gösta Phragmén in Stockholm, Bradley returned as an expert on powder diffraction techniques, solving complex structural problems. His most outstanding achievement was the determination of the structure of the alloy γ-brass (Cu_5Zn_8), which has 52 atoms in a cubic unit cell (Bradley and Thewlis, 1926). The technique for deriving the molecular structures of very complex molecules was at hand and was eventually to lead to the possibility of determining the structures of biological molecules. Among the graduate students who joined Bragg during this period was W.H. (Will) Taylor, who became a University lecturer in 1928 and who would eventually join Bragg in Cambridge as Head of the Crystallography Laboratory.

Bragg had been tempted to return to Cambridge in the early 1930s, feeling that he wanted to maintain the physics connections of his research work rather than becoming a pure crystallographer, but various discussions with Rutherford came to nothing. This

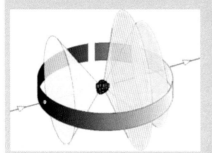

The powder camera greatly extended the range of substances which could be examined by X-ray diffraction, as this technique does not require a large single crystal. This was particularly useful in the analysis of metals and their alloys. A powder camera uses a sample consisting of several crystals (crystallites) oriented at random in all directions. The diffraction effect is therefore the same as rotating a single crystal in all directions, similarly producing a series of concentric cones around the sample. The film, which lines the circumference of the camera, intersects these cones, producing a series of arcs.

disappointment, and the many pressures on him as Head of Department, resulted in a breakdown in 1930, from which he recovered in 1931.

In 1937, the Directorship of the National Physical Laboratory (NPL) became vacant and Bragg was appointed to the post, taking his collaborators Bradley and Henry Lipson with him. He soon found that he had little taste for the work, which involved management under civil service administrative practices and little original science. Before he moved to the NPL, however, Rutherford died and this opened up the possibility of a move back to Cambridge from the NPL. The offer of the Cavendish Chair was made by the electors and Bragg had no hesitation in accepting the position. He was soon offered a Professorial Fellowship at Trinity College as well. Prophetically, a *Nature* editorial commented that:

> The Cavendish Laboratory is now so large that no one man can control it all closely and Bragg's tact and gift of leadership form the best possible assurance of the happy cooperation of its many groups of research workers. (Phillips, 1979)

Bragg fully acknowledged his debt to Manchester University in providing experience of running a large laboratory. In his final report, he wrote:

> Manchester University has taught me all I know about the running of a department, and the fascinating and intricate life of a modern university. I will always remember with gratitude the kindness of my colleagues and the inspiring atmosphere of this University. (Phillips, 1979)

11.2 Changing directions: Bragg and the immediate pre-war years

Bragg took up the Cavendish Professorship in October 1938 with the shadow of war looming. His tenure of the Cavendish Chair was to be a period of major change in the management, administration and scientific direction of the Laboratory. Rutherford's reluctance to raise resources for nuclear physics had delayed investment in the next generation

Box 11.2

The principles of the X-ray photogoniometer

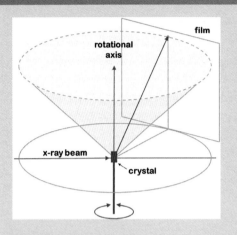

The X-ray photogoniometer is a powerful instrument for indexing the diffraction spots of complex crystals. The crystal is mounted in front of an X-ray beam and is rotated about a prominent crystal axis. As it rotates, the X-rays are diffracted from each layer of atoms within the crystal, resulting in diffracted beams in the shape of a series of cones. A flat photographic plate intercepts these cones, producing an image of the diffraction spots which lie along a series of hyperbolae. The repeat distance of atoms along the axis can be found by measuring the spacings between the hyperbolae. Sometimes a cylindrical camera is used with its axis coinciding with the rotation axis, so that the reflection spots lay on horizontal rows. These principles were built into Bernal's universal X-ray photogoniometer.

of accelerators and resulted in the departure of many of his most experienced colleagues in this field to other universities, where they were now well established (see Section 9.9). Leadership in nuclear physics was passing to other universities. Of the radioactivists, only Dee and Feather remained in Cambridge, although nuclear physics was still the dominant discipline in terms of numbers of staff and research workers. Despite his intention to concentrate upon physics rather than crystallography, Bragg could not escape the fact that his major achievements had been in the latter field. If the Rutherford legacy described in Chapter 9 was in decline, the topics reviewed in Chapter 10 were to be the growth areas of the future, and many of these were well matched to Bragg's major interests and enthusiasms.

11.2.1 Bernal and the growth of crystallography

A particular interest was the Crystallography Laboratory, which had been incorporated into the Cavendish Laboratory under the leadership of Desmond Bernal. Bernal had graduated in physics from Emmanuel College in 1923 and in his last year as a student had made a major study of what turned out to be the 230 space groups of crystallography, when he should have been concentrating on his studies for the Tripos examinations. For his endeavours he won the college Sudbury Hardyman Prize for his essay, 'On the analytic theory of point group systems'. This was the origin of his lifelong interest in crystallography. Arthur Hutchinson, then Lecturer in Mineralogy at Cambridge, recommended that he carry out research with William Bragg, who was about to take up the Directorship of the Davy Faraday Laboratory at the Royal Institution. The topic suggested by Bragg was the structure of graphite, which Bernal solved using the rotation method which had recently been developed by Michael Polanyi in Germany (Box 11.2). In collaboration with his colleagues

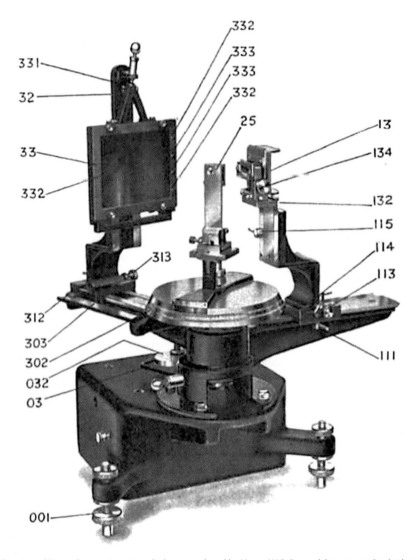

Bernal's universal X-ray photogoniometer, which was marketed by Messrs W.G. Pye and Company at Cambridge. The **Fig. 11.1** construction and operation of the device is described in a series of four papers in the *Journal of Scientific Instruments*. This image is taken from the fourth paper, which describes in detail the operation of the goniometer (Bernal, 1929).

Bernal determined that in graphite the atoms lie in planes which are defined by hexagons of carbon atoms (Bernal, 1924). He was also a pioneer in creating charts in which it was possible to read directly the coordinates of the reciprocal lattices for the reflections which appear on X-ray photographs (Bernal, 1926). The next step was the design and manufacture of the X-ray equipment which would eliminate the need for every laboratory to built its own apparatus. The result was the universal X-ray photogoniometer, which was marketed by Messrs W.G. Pye and Company at Cambridge (Figure 11.1).

In 1926, Hutchinson was promoted to Professor of Mineralogy at Cambridge and in 1927 created the new post of Lecturer in Structural Crystallography, to which Bernal was appointed. At the same time, William (Peter) Wooster, a graduate of the Laboratory, was appointed Demonstrator in Crystal Physics, and a laboratory technician was employed. The Mineralogy Department was small, but was ideally located for someone of Bernal's broad and exuberant interests. It was next to the Cavendish Laboratory, the Mond Laboratory and the Philosophical Library, as well as being close to the Chemistry, Zoology and Anatomy departments. With Bernal's extraordinary energy and personality, the Crystallography Laboratory and the numbers of graduate students grew rapidly, despite Rutherford's antipathy to Bernal's approach and political views.[2] During Bernal's period in Cambridge, his graduate students included Helen Megaw, Dorothy Crowfoot (Hodgkin) and Max Perutz. On Hutchinson's retirement in 1931, Bernal was promoted to the position of Assistant Director of Research and the Crystallography Laboratory was transferred formally to the Cavendish Laboratory, although it retained its own identity. Of particular significance was Bernal's extension of his crystallographic studies to include not only metals and inorganic materials, but also organic and biological molecules in their crystalline state.

A major breakthrough took place when Bernal was given a crystal of the enzyme pepsin which had been grown by John Philpot in Theodor Svedberg's laboratory in Uppsala. Bernal found that if the crystal was kept wet in a sealed capillary, it displayed a huge number of X-ray reflections. He left the rest of the crystallographic analysis to Crowfoot, who found that the pepsin molecule was

> an oblate spheroid with diameters of ca. 35 and 25 Å arranged in hexagonal nets, and largely hydrated in the crystal. In spite of the molecule's size and complexity, the extent of the diffraction pattern showed that the arrangement of atoms inside the protein molecule is also of a perfectly definite kind. (Bernal and Crowfoot, 1934; Dodson, 2002)

This was the beginning of protein X-ray crystallography.[3]

Also during this period, Hodgkin began her work on insulin, the structure of which she only completed 35 years later. Perutz began his analysis of haemoglobin and in 1938 published the first X-ray photographs of haemoglobin crystals (Figure 11.2) (Bernal *et al.*, 1938), another very difficult problem of crystallographic analysis which was to take over 20 years to solve. Both were to be awarded Nobel Prizes in Chemistry, Perutz in 1962 and Hodgkin in 1964.

With Bragg's move to Cambridge, Blackett took up the Langworthy Chair at Manchester and Bernal replaced Blackett at Birkbeck College, London. Crowfoot moved to Oxford in 1934 to take up a fellowship at Somerville College and continued her ground-breaking crystallographic studies of insulin, penicillin and vitamin B_{12}. Thus, Bragg inherited in the Laboratory a distinguished Crystallographic Laboratory which was carrying out cutting-edge research, but which had lost a number of its leading lights. Following Bernal's departure, only Perutz remained in Cambridge, but from the National Physical Laboratory Bragg brought with him Bradley to head the Crystallography Laboratory and Lipson as his assistant. In addition, Paul Peter Ewald, the distinguished German crystallography, forced out of his post at Stuttgart University in 1937 by the Nazi administration, emigrated to

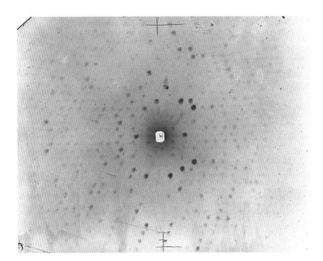

An X-ray crystallographic image of the diffraction spots of a wet haemoglobin crystal obtained by Bernal, Fankuchen and Perutz (Bernal *et al.*, 1938).

Fig. 11.2

England where he was given a research position in the Cavendish Laboratory for a couple of years until he was appointed to a lectureship at Queen's University, Belfast in 1939.

11.2.2 Bragg and the study of proteins and biomolecules

Despite these many changes and the threat of war, Bragg's first year in the chair was remarkably productive. Probably the most important development was his appreciation of the role which crystallography could play in understanding yet more complicated molecules, in particular proteins and biomolecules. The stimulus was provided by Perutz who, soon after Bragg's arrival in Cambridge, showed him his excellent X-ray diffraction images of haemoglobin (see Figure 11.2). As Perutz recalls in his memoirs,

> When I showed him my X-ray pictures of haemoglobin his face lit up. He realized at once the challenge of extending X-ray analysis to the giant molecules of the living cell. Within less than three months he obtained a grant from the Rockefeller Foundation and appointed me as his research assistant. Bragg's action saved my scientific career and enabled me to bring my parents to Britain. (Perutz, 1970)

This meeting with Perutz was to determine Bragg's principal research interests for the rest of his career. There immediately followed a paper on protein structure analysis (Bragg, 1939b). In the same year, he came up with the ingenious idea of using Young's fringes as a means of carrying out Fourier analysis, with the holes in the plate corresponding to the observed diffraction spots. Then he realised that the locations of the holes corresponded to the distribution of points in the reciprocal lattice of the crystal. The experiment was successfully carried out on a diopside crystal and demonstrated at the Royal Society Conversazione in 1939 (Bragg, 1939a). The stage was set for the remarkable developments in protein crystallography after the war.

11.2.3 Other pre-war movements

Geoffrey Taylor had carried out pioneering research into the role of dislocations in explaining shearing in metals, another subject close to Bragg's heart (Section 10.5.1). In 1939, Bragg attracted Egon Orowan, who had written a contemporaneous and independent paper on dislocations (Orowan, 1934), from Birmingham University to Cambridge, and his arrival significantly strengthened the Laboratory's activities in metal and condensed matter physics in general.

Appleton's departure to become Secretary of the Department of Scientific and Industrial Research left vacant the Jacksonian Chair of Natural Philosophy, which was filled by Cockcroft. Appleton's research programme into radio propagation in the ionosphere (Section 10.2) was continued by John (Jack) Ratcliffe, who not only was to provide essential support for the development of radio astronomy after the war, but was also a very able administrator who took much of this burden from Bragg's shoulders, particularly in understanding and coping with the complexities of Cambridge University's somewhat arcane administrative and management practices. As it turned out, Ratcliffe's expertise in all aspects of radio science was ideally matched to Bragg's expertise in optics, Fourier transforms and their application, tools which were to be essential in the post-war years.

Bragg fully appreciated the need for reform of the management structure of the Laboratory, but this had to be put on hold until after the war. In the meantime, a priority was completion of the construction of the Austin Wing (Section 8.5), which was supervised by Cockcroft. The understanding was that the Austin Wing would be used by the services in the event of war, which indeed broke out in September 1939. As recounted in the next section, many members of the Cavendish staff left to support the war effort as soon as war was declared, and the research carried out at the Laboratory during the war was largely related to support of these efforts.

11.3 The war years

It is often said that the First World War was the 'chemist's war', with the need for increasing quantities of explosives, poison gases, optical glass, synthetic dyes and pharmaceuticals. In contrast, the Second World War was undoubtedly the 'physicist's war'. Essentially the whole of physics was involved in meeting the challenge of the Axis Powers in Europe, who themselves were exploiting the skills of many of their best physicists and engineers to support their war efforts. This continuing game of cat-and-mouse is splendidly recounted in the case of scientific intelligence and deception in R.V. Jones' classic book, *Most Secret War* (Jones, 1978). There was a desperate need for trained physicists with expertise ranging from nuclear physics, explosives and materials science to radio transmitters and receivers, to mention only some of the most obvious examples.

Bragg's immediate response was to ensure that the Cavendish Laboratory made a full contribution in support of the war effort. The teaching of physics had to be adapted to

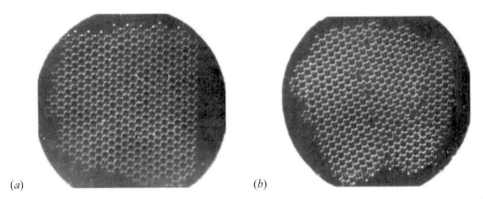

(*a*)   (*b*)

Illustrating the role of dislocations in understanding the plastic properties of materials. (*a*) A freely floating raft of bubbles, representing a perfect crystal structure. (*b*) 'A raft consisting of three "crystals". The inter-crystalline boundaries are marked by a chain of dislocations. The rows in each crystal are slightly warped, but the crystalline array extends right up to the boundary.' The raft shown in (*a*) is relatively rigid, whereas the second (*b*) can shear by movement of the dislocations (Bragg, 1942a).

Fig. 11.3

the needs of war. Queen Mary College and Bedford College were evacuated from London to Cambridge and the teaching efforts of these colleges were combined with those of the Laboratory. The teaching was concentrated into a two-year honours course including special courses on electronics which were needed to support work on radar, which was essential for defence against air attacks by the Luftwaffe and for guiding allied fighters and bombers to their targets. The redoubtable Searle was brought out of retirement to teach his renowned courses on practical physics.

Bragg also contributed to the national efforts through his membership and chairmanship of various wartime committees such as the Admiralty's committee on the development of underwater sonar detection of submarines. This followed on naturally from his researches into the sonic location of enemy gun emplacements during the First World War (Section 7.8). He made regular visits to the Admiralty research station at Fairlie on the River Clyde as an advisor on sonar location. He was also a member of the Metallurgy Committee set up by the Ministry of Supply, and Chairman of the General Physics Committee. From 1942, he was a member of the Advisory Council of the Department of Scientific and Industrial Research at the invitation of Appleton. He also served in Ottawa as liaison officer between Canada and the UK for six months in 1941, laying the groundwork for future collaboration in nuclear research relevant to the atomic weapons programme.

Bragg did not carry out research related to the war effort during these years, but, as a result of discussions with Orowan, he came up with the delightful invention of the bubble raft as a means of visualising the behaviour of dislocations in solids and their plastic deformation in two dimensions (Figure 11.3) (Bragg, 1942a). The model bears more than passing resemblance to Bragg's concept of assigning radii to atoms in crystal structures. This model would prove to be important after the war in promoting the theory of dislocations. It also illustrates Bragg's vivid pictorial imagination and his ability to explain concepts in physics in terms which could be understood by a lay audience.

11.4 The Tube Alloys and Manhattan projects

In 1939, the discovery of nuclear fission with the release of sufficient neutrons to initiate a chain reaction indicated that nuclear reactors and nuclear weapons could in principle be constructed (Section 9.8). The complex scientific and political history of the development of nuclear weapons through the Tube Alloys and Manhattan projects has been told extensively in, for example, the books by Clark (1961), Gowing (1964) and Rhodes (1986). In this brief account, we concentrate upon the activities during the war years of those who were past, present and future members of the Laboratory. Much of the UK expertise in experimental and theoretical nuclear physics was the province of Rutherford's former research associates, Blackett,[4] Chadwick, Oliphant and their students who had now moved to other universities, while Cockcroft, Dee and Norman Feather, who was appointed to a University Lectureship in 1936, continued the work at the Laboratory.

The French group led by Joliot had realised that a runaway nuclear reaction of ^{238}U would take place more effectively if the fast neutrons were slowed down by a suitable moderator, since the nuclear cross-section for the absorption of neutrons was much greater for slow than for fast neutrons. The preferred moderator was heavy water, D_2O, the sole source of which was at the Norsk Hydro hydroelectric station at Vemork in Norway. It was discovered that Germany had already made a bid for the entire stock of Norwegian heavy water, suggesting that their scientists might also be developing a nuclear weapon. Before that could happen, the French scientists were successful in obtaining the stock of heavy water from the Norwegians. Germany invaded France in May 1940 and so, at the invitation of the Churchill government, Hans von Halban and Lew Kowarski of the Paris group moved their research activity to the Cavendish Laboratory, bringing with them 188 litres of heavy water. They joined Nicholas Kemmer, Allan Nunn May,[5] Anthony French, Denys Wilkinson and Samuel Curran, who were working on nuclear physics projects relevant to the war effort.

Otto Frisch had left Germany in 1933 because of the Nazi oppression of those of Jewish origin to join Blackett's new group at Birkbeck College. While on a five-year fellowship at Copenhagen with Bohr, Frisch wrote his famous paper with Lise Meitner interpreting Hahn and Strassmann's nuclear physics experiments in terms of nuclear fission induced by neutrons (Meitner and Frisch, 1939) (Section 9.8). At the outbreak of the war, Frisch made a visit to Birmingham where Chadwick's physics group now included Rudolf Peierls. It was no longer advisable for Frisch to return to Copenhagen. Together, in 1940, Frisch and Peierls wrote their famous memorandum showing that the critical mass at which a nuclear chain reaction would occur in pure ^{235}U, the rarer isotope of uranium, was only about 10 kg, light enough to be delivered by a long-range bomber.

Oliphant informed Henry Tizard, Chairman of the Committee on the Scientific Survey of Air Defence of the British Army, about these calculations. As a result of the Frisch–Peierls memorandum, Tizard set up the MAUD Committee in 1940 to investigate the practical feasibility of a nuclear weapon. The MAUD Committee consisted of many of the Cavendish pioneers: the chair was George Thomson and the other committee members were

Oliphant, Blackett, Chadwick, Philip Moon – who had been Oliphant's graduate student in Cambridge and was now at Birmingham – and Cockcroft.

In the same year, there was a major breakthrough thanks to theoretical work by Feather and Egon Bretscher, who had joined the Laboratory in 1936 as a Rockefeller Scholar.[6] They realised that a uranium reactor which created slow neutrons would also create as a by-product an element with atomic number 94, which became known as plutonium ^{239}Pu.[7] The reaction chain, involving two β-decays following the absorption of the slow neutron, is as follows:

$$^{238}_{92}\text{U} + {}^{1}_{0}\text{n} \longrightarrow {}^{239}_{92}\text{U} \xrightarrow[23.6 \text{ min}]{\beta^-} {}^{239}_{93}\text{Np} \xrightarrow[2.3565 \text{ d}]{\beta^-} {}^{239}_{94}\text{Pu,} \qquad (11.1)$$

They predicted that ^{239}Pu would be readily fissionable and, because of its quite different chemical constitution from uranium, would be easily separable from ^{238}U. The critical mass for a plutonium bomb would be less than that for a ^{235}U weapon. The first samples of plutonium were isolated by Glenn Seaborg in the following year by bombarding ^{238}U with neutrons, using one of the advanced cyclotrons constructed by Lawrence at Berkeley. In March 1941, Seaborg, Emilio Segrè and Joseph Kennedy showed that plutonium was indeed fissile, as predicted by Bretscher and Feather (Kennedy et al., 1946).[8] Feather was to remain the only senior physicist leading the nuclear research programme at the Laboratory throughout the war years.

These considerations concerning the feasibility of a nuclear weapon and a civilian nuclear reactor were set out in detail in the two famous MAUD reports, after which the committee was disbanded. Although a copy of these reports had been sent to the Director of the US Uranium Committee, they were simply filed. Oliphant went to the USA in 1941, discovered that the Uranium Committee had never seen the MAUD reports and was able to present their contents to the full committee. He pressed upon the US scientists the urgency of developing a nuclear weapon, pointing out that the German scientists had discovered nuclear fission three years earlier and had set up a Uranium Club in 1939 whose activities were being managed by the Kaiser-Wilhelm Institut für Physik in Berlin-Dahlem. The report was escalated to the highest political level in the USA and the proposal to develop nuclear weapons received presidential approval.

Attempts were made to develop a collaborative nuclear weapons programme between the US and the United Kingdom, but nothing came of the collaboration until much later – the UK and the US developed their own secret nuclear programmes, the UK *Tube Alloys Project* and the US *Manhattan Project*. It soon became apparent that the US endeavour was on a very much larger scale than could possibly be afforded in the UK, which was in the middle of a vastly expensive war which consumed a large percentage of the gross domestic product. The opportunity for collaboration was lost, despite the fact that the UK scientists had already contributed substantially to the development effort.

Although both sides had originally rejected the plutonium–slow neutron approach to nuclear chain reactions, interest revived in this possibility thanks to the efforts of Oliphant, and the Laboratory's research into slow neutrons assumed new importance. Because of their vulnerability in Cambridge to German bombing, Churchill requested the transfer of the Cambridge research team to Chicago, but US–UK relations over the security of the nuclear programme were at a low ebb, and only one of the Cavendish team was a UK national. Instead, the activity was transferred to Canada, first to Montreal and then to Chalk River. In 1944, Cockcroft took charge of the joint UK–Canadian atomic energy project and became Director of the Montreal Laboratory and Chalk River Laboratories. The Manhattan Project developed at an extraordinary pace and employed a galaxy of the best American physicists, and soon the US efforts had far outstripped those in the UK. Security concerns also limited the opportunities for cooperation. Eventually, thanks to the Quebec Agreement of 1944, collaboration was re-established, Chadwick playing a major liaison role in the selective sharing of nuclear data and information between the US, the UK and Canada.

In the end, the US developed two nuclear weapons. The first, a ^{235}U gun-type atomic bomb (Little Boy), was dropped on Hiroshima on 6 August 1945 and the second, a ^{239}Pu implosion-type bomb (Fat Boy), on the city of Nagasaki on 9 August 1945, ending the Second World War.

The UK government was determined that it should develop its own independent nuclear weapon, but the UK was many years behind the US efforts. Cockcroft returned to the UK in 1946 to become the first director of the UK's Atomic Energy Research Establishment (AERE) at Harwell, where weapons-grade ^{239}Pu was produced. This led to the formation of the UK's Atomic Weapons Research Establishment (AWRE), later the Atomic Weapons Establishment (AWE), at Aldermaston in April 1950. The first explosion of a UK nuclear weapon, which adopted the ^{239}Pu implosion-type technique, took place off the Monte Bello Islands in western Australia in October 1952.

11.5 G.I. Taylor and high-energy explosions

As early as 1938, Geoffrey Taylor had been invited to join the UK Civil Defense Committee to assess and advise on the effects of the blast waves of bombs. He was undoubtedly the UK expert in this field, having written his first published paper in fluid dynamics on the physics of strong shock waves (Taylor, 1910). These studies resulted in an assessment of the protection measures which should be taken to minimise the impact of the Blitz. In 1941, he was also involved in understanding the physics of underwater explosions, which was important in minimising the impact of the German U-boat threat.

In 1950, Taylor published his famous analysis of the dynamics of shock waves associated with very high-energy explosions, although this work had been carried out in 1941 and 1942 at the request of the UK government. Indeed, much of his work on shock waves during the war was in the form of classified internal reports and first appeared in his collected scientific papers (Taylor, 1958, 1960, 1963, 1971). In a high-energy explosion, a huge amount of energy E is released within a small volume, which results in a strong spherical blast wave and shock front which propagate through the surrounding air. The internal pressure is enormous and very much greater than that of the surrounding air. The dynamics of the shock front depend, however, upon the density ρ_0 of the surrounding air which is swept up by the expanding shock front and causes its deceleration. The compression of the ambient gas by the shock front plays a role in its dynamics and depends upon

---

| **The Buckingham Π theorem and nuclear explosions** | Box 11.3 |

Edgar Buckingham enunciated his theorem in 1914, following some derogatory remarks about the method of dimensions by the distinguished theorist Richard Tolman (Buckingham, 1914). Buckingham made creative use of his theorem in understanding how to scale from model ships in tanks to the real thing. The procedure is as follows:

- First, guess what the important quantities in the problem are.
- Form from the important quantities all the possible independent dimensionless combinations – these are known as *dimensionless groups*.
- Then apply the Buckingham Π theorem,[9] which states that a system described by n variables, built from r independent dimensions, is described by $(n - r)$ independent dimensionless groups.
- The most general solution to the problem can be written as a function of all the independent dimensionless groups.

Hence, the table of variables and their dimensions for a high-energy explosion is:

Table 11.1 G.I. Taylor's analysis of explosions

Variable	Dimensions	Description
E	$[M][L]^2[T]^{-2}$	energy release
ρ_0	$[M][L]^{-3}$	external density
γ	–	ratio of specific heat capacities
r_f	$[L]$	shock front radius
t	$[T]$	time

From the Buckingham Π theorem, there are five variables and three independent dimensions and so we can form two dimensionless groups. One of them is $\Pi_1 = \gamma$. We find the second dimensionless quantity by elimination. From the quotient of E and ρ_0, we find

$$\left[\frac{E}{\rho_0}\right] = \frac{[M][L]^2[T]^{-2}}{[M][L]^{-3}} = \frac{[L^5]}{[T^2]} = \left[\frac{r_f^5}{t^2}\right] \quad \text{and so} \quad \Pi_2 = \frac{Et^2}{\rho_0 r_f^5}.$$

We can therefore write the solution to the problem as

$$\Pi_2 = f(\Pi_1), \quad \text{that is,} \quad r_f = \left(\frac{E}{\rho_0}\right)^{1/5} t^{2/5} f(\gamma). \tag{11.2}$$

the ratio $\gamma = C_p/C_V$ of specific heat capacities of air, which is $\gamma = 1.4$ for the molecular gases oxygen and nitrogen. The only other parameters in the problem are the radius of the shock front r_f and the time t. Taylor was an expert on dimensional and similarity methods in fluid dynamics and made full use of these skills in his beautiful analysis (Box 11.3).

In his detailed analysis of the expansion of the very strong shocks associated with explosions, he derived equation (11.2) and showed that the constant $f(\gamma) = 1.03$ for air. He

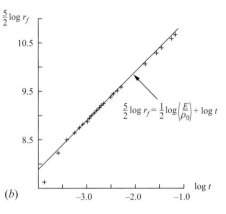

$\frac{5}{2}\log r_f$

10.5

9.5

$\frac{5}{2}\log r_f = \frac{1}{2}\log\left(\frac{E}{\rho_0}\right) + \log t$

8.5

$\log t$

−3.0   −2.0   −1.0

(a)   (b)

Fig. 11.4   (a) A frame from Mack's film of a Nevada nuclear explosion taken 15 ms after ignition. (b) G.I. Taylor's analysis of the dynamics of the shock front, showing that its radius is proportional to the $2/5$ power of the time, $r_f \propto t^{2/5}$ (Taylor, 1950b).

compared his results with high-speed films of chemical explosions available at that time and found that his theoretical predictions were in reasonable agreement with what was observed (Taylor, 1950a). In 1949, he compared his calculations with the results of higher-energy explosions using TNT-RDX and again confirmed the validity of equation (11.2) (Taylor, 1950b). In 1947, the US military released Julian Mack's movie of the first atomic bomb explosion, the Trinity test of a ^{239}Pu implosion-type bomb, which took place in the New Mexico desert in July 1945, under the title *Semi-popular motion picture record of the Trinity explosion* (Mack, 1947) (Figure 11.4(a)). Taylor had been present at the Trinity test during one of his two visits as advisor to the Manhattan Project on strong shock waves. He was able to determine, directly from the movie, the relation between the radius of the shock wave and time (Figure 11.4(b)). To the annoyance of the US government, he was then able to work out rather precisely the energy E released in the explosion and found it to be about 10^{14} J, equivalent to the explosion of about 20 kilotons of TNT.

11.6 Radar

While the UK could not keep up with the USA in the development of nuclear weapons, it played the leading role in the development of radar.[10] Early experiments in radar, which stands for *radio detection and ranging*, had been developed in secret by many nations during the 1930s. With the rise of Nazi Germany and the threat of attacks on the UK, which was well within the range of German bombers, the detection of incoming aircraft in good time for fighter aircraft and anti-aircraft batteries to be deployed became a national priority. In addition, radar would provide a means of guiding UK fighters and bombers to their targets in continental Europe. In February 1935, Robert Watson-Watt wrote a secret report for the Air Ministry entitled *Detection and location of aircraft by radio methods*. This was followed in June of the same year by a successful demonstration of the technique, detecting aircraft at distances of up to 27 km. By the end of the year, the range had extended to 100 km. The decision was taken to construct a network of radar defence antennae along the south-east coast of the UK, which eventually became known as the Chain Home. Watson-Watt became Superintendent of the new British Air Ministry Bawdsey Research Station in Suffolk in September 1936. Because of the urgency of setting up the defence radar system, Watson-Watt had to make do with what was available to him. He described his approach as the 'cult of the imperfect', remarking:

> Give them the third best to get on with; the second best comes too late, the best never comes. (Brown, 1999)

In 1936, the first five stations were commissioned. By the time of the Battle of Britain in 1940 there were 19 operational Chain Home stations; by the end of the war over 50 had been built. The original antennae operated at the long radio wavelength of about 10 m (20–30 MHz), for which conventional radio transmitters and receivers could be purchased.

With the outbreak of war, there was an urgent need to develop radar techniques and the associated electronics to produce the pulsed and encoded signals necessary to avoid countermeasures by the German military. R.V. Jones' book *Most Secret War* tells the remarkable story of the measures and countermeasures taken by both the UK and German sides in their intense war of deception and disinformation (Jones, 1978).

The Cavendish Laboratory had already made major contributions to radio studies of the ionosphere through Appleton's pioneering researches and had a large body of experience relevant to the radar programme. Among the most important of the leaders of the research and development programmes was Ratcliffe who, early in the war, was appointed Group Leader at the Telecommunications Research Establishment (TRE), where he supervised the activities of Martin Ryle, Bernard Lovell and Antony Hewish, later to become leaders of UK radio astronomy. It would take us too far off the main track to describe in detail all the work which the Cavendish researchers carried out during the war in support of the radar programme, but brief descriptions of the activities of Martin Ryle and Brian Pippard illustrate the impact which their experiences had upon the development of the future programme of the Laboratory.

11.6.1 Ryle at TRE

Martin Ryle's interest in radio was already well developed by the time he left school. He had built his own radio transmitter and acquired a Post Office licence to operate it. At Oxford, he and Edmund Cooke-Yarborough, a fellow undergraduate, set up the university amateur radio station. In 1939, Ryle joined Ratcliffe's Ionospheric Research Group at the Cavendish Laboratory. On the outbreak of the Second World War, Ratcliffe joined the Air Ministry Research Establishment, later to become the Telecommunications Research Establishment (TRE), and Ryle followed in May 1940. For the first two years, he worked mainly on the design of antennae and test equipment. In 1942, Ryle became the leader of Group 5 of the newly formed Radio Countermeasures Division, whose task was to provide jamming transmitters against the German radar defense system and to devise radio-deception operations.

Among the latter was the electronic 'spoof invasion' on D-day, which led the German High Command to believe that the invasion was to take place across the Straits of Dover, rather than at Normandy.

Radar techniques developed at an astounding pace during these years, and Ryle and his colleagues worked in a frantic atmosphere, constantly having to find immediate practical solutions for the electronic defence of the RAF's bomber fleet. According to Francis Graham-Smith,[11] his future colleague in the fledgling Radio Astronomy Group, perhaps the greatest achievement of his war years was the discovery of a vulnerable element in the V-2 rocket radio guidance system (Graham-Smith, 1986). The system developed by Ryle and his old college friend Cooke-Yarborough successfully disrupted the accurate aim of the V-2 rockets and probably contributed to the abandonment of radio control only a few weeks later.

Besides the urgency of the need for innovation and imaginative solutions to technical problems, on which peoples' lives would depend, the war years left an indelible mark on Ryle's leadership role. Bernard Lovell remarked,

> As head of this group, Ryle's extraordinary inventiveness and scientific understanding soon became evident. Under the stress of urgent operational requirements he became intolerant of those who were not blessed with his immediate insight. (Lovell, 1985)

But Ryle learned how to motivate groups of research workers. On his retirement from the Laboratory in 1982, Peter Scheuer reflected on Ryle's huge contributions to the development of aperture synthesis in radio astronomy, the legacy of his wartime experiences:

> [It] is very much the story of one remarkable man, who not only provided the inspiration and the driving force but actually designed most of the bits and pieces, charmed or savaged official persons according to their deserts, wielded shovels and sledgehammers, mended breakdowns, and kept the rest of us on our toes. (Scheuer, 1984)

In a personal letter to me, written only two months before his death in 1984, Ryle wrote:

> Presumably I knew some physics in 1939 – but this evaporated during the six succeeding years, though it was replaced by other things. But six years of designing/installing/flying boxes of electronics gave one 'state of the art' electronics, a fair intimacy with aircraft and the ability to talk constructively with Air Vice-Marshals – or radar mechanics – and above all gave one the privilege of flying with the in-between-operational-tours air-crew who flew our aircraft.

11.6.2 Pippard at ADRDE and RRDE

In spring 1941 Pippard, along with all the physicists in his year at Cambridge, was interviewed by Charles (C.P.) Snow to judge his suitability for scientific research in support of the war effort. He was next interviewed by Cockcroft, then the Chief Superintendent of the Air Defence Research and Development Establishment (ADRDE) at Christchurch, Hampshire, and was assigned to the group working on antennae and transmission lines. A year later, the group moved from the south coast to Malvern for reasons of security. The ADRDE and the Telecommunications Establishment were consolidated at Malvern, where the combined organisation was renamed the Radar Research and Development Establishment (RRDE). The four-year period during which he was employed as a Scientific Officer was devoted to mastering and developing microwave radar techniques. During these years, the development of klystrons and particularly cavity magnetrons provided new powerful sources of centimetre radiation which, combined with parabolic antennae, had higher angular resolution than the existing radar arrays, which operated at metre wavelengths.

At that time, the understanding of the physics of radio propagation and microwave components was rudimentary and had to be developed quickly. Soon after he joined the ADRDE, Pippard attended what he considered an exceptional course of special lectures on the physics of waves and radio by Ratcliffe. For most of his career as a Scientific Officer, his Head of Section was John Ashmead, who had just completed his PhD at the Cavendish Laboratory and who, after the war, would return there as a University lecturer. The four years devoted to uninterrupted research in radio and microwave techniques at the RRDE provided Pippard with an ideal training in research techniques. In particular, he developed the discipline of performing experimental work with economy and imagination. In his words,

> In the ten post-war years during which my Cambridge research was largely on microwaves and metals, I stuck to the same basic technique and needed to waste no time on fancy electronics; my war years were well spent, learning and applying the most economical methods. Much of the effort was devoted to determining the beam patterns of microwave antennae in different configurations and to developing flexible systems that would scan rapidly across targets of interest. Little was known about waveguides at that time. (Longair and Waldram, 2009)

Examples include his experiments to improve the polar diagram of a parabolic reflector used to detect the returning signals at the focus of a radar antenna. His understanding of optics enabled him to realise that placing a circular reflector of the right size but shifted forward by one-sixth of a wavelength would eliminate unwanted sidelobes. This elegant solution worked beautifully. A second example of Pippard's ingenuity concerned Ashmead's idea of using a spherical rather than a parabolic reflector so that the feed could be moved in the focal plane and scan the area about the radar target. Pippard was responsible for the feed design and, because of its similarity in appearance to a pig, the device was christened a Hoghorn.

11.7 De Bruyne and glues for aircraft structures

A key contribution to the War effort was made by Norman de Bruyne, a remarkable experimenter and inventor who joined the Laboratory in 1927 to study for a PhD under Rutherford.[12] He did not, however, work on radioactivity but rather on field emission, the ejection of electrons from the surfaces of materials by the action of an electric field. Part of the reason he chose this topic was that it could be completed in a year and so he could be a candidate for a Trinity Prize Fellowship in 1928. This was a clever choice since he had worked for the General Electric Company under B.S. Gosling, one of the leading experts on field emission, during the previous summer and brought with him much of the equipment needed to complete his study, which was published in 1928 (de Bruyne, 1928). He was successful in winning his Trinity Fellowship in September 1928 and his PhD in 1930. The following year he worked on the first thyratron to be made available to the Laboratory. Having been appointed Junior Bursar of Trinity College, he decided to leave the Laboratory in 1931 to pursue his great passions, which were flying, aircraft materials and structural engineering. He learned to fly at the newly opened Marshall's aerodrome at Cambridge, where he was the first student to obtain his pilot's licence in Arthur Marshall's flying school.

His major contributions, which were to have a significant impact on the war effort, were in the use of stressed plywood skins bonded by synthetic glues for all parts of the aircraft structure. The strength of these materials enabled the development of single-wing aircraft without the struts and wires used in conventional aircraft construction. Furthermore, the plane was only half the weight of one using conventional materials. The key to de Bruyne's success was the development of artificial resins based on phenol-formaldehyde to replace natural resins. Unlike the natural resins used in making plywood by the conventional method, the new adhesives did not lose their adhesive properties on exposure to moisture. The development of new generations of adhesives and his innovations in structural design were major achievements. De Bruyne built his pioneering single-wing aircraft himself and in December 1934 his plane, the Snark, made its maiden flight from Marshall's aerodrome with de Bruyne at the controls – and obtained its airworthiness certificate.

This success led to the award of a contract from the de Havilland Aircraft Company for the development of reinforced plastics for aircraft propellers. By 1937, he had demonstrated the use of phenolic resins with suitable reinforcement as a means of producing strong, lightweight aircraft components. He introduced the term Aerolite for the phenolic resins, and a version of this called Gordon Aerolite was used in the wings of the Bristol Blenheim aircraft. He went on to develop synthetic urea-formaldehyde-based adhesives, and these were used in the construction of the Mosquito fighter-bomber, nicknamed *the wooden wonder*, which came into service in 1942 as the fastest and most versatile aircraft of the war.

De Bruyne went on to solve the problem of bonding plywood to aluminium with the invention of the adhesive called Redux – standing for Research at Duxford, the airfield outside Cambridge where his company was based. This material was first used in bonding clutch plates in the Cromwell and Churchill tanks. Redux was first used in aircraft in the de Havilland Hornet and Sea Hornet in 1944 and then in the world's first jet airliner, the de Havilland Comet. Although the Comet suffered a disastrous structural failure, the Redux adhesives were exonerated from any blame. De Bruyne's Redux adhesives were to be used in many aircraft. In 1947, de Bruyne's company was taken over by the Swiss CIBA Company.

11.8 Impact of the war years

These few examples of research stimulated by the needs of war led to the development of new technologies which were to be fully exploited by the physicists when they returned to peacetime duties. But they also learned about carrying out research under pressure and using ingenuity rather than brute force to solve pressing scientific and technical problems. These characteristics, very much in the Rutherford tradition, were to be influential in their approach to post-war research in the Cavendish Laboratory.

After the war, there was a change in attitude of governments to the role of basic research. There was a very large increase in investment in basic research in the USA, largely stimulated by the huge contributions which the very best research scientists had made during the period of hostilities and by the realisation of the enormous potential for economic growth, as well as strategic defence capabilities, which the fruits of basic research could bring. In Europe, recovery from the ravages of the war took somewhat longer than in the USA but, in due course, the European countries began to invest heavily in pure and applied research.

But equally important was the change in attitude of many of the best research workers as a result of their wartime experiences. To quote Bernard Lovell, they adopted an approach to research

> utterly different from that deriving from the pre-war environment. The involvement with massive operations had conditioned them to think and behave in ways which would have shocked the pre-war university administrators. All these facts were critical in the large-scale development of astronomy. (Lovell, 1987)

Lovell's remarks concerned the development of astronomy facilities, but they apply equally to all the sciences, no more so than in physics. How Lawrence Bragg and the Laboratory adapted to this new era is the subject of the next chapter.

12 Bragg and the post-war years

12.1 Restructuring the Laboratory: the immediate post-war years

With the end of hostilities, Bragg faced the challenge of converting the Laboratory from an organisation supporting the war effort to a physics department which had to be restructured in the light of a number of changing circumstances.[1]

- The Austin Wing could now be returned to its original purpose, which was to relieve the serious overcrowding and lack of adequate laboratory space, a legacy of the Rutherford era.
- Just as at the end of the First World War, the demands of the Second World War had resulted in a backlog of undergraduate and postgraduate physics students to be trained in the physical sciences. In physics, the number of undergraduates was 50% greater than it had been before the war and the numbers in the final (Part II) year almost trebled to about 120. But the greatest increase was in the number of graduate students, which now totalled 160, of which 110 were studying for the PhD degree. The result was that, even with the addition of the new Austin Wing, the Laboratory was already full to capacity by 1948 and indeed overcrowded in some areas (Bragg, 1948).
- The huge role which physics had played in the war, in particular in national security, meant that more physicists were needed for military and civilian research and development. The Barlow Report had recommended that there was a national need to maintain this larger number of qualified physics students. These conclusions were broadly in line with Bragg's own estimate, namely that 200 to 300 professional physicists would need to be trained each year in the UK after the war (Bragg, 1942b).
- As alluded to in Section 11.8, the physicists returned from the war with attitudes quite different from pre-war assumptions about how research should be carried out. The urgency of research and development in all areas of support for the war effort meant that physicists had become used to essentially unlimited financial and administrative support for their activities.
- The increasing diversity of activity in the Laboratory meant that there could no longer be a Head of Department who could oversee all the research. Rutherford, with his larger-than-life personality, charisma and sheer personal energy, had been able to cope with the old-style extensive approach to research with a single motivating director, but Bragg had a very different personality and recognised the need for change.

- Bragg was also well aware that his appointment did not command universal acclaim. There was certainly an impression that the Laboratory had declined from the triumphant years of the early 1930s. As Mott expressed it,

> Bragg was offered the job by the electors, I have to assume because they [felt] the Cavendish needed a new line.... I know of the tendency at the time of the nuclear fraternity to feel and express the view that what isn't nuclear isn't (fundamental) physics. (Mott, 1990)

This sentiment is repeated by Pippard when he wrote:

> W.L. Bragg's appointment to the Cavendish chair of experimental physics was taken by many as a threat to the great tradition of fundamental physics research established by J.J. Thomson and, especially, Rutherford... The choice of a crystallographer, however distinguished, was a blow to many hopes. (Pippard, 1990)

Bragg faced a formidable set of challenges, but there can be no doubt that, in his much less flamboyant way, he foresaw how university physics research would develop in the future. His problems were compounded by the continuing exodus of Cavendish nuclear physicists to other universities, disillusioned by the course of events. The immediate post-war years were undoubtedly very difficult for Bragg as he sought to revitalise the department and take it in somewhat different directions. As remarked by Ratcliffe, however, Bragg's outstanding personal qualities enabled the necessary changes of direction to take place:

> Bragg had the admirable gift of supporting people without running them, for example, Perutz and Kendrew, and he did the same thing with Ryle. (Crowther, 1974)

During the war years, Bragg had given serious consideration to the need for physicists after the war and the organisation of scientific research and teaching in the universities, publishing two papers on how these could be provided (Bragg, 1942b, 1944). The structure of the Laboratory had to change, his view being that

> the ideal research unit is one of six to twelve scientists and a few assistants, together with one or more first-class mechanics and a workshop in which the general run of apparatus can be constructed. (Bragg, 1942b)

By 1948, Bragg's reforms were well in place, as summarised in his paper in *Nature* describing the organisation and work of the Laboratory (Bragg, 1948). The research groups and their leaders were in the areas of nuclear physics (Frisch), radio (Ratcliffe), low-temperature physics in the Mond Laboratory (Shoenberg), crystallography (W.H. Taylor), metal physics (Orowan) and mathematical physics (Hartree). There were also smaller groups in areas such as fluid mechanics. Other activities were to become associated with the Laboratory, including the initiatives in electron microscopy led by Ellis Cosslett, the Meteorological Physics Group led by Thomas Wormell which was to be transferred from the Observatories to the Laboratory, and later the group which was to become known as Physics and Chemistry of Solids, led by Philip Bowden and David Tabor. How these groups came about and how they fared under the new arrangements is detailed in the following sections and chapters. In Bragg's paper of 1948, he estimated that the total cost of research per year in 1946–47 was about £60,000, including manpower and administration costs, but not rental of the buildings. Of this, £10,000 came from outside sources.

But the changes ran more deeply. Devolving responsibilities onto the groups meant that administration and management of what was one of the largest physics laboratories in the UK had to become better organised. The senior academic staff were now allocated their own offices and the groups each had their own secretaries, support staff and telephones, unlike the pre-war arrangements. To coordinate the overall administration of the department, Bragg appointed first I.T. James and then Kenneth Dibden, who was to be in post from 1948 to 1960, as Laboratory Secretaries, supported by nine administrative staff to carry out the many functions which had previously been largely dealt with by the academic staff members themselves. In his *Nature* article, Bragg described the role of the Laboratory Secretary as follows:

> This officer has in his charge finance, appointments of assistant staff and rates of pay, the formal work concerning admissions, structural alterations and upkeep of buildings, preparations of agenda for meetings, and other materials of this kind. (Bragg, 1948)

Dibden, a former naval officer, was particularly skilled at managing the support staff and was eventually to look after all the assistant staff in the University. Ratcliffe, who had returned early from the war and was Bragg's trusted advisor and facilitator of the many changes being undertaken, recalled that

> [Bragg] modernised the very antiquated notepaper, introduced a departmental secretary to help him run the Laboratory and the place worked in a completely different way from the old one. (Phillips, 1979)

Brian Pippard, who became Cavendish Professor in succession to Mott, wrote:

> Bragg performed a notably excellent job in decentralising the work of the Cavendish, and thus effectively breaking away from what would have ultimately become the dead hand of the Rutherford tradition. His decision to give each research section as near as possible autonomy, consistent only with very general central principles and of course financial control, has played a significant part in subsequent developments. Ever since then, the Cavendish has been notable among Cambridge departments for the democratic way in which it conducts its business. (Pippard, 1990)

The same basic structure has been maintained over the years since Bragg introduced these reforms with the flexibility to allow the nature of the group structure to evolve in response to changing priorities.

The group structure of the Laboratory in 1950, the middle of Bragg's post-war tenure as Head of the Laboratory, is summarised in Table 12.1. According to his Report to the University for 1950, published in 1951, the staff comprised three Professors, four Readers, eight Lecturers in Experimental Physics, two Lecturers in Theoretical Physics, one Assistant Director of Research, eight University Demonstrators (Assistant Lecturers) and two Assistants in Research – not only was the scope of the research activity significantly increased, but the need for more teaching necessitated an increase in staff. The distribution of these staff among the research groups is also indicated in Table 12.1. Figure 12.1, the

Fig. 12.1 The staff and graduate students in the Laboratory in 1950. Many of the principal players during the post-war Bragg years are in the front row. From left to right: E.H.K. Dibden, A.S. Baxter, H.N.V. Temperley, J.M.C. Scott, A.B. Pippard, H.D. Megaw, V.E. Cosslett, D. Shoenberg, J.A. Ratcliffe, Prof. Sir Lawrence Bragg, Prof. O.R. Frisch, E.S. Shire, W.H. Taylor, A.E. Kempton, W.E. Burcham, J. Ashmead, W. Cochran, D.H. Wilkinson, J.W. Findlay, A.P. French and D.E. Nagle.

			Numbers of registered
Table 12.1 Research groups, staff and numbers in the Cavendish Laboratory in 1950			
Research group	Head of group[a]	Other staff members[a]	graduate students
Nuclear Physics	Frisch (P)	Shire (R), Burcham (L), Kempton (L), French (D), Wilkinson (D)	37
Radio	Ratcliffe (R)	Ryle (L), Weekes (ADR), Budden (D), Findlay (D), Gold (D), Falloon (AiR)	20
Crystallography	Taylor (R)	Bragg (P), Megaw (ADR), Cochran (D)	17
Low Temperature Physics	Shoenberg (L)	Ashmead (L), Pippard (L)	10
Metal Physics	Orowan (R)	–	9
Fluid Dynamics	Townsend (ADR)	–	3
Theoretical Physics	Hartree (P)	Scott (L), Temperley (L)	3
Electron Microscopy	Cosslett (L)	Bradfield (AiR), Baxter (D)	
Meteorological Physics	Wormell (L)	Pierce (AiR)	
Others		Mrs Horton (L), D. Tabor (ADR, Lubrication & Friction)	

[a] Notation: P = Professor, R = Reader, L = Lecturer, ADR = Assistant Director of Research, D = Demonstrator (Assistant Lecturer), AiR = Assistant in Research.

annual laboratory photograph of the academic staff and graduate students for 1950, gives a vivid impression of the growth in size of the research activity in the Laboratory from the pre-war years, as represented by the corresponding photograph from 1932 (see Figure 9.16). Equally striking is the very different set of personalities and specialisms of the staff from the pre-war era. These changes of direction were far from easy, none more of a headache for Bragg than the future of nuclear physics.

But if Bragg was somewhat ambivalent in his support of nuclear physics, there can be no doubt about his enthusiasm and strong support for the opportunities offered by developments in crystallography, biological physics, electron microscopy, condensed matter physics and radio astronomy. In their different ways, many of these experimental disciplines were closely related to Bragg's deep interests in optical and X-ray systems, the problems of phase determination and the applications of Fourier optics.

12.2 Teaching

The structure of the Tripos courses in physics maintained the same overall format of Part I and Part II physics as shown in Table N.1 of the notes to Chapter 2. An important innovation, which resolved a problem which had been of concern since the Thomson era, was the appropriate level of instruction in theoretical physics for experimentalists and for those theorists who wished to maintain a close involvement with experimental work. A Part II course in theoretical physics was introduced in the academic year 1949–50, directed by Hartree and Temperley. The course was combined with the normal Part II physics course, with a reduced amount of practical class work. Bragg was quite clear about the objectives of the course:

> [T]he course is specially designed for physicists and not for mathematicians with a bent for mathematical physics. (Bragg, 1951)

The course was a success, 14 students taking the course in the first year of its delivery; it was undoubtedly filling a real need. This basic structure has been maintained ever since.

An issue which Bragg addressed was the problem of the differing levels of preparation of students taking Part I of the physics examinations. Many of the best students found that the first year repeated much of what they knew already, so the second year of the course was modified so that some Part II topics were covered in Part I. Furthermore, use was made of the Long Vacation terms between Parts I and II for further courses so that the three-year course could be reduced to two for the best prepared students.

Another of Bragg's innovations was a series of *arts lectures*, given by distinguished members of the University in the Arts and Humanities. These lectures were given in the Michaelmas and Lent terms at 10 a.m. on Monday mornings and attracted large audiences, typically about 200 persons. The courses from 1949 to 1953 were entitled *The Ancient World, The Growth of English Literature, Economic and Social Development since the Middle Ages, Art, Music, Man and his Environment, The Novel* and *The History of North America*. Bragg recorded his appreciation for the efforts of the lecturers and the enthusiasm of the audience for the value the lectures added to their studies.

12.3 Electronic computing: EDSAC and EDSAC 2

The use of computing engines to carry out arithmetic calculations had a long Cambridge history, from Babbage's difference engine of 1822 to Wynn-Williams' scale-of-two counter of 1932 and the pioneering code-breaking researches of Alan Turing during the Second World War. The history of computing in Cambridge is splendidly told in Haroon Ahmed's highly illustrated *Cambridge Computing: The First 75 years* (Ahmed, 2013). The leader of the initiative to form a computer laboratory for academic research was John Lennard-Jones, who was appointed John Humphrey Plummer Professor of Theoretical Chemistry in 1932. The proposal to set up a computer laboratory capable of carrying out lengthy computations was supported by the School of Physical Sciences, as well as by the faculties of Economics and Politics, of Agriculture and the Biological Faculties. The machines were to be the best mechanical calculators available. The project was approved by the University in 1937, although the name of the department was changed to the Mathematical Laboratory.

Lennard-Jones was appointed Director and Maurice Wilkes, one of Ratcliffe's former research students in the Cavendish Radio Group, was appointed a University Demonstrator in the Mathematical Laboratory. With the assistance of Douglas Hartree, then at Manchester, a number of mechanical computers were purchased, including a differential analyser built by Metropolitan-Vickers. Events were overtaken by the Second World War, during which Wilkes was seconded to the Telecommunications Research Establishment (TRE), where he honed his skills in electronics.

After the war, Wilkes returned to Cambridge and before long was appointed Director of the Mathematical Laboratory. Inspired by the ENIAC and EDVAC projects in the USA, Wilkes determined to develop a Cambridge version of these electronic computers. The role of the Mathematical Laboratory was to provide computational services for the whole University and not act as a research and development activity for electronic computing. Wilkes took the strategic decision that the Laboratory would not devote effort to developing new hardware or cutting-edge technology, but use existing proven technologies to enable a practical service to be delivered to the University. The EDSAC computer, the acronym standing for Electronic Delay Storage Automatic Calculator, was the first practical electronic stored-program computer for general use (Figure 12.2). Its first programs were run on 6 May 1949.

As described by Joyce Wheeler (Wheeler, 1992), the project was a great success and members of the sponsoring departments made full use of the computational capabilities on offer. There were mathematical projects involving the solution of non-linear differential equations, as well as some of the very first stellar structure and evolution computations by Brian Hazelgrove (Haselgrove and Hoyle, 1956) and Joyce Blackler (Blackler, 1958), the future Joyce Wheeler, under the supervision of Fred Hoyle.[2] The economists made full use of the machine, Richard Stone carrying out pioneering economic modelling for which he was awarded the 1984 Nobel Prize in Economics.

From the perspective of research in the Cavendish Laboratory, the computational facilities were essential for the advance of the research programmes. To mention just a few of these areas:

- Douglas Hartree and the computation of atomic structure by the Hartree–Fock self-consistent field method (Hartree, 1957) (Section 12.10.4)
- Molecular biology and the determination of the structures of myoglobin and haemoglobin by Kendrew and Perutz (Sections 12.6.1 and 15.3.1)
- Radio astronomy and the practical application of the principles of aperture synthesis (Section 14.3). The first prototype Earth rotation aperture synthesis observations by Blythe (1957a, 1957b) required the use of EDSAC to reconstruct the map of the sky. In his Nobel Prize lecture, Ryle makes the point that the sequence of radio telescopes which he constructed were only feasible because of the parallel increase in computing power provided by the Mathematical Laboratory
- The application of computing to the theory of condensed matter physics by Heine and his colleagues (Section 15.5.1)

The EDSAC computer, which was use extensively by the theoretical physicists, by the Radio Group and by the MRC Unit for Biological Systems. The physicists made pioneering uses of computers to analyse the data from radio telescopes and X-ray crystallographic images and to carry out theoretical atomic and condensed matter physics computations. Maurice Wilkes is on the left and Bill Renwick on the right (courtesy of the University of Cambridge Computer Laboratory).

Fig. 12.2

- The pioneering computations of Howie and Whelan using EDSAC 2 to interpret quantitatively electron microscope and electron diffraction images of dislocations (Section 15.2.1).

The memory of the early computers was very limited and so many innovations were adopted to make optimum use of their capacities. A good example, which was essential for many of the applications involving large data sets, is the fast Fourier transform, which reduces the number of computations needed to take a discrete Fourier transform from $\mathcal{O}(n^2)$ to $\mathcal{O}(n \log n)$ computations, where n is the size of the data set. This technique, later to be known in more general form as the *Cooley–Tukey method* (Cooley and Tukey, 1965), was crucial for analysing Fourier data obtained in molecular biology, radio astronomy and many other disciplines.

EDSAC was closed down in July 1958, when it was replaced by EDSAC 2, which remained in operation until 1965.

12.4 Nuclear physics

We need to return to the pre-war years to recount the history of developments in nuclear physics during the last years of Rutherford's tenure of the Cavendish Chair and the beginnings of Bragg's. Cockcroft and Walton made their key discovery of the transmutation of the nucleus in 1932 but other investigators were hot on the trail, each taking a different approach to the creation of intense beams of high-energy protons and light nuclei for nuclear physics experiments. In 1931, Robert van de Graaff invented the van de Graaff electrostatic accelerator for creating megavolt potential differences (van de Graaff, 1931), while Merle Tuve was developing the Tesla transformer with which he achieved voltages of 3 MV. Most significantly, Ernest Lawrence at Berkeley, California began his development of the cyclotron, based upon acceleration concepts developed by Rolf Wideröe (1928), which were to lead to the invention of the betatron. In 1931, Lawrence and M.S. Livingston had their first success in accelerating electrons to energies of 80 kV in a 4.5-inch diameter cyclotron (Lawrence and Livingston, 1931). By 1935, Lawrence had developed a successful 27-inch cyclotron, producing currents of 20 μA of 6.2 MeV deuterons. By early 1936, this cyclotron had become the pre-eminent source of radioactive elements for medical and therapeutic purposes, as well as for nuclear physics research. This was followed by a 37-inch cyclotron which eventually produced 74 μA fluxes at 8 MeV. The culmination of the pre-war activity was the construction of a 60-inch cyclotron with a 220-ton magnet, which became operational in 1939. By late 1939, the machine was producing 100 μA beams of 16 MeV deuterons.[3] Lawrence and his colleagues saw no reason why much higher-energy particles could not be produced:

> We ... see no difficulties in the way of producing with the present equipment 25 million volt deuterons and 50 million volt alpha-particles, and moreover we are convinced that much higher energies could be obtained from a cyclotron of larger dimensions. (Lawrence *et al.*, 1939)

As already discussed, Rutherford was reluctant to seek the significant funds needed to build the next generation of particle accelerators, believing that there was still a great deal of useful research to be carried out with the energies of machines similar to the Cockcroft–Walton accelerator. This led to the departure of many of his most distinguished collaborators to other universities, which were more sympathetic to investment in more powerful accelerators (Section 9.9). Nuclear physics research in the Laboratory lost its momentum for a few years until Lord Austin made his gift for the construction of the Austin Wing and a suite of more powerful accelerators. Three different types of accelerator were constructed.

The High Tension Laboratory was built in 1937 on a site adjacent to the Austin Wing to house the 1 MeV and 2 MeV accelerators (see Figure 5.3). It was decided that, rather than building these accelerators in-house, commercial versions of the Cockcroft–Walton accelerator built by Philips of Eindhoven would be purchased (Figure 12.3). According to Otto Frisch, these machines

(a) (b)

(a) Construction of the 1 MeV Cockcroft–Walton machine. (b) The completed High Tension Laboratory, the 2 MeV machine in the foreground and behind it the 1 MeV machine.

Fig. 12.3

 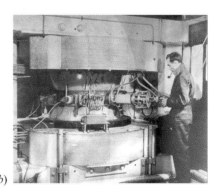

(a) (b)

(a) The circular Ds of the Cavendish cyclotron with the electrostatic accelerating gap between them. (b) The completed cyclotron.

Fig. 12.4

were well beloved by journalists and TV producers, being their idea of the shape of science to come, with their tall columns of polished metal electrodes and the crashing sparks they could be provoked to generate. (Frisch, 1974)

There was a need to develop other means of accelerating particles to high energies in addition to the Cockcroft–Walton machines. The department invested in a cyclotron, in which particles are bent in a circle by a strong uniform magnetic field and their energies are increased each time they pass over the accelerating gap between the two 'Ds' (Figure 12.4). Cockcroft was in charge of the project to construct a cyclotron, which was essentially a copy of the Berkeley 37-inch cyclotron developed by Lawrence; Chadwick's cyclotron

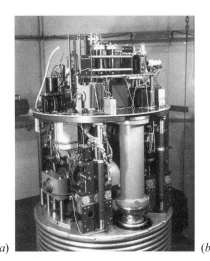 

(a)   (b)

Fig. 12.5   (a) The ion source of the van de Graaff accelerator for spraying electric charges onto the fast-moving belt and carrying them to the high-voltage electrodes. (b) The moving belt at the top of the machine.

at Liverpool was another copy of the Berkeley accelerator. A 36-inch diameter magnet, weighing 50 tons, was built by Metropolitan-Vickers and Brown-Firth of Sheffield. The first weak deuteron beams were observed on 1 August 1938 and a 13 μA beam of 5 MeV protons produced by Christmas of the same year (Dahl, 2002). In the end the cyclotron could accelerate deuterons to about 10 MeV.

Higher voltages and greater stability than those of the 1 MeV and 2 MeV Phillips accelerators could be produced by a van de Graaff generator, in which a fast-moving electrified belt inside a large steel tank filled with compressed gas carried the charges to the high-voltage electrode (Figure 12.5). Such a machine was built by Edward Shire,[4] and housed in the Electrostatic Generator Building (see Figure 5.3). It came into operation in the early 1950s and could accelerate particles to 3 MeV. All three machines were used for studying the nuclear energy levels of light nuclei.

On Bragg's appointment to the Cavendish Chair in 1938, nuclear physics, still the largest activity in the Laboratory, was run by Philip Dee and Norman Feather. But they were both to depart from Cambridge, Dee to the University of Glasgow in 1943 to take up the Regius Professorship of Natural Philosophy and Feather, who had acted as Assistant Head of the Department, to the Professorship of Natural Philosophy at the University of Edinburgh in 1945.

Bragg faced the problem that the resources needed to support the next generation of nuclear physics facilities far exceeded what could be afforded or managed by a single university department. Nor were the resources needed to compete with the USA in the construction of large cyclotrons or linear colliders likely to be made available – the UK was bankrupt as a result of a devastatingly expensive war. Bragg was of the firm view that such large-scale projects were not well adapted to the university environment, but required a national effort in an institute dedicated to the construction and operation of powerful accelerators for use as tools for basic research by the community of nuclear physicists. As Perutz remarked,

> This was an inevitable consequence of the War and the transformation of atomic physics to 'Big Science', to which the tradition and structure of Cambridge University was ill-adapted. (Perutz, 1970)

None of the former leaders of the nuclear physics activities in the Laboratory could be persuaded to return to Cambridge. Somewhat to his surprise, Otto Frisch was appointed to the Jacksonian Chair of Natural Philosophy in succession to Cockcroft who, after some post-war delays, had been appointed as the first Director of the Atomic Energy Research Establishment (AERE) at Harwell, charged with developing Britain's nuclear power programme.

Frisch was a very different character from Rutherford and there was some disappointment that a nuclear physicist of the stature and personality of a Rutherford could not be attracted to Cambridge. Bragg was not particularly encouraging about the future of nuclear physics in Cambridge. By the time he wrote his *Report of the Head of Department for the year 1949–50*, the section on nuclear physics reads:

> Although these sets [of four accelerators] are on a scale which would have seemed immense in a physics laboratory a generation ago, and their maintenance is a matter of engineering rather than of physics, they are of course modest compared with the vast units in some other centres in this country and America. (Bragg, 1951)

Nonetheless, Bragg supported plans to develop a linear accelerator which would produce electrons of several hundred MeV energy, in collaboration with the atomic energy establishments at Harwell and Malvern, the Cambridge efforts being led by Albert Kempton. Preliminary experiments with large magnetrons were carried out in a former Madingley Road factory which was taken over for use as a laboratory. But this was not to lead to the accelerator the nuclear physicists had hoped for. Frisch was remarkably cool about what he perceived to be the future of nuclear physics:

> By the 1950s nuclear physics had become a mopping-up operation; while it was still possible to attack a great many questions, and to pin down more precise knowledge was important, there was no longer the feeling of worlds to be conquered. . . . The construction of a linear electron accelerator was begun by [Samuel] Devons and [Hugh] Hereward and continued by [William] Burcham and [Albert] Kempton. The planned energy of 0.4 GeV would have been enough to make π-mesons. But when Mott succeeded Bragg as Cavendish Professor in 1953, one of his first actions, after consulting various people concerned, was to terminate the project, having concluded that it was 'too little and too late'. (Frisch, 1974)

This decision led Denys Wilkinson, then a Reader in Nuclear Physics, to accept the offer in 1956 of the Professorship of Nuclear Physics at Oxford University, where the prospect of collaborating with the Harwell and Aldermaston laboratories seemed to offer more opportunities for gaining access to state-of-the-art accelerators. In the late 1950s the decision was taken to get rid of the Cavendish machines, which were no longer competitive with bigger and better facilities elsewhere. The two Philips machines were shipped to South Africa and the van de Graaff machine was scrapped. The cyclotron was given to William Burcham, who had moved to Birmingham University, where he gave it a new lease of life as a strong-focussing machine with much higher energy and intensity, and with a source of polarised protons and deuterons which allowed new types of research to be carried out. In the early 1960s, the High Tension Laboratory was converted to create the Cockcroft Lecture Theatre and a suite of practical class laboratories.

Frisch realised that the way to make a serious impact in particle physics was to enhance the capabilities for data analysis for the new generation of bubble chambers and to make technological contributions to the large international projects about to be undertaken at CERN in Geneva – as he made clear in his reminiscences, building ingenious instruments and detectors was what he really enjoyed doing (Frisch, 1980). The development of the Sweepnik high-speed measuring machine for analysing bubble-chamber particle tracks was to play a major role in analysing the data from the CERN and US accelerators. These changes of direction were to come to fruition during the Mott era.

12.5 Crystallography

The origins of the Crystallography Laboratory, as well as Bragg's appointment of Albert Bradley to lead these activities on his assumption of the Cavendish Chair, have been de-scribed in Sections 11.2.1 and 11.2.2. Sadly, Bradley suffered a breakdown during the war years, but Bragg was fortunate enough to attract his former student and colleague W.H. (Will) Taylor, who had been Head of Physics at the Manchester College of Science and Technology, to a Readership in Crystallography and to become Head of the Crystallog-raphy Group. Among the sceptics, crystallography was scarcely regarded as part of physics, but Bragg was quite clear about its fundamental significance for condensed matter physics. Two quotations illustrate his views on physics and the role which crystallography would play. Crowther quotes Bragg as stating that modern physics was

> the study of the fundamental particles of which matter is made, as contrasted with the study of matter in bulk and of large scale forces. (Crowther, 1974)

In his lecture to the Institute of Metals in 1948, Bragg made his vision very clear:

> The department which we call crystallography would perhaps be better described as the department for discovery of the structure of the solid state . . . Mainly by X-rays we seek to discover the way the atoms are arranged in crystals and in other forms of solids. The scope of the work is very considerable. At the one end we are investigating such substances as minerals and alloys in the inorganic field; other researchers are examining complex or-ganic compounds . . . ; finally at the other extreme we have a little group which is financed by the Medical Research Council under the direction of Perutz, which is engaged in a

gallant attempt to work out the structure of the highly complex molecules which build up living matter, the proteins . . .

This section of the laboratory is closely linked to metal physics under Orowan. His students are particularly studying the mechanical properties of metals and relating them to their structure. The effects of cold work on a metal, recrystallisation, yield point and plastic flow, brittle fracture, distortion under rolling and drawing and so on, are being investigated as physics phenomena.

Bragg's vision set the path for the development of what would come to be called *solid state* or *condensed matter physics*.

By the date of Bragg's 1952–53 Report to the University, Taylor was able to report on many significant advances across the broad front of crystallographic activities. To quote Taylor's contribution to the report (Bragg, 1954):

- The success of Mr. Cochran and his collaborators in analysing really complex organic structures of great chemical interest, some of them related to nucleic acids; and in particular their development of experimental and interpretative techniques for the location of hydrogen atoms.
- The growing importance of the research on complex alloy phase structures involving transition metals, not only in the provision of accurate factual information on a difficult section of alloy chemistry, but also in the stimulus it is providing for more fundamental theoretical treatment of electron distributions both in the atoms and in the crystal as a whole.
- The development of a model of the cold-worked state in a metal, based on the data obtained with the help of the X-ray microbeam diffraction technique. The model not only provides a self-consistent interpretation of observations of particle-size and strains for a considerable number of metals, both hard and soft, but also proves capable of accounting satisfactorily for observations made by other workers using different techniques.
- The analysis of the structure of coal, which has for the first time provided a soundly-based and quantitative picture of the physical nature of the coal substance. The work has been supported by the National Coal Board, and we have had the benefit of co-operation with a number of their research teams.
- A renewal of activity in mineral structure analysis and the use of X-ray diffraction techniques for the examination of mineral systems both natural and synthetic; examples include the feldspars, the kaliophilite-nepheline family, and the hydrated calcium silicates.
- A completely new approach to the problem of ferro-electricity has stimulated much discussion, particularly in the U.S.A.

The Crystallography Laboratory was providing data for investigators in chemistry, metallurgy and mineralogy, as well as maintaining useful relations with industry. In his report, Bragg describes the analysis of crystal structure by X-rays as a 'typical borderline subject' and yet these techniques were to inform some of the Laboratory's next generation of basic discoveries in the physical sciences. The discoveries of the structure of haemoglobin, myoglobin and DNA depended upon the techniques developed by the Crystallography Laboratory.

12.6 The MRC Research Unit for the Study of the Molecular Structure of Biological Systems

12.6.1 Perutz and Kendrew

Despite the enormous challenge of disentangling the molecular structure of biological molecules, Bragg was unswerving in his support for Perutz's slow but steady progress to achieve this goal in his studies of haemoglobin, which was known to be a key molecule in human and animal biology (Box 12.1). As an alien refugee, Perutz was interned in the early years of the war but released in 1941. He returned to Cambridge, but was set to work on research associated with the war effort[5] and only returned to research on biological molecules in 1944, supported by an ICI fellowship. In January 1946, John Kendrew, who had trained as a chemist, joined the Laboratory with the encouragement of Bernal. Bragg suggested that he work with Perutz on proteins, unaware of the great difficulty of the problem. As Kendrew later recounted,

> Bragg was the *only* crystallographer in Cambridge – apart from Perutz – who did not believe we were wasting our time on a project much more complicated than had previously been attempted by the methods of X-ray crystallography. (Kendrew, 1990)

Perutz had first been supported by a grant from the Rockefeller Foundation and then by an ICI fellowship, but now that was coming to an end. David Keilin, the Quick Professor of Biology and Head of the Moltano Institute of Parasitology on the neighbouring Downing site, provided space in his laboratory for Perutz and Kendrew to prepare their crystal specimens, which were then studied by X-ray crystallography in the Laboratory. As Perutz wrote in 1963 following Keilin's death,

> J.C. Kendrew and I owe Keilin a tremendous debt, for he was one of the first to see the potentialities of our physical approach to biochemistry. Until the 1950s we had no facilities for biochemical work in the Cavendish Laboratory; Keilin gave us bench space in his institute even though he was short of space himself, and he helped us to grow protein crystals. He also persuaded his Faculty Board to let me lecture on Molecular Biology long before that subject had even acquired its name. When Kendrew's and my research was in danger of closing down for lack of support by the university Keilin suggested and supported Sir Lawrence Bragg's approach to the Medical Research Council which saved us. (Mann, 1964)

Bragg met with Edward Mellanby, the secretary of the Medical Research Council (MRC), in May 1947 to explore how the research of Perutz and Kendrew could be supported and, somewhat to their surprise, the council accepted the proposal with alacrity, although it had only been intended as a preliminary exploration of possible ways ahead. The official title of the new unit was the MRC Research Unit for the Study of the Molecular Structure of Biological Systems. In subsequent years, the unit grew with the arrivals of

Haemoglobin and myoglobin

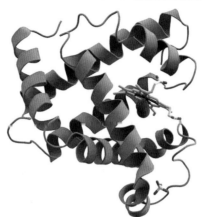

The structure of human haemoglobin (left); the protein's α and β subunits are shown in red and blue, and the iron-containing haem groups in green. The structure of myoglobin (right) (courtesy of Creative Commons).

Haemoglobin is the iron-containing oxygen-transport metalloprotein in the red blood cells of verte-brates. Haemoglobin in the blood carries oxygen from the lungs or gills to the rest of the body where it releases the oxygen to burn the nutrients which provide energy to power the functions of the organism in the process of metabolism. Haemoglobin has an oxygen-binding capacity of 1.34 mL O_2 per gram of haemoglobin, which increases the total blood oxygen capacity 70-fold compared with dissolved oxygen in blood.

Haemoglobin is responsible for the colour of the red blood cell. It contains a haem group with an iron atom at its centre. When the iron is bound to oxygen, the haem group is red in colour (oxyhaemoglobin), and when it lacks oxygen (the deoxygenated form) it is blue-red. As blood passes through the lungs, the haemoglobin picks up oxygen because of the increased oxygen pressure in the capillaries of the lungs, and it can then release this oxygen to body cells where the oxygen pressure in the tissues is lower. In addition, the red blood cells can pick up the waste product, carbon dioxide, some of which is carried by the haemoglobin at a different site from where it carries the oxygen, while the rest is dissolved in the plasma. Haemoglobin is composed of four polypeptide chains, which in adults consist of two alpha (α) globin chains and two beta (β) globin chains. Each polypeptide has a haem group attached, in which each haem can bind one oxy-gen molecule – there are thus four haem groups per haemoglobin molecule that together bind four oxygen molecules.

Myoglobin is a protein found in the muscle cells of animals and acts as an oxygen-storage unit, providing oxygen to the working muscles. There is a close chemical similarity between myoglobin and haemoglobin. Both proteins contain a molecular constituent called a *haem*, which enables them to combine reversibly with oxygen. The bond between oxygen and haemoglobin is more complex than that between oxygen and myo-globin and accounts for the dual ability haemoglobin has to transport oxygen as well as to store it. Myoglobin has a molecular weight of 16,700, about one-fourth that of haemoglobin.

Hugh Huxley (1948), Francis Crick (1949), James Watson (1951) and Vernon Ingram (1954).[6]

The magnitude of the task facing Perutz was already apparent in his paper with Bernal and Fankuchen (Bernal *et al.*, 1938). It had been established by Gilbert Adair, Assistant Director of Research in the Physiological Laboratory and an expert on the physical chemistry of haemoglobin, that the atomic weight of the molecule was enormous, about 67,000 daltons or atomic mass units (Adair, 1925). In the same sequence of papers, Adair showed from oxygen binding measurements that the molecule contained four haem groups (Box 12.1). Adair kindly provided Perutz with a number of his 'beautiful haemoglobin crystals', the crystallographic measurements showing that the unit cell must have two identical halves and that the four haem groups lay roughly normal to the crystallographic *a*-axis. Perutz's X-ray crystallographic measurements also agreed with the very large atomic weight of haemoglobin (Bernal *et al.*, 1938).

After he returned to Cambridge, Perutz resumed his studies of the intensities of diffraction patterns of crystals of horse methaemoglobin and measured about 7000 Bragg reflections. He carried out the Fourier synthesis of these measurements to determine the Patterson functions, exploiting all the tricks of the trade of X-ray crystallography – for example, the use of differences in the X-ray pictures between the hydrated (expanded) and unhydrated (contracted) states of the molecule. Exploiting the symmetry of the molecule, these measurements provided a two-dimensional reconstruction of the molecule, but a three-dimensional inversion was beyond computation at that time. On the basis of the information available to him, Perutz came up with various models for the overall structure of the haemoglobin molecule, including bundles of α-keratin fibres and a 'pill-box' model of the structure of the four major groups of the molecule. Soon after he arrived in the Laboratory in 1949, Crick reviewed critically all the work carried out by Perutz on the structure of haemoglobin and quickly found flaws in the models. Crick soon turned his attention to the structure of the DNA molecule.

Meanwhile, Kendrew decided to tackle the structure of the myoglobin molecule, which was very similar to one of the four components of the haemoglobin molecule and so potentially a simpler molecule for analysis. Kendrew was one of the first to appreciate the importance of computers for the analysis of X-ray crystallographic data, and developed programs for the EDSAC computer to speed up the tedious Fourier inversions.

The breakthrough occurred in the last year of Bragg's tenure of the Cavendish Chair. It was realised that the way to determine the phases needed for the three-dimensional reconstruction of the structure of these molecules was the technique of isomorphous substitution, in which a heavy metallic atom is attached to the molecule, without changing its molecular structure (see Section 12.6.2).[7] The consequent changes in the intensities of the spots on the X-ray photographs enabled the phases to be determined and thence the three-dimensional structure of the molecule. Initially only one type of atom could be attached, but once more than one different type of atom could be added, the full reconstruction could take place. This was still some years in the future because of the large amount of data and computation needed to carry out the programme successfully. But Bragg's tenure of the Cavendish Chair in 1953 was to end with Crick and Watson's brilliant solution to the structure of the DNA molecule.

The outline structure of a nucleic acid (left); a sugar-phosphate unit in the DNA molecule (right).

The DNA molecule is a very long chain polymer, the monomers of which are called *nucleotides* (left diagram). Each nucleotide consists of three units: a sugar known as deoxyribose, a negatively charged phosphate and a base. The nucleotides are linked along the sugar–phosphate units, which form the backbone of the molecule. A base is then attached to each sugar (right diagram).

The atomic arrangement of each sugar–phosphate unit along the backbone of the molecule is identical. The attached bases, however, are not. The location of the four possible bases – thymine, cytosine, adenine and guanine – are shown in the right diagram. The sequence of the bases along the sugar–phosphate chain is irregular, and it is this sequence that contains the genetic information in DNA.

12.6.2 Crick, Watson and the discovery of the structure of the DNA molecule

Francis Crick was already 33 years old when he joined Perutz's group in the Laboratory as a graduate student. His studies had been interrupted by the war and then, after a few false starts, he began his research on the physical basis of biological systems and genetics. His PhD project under Perutz concerned the structure of polypeptides and proteins.

James Watson was 12 years younger than Crick and had already completed his PhD on bacterial viruses in 1950 under the supervision of Salvador Luria at the University of Indiana. Luria sent him to Copenhagen to learn biochemical and genetic techniques but he changed direction in May 1951 when he heard Maurice Wilkins' report at a conference in Naples on his X-ray diffraction images of fibres of DNA molecules. Luria arranged for Watson to join Kendrew in Cambridge in August 1951 to work on the structural chemistry of nucleic acids and proteins.

The Crick–Watson collaboration began as soon as they discovered their common interest in understanding the nature and role of DNA molecules in biology and genetics (Box 12.2).

They were brilliant and outspoken scientists with complementary expertise – it was an ideal collaboration, but one which could easily ruffle feathers. As Perutz wrote long after these events took place,

> They shared the sublime arrogance of men who have rarely met their intellectual equals. . . . To say that they did not suffer fools gladly would be an understatement: Crick's comments would hit out like daggers at *non sequiturs* and Watson demonstratively unfolded his newspaper at seminars that bored him. (Perutz, 1993)

The evidence suggested that the molecules which carry the genetic information were likely to be some form of macromolecule such as a protein. The proteins were known to be structural and functional macromolecules, some of which are involved in enzyme reactions in cells. In 1944, Oswald Avery and his collaborators had shown that heritable differences could be caused in bacteria by specific DNA molecules, suggesting their role in transferring genetic information (Avery *et al.*, 1944).

The study of helices in biological systems began in the 1930s with the work of William Astbury at the University of Leeds, who showed that there were drastic changes in the X-ray diffraction patterns of moist wool or hair protein fibres upon significant stretching. The data suggested that the unstretched fibres had a coiled molecular structure with a characteristic repeat distance of 0.51 nm. This led Maurice Huggins to propose that the unstretched protein molecules formed a helix, the α-form, while stretching caused the helix to uncoil, the β-form (Huggins, 1943).

Bragg was strongly attracted by the idea of the helical structure of proteins and he, Kendrew and Perutz made various attempts to create such chemical models, but they failed to find any particularly convincing solution (Bragg *et al.*, 1950). Within a year, Linus Pauling had found the solution to the structure of the α-helix, much to Bragg's consternation (Pauling *et al.*, 1951). Bragg lacked Pauling's insight into the nature of chemical bonding which implied that the peptide units would be planar. The Cavendish physicists had also been misled by the apparent repeat distance of the α-keratin molecule and had not considered the possibility of non-integral helices, perhaps Pauling's most imaginative innovation. Bragg never forgave himself for these oversights which led to Pauling winning the race to discover the α-helix structure of proteins. As he wrote in 1965 of his 1950 paper with Kendrew and Perutz,

> I have always regarded this paper as the most ill-planned and abortive in which I have ever been involved. (Bragg, 1965)

As Phillips (1979) remarked, it was especially aggravating (for Bragg) to have asked the right question only to have Pauling provide the answer.

Having been scooped by Pauling, Bragg wondered whether they should change tack and study the structure of DNA. In his gentlemanly way, however, Bragg did not want to enter into competition with Maurice Wilkins, whose group at King's College, London was well ahead of the Cambridge group in DNA studies. Wilkins was the senior member of the King's College MRC Biophysics Unit which had been set up in 1946 by John Randall, a former student of Bragg's from his Manchester days, about a year before the creation of

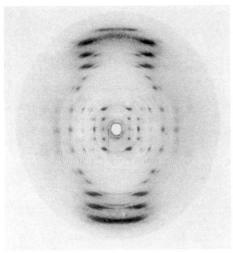 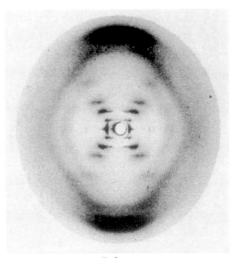

A form   B form

The B51 X-ray pictures taken by Franklin and Gosling of the A and B forms of DNA in May 1952 (Franklin and Gosling, 1953c,d) (reproduced by kind permission of *Acta Crystallographica* and the International Union of Crystallography).

Fig. 12.6

the Cambridge MRC Unit. Bragg therefore discouraged active research in DNA, although the group maintained a sideline interest in the subject.

Rosalind Franklin joined Wilkins' group in January 1951 and within a year she had transformed the field. She was a brilliant experimental physical chemist who was able to create thinner fibres of DNA than those used by Wilkins, resulting in much sharper X-ray diffraction patterns. In October 1951, Franklin and Raymond Gosling established that there are two different patterns of DNA X-ray diffraction images, depending on the relative humidity of the environment of the sample (Figure 12.6) (Franklin and Gosling, 1953c,d). Each pattern corresponded to a different structure, which they termed A and B. The A form occurs at a relative humidity of 75% and the B form at a relative humidity of 90%. The A form is crystalline, while the B form exists as fibres – the two forms are readily interchanged by simply varying the relative humidity. The DNA in living material is the B form, but the A form is important in determining the structure of DNA because it could be studied by X-ray crystallography. From their X-ray measurements of the A form, Franklin and Gosling determined the dimensions and hence the volume of the unit cell in the crystal. This, combined with the known density and water content of the samples, gave the result that there are two chains in the unit cell.

Franklin and Gosling's X-ray data, however, provided no information about the structure of the molecule – for example, whether it is helical or otherwise. To obtain this information, details of the X-ray diffraction images were needed. Bragg returned to the chase and asked William Cochran, a postdoctoral fellow from Edinburgh who was to become a University Demonstrator in the Laboratory, to carry out theoretical calculations of the X-ray diffraction pattern of thin molecular helices. In collaboration with Vladimir Vand, who had already considered the case of uniform helices, and with Crick, who elbowed his way into

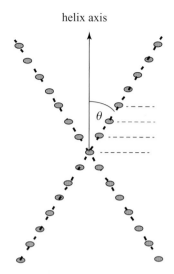

helix axis

θ

The basic diffraction pattern of a helix.

these calculations as a distraction from his PhD project, the analysis was carried out in October 1951 (Cochran and Crick, 1952; Cochran *et al.*, 1952).

Cochran had already made key contributions to the understanding of the structure of molecules of biological importance, working on the borderline between crystallography and physics. In his PhD dissertation, he had introduced the technique of isomorphic replacement to determine the molecular structure of sucrose (Beevers and Cochran, 1946; Cochran, 1946). He used counter techniques to measure scattered X-ray intensities which, combined with his development of Fourier techniques (Cochran, 1948), enabled chemical bonding to be determined and, for the first time, the location of hydrogen atoms in, for example, hydrogen bonding. These are only two examples of Cochran's pioneering contributions to the rigorous determination of the molecular structure of complex molecules – these were soon to be exploited by Perutz, Kendrew, Watson and Crick.[8]

The results of the calculation showed that the X-ray spectra lie on a series of 'ley lines' in Fourier space. The points corresponding to the first maxima form a cross-like pattern, a characteristic feature of diffraction by a helix (Figure 12.7). There are further, less intense maxima outside the first maxima, and there is some intensity between them. From the separation of the ley lines, the pitch of the helix can be deduced and this, combined with the angle θ of the cross, gives the radius of the molecule. The calculations of Cochran, Crick and Vand were extended to the case of two identical helices – each helix scatters the X-rays as before, but there is interference between the two sets of scattered waves. The bases which make up the DNA molecule (Figure 12.8(*a*)) give rise to a different set of spectra which also occur on ley lines, the separation of which depends inversely on the separation of the bases along the direction of the helix axes. The spectra of the bases form the dark areas at the top and bottom of the X-ray pictures (Figure 12.6) and are easily distinguished from those of the sugar–phosphate backbones.

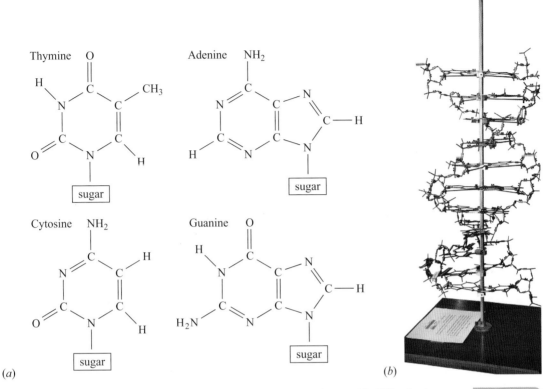

(a) The four DNA bases which bind in pairs by hydrogen bonds to link the two backbones of the DNA molecule. Adenine pairs with thymine, and cytosine with guanine. (b) Crick and Watson's half-size model of the DNA molecule constructed at the time of their discovery of the structure of the molecule.

Fig. 12.8

Inspired by Franklin's work on the A form of DNA, Watson and Crick began work on the physical and chemical structure of DNA. In November 1951, Franklin described the two forms of DNA at a colloquium in London attended by Watson. Watson and Crick concluded that X-ray diffraction data indicated that the molecule had a helical structure. Relying on Watson's recollections of the talk given by Franklin, they produced a triple-stranded model of DNA in which the phosphate groups were on the inside and the bases pointed outwards. In a brief visit to Cambridge by Wilkins and Franklin, the model was quickly dismissed as incompatible with the data since it contained no water molecules – Watson had misunderstood the water content of the unit cell. Bragg was incensed that Watson and Crick had attempted to elbow their way into studies of DNA molecules, which he regarded as the province of the King's College group. What made matters worse was that they had come up with an incorrect model. They were instructed to discontinue these studies.

There the matter rested until the autumn of 1952 when the Cavendish group learned that Pauling was working on the structure of DNA. As soon as Bragg learned that Pauling had taken up this challenge, he gave Crick and Watson permission to return to their work on the DNA molecule – Bragg did not want to be scooped by Pauling yet again. Pauling's paper on a triple-stranded model for the molecule bore similarities to the abortive attempt which Crick and Watson had made in the previous year. As soon as the manuscript of Pauling's paper arrived in Cambridge in January 1953, Watson recognised that the model was chemically impossible but was concerned that, as the world's leading chemist, Pauling would soon find his mistake.

In November 1952 a report by Franklin and Gosling was handed to Crick which contained new crystallographic data on the A form of DNA. It stated that the lattice is monoclinic face-centred and the space group is C2, and gave the dimensions of the unit cell. Crick realised that the C2 space group implied a diad, that is, a two-fold axis of symmetry, in which the two sugar–phosphate chains run in opposite directions. Wilkins showed Watson the superb new diffraction images which Franklin had obtained of the B form of the DNA, the famous B51 image obtained in May 1952 (Figure 12.6). In addition, Perutz gave Crick a copy of an MRC report on the work of the King's College group which contained information about the unit cell of the B form. With these clues, Crick and Watson devoted all their considerable energies to unravelling the structure of the DNA molecule. Here the computations of Cochran, Crick and Vand came into their own since these were the tools needed to interpret Franklin's images. Another key piece of information was provided by Jerry Donohue, who was working in the same office as Crick and Watson. He pointed out that Watson had used the wrong tautomeric forms for the bases in trying to put the molecules together – he should have used the keto form, which Donohue said was the common one, rather than the enol form, despite what it said in the textbooks.

On the morning of 28 February 1953, Watson came up with the crucial finding that in pairing adenine with thymine and cytosine with guanine, each pair joined by hydrogen bonds, the pairs had exactly the same length and fitted perfectly between the helical backbones provided by the sugar–phosphate chains. Furthermore, these pairings accounted for Chargaff's rule that the number of adenine bases in any sample of DNA was roughly equal to the number of thymines and that the number of cytosines equalled the number of guanines (Figure 12.8(a)). Watson appreciated that this structure overcame all the objections to his earlier models. At lunch the same day in the Eagle pub in Bene't Street, Watson and Crick announced to all present that they 'had found the secret of life'.[9]

The discovery of the structure of the DNA molecule and the data on which it was based were published in three consecutive articles in the 25 April 1953 edition of *Nature* (Franklin and Gosling, 1953b; Watson and Crick, 1953b; Wilkins *et al.*, 1953). The Watson and Crick paper concludes with the famous remark:

> It has not escaped our notice that the specific pairing we have postulated immediately suggests a possible copying mechanism for the genetic material.

Aaron Klug pointed out that the uncharacteristic element of caution in their statement was a result of the fact that they had not seen the King's College papers before publication. Their ideas were much more fully explored in their subsequent paper of 30 May 1953, entitled *Genetical implication of the structure of deoxyribonucleic acid* (Watson and Crick, 1953a). The double-helix model (Figure 12.8(b)) still had to be tested quantitatively against the King's College data. Given the model, Patterson functions could be predicted, and these provided a satisfactory match to the X-ray diffraction patterns (Franklin and Gosling, 1953a,e).

The subsequent story is well known. Tragically, Franklin died of ovarian cancer in 1958 at the age of only 37. Nobel Prizes were awarded to Crick, Watson and Wilkins in 1962.

12.7 Ratcliffe, the Radio Group and the birth of radio astronomy

During the post-war years, Jack Ratcliffe played a central role in the organisation of the Laboratory, its teaching programme and its interactions with the University and external bodies. Bragg was happy to leave these responsibilities to Ratcliffe since he had little taste for the committee and management activities which necessarily accompanied the far-reaching changes he made during the post-war years. In contrast, Ratcliffe was a master at these aspects of University management, administration and politics. His familiar utterance at departmental meetings was, 'We think, Professor, don't we . . . ' Despite Ratcliffe's modesty about his contributions, he was an outstanding scientist and teacher and had a clear vision of the future, nowhere more so than in the evolution of the pre-war Radio Group into the Radio Astronomy Group under Ryle.

12.7.1 The Radio Group

The pre-war research activities of Appleton and his colleagues were described in Section 10.2. In the early 1930s, a radio receiving station was set up behind the Cambridge University rugby ground, on a site which was used by the Officer Training Corps as a rifle range. This was replaced by a brick building in 1933 and was used for experimental ionosphere research. With Appleton's departure in 1938, Ratcliffe took over the running of the group. He led the research programme as a theorist, leaving the graduate students to develop the electronics and technology needed for experiments in ionospheric research. A particular area of expertise was in ionosonde measurements which involved the use of pulsed high-frequency radio signals to probe different regions of the ionosphere. This was the breeding ground for a number of future leaders of radio science and related technologies, including John Findlay, Joseph (Joe) Pawsey and Maurice Wilkes. The activities involved many aspects of radio wave propagation and the physics of the ionosphere. In 1939, Ratcliffe persuaded Martin Ryle to join the Radio Group but, with the onset of the war, Ryle soon became caught up in the development of radar, eventually ending up at the Telecommunications Research Establishment (TRE) as recounted in Section 11.6.1.

During the war, members of the Radio Group were in high demand by the TRE, Ratcliffe becoming the head of TRE Development Services in 1941. The main area in which Ratcliffe was involved was the development of airborne radar for the location of enemy aircraft by night fighters and bombers. In addition, Development Services was responsible for providing training courses in the use of all aspects of radar. It is estimated that during the three and a half years of its existence from August 1941, 6000 persons attended these courses.

With the end of hostilities, Ratcliffe returned to Cambridge and Ryle soon joined him. Ryle preferred to carry out research on the extraterrestrial radio signals which had been discovered during the war rather than ionospheric research, and so Ratcliffe split the Radio Group into two parts: the Radio Ionosphere Section continued radio studies of the atmosphere and ionosphere, while Ryle became head of the Radio Astronomy Section, a discipline which barely existed at the time. Both groups expanded rapidly. In 1945 Weekes, Findlay and Booker became Cavendish staff members in the Ionosphere Section and in 1947 they were joined by Budden and Briggs. A corresponding expansion occurred in the Radio Astronomy Section, as described in Section 12.7.2. Soon there were about 20 to 30 graduate students working in the two sections and the activity had become the second-largest research area in the Laboratory. Initially most were in the Radio Ionosphere Section, but by 1956 the majority were members of the Radio Astronomy Section.

The Radio Ionosphere Section was a hive of activity during the immediate post-war years. Budden (1988) summarised the achievements of the Section under Ratcliffe's leadership under five headings:

- *The propagation of radio waves of very low frequency*. These observations and their theoretical interpretation gave information about the lowest part of the ionosphere. Hartree derived the necessary differential equations and he and Budden used the EDSAC computer to solve simple functions for $N(z)$, the variation of the electron concentration with height z above the Earth's surface.
- *The theory of diffraction applied to the fading of radio waves and irregularities and movements of the plasma in the ionosphere* (Ratcliffe *et al.*, 1950). These techniques could be used to determine the motions and sizes of irregularities in the ionosphere and were to be important in understanding the fluctuations of the intensities of signals from sources of radio waves from beyond the atmosphere (see Section 12.7.2).
- During a period of sabbatical leave in 1952, Ratcliffe developed techniques for determining *the electron distribution in the F-layer of the ionosphere*, the uppermost and densest region where the ultraviolet radiation from the Sun is absorbed by ionising the neutral component of the atmosphere.
- One of the most remarkable pieces of research from the early post-war years concerned the origin of *'whistlers'*. These are low-frequency electromagnetic radio signals which can be made audible in the 1–10 kHz waveband by a suitable radio-to-audio amplifier and loudspeaker. The characteristic whistling sound starts at a high frequency and then decreases in frequency until it passes below the threshold of hearing. L.R.O. Storey began his researches into whistlers in 1950–51 at Ratcliffe's suggestion. Storey recorded the electromagnetic signals on magnetic tape and then passed the signal through a simple audio frequency analyser. He established that whistlers are initiated by lightning discharges which produce radio waves of all frequencies. The very low-frequency waves are guided along magnetic field lines in the ionosphere, but because of dispersion, the different frequency components travel at different speeds, the high-frequency signals arriving first and the lower frequencies later. Storey deduced that the waves travelled along the Earth's dipole magnetic field lines from one hemisphere to another at distances up to four times the Earth's radius, much greater distances than anyone had expected

(Storey, 1953). These observations and their interpretation marked the beginning of the study of the Earth's magnetosphere, and the technique was used to explore the electron concentration at great heights above the surface of the Earth. Occasionally whistlers of twice the duration were observed, associated with the reflection of whistler signals by converging field lines in the dense plasma in the opposite hemisphere to the observer.

- *Studies of the cross-modulation of radio signals.* This turned out to be a non-linear effect associated with the heating of the ionospheric plasma by the radio waves themselves, resulting in coupling with other radio waves. Ratcliffe and his colleagues clarified the physics of these interactions (Ratcliffe and Huxley, 1949) and then used the measurements to study the height at which the interactions took place and the frequency of electron collisions, which determines the rate at which energy is transferred from the heated electrons to the neutral molecules.

By 1960, Budden was the only staff member left in the Radio Ionosphere Section besides Ratcliffe, the others having moved to other departments of the University or taken up appointments elsewhere. Posts were needed for the expanding activities in the Radio Astronomy Section and in Solid State Physics. In October 1960, Ratcliffe took up the posts of Director of Radio Research in the Department of Scientific and Industrial Research and Director of the Radio Research Station at Slough, where he made full use of the expanded opportunities for ionospheric and magnetospheric research.

12.7.2 The birth of radio astronomy

Working at the Bell Telephone Laboratories at Holmdel, New Jersey before the war, Karl Jansky had been assigned the task of identifying naturally occurring sources of radio noise which could interfere with radio transmissions. In what turned out to be a classic series of observations made at the long wavelength of 14.6 m (20.5 MHz), he discovered the radio emission from the Galaxy (Jansky, 1933). This discovery was confirmed by Grote Reber, a radio engineer and enthusiastic amateur astronomer, with his home-built radio antenna and receiving system operating at a wavelength of 1.87 m (160 MHz) (Reber, 1940, 1944). Comparison of Jansky's and Reber's observations showed that the emission could not be black-body radiation. In the paper immediately following Reber's in the *Astrophysical Journal*, Henyey and Keenan (1940) showed that the intensity observed by Jansky at the longer wavelength was far too great for the emission process to be *bremsstrahlung*, the characteristic thermal radiation of an ionised plasma. Other than these negative conclusions, these observations attracted little attention from professional astronomers.[10]

The development of radar during the Second World War had two consequences for radio astronomy. First, sources of radio interference which might confuse radar location had to be identified. In 1942, James Hey and his colleagues at the Army Operational Research Group in the UK discovered intense radio emission from the Sun which coincided with a period of unusually high sunspot activity (Hey, 1946). They continued to improve the sensitivities of the receivers but discovered that the noise performance of the telescope system did not improve. They soon realised that background radio emission from the Galaxy itself, and not the receivers, was limiting the sensitivity of the telescope system. At the end of hostilities, Hey and his colleagues began mapping the sky at 5 m wavelength and in 1946 they discovered the first discrete source of radio emission, which lay in the constellation of Cygnus – the source became known as Cygnus A (Hey *et al.*, 1946).

The second consequence was that the extraordinary research efforts to design sensitive receivers and improved radio antennae resulted in new technologies which were to be exploited by the pioneers of radio astronomy, all of whom came from a background in radar. The three main groups were to be headed by Martin Ryle at Cambridge, Bernard Lovell at Manchester and Joseph Pawsey at Sydney.

Ryle returned to Cambridge supported by an Imperial Chemical Industries (ICI) fellowship and joined Derek Vonberg. Their first project was to measure the properties of the radio emission from the Sun. There was scarcely any money for equipment, but they were able to buy considerable amounts of surplus war electronics very cheaply and also acquire large amounts of high-quality German radar equipment which had been requisitioned after the war. They took away five truckloads of surplus equipment from the Royal Aircraft Establishment (RAE) at Farnborough, including two 7.5 m Wurzburg antennae, several 3 m dishes and a vast amount of high-quality German coaxial cable – they were all superior to the UK equipment and were to be used for many years.

The angular resolving power of the radio antennae available at that time was not sufficient to resolve the disc of the Sun. Ryle and Vonberg therefore adapted the surplus radar equipment and developed new receiver techniques for metre wavelengths to create a radio interferometer, the antennae being separated by several hundred metres in order to provide high enough angular resolution. Only later was it realised that they had invented the radio equivalent of the Michelson interferometer. A massive sunspot occurred in July 1946, and their observations showed conclusively that the radio emission originated from a region on the surface of the Sun similar in size to the sunspot region (Ryle and Vonberg, 1946, 1948). The short *Nature* paper is remarkable in that it foreshadows many of the features of the future discipline of radio astronomy: the first radio Michelson interferometer with variable antenna spacing, the demonstration that the brightness temperature of radio emission was greater than 2×10^9 K and that the radio emission was circularly polarised, the latter two observations implying that some non-thermal emission mechanism was involved.

Francis Graham-Smith joined Ryle as a graduate student in 1946 and they realised that the interferometer could be modified to study the source Cygnus A. The interferometer traces revealed not only strong signals from Cygnus A, but also another source, which was located in the constellation of Cassiopaeia – it became known as Cassiopaeia A (Ryle and Graham-Smith, 1948). It was realised that interferometric techniques could be used not only to measure accurate positions, but also to analyse the brightness distribution of the radio emission of the sources. In 1948, midway through his ICI fellowship, Ryle was appointed a University lecturer.

In 1949 further discrete sources of radio emission were discovered by the Australian radio astronomers John Bolton, Gordon Stanley and Bruce Slee, who succeeded in associating three of them with remarkable nearby astronomical objects (Bolton *et al.*, 1949). One was associated with the supernova remnant known as the Crab Nebula and the others, Centaurus A and Virgo A, were associated with the strange nearby galaxies NGC 5128 and M87 (NGC 4486), respectively.

The 1C radio telescope with which the first Cambridge survey of the radio sky was carried out. In the background one of the 7.5 m Wurzburg antennae can be seen.   Fig. 12.9

It was evident that radio telescopes of larger aperture would detect more and fainter radio sources and so Ryle built the first of a series of radio telescopes designed to make surveys of the northern sky. What eventually became known as the 1C (first Cambridge) survey was carried out with what was referred to as a 'long Michelson' interferometer, which was built on the rifle range site (Figure 12.9). The telescope was a transit instrument, meaning that it had a fixed polar diagram and measured the radio intensity as the sky passed over the telescope. The telescope had good resolution in right ascension, but poor resolution in declination (Figure 12.10(a)). One of the key inventions made by Ryle early in the history of the new discipline was that of the *phase shift receiver*. In this device, an extra half-wavelength delay could be inserted into one of the two arrays of the interferometer so that the interferometric beam pattern was shifted by half a fringe, with the result that the maximum in one configuration lay on top of the minimum of the other. By subtracting these measurements, many of the problems of measurement could be eliminated (Figure 12.10(b)). For example, the diffuse galactic background emission was subtracted out, as well as man-made interference, receiver gain variations, and inequalities between the arms of the interferometer (Ryle, 1952). The receiver detected only compact radio sources and this arrangement soon became the standard technique for all radio interferometers.

So many sources were detected that they could not be separated out, and so Ryle doubled the length of the elements of the interferometer, providing better resolution in right ascension. Operating at a wavelength of $\lambda = 3.7$ m, the final interferometer consisted of two long elements, each $20\lambda \approx 75$ m long, their centres being separated by $110\lambda \approx 400$ m.

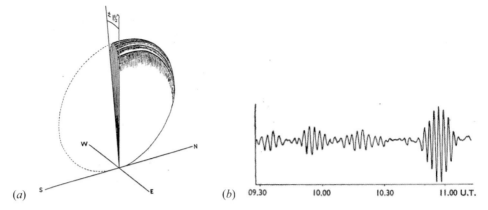

(a)   (b)   09.30   10.00   10.30   11.00 U.T.

Fig. 12.10   (a) The reception pattern, or polar diagram, of the 1C aerial-spaced interferometer, showing the high resolution in right ascension (E–W), 1.5°, and the poor resolution in declination (N–S). (b) A section of the 1C record obtained with the phase-switching technique, showing a number of weak radio sources. The sensitivity is about 100 times greater than the total power records (Ryle *et al.*, 1950).

In their paper *A preliminary survey of the radio stars in the northern hemisphere*, about 50 discrete radio sources were listed (Ryle *et al.*, 1950). The positional accuracy of the radio sources was too poor to allow them to be identified securely with known astronomical objects. A few appeared to be associated with nearby galaxies, but most remained unidentified without any optical counterparts. Ryle and his colleagues favoured the view that the majority of the sources were 'some hitherto unobserved type of stellar body, distributed widely throughout the Galaxy'. The interferometer system had cost about £350.

In 1951, Graham-Smith measured interferometrically the positions of the four brightest sources in the northern sky with an accuracy of about 1 arcmin (Graham-Smith, 1951). He confirmed the association of Taurus A with the well-known supernova remnant, the Crab Nebula, and Virgo A with the galaxy M87 (NGC 4486), which was known to possess a strange optical jet. The observations of Cygnus A and Cassiopaeia A led to their optical identification by Walter Baade and Rudolph Minkowski, who made their observations with the Palomar 200-inch telescope (Baade and Minkowski, 1954). Cassiopaeia A was associated with a young supernova remnant in our own Galaxy, while Cygnus A was associated with a faint, distant galaxy with a redshift $z = 0.0561$. When this result was communicated to Ryle, he quickly changed his view about the nature of the radio source population. Cygnus A, the second-brightest radio source in the northern sky, was associated with a faint galaxy at cosmological distances; fainter radio sources should lie even further away and so could be used for cosmological investigations.

One of the issues of radio observations at long radio wavelengths concerned the origin of the strong fluctuations in intensity observed from unresolved radio sources. Antony Hewish joined the Radio Astronomy Section in 1948 and his research included investigations of the nature of these fluctuations, or *scintillations*. The cause of the radio scintillations was the deflection of radio rays when they pass through irregularities in the ionospheric plasma, a subject on which Ratcliffe and his colleagues had already written pioneering papers

(Ratcliffe *et al.*, 1950). In 1951, the theory of the scintillation process as applied to observations of unresolved radio sources was worked out in detail by Hewish (1951). The same concepts and techniques could be applied to the physics of fluctuations due to ionospheric, interplanetary and interstellar electron density fluctuations. Hewish showed that the scale of the ionospheric irregularities ranged from 2 to 10 km, that the variation of electron content was about 5×10^9 electrons cm^{-2}, and that the irregularities are located at a height of about 400 km above ground level. These irregularities moved with a steady wind-like motion at a velocity of the order 100 to 300 m s^{-1}. The same technique could be used to study the solar corona, the region of hot plasma surrounding the Sun. The radio source Taurus A (the Crab Nebula) was observed at varying angular distances from the Sun, and the variability of the signal could be accounted for by scattering by fluctuations of the electron density in the solar corona (Hewish, 1955). Hewish derived the sizes and electron densities of coronal irregularities in the distance range 5 to 15 solar radii. Throughout his career, Hewish was to pursue these studies, which involved observation of the short-time-scale variability of radio sources. These were to result in an unexpected key discovery, the radio pulsars, which turned out to be magnetised, rotating neutron stars (Section 14.4).

Two further discoveries of the early 1950s were to be crucial for the development of radio astronomy. The nature of the galactic radio emission and of the discrete radio sources was resolved. In 1950 Hannes Alfvèn and Nicolai Herlofson proposed that the emission of the 'radio stars' was the *synchrotron radiation* of high-energy electrons gyrating in magnetic fields with flux density $10^{-10} - 10^{-9}$ T within a 'trapping volume' of about 0.1 light year radius about the star (Alfvén and Herlofson, 1950). Then Karl-Otto Kiepenheuer and Vitali Ginzburg made the much better suggestion that the galactic radio emission is the synchrotron radiation of ultra-relativistic electrons gyrating in the interstellar magnetic field (Kiepenheuer, 1950; Ginzburg, 1951). By the mid 1950s, the power-law spectrum of the galactic radio emission and its high degree of polarisation convinced everyone of the correctness of the synchrotron hypothesis. Radio emission is observed throughout the disc of the Galaxy and so provides direct evidence for an interstellar flux of very high-energy electrons and of the interstellar magnetic field present throughout the disc of the Galaxy. A consequence was that the total energy requirements of some of the most luminous radio emitters such as Cygnus A, which is about 10^8 times more luminous as a radio emitter than our own Galaxy, must be enormous – it turned out that the energy needed corresponded to the rest mass energy of about a million solar masses of matter.

The second discovery was that the radio emission of Cygnus A did not originate from the body of the galaxy. In 1953, Roger Jennison and Mrinal Kumar Das Gupta at Jodrell Bank used interferometric techniques to show that the radio emission originated from two huge radio lobes (Jennison and Das Gupta, 1953). Once the optical identification of the source was made with a massive distant galaxy in the following year, it turned out that the lobes were located on either side of the radio galaxy (Figure 12.11). Thus, not only must the radio galaxy accelerate an enormous amount of material to relativistic energies, but this material also has to be ejected into intergalactic space in opposite directions from the parent galaxy. These observations and their interpretation in terms the properties of relativistic plasmas and magnetic fields provided a powerful stimulus for the new discipline of high-energy astrophysics.

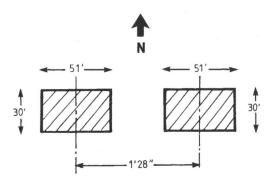

Fig. 12.11 A reconstruction of the radio structure of the radio source Cygnus A from radio intensity interferometric observations by Jennison and Das Gupta at a frequency of 125 MHz (Jennison and Das Gupta, 1953). The parent galaxy is located between the radio lobes.

By the time of Bragg's resignation from the Cavendish Chair in 1953, many of the pioneering advances which would result in the spectacular technical and scientific development of radio astronomy and high-energy astrophysics in the succeeding years were already in place. Bragg was rightly proud of the way in which radio astronomy had developed, as is evident from his last Report to the University:

> This new branch of astronomy is one which, it would seem, would repay development at Cambridge . . . The new knowledge of the Universe which it is yielding is proving to be of intense interest, and as the Cambridge unit under Mr Ryle has already established a leading position, the opportunity to develop this new science should be exploited vigorously. (Bragg, 1954)

Ryle's inspiring leadership and total involvement in all aspects of the work were crucial to these developments, but he also benefitted enormously from Bragg's unswerving enthusiasm for the research and Ratcliffe's outstanding administrative support as Head of the Radio Group, both of which relieved Ryle of many of the more tedious aspects of management and administration, for which he had little patience or sympathy. As Sullivan (2009) points out, the period 1953–54 marked a turning point in the development of radio astronomy. The succeeding generations of radio telescope were to be expensive instruments, as the discipline changed from being a cheap 'string-and-sealing-wax' activity to a major branch of astronomy and a 'Big Science' with correspondingly much greater budgets than had been conceivable in the early pioneering days. Ryle came into his own in the development of the subsequent generations of radio telescope, the story of which is taken up in Chapter 14.

12.8 Electron microscopy

The early development of electron microscopy was almost entirely carried out by physicists working in Berlin.[11] In 1924 Louis de Broglie showed that the quantum waves associated with electrons have wavelength $\lambda = h/p$, where p is the momentum of the electron. This made it clear that, if electron waves were to be used in microscopes instead of light waves, very much higher resolution could be obtained.

The idea that a short solenoid could act as a magnetic lens was first worked out in 1926 by Hans Busch, who calculated the trajectories of electrons in short rotationally symmetric electric and magnetic fields (Busch, 1926). He found to his surprise that the electron beams behaved almost exactly like light waves in an optical lens system and obeyed the same optical lens equation. Max Knoll and Ernst Ruska developed the first electron microscope prototype at the Technical University of Berlin in 1931. Ruska, his brother Helmut and Bodo von Borries were financed by Siemens in 1937 to develop applications for the electron microscope, focussing their efforts on its use for biological specimens. During the period 1938–39, Ruska at Siemens produced the first commercial transmission electron microscope.

The typical resolution of an optical microscope is about $\lambda/2 \approx 200$ nm. For 100 keV electrons, the de Broglie wavelength is $\lambda \approx 0.04$ nm, but the electron beam was focussed by magnetic lenses which suffered from severe spherical and chromatic aberration. This restricted focussing to a small angular range and to a much degraded resolution of about $100\lambda = 4$ nm. Nonetheless, this resolution is a very significant improvement over the optical microscope. By the mid 1940s electron microscopes were designed and manufactured with theoretical resolutions of 2 nm.

Ellis Cosslett's lifelong involvement in the development of electron microscopes and their applications began in Bristol in 1933 while studying the papers by Knoll and Ruska on the electron microscope (Knoll and Ruska, 1932a,b). The papers described the construction of the first two-stage transmission electron microscope (TEM) (Box 12.3). Cosslett was probably the first person in the UK to appreciate the profound significance of this work. After completing an MSc degree at the University of London, he moved to Oxford University on the outbreak of war in 1939, where he set up a primitive electron-optical bench in the Electrical Laboratory to study the aberrations of magnetic lenses.

In 1942 six RCA electron microscopes were imported to the UK, one of them coming to Cambridge University, much to the chagrin of Oxford. With the encouragement of Bragg, Cosslett moved to Cambridge in 1946 to take charge of the RCA TEM (Figure 12.12(a)), supported by a three-year ICI fellowship; he was destined to spend the rest of his life in Cambridge. In the same year, a 1943 Siemens 'Übermikroscop' electron microscope was acquired as war booty from the Krupps armament factory and presented to the Laboratory (Figure 12.12(b)). Cosslett was joined by Robert Horne, an expert in electronics from the Royal Air Force. Cosslett's achievements were to be in the field of electron optics, the development of electron microscopes for X-ray microanalysis and his management of electron microscope services for use by the research groups in the Laboratory and other University departments. He left the scientific exploitation of the facilities to others. Within the Laboratory, these were to be particularly developed by members of the Crystallography and Metal Physics groups. In his 1950 Report to the University, Bragg reported that 2560 photographs had been taken by the two machines, including images for researchers in animal pathology, biochemistry, botany, colloid science, geology, metallurgy, parasitology, pathology, physiology and zoology (Bragg, 1951). Prominent workers in Cosslett's

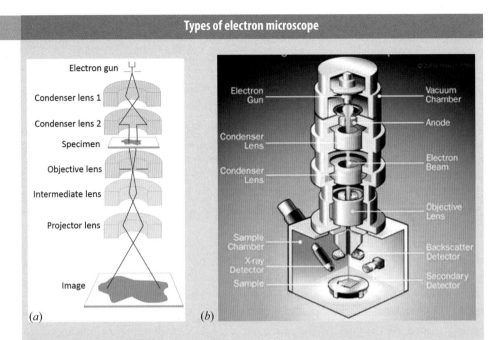

Box 12.3   **Types of electron microscope**

Transmission Electron Microscope (TEM) The TEM (*a*) uses a high-voltage electron beam directed by the condenser lens at a thin slice of the sample. The image is then formed from the transmitted electrons by a series of magnetic lenses.

X-ray Microanalyser Two sets of deflection coils (not shown) scan the electron beam across the specimen. and a scanned image is constructed as in a scanning electron microscope (SEM) (*b*), but an X-ray detector is used rather than an electron detector. The interaction of the electron beam with atoms on the sample's surface cause X-ray line and continuum emission, allowing local variations in the concentration of the chemical elements to be measured (diagram courtesy of the World Wide Web and Jonathan Atteberry).

Scanning Transmission Electron Microscope (STEM) The scanning transmission electron microscope (STEM) combines the high-resolution advantage of a TEM with the SEM's ability to handle many signals. It requires a very bright X-ray source, usually a field emission gun, and was pioneered by Albert Crewe in Chicago. Between 1975 and 2000 the applications of STEM were greatly developed by the Microstructural Physics Group at the Laboratory by A. (Archie) Howie and L.M. (Mick) Brown, culminating in the correction of spherical aberration in the STEM by Ondrej Krivanek.

Electron Microscope Group included William Nixon, Peter Duncumb, Ray Dolby, who was to gain fame and fortune as the inventor of the Dolby noise-reduction system for high-fidelity musical reproduction, Ken Smith, who also worked on nuclear magnetic resonance (NMR), and David Smith.

In the earliest versions of the TEM, the contrast in the images resulted from the different absorption coefficients of the materials in the specimen under study, as illustrated in the

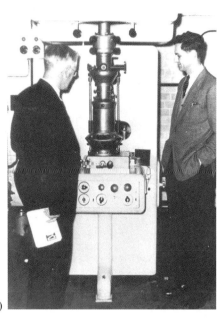

(a) The original RCA electron microscope. At voltages up to 60 kV, it gave a resolving power of 5–6 nm. (b) The 1943 Siemens electron microscope. It was requisitioned from the Krupps armament factory after the war and presented to the Laboratory in 1946. Cosslett is on the left and Horne on the right.

Fig. 12.12

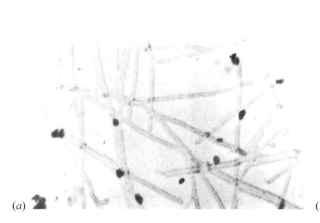
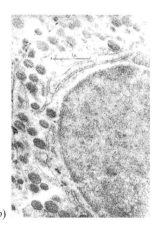

(a) A photograph of the tomato aucuba mosaic virus taken with the RCA TEM in 1947. (b) One of a series of TEM images of ultra-thin slices through the pituitary gland of a mouse, taken in 1959.

Fig. 12.13

schematic illustration of a TEM in Box 12.3. An early example of what could be achieved by the RCA electron microscope in 1947 is shown in Figure 12.13(a), a photograph of the tomato aucuba mosaic virus. Only a few viruses are big enough to be seen with an optical microscope, but with the electron microscope their exact size and shape could be studied.

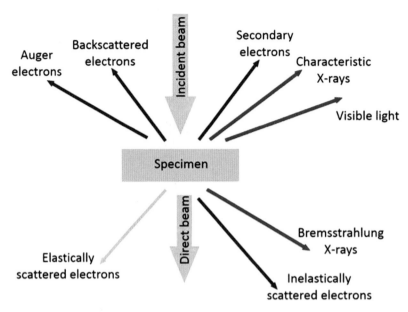

Fig. 12.14   A schematic diagram illustrating the types of interaction which a focussed beam of electrons makes with a target.

From the beginning, the major role which electron microscopy had to play in biology was fully appreciated. Most biological and organic specimens are radiation-sensitive and so only a low intensity of electrons can be used. Furthermore, only very thin samples of the biological material can be studied in this way. The 1959 photograph in Figure 12.13(*b*) shows an image of one of a series of ultra-thin slices through the pituitary gland of a mouse. The large, almost featureless area is the cell nucleus. The pores in the wall of the nucleus are clearly shown, connecting the nucleus to the rest of the cell; previously, the existence of these pores had only been postulated.

The electron microscope can be operated in many different modes. Figure 12.14 illustrates the various types of emission which occur when a focussed beam of electrons interacts with a target. The physical processes involved can be summarised as follows:

- Absorption of the electrons in the incident beam by the material of the specimen, the original mode of operation of the transmission electron microscope
- Backscattering of electrons from the incident beam
- Emission of secondary electrons caused by the interaction of the electron beam with the atoms of the specimen
- Emission of characteristic X-rays as a result of the excitation of atoms in the surface layers of the specimen. The X-ray spectra include the K and L X-ray emission lines, which can be used to characterise the chemical composition of the surface material at high resolution. By scanning the beam across the sample, a map of the chemical composition over the surface can be created
- Emission of bremsstrahlung X-rays in encounters between the electron beam and the nuclei of atoms in the specimen

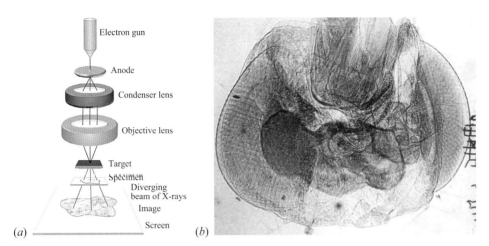

(*a*) The layout of the point-projection X-ray microscope built by Cosslett and Nixon. (*b*) A detail of the eye of the fruit fly, showing clearly that it is made up of a hexagonal network of small lenses (photography by W. Nixon).

Fig. 12.15

- Detection of elastic and inelastic scattering of electrons which pass through the sample
- Electron beam induced conductivity and cathodoluminescence, particularly in insulators and semiconductors.

These types of electromagnetic and electron processes can be studied by placing different types of X-ray, optical and electron detectors about the sample, as indicated schematically in the sketch of a scanning electron microscope in Box 12.3. There are consequently many variants on the theme of electron microscopes; three examples are described in Box 12.3.

The optical system of an electron microscope may also be used to set up a point-projection X-ray microscope (Figure 12.15(*a*)). An intense point source of X-rays is created by concentrating an electron beam to a focus 1 μm or less in diameter on a target. This was shown to be practicable using magnetic lenses developed by Cosslett and Taylor in 1947. The specimen is placed close to a point source of X-rays and the image is magnified as the X-ray beam diverges. Such a point-projection X-ray microscope was built by Cosslett and Nixon in 1951 and was the first X-ray microscope with reasonably high power with resolution as good as an optical microscope (Cosslett and Nixon, 1951). This type of X-ray microscope can be used to examine materials which are opaque to light, such as metals, ores, rocks, bone, teeth and wood, as well as the structure of transparent materials. It is very useful in studying living materials, which have to be sectioned to be studied in an optical microscope. An example of the imaging of the eye of a fruit fly is shown in Figure 12.15(*b*).

The scientific exploitation of the full capabilities of electron microscopy was to be undertaken by the next generation of electron microscopists, with major developments in the resolution of the microscopes, thanks to the development of improved aberration correctors and of new techniques for the detection of electrons and X-rays. Imaging studies of dislocations in metals were to prove to be one of the highlights of the science produced under Bragg's successors (see Sections 15.2 and 18.3).

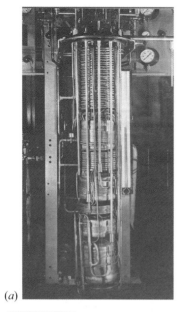

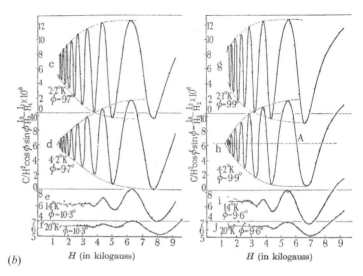

(a)   (b)

Fig. 12.16   (a) Ashmead's helium liquifier (Ashmead, 1950). (b) Shoenberg's measurements of the de Haas–van Alphen effect in bismuth at different temperatures and crystal orientations relative to the axis of the torsion pendulum and magnetic field direction (Shoenberg, 1939). These data were obtained in 1938 while he was a visitor at Kapitsa's institute in Moscow.

12.9 Low-temperature physics

Soon after the end of the war Jack Allen accepted a professorship at St Andrews University, and David Shoenberg took over the headship of the Mond Laboratory. By 1947 there was a need for larger supplies of liquid helium than could be provided by Kapitsa's liquefier, which had done noble service for ten years. The challenge was taken up by John Ashmead, and his new machine was completed in 1949 (Ashmead, 1950). He put great emphasis on simplicity of operation and ease of servicing. Liquefied helium was produced by Joule–Thomson expansion rather than by adiabatic cooling, as favoured by Kapitsa (Figure 12.16(a)). Another difference was that Ashmead enclosed the helium liquefier and the closed-cycle hydrogen liquefier within a single container, the helium being pre-cooled to 14 K by liquid hydrogen. This combination had three advantages: the hydrogen seldom had to be purified, the hydrogen explosion hazard was much reduced and the time taken to start the machine was short. In normal operation, the Ashmead liquefier produced 3.8 litres per hour, the helium part of the apparatus usually attaining about 75% of the theoretical maximum efficiency.

The Mond Laboratory kept the Laboratory glassblower well occupied, for the research students normally had their own cryostats, each comprising an outer glass Dewar flask containing liquid nitrogen and an inner one filled with liquid helium which also contained the experimental equipment, usually constructed out of thin-walled Cu–Ni tubing to reduce heat losses. The apparatus would be pre-cooled with liquid nitrogen and then carried bodily to the liquefier to be filled with liquid helium, the gas boiled off in this process being carefully collected. On return to the research laboratory the cryostat would be connected to an elaborate helium recovery system, which contained high-volume pumps that could be used to reduce the boiling point of the helium from 4.2 K to about 1.2 K. Lower temperatures could be reached using booster diffusion pumps.

All this activity was supported by the dedicated workshop set up by Kapitsa, and by a team of remarkably skilled instrument makers under the eagle eye of chief assistant Frank Sadler, who also ran the liquefier.

12.9.1 Shoenberg and the de Haas–van Alphen effect

Shoenberg began his studies in low-temperature physics in 1933, as Kapitsa's first graduate student (Section 10.3). The initial experiments suggested by Kapitsa concerned the magnetostriction of a single bismuth crystal, which introduced Shoenberg to this remarkable material. He then turned his attention to the de Haas–van Alphen effect (de Haas and van Alphen, 1931). The effect had been predicted by Landau (1930) and is a result of the quantisation of electron energies in an applied magnetic field. In the plane normal to the magnetic field direction, the energy of a free electron takes quantised values which are odd multiples of $\mu_B H$ and also of the magnetic moment associated with electron spin, where μ_B is the Bohr magneton. As expressed by Pippard (1995),

> [In] semi-classical terms, the orbits permitted to the electrons are such as to enclose a half-integral multiple of the flux quantum h/e. Landau saw that as H was changed and the energy levels moved through the Fermi level the energy of the whole electron assembly would undergo periodic fluctuations and consequently the magnetic moment would fluctuate with field strength.

Because of thermal smearing, the effect is visible only at low temperatures and in high magnetic fields. Since a level passes through the Fermi surface whenever $(2n+1)\mu_B H = \epsilon_F$, the fluctuations are periodic in $1/H$, as may be seen in Shoenberg's early results, illustrated in Figure 12.16(b).

As described in Section 10.4.2, Shoenberg's Moscow experiments of 1937–38 were far superior to those of de Haas and van Alphen, and stimulated Landau to carry out further theoretical investigations. The intention was that he and Shoenberg should write up the results in a joint paper, but Landau was incarcerated before the work could be completed. In the end, Landau's work was put together by Peierls and published as a short appendix to Shoenberg's paper with appropriate acknowledgements (Shoenberg, 1939). Shoenberg found that the electrons rather precisely obeyed Fermi–Dirac statistics with a tensorial effective mass, as predicted by Landau. These studies were the beginning of Shoenberg's lifelong involvement with the de Haas–van Alphen effect.

After the war, Jules Marcus (1947) discovered the effect in zinc which, as remarked by Pippard, 'is by no means an eccentric metal like bismuth'. It turned out that the effect was common in metals, and Shoenberg immediately rose to the challenge. Using the torsional balance technique, he found the effect in gallium, tin, graphite, cadmium, aluminium, mercury and thorium (Shoenberg, 1952).

12.9.2 Pippard and non-local theories for normal metals and superconductors

Following his return from wartime radar research, Brian Pippard joined the Mond Laboratory as a graduate student under the supervision of Shoenberg. Ashmead, who had been Pippard's section leader at RRDE, persuaded Shoenberg that Pippard should repeat some measurements made by Heinz London in 1940 of the microwave surface resistance of superconducting tin (London, 1940).

It was well known that there is a *skin effect* at high frequencies, meaning that microwaves can only penetrate a short distance into a metal. The analysis of this effect is similar to London's analysis for the superconducting penetration depth at zero frequency, and so the question naturally arose as to what would happen if microwaves were applied to a superconductor. At finite frequency, according to the two-fluid model, it would be expected that an effective complex conductivity would take the form $\sigma = \sigma_1 - i\sigma_2$, where σ_1 represents the real ohmic response of the normal electrons and the imaginary part σ_2 arises from the response of the superelectrons according to the first London equation, which describes their acceleration by an electric field (equation (10.2) in Section 10.4.1). According to the theory of the skin effect, a superconductor should present to its surroundings a microwave surface impedance

$$Z_s = R_s + iX_s = \left(\frac{i\omega\mu_0}{\sigma_1 - i\sigma_2} \right)^{1/2}. \tag{12.1}$$

Heinz London had obtained some useful data on R_s at 1.5 GHz by measuring the microwave heating of a sample of superconducting tin, his idea being that this in turn would give useful information about the conductivity σ_1 of the normal electrons.

Pippard took full advantage of the opportunity. He had the tools available, since he had brought state-of-the-art klystrons, crystal rectifier cartridges and bits and pieces of waveguides back to Cambridge – the authorities apparently had no objection to the recycling of the fruits of wartime research in a civilian context. With his experience of microwave circuits and the new technologies, he found it an 'easy task' to repeat the experiments with much enhanced precision. He achieved this, not by measuring the microwave heating as Heinz London had done, but by allowing the superconductor to form part of a microwave resonator. R_s could be obtained from measurements of the bandwidth of the resonance response. Later he realised that by observing changes in the frequency of resonance he could also simultaneously obtain values of the surface reactance X_s. This vital development made it possible to determine simultaneously and independently both the normal conductivity σ_1 and the superelectron response σ_2, from which the penetration depth λ could be deduced. This opened the way for a series of measurements by Pippard and his students on superconductors of many types and in various limits, which extended over many years. The experiments did much to clarify the dynamics of both normal electrons and superelectrons.

In the course of this work on superconductors, Pippard also made highly significant measurements of the surface impedance of normal metals, reaffirming Heinz London's

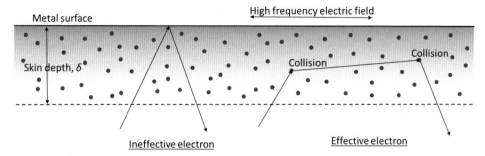

Metal surface

High frequency electric field

Skin depth, δ

Collision

Collision

Ineffective electron

Effective electron

Illustrating Pippard's concept of 'effective' and 'ineffective' electrons.

Fig. 12.17

observation (London, 1940) that in normal metals the surface resistance did not continue to decrease with falling temperature as the direct-current resistance did, but reached a limit. London had suggested that this *anomalous skin effect* occurred when the mean free path of electrons became greater than the skin depth. Ohm's law $J = \sigma E$ would then no longer be valid because electrons arriving at a particular point r would, after their velocity had last been randomised by scattering, have been accelerated by the electric fields at other points r' which might be some distance from r. Pippard worked out the details of this effect, and his conclusions were re-expressed by his fellow student Robert (Bob) Chambers in the form

$$J(r) = \frac{e^2 k_F^2}{4\pi^3 \hbar} \int \frac{[E(r') \cdot R]R}{R^4}\, e^{i\omega R/v_F}\, e^{-R/\ell}\, d^3 r', \qquad (12.2)$$

where $R = r - r'$ and ℓ is the electronic mean free path. When, however, this non-local form of current–field relation was included in the analysis of the skin effect, it led to an apparently intractable integro-differential equation.

Pippard was, however, able to develop a semi-empirical theory of this anomalous skin effect using a simplified model in which he regarded as 'effective' only those electrons moving at such a small angle to the surface that they remained within the skin depth between scatterings (Figure 12.17). These electrons would experience a more or less constant field and so make a full contribution to the surface current. He predicted correctly that in the limit of long free paths, the *extreme anomalous limit*, which for pure metals was also the low-temperature limit, the surface resistance should be independent of the free path and a simple function of the density of conduction electrons, varying as the two-thirds power of the frequency.

Investigations of the anomalous skin effect were taken up with enthusiasm by Chambers, who extended them to a wide range of metals. The integro-differential equations were eventually solved with the help of Pippard's theorist friend Ernst Sondheimer, with the assistance of the pure mathematician Harry Reuter. This exact theory of the anomalous skin effect was subsequently shown to fit all the data remarkably closely (Reuter and Sondheimer, 1956).

In 1947, Pippard was appointed a University Demonstrator, and in 1950 promoted to a University Lectureship. At about the same time, he made two significant new observations of the superconducting penetration depth, which he was able to deduce from the microwave surface impedance (Pippard, 1953). First, he measured the skin depth of tin alloyed with indium to reduce the free path of the electrons. He observed that the alloying not only increased the normal state skin depth as expected, but also, and in much the same way, increased the superconducting penetration depth. Second, he measured the anisotropy of the penetration depth in pure tin and found that it varied with surface orientation in a non-tensorial manner that was not possible if the superelectrons obeyed the second London equation. The latter can be rewritten in the local form $\Lambda J_s = -A$, where A is the magnetic vector potential expressed in a suitable gauge. It was, however, very similar in form to the anisotropy of the normal state skin depth, governed by Chambers' non-local equation.

Pippard (1953) drew the bold conclusion that he must be observing non-local effects in the supercurrent, and in his paper proposed that the second London equation should be replaced by

$$\Lambda J_s(r) = -\frac{3}{4\pi \xi_0} \int \frac{\left[A(r') \cdot R\right] R}{R^4} \, e^{-R/\xi_0} \, e^{-R/\ell} \, d^3 r', \qquad (12.3)$$

a non-local equation analogous to Chambers' formula (12.2). In this picture, the supercurrent depended on distant values of the vector potential, the range being limited by the electronic mean free path as in the normal metal. Pippard decided, however, that even in pure material with no scattering there must be some limit to the range over which the vector potential could affect the supercurrent and so he added the extra term e^{-R/ξ_0}, where ξ_0 was to be interpreted as an *electromagnetic coherence length*, which he suggested might be of order $\hbar v_F / k T_c$, the range over which the one-electron states involved in the superconducting condensation would get out of phase with each other. This proposal was initially controversial, but it fitted the surface impedance data very well and was eventually completely vindicated.

12.9.3 Magnetic fields in superconductors

According to the Meissner effect, if a magnetic field is applied to a long rod of a superconductor, the field is excluded from the rod, apart from the very small penetration depth. But the Leiden group had discovered that the common superconductors such as tin and lead had a *critical magnetic field* B_c, typically a few hundred gauss. When the parallel field exceeded B_c, a phase transition occurred, flux entered the rod and the normal state was restored.

But if, on the other hand, the field was applied normal to the rod, something different happened. Initially the magnetic flux was excluded as before by screening supercurrents flowing around the surface of the rod. But when the applied field reached $\frac{1}{2} B_c$, the magnetic flux gradually began to penetrate; penetration was not complete until the applied field reached B_c. Superconductors that behave in this way are known as *type I superconductors* and are said to be in the *intermediate state*. The essential theory had been elucidated in the 1930s by Peierls and by Fritz London, and later by Landau: the state consists of normal and superconducting domains parallel to the field, the local field inside the normal domains being equal to B_c; as the applied field increases so does the proportion of normal material. The domain size is typically about 0.15 mm.[12]

The original analyses of the intermediate state by Peierls and London were based on the idea that the normal–superconducting (NS) interface had a positive surface energy, and in 1951 Pippard suggested that the variation in superconducting order with position at the interface could not be immediate but must be spread over a characteristic *coherence length* ξ. He also pointed out that the boundary energy would only be positive if this coherence length were greater than the penetration depth λ. Since λ was known to diverge at T_c, it was evident that the same must be true of ξ.[13]

The behaviour of the system at the normal–superconducting boundary was soon illuminated by the remarkable theory produced by Vitali Lazarevich Ginzburg and Lev Landau (1950).[14] They proposed a semi-empirical expression for the free energy of a superconductor, written in terms of an order parameter $\Psi(r)$ which they boldly assumed, following the ideas of Fritz London, was *complex*: the square of the amplitude of the order parameter $|\Psi|^2 = \Psi\Psi^*$ represented the density of superelectrons n_s, but its phase was assumed to behave under a gauge transformation like the phase of a single-particle wave function. By minimising the free energy with respect to Ψ they obtained the remarkable *Ginzburg–Landau* equation,

$$\frac{1}{2m}(-i\hbar\nabla + 2eA)^2\Psi + (\alpha + \beta\Psi\Psi^*)\Psi = 0, \qquad (12.4)$$

where α and β were parameters to be fitted to the known variation of the bulk free energy, penetration depth, and so on. This remarkable blend of quantum theory and thermodynamics could be solved to show how Ψ varied with position at a normal–superconducting boundary – and their solution had a coherence length, the Ginzburg–Landau coherence length ξ_{GL}, that diverged at T_c just as Pippard had expected.

Ginzburg and Landau could also minimise the free energy with respect to the magnetic vector potential, but this simply led back to the second London equation for the supercurrent, which was a local equation. Because of this, and for other reasons, Landau rejected Pippard's proposal that the supercurrent showed a non-local response to the vector potential, as he made clear forcibly in early 1957 when Pippard encountered Landau, Ginzburg and Abrikosov in Moscow. However, after the microscopic BCS theory of superconductivity was published later the same year and the Moscow theoreticians had studied it, they generously sent Pippard a private message conceding that he had been right.

Ginzburg–Landau theory rapidly became extremely important in understanding the behaviour of *type II superconductors* (Section 15.1.3).

12.10 Other research activities

In addition to the large groups in the Laboratory, there were a number of smaller activities, which would eventually be absorbed into new groups, be transferred to other departments or end their natural research life-spans.

12.10.1 Fluid dynamics

G.I. Taylor's pioneering work in fluid dynamics continued throughout the period of Bragg's tenure as Cavendish Professor and his activities, particularly in the area of turbulence, were fostered by the arrival from Australia of George Batchelor and Alan Townsend in 1945 as graduate students. Townsend has given a delightful account of their activities in the period 1945 to 1956 (Townsend, 1990). During the war, both of them had worked at the CSIR Aeronautical Laboratory in Melbourne, Batchelor in the Aerodynamics Section and Townsend in the Instruments Section. Batchelor was to devote his research to the theory of turbulence, particularly to the statistical theory which had been pioneered by Taylor in his classic pre-war papers (Taylor, 1935a,b,c,d), with which Batchelor was already familiar. Townsend was to develop experimental techniques for the study of turbulence under a variety of different experimental arrangements using hot-wire anaemometers.

These activities culminated in two books, Batchelor's monograph *The Theory of Homogeneous Turbulence* (Batchelor, 1953) and Townsend's *The Structure of Turbulent Shear Flow* (Townsend, 1956), both of them representing the state of the art at the time. Batchelor subsequently abandoned active research and devoted his energies to establishing and directing the Department of Applied Mathematics and Theoretical Physics.

Townsend became an internationally recognised leader for his experimental and analytic studies of turbulent motion in fluids. He was the first to design apparatus to measure the triple and quadruple correlations in turbulent flow, the mean squares of first and higher time derivatives of velocity fluctuations and many other important statistical parameters of turbulence. He went on to make pioneering measurements of turbulent diffusion, all these researches requiring outstanding experimental skills.

12.10.2 Metal physics

Egon Orowan was one of the three pioneers of the role of dislocations in the theory of plasticity and the strength of materials, along with Geoffrey Taylor and Michael Polanyi (see Section 10.5.1). Orowan was appointed to a post in the Laboratory in 1939 and in 1947 was promoted to a readership. During the war years, he made several notable contributions to the technology of munitions production. Perhaps the most impressive paper was entitled *The calculation of roll pressure in hot and cold flat rolling* (Orowan, 1943), which Nabarro and Argon (1995) describe as a 'formidable sustained effort of applied mechanics' and which elucidated the processes involved in plastic flow during metal rolling. Another major achievement was the understanding of the nature of brittle fracture in welded-steel 'Liberty' ships, which led some of them to sink during transatlantic crossings (Orowan *et al.*, 1944). The group had excellent graduate students, many of whom became leaders in the fields of metallurgy and materials science rather than as physicists.

Curiously, although Orowan was highly regarded, he published remarkably few papers during the period from 1946 until he left the Laboratory in 1950 to take up a professorship at the Massachusetts Institute of Technology. In addition, although of high quality, these papers tended to be published in obscure journals. Bragg's report for the academic year

1949–50 lists eleven papers from the group in major journals, only two of them with Orowan as an author. Perhaps surprisingly, Bragg makes no mention of the Metal Physics Group in his report for the academic year 1952–53 (Bragg, 1954). The activity was, however, about to flourish through advances in electron microscopy techniques (Section 15.2) and reappeared in Mott's biennial reports. Eventually, the group was renamed the Microstructural Physics Group.

12.10.3 Meteorological physics

The Laboratory had maintained an interest in meteorological physics from the Thomson era through the activities of William Napier Shaw, who became Director of the Meteorological Office in 1905 (Section 4.4) and, more significantly, by C.T.R. Wilson, whose cloud chamber experiments had been designed to provide answers to meteorological questions, particularly concerning the origin of clouds (Section 6.3). Wilson also had a long-standing interest in atmospheric electricity which he pursued after he moved his experimental apparatus from the overcrowded Cavendish to the Cambridge Observatories. In 1925 Thomas Wormell was appointed Observer in Meteorological Physics and began his graduate studies under Wilson in the area of the electrical effects of thunderstorms. In 1950, he was appointed Lecturer in Meteorological Physics and transferred his activities from the Observatories to the Cavendish Laboratory. At the same time, the books bequeathed to the University by Napier Shaw were transferred to the Laboratory. With the move of the Cavendish Library to the Austin Wing, Bragg created the Napier Shaw Library in the old Cavendish Library, where it also acted as a study area for graduate students.

Wormell's interests continued to be principally in the area of lightning and the resulting 'atmospherics', which were of interest to Ratcliffe and Budden in the Radio Group. The group also carried out studies of radiative transfer in the atmosphere, supported by laboratory experiments on the infrared absorption spectrum of ozone and its dependence on pressure. These studies were used to estimate the mean height and properties of the ozone layer and how it changed with the seasons and with day-to-day atmospheric conditions. Members of the group also used radar techniques to study the structure and development of rain clouds, including Doppler measurements of the vertical motions of the precipitation particles. The group continued these studies until Wormell's retirement in 1971.

12.10.4 Theoretical physics

Rutherford had an ambivalent attitude towards theorists, but had benefitted enormously from the presence of Ralph Fowler. Fowler had an enormous influence on the development of theoretical physics research – it is estimated that, during the period 1922 to 1939, he supervised 64 graduate students, about 11 at any one time. With Fowler's untimely death in 1944, a large gap was left in the theoretical physics expertise directly available to the Cavendish experimentalists. In 1945, the appointment of Douglas Hartree to the John Humphrey Plummer Professorship of Mathematical Physics, which was assigned to the Cavendish Laboratory, helped solve the problem. In addition, two lectureships in theoretical physics were established and were filled by Harold Temperley and John Scott, the latter in the field of theoretical electrodynamics and radio.

The appointment of Hartree was particularly important for members of the Laboratory and for the University as a whole. His area of expertise was numerical analysis, in which he was regarded as a world authority (Darwin, 1958). He was the originator of the self-consistent field approach to solutions of the Schödinger equation for many-electron atoms. The inclusion of exchange forces by Vladimir Fock resulted in the Hartree–Fock self-consistent field model for determining the electron distribution in complex atoms by numerical analysis.

In the 1930s, these iterative calculations were carried out using mechanical calculators, in particular by the differential analyser built by Vannevar Bush in the USA. Hartree built the first such machine in the UK and used it to solve differential equations applicable in many fields. After the war, he realised that such machines would be superseded by electronic calculating machines. On his appointment to the Plummer Chair in Cambridge, he devoted a great deal of effort, working with Maurice Wilkes, to develop the algorithms and procedures which would be at the heart of the solution of ordinary and partial differential equations by the EDSAC and EDSAC 2 computers. In addition, he was notably generous in helping his colleagues with the solution of theoretical problems which required his particular skills in numerical analysis. The huge range of topics in which he became involved can be appreciated from the list of papers in Darwin's biographical memoir of Hartree (Darwin, 1958).

Bragg's clear vision of the role of theorists at the Laboratory is best summarised in his own words:

> Most of the work in theoretical physics is closely related to one or another of the investigations in the experimental sections. It is becoming increasingly difficult to find space for the theoreticians who like to have a base in the Cavendish Laboratory because of the contact with the experimental work which it makes possible. In all, some 20 have been accommodated with working facilities. Their lines of interest are so varied that it is difficult to classify them. Broadly speaking they fall into the following groups:
>
> - Electronic structure of atoms.
> - Application of the EDSAC, the electronic computer in the Mathematical Laboratory, to various calculations in mathematical physics.
> - Statistical mechanics. Ferromagnetism, ferroelectricity, order-disorder phenomena, superconductivity, electron theory of metals, theory of liquids.
> - Fluid dynamics.
>
> Each of the sections concerned with nuclear physics, radio waves, and low-temperature physics has associated with it a member of the University teaching staff who is a theoretical physicist and can direct the Research Students who are doing theoretical work and advise the experimentalists. (Bragg, 1951)

There would be major developments in theoretical physics, but Bragg's philosophy set the basic pattern within the Laboratory for the interaction between theory and experiment. This would be formalised under Mott with the formation of the Solid State Theory Group, while there remained theorists primarily associated with the experimental

The Cavendish Collection of Historic Scientific Instruments in the Bragg Building of the Cavendish Laboratory at West Cambridge. The display cabinets are the same as those in the Austin Wing of the old Laboratory and contain pieces of apparatus up to about 1939.

groups. An important development initiated by Bragg was the formal introduction of undergraduate courses in theoretical physics into the physics syllabus (Section 12.2).

12.11 The Cavendish Collection of Historic Scientific Instruments

Bragg was particularly concerned about the preservation of the historic instruments which had been used by members of the Laboratory in their pioneering researches, and made a determined effort to locate and preserve the more important of these, as well as numerous letters, manuscripts and papers. Derek de Solla Price, later to become a distinguished historian of science, undertook the cataloguing and organisation of the Cavendish archives and the collection of scientific instruments. Bragg found space in the Austin Wing for four large display cases in which the more important pieces of apparatus used by members of the Laboratory from the time of Maxwell onwards were displayed. This remarkable collection was transferred to the Bragg Building when the Laboratory moved to its new site at West Cambridge in 1974 (Figure 12.18).

12.12 The end of the Bragg era

Bragg had had an association with the Royal Institution of Great Britain as a non-residential Professor of Natural Philosophy since 1937. In this position, he lectured each year at the

Friday Evening Discourses. His father was the resident Professor of Natural Philosophy until his death in 1942. As a non-governmental organisation, the financial position of the Royal Institution was always somewhat precarious, and matters worsened during the early 1950s as a power struggle developed between the President of the Institution, Lord Brabazon, and the incumbent resident Professor of Natural Philosophy, Edward Andrade. The tortuous history of the events of 1952–53 resulted in the departure of Andrade and the offer of the post of residential Professor of Natural Philosophy to Bragg in April 1953. Out of regard for his father's legacy and loyalty to the Royal Institution, he accepted the post and agreed to take up the position from 1 January 1954. When he was offered the post, he was already 63 and approaching the end of his tenure as Cavendish Professor. For the third time in his career, he faced the arduous task of regenerating a major academic institution. He devoted his energies to restoring the fortunes of the Royal Institution and promoting science for young people, particularly through his much-loved and brilliant Christmas Lectures which, through the medium of television, reached a huge audience.

Bragg could be justly proud of his achievements as Cavendish Professor. Comparison of the profile of the Laboratory under his great predecessor and its complete restructuring in the post-war years demonstrates his astonishing achievement. By the time of his departure, the structure of DNA had been solved by Crick and Watson, radio astronomy under Ryle was already making a major international impact, low-temperature physics was at the forefront of research internationally and electron microscopy was adding completely new dimensions to studies in condensed matter physics. This was achieved by changing the management structure of the scientific programme from a monolithic entity to a group structure which allowed the individual groups to determine their own priorities for research, with a light but encouraging touch from the Director of the Laboratory. But there was more to the success than that. Bragg's appointment was originally seen as a disappointment by many and yet he had the extensive vision to realise that the new physics was not to develop from well-trodden paths but from quite new directions, often involving cognate disciplines such as chemistry, metallurgy, biology and engineering. This may seem obvious with hindsight, but to have the foresight to realise this and then demonstrably deliver the new science was a great achievement.

PART VII

1953 to 1971

Nevill Francis Mott (1905–1996)
Oil painting by Paul Gopal-Chowdhury

The Mott era: an epoch of expansion   13

Nevill Mott's appointment as Cavendish Professor in 1954 marked the beginning of further changes of emphasis in the scientific direction of the Laboratory. While he worked closely with experimentalists, he carried out no experimental physics research himself, despite being Cavendish Professor of Experimental Physics. He was unquestionably the leading UK theorist in solid state physics and he was to consolidate and expand considerably both theory and experiment within the Laboratory.

13.1 Mott's pre-Cavendish days

Mott's pedigree in physics was impeccable. His parents, Charles Francis Mott and Lilian Mary Reynolds, had been research students in the Cavendish Laboratory, both appearing on the 1904 photograph of Cavendish graduate students. Mott read mathematics at St John's College, Cambridge in 1924, which he completed with high distinction. His undergraduate studies spanned the dramatic years when quantum mechanics was discovered with its profound implications for physics at the atomic level – Mott was determined to master the new discipline. He was in the same college as Dirac and, despite Dirac's notorious reticence, benefitted from the presence of the only quantum theorist in Cambridge on the same level as the great European pioneers. Mott's PhD supervisor was Ralph Fowler, who arranged for him to spend the autumn of 1928 working with Bohr in Copenhagen. In the same year, he made his important contribution on the scattering probabilities of particles which obey Bose–Einstein statistics (Mott, 1928). Mott's predictions were confirmed by Chadwick's experiments (Blackett and Champion, 1931), which greatly impressed Rutherford (Section 9.3).

In 1929, Mott accepted Lawrence Bragg's offer of a lectureship at the University of Manchester, remarking in his autobiography that 'Cambridge and Manchester were the only worthwhile schools of physics in the UK (with Bristol just beginning . . .)' (Mott, 1986). Although he only stayed a year at Manchester, it was an important year for Mott intellectually. As he put it,

> In fact, [the Braggs] were the originators of a new science, the scientific investigation of the structure of crystals, and that meant of most materials . . . [Lawrence] Bragg was the world leader in this subject. It was a change from Rutherford's Cavendish and indeed Bohr's Copenhagen, where the atom, the nucleus and the electron were the thing. Here one asked, and found out, how atoms were put together to form real materials. It was

321

an introduction to 'solid state physics', in which I made my career from 1933 onwards. (Mott, 1986)

During the year, Mott delivered a course of lectures on wave mechanics, which led him to write his first book, *An Outline of Wave Mechanics* (Mott, 1930), at the age of only 25.

In 1930 Mott took up the posts of University Lecturer in the Faculty of Mathematics and Fellow of Gonville and Caius College at Cambridge, where he was to stay for the next three years. In fact, he was just as deeply interested in the work going on in the Cavendish Laboratory as the activities in the Mathematics Faculty and was present throughout the *annus mirabilis* of 1932 (Sections 9.5 and 9.6). In 1933, he accepted the Chair of Theoretical Physics at the University of Bristol, where he was to remain for 21 years until he took up the Cavendish Chair.

At Bristol he established his preferred mode of research, opening up new fields in the general area of solid state physics, working closely with experimentalists. He remained in close touch with the experimentalists throughout his career, even after he had created the Solid State Theory Group in the Cavendish Laboratory. Once he had opened up a new field, he tended to move on to other topics rather than become a specialist in a particular area. An example of this is the follow-up to his pioneering work on the theory of atomic collisions. Fowler invited him to write a book on atomic collisions for the Clarendon Press and this was published jointly with Harrie Massey (Mott and Massey, 1934). This authoritative text, *Theory of Atomic Collisions*, went through several editions and expanded enormously in scale, but Mott made no contribution to the subsequent editions beyond his first efforts. Other influential books before his return to Cambridge in 1954 included *The Theory of the Properties of Metals and Alloys* (1936) with Harry Jones and *Electronic Processes in Ionic Crystals* (1940) with Ronald Gurney. Perusal of Mott's complete list of publications reveals his broad-ranging interests across the complete range of topics in solid state physics (Pippard, 1998). Given the major changes of direction initiated by Bragg, there could be no doubt that Mott was the ideal person to fill the Cavendish Chair. At the same time, he became a Professorial Fellow of Gonville and Caius College, where Chadwick, having returned from Liverpool, had become Master in 1947.

13.2 Strategic decisions in research

Mott was an effective scientific manager and administrator and well able to cope with the complexities of Cambridge University's advisory and management procedures. He quickly realised that, to be able to make any headway with the reforms and changes he wanted to implement, he had to become involved at a high level in the decision-making procedures of the University. To this end, he became a member of the General Board of the Faculties and of its Needs Committee, where priorities for posts and funding are thrashed out.[1] He faced many challenges; during his tenure of the increasingly demanding role of Cavendish Professor and Head of Physics, the Laboratory expanded considerably and plans were completed for the next phase of its development with the move to West Cambridge.

Throughout this period, his scientific creativity continue unabated and, after his retirement, he completed some of his very best work. He was working at his desk until three days before his death in 1996 at the age of 90.

Mott was immediately faced with the issue of what to do about the proposed linear accelerator which was being developed by the Nuclear Physics Group. Even before he arrived in Cambridge, his mind was made up – the accelerator was 'too little, too late'. A university like Cambridge could not compete with the major investments being made in the USA. In his own words,

> I felt this machine was a profound mistake. The Cavendish must excel or nothing. What chance had we with this machine, miles behind the Americans? And yet I knew men I admired thought otherwise. I remember a night in a bitterly cold guest room in Caius College, worrying about it in a sleepless kind of way, and deciding that I must stop it. I am not sure that this was a wholly rational decision – just something that I had to do to prove myself. Certainly, it cost me something. But in the event only a few people seemed to mind, and others admired decisive action for its own sake. I think it was the right decision. (Mott, 1986)

The General Board of the University was persuaded of his case and provided the Laboratory with a new Lectureship in Theoretical Physics, which was to be filled by the theoretical solid state physicist John Ziman, and the opportunity to bring Hans Bethe to the Laboratory for a year. This was a somewhat different management approach from that of the conciliatory Bragg.

Mott's next problem was what to do about the MRC Unit (Section 12.6). The success of Watson and Crick in disentangling the structure of the DNA molecule had been an undoubted triumph. Furthermore, Kendrew and Perutz, with their understanding of how to use isomorphous replacement to determine the phases of X-ray images, were on the brink of determining the structures of myoglobin and haemoglobin. Not unnaturally, there was an increasing demand for space and resources within the Laboratory. This in itself was a major issue, but there was the problem that the MRC Unit was a pure research organisation dedicated to molecular biology, with staff of the highest calibre who did not teach, or refused to teach in the case of Crick, and whose research students were ill-prepared to work in a physics laboratory. The MRC Unit, despite its undoubted distinction, was a burden on the Laboratory. At the same time, the numbers of students admitted by the colleges to read physics continued to increase and there was a corresponding increase in the physics teaching staff. Mott needed the space and so, in the summer of 1957, the MRC Unit was decanted into what became known as the *MRC hut* (Figure 13.1). This prefabricated temporary accommodation was located just to the east of the Austin Wing (see Figure 5.3) and had been occupied by the Metal Physics Group under Orowan. When Orowan departed for the USA, the Metal Physics activities were relocated to the Austin Wing. The intention of the University was that the hut should be demolished, but Mott was able to get the demolition order overruled. The MRC agreed to refurbish the building and Kendrew and Perutz moved into the hut – it was still there in 2016. This was where the structures of myoglobin and haemoglobin were finally solved.

Fig. 13.1 The MRC hut, to the east of the Austin Wing (courtesy of Hans Boye, MRC Laboratory of Molecular Biology).

These arrangements provided only temporary respite for the expansion of all aspects of the Laboratory's activity. With the tremendous success of the programme, the molecular biologists continued to expand in numbers and they sought space wherever they could, including further vacant rooms in the Zoological Laboratories. Mott tried to find some other department which would house them, but nothing was forthcoming. During 1957–58, Perutz and Crick made the case to the Medical Research Council that new accommodation was needed for molecular biology, and in 1958 the MRC agreed to the foundation of the MRC Laboratory of Molecular Biology. Mott was convinced that they had to find some other home. As he wrote in his autobiography,

> I judged that they would develop into a bigger thing than we could cope with and encouraged the General Board to find them a new site. (J.G. Crowther's book on the Cavendish says that I was very sad when they left, but this was not so). The space they left enabled me to build up electron microscopy, and develop the work on dislocations . . . (Mott, 1986)

There was no space left in the centre of Cambridge and so a site on the new Addenbrooke's Hospital site, next to the Radiotherapeutics Department and two miles from the centre of Cambridge, was selected. The Laboratory for Molecular Biology was opened by Queen Elizabeth in May 1962 – later that same year, Nobel Prizes were awarded to Crick, Watson, Kendrew and Perutz.

Space was still a headache, particularly for the theorists, many of whom were members of the Mathematics Faculty but who found the atmosphere in the Laboratory conducive to their research programmes. Mott was personally all in favour of encouraging theorists to work closely with the experimentalists, as he had done successfully in Bristol. In 1959 he obtained a grant of £45,000 from the University for the construction of an additional floor on the top of the Austin Wing to house the theoretical physicists and applied mathematicians.

But trouble was brewing. The Mathematics Faculty had no departmental structure, no sense of community and no offices provided by the University – the assumption was that, with the University providing lecture space and a faculty office, supervision of students and their own research would be carried out in the mathematicians' college rooms. Batchelor was the leading figure in promoting the concept that there should be a departmental structure for mathematics to enable the students and staff to interact more effectively and have their own University offices. In 1959, Batchelor's vision was fulfilled with the establishment of the Department of Applied Mathematics and Theoretical Physics (DAMTP), of which he became the first Head, and the Department of Pure Mathematics and Mathematical Statistics (DPMMS), the Head of which was William Hodge. Most of the theorists in high-energy physics, theoretical astronomy and fluid and solid mechanics became members of DAMTP, Dirac and Fred Hoyle becoming the senior professors. On its formation, DAMTP was housed in various locations within the Cavendish Laboratory but, in 1963, Cambridge University Press moved out of the Pitt Building in the centre of Cambridge and Batchelor quickly secured the old Press machine shop for the new department. At the same time, Hodge made a bid for a second building vacated by the Press for DPMMS. Both proposals were supported by the University and over the next few years these departments were consolidated in the old Press buildings.

Mott's and Batchelor's approaches to the role of theory could not have been more different. Most of the particle theorists became members of DAMTP but, on his promotion to a readership in 1964, Richard Eden transferred from the Mathematics Faculty to the Cavendish Laboratory, bringing with him his research group. He was clear that he wished to maintain close relations with the experimentalists. Eventually his group joined up with the High Energy Physics Group. From Mott's perspective, the field was open for the Laboratory to take the lead in solid state theory and he actively promoted the interests of the solid state theorists, who had the benefit of remaining close to the experimentalists. At the same time, he ensured that there were theorists, such as Eden in High Energy Physics and Scheuer in Radio Astronomy, associated with the experimental research groups.

Another opportunity to reinforce solid state physics came with the move of the Chemistry Department to its new buildings in Lensfield Road in 1958. Philip Bowden had built up a large and impressive Surface Physics Group in the Department of Physical Chemistry, as recounted in Section 15.4. Their topics of research were remarkably close to Mott's interests and often closer to physics than to chemistry. Another distinctive feature of the Surface Physics Group was that its interests were closely aligned with those of industry and much of its support was provided by industrial funding. Bowden and Mott agreed that the research of the Surface Physics Group would be best served by being transferred to the Laboratory and, after what Mott describes as 'a good deal of fixing', their aim was achieved – Mott's membership of the General Board and its Needs Committee undoubtedly helped in effecting this move. The Surface Physics Group took over most of the space vacated by the Department of Physical Chemistry in Free School Lane (see Figure 5.3). Eventually Bowden's group was renamed the Physics and Chemistry of Solids Group within the Cavendish Laboratory. This development was very much to Mott's taste. Writing in 1986, he stated:

Table 13.1 Research groups, staff and numbers in the Cavendish Laboratory in the Lent term, 1970

Research group	Head of group	Other staff members[a]	Numbers of persons in research group[b]
High Energy Physics	Frisch (P)	Shire (R), Eden (R), Kempton (L), Riley (L), Rushbrooke (L), Neale (D)	57
Radio Astronomy and Ionosphere	Ryle (P)	Budden (R), Hewish (R), Kenderdine (L), Baldwin (ADR), Scheuer (ADR), P. F. Scott (ADR), Shakeshaft (ADR)	71
Crystallography Laboratory	Taylor (R)	Megaw (L), Brown, P.J. (ADR),	31
Mond and Magnetic Laboratories (Low Temperature Physics)	Shoenberg (R)	Pippard (P), Ashmead (L), Waldram (L), Adkins (L), Josephson (ADR)	37
Metal Physics (Microstructural Physics)	Howie (L)	Brown, L.M. (L), Metherell (L),	32
Fluid dynamics	Townsend (R)	–	6
Solid State Theory	Heine (R)	[Mott (P)],[c] Anderson (P), J.M.C. Scott (L), Lekner (D)	32
Electron Microscopy	Cosslett (R)	Ferrier (ADR)	38
Meteorological Physics	Wormell (L)		10
Liquid Metals	Faber (L)		5
Slow Neutron Physics	Squires (L),		4
Surface Physics	Tabor (R)	Yoffe (R), Field (D)	67

[a] Notation: P = Professor, R = Reader, L = Lecturer, ADR = Assistant Director of Research, D = Demonstrator (Assistant Lecturer).

[b] These numbers include staff members, postdoctoral workers, research students, research assistants, computing assistants and technicians.

[c] As Mott writes in his 1970 review of the work of the Laboratory: 'Several projects in the Laboratory involve more than one section; for instance, work on electrical and optical properties of non-crystalline materials involves the sections Liquid Metals, Surface Physics, Electron Microscopy, Solid State Theory, and the Mond, and this is also the main research interest of Professor Mott.'

I think myself that this is one of the best things I did for the Cavendish. Philip Bowden died in 1968, and his unit still retains a touching loyalty to his memory. His achievement was to create a unit, within a physics department, where chemists and physicists worked together. This is the way to do solid state science, and the unit in the Cavendish has lasted till today [1986] and is very highly regarded. Since my retirement [in 1971] I have chosen to have my office in this unit. (Mott, 1986)

By the last year of Mott's tenure of the Cavendish Chair, the balance of effort between the research groups had shifted markedly. Table 13.1 gives some indication of the changes in the size of the groups and can be compared with the corresponding figures for 1950, towards the end of the Bragg era (see Table 12.1); note that the figures in the last column of Table 12.1 refer to registered graduate students, whereas those in Table 13.1 include staff members, postdoctoral workers, research students, research assistants, computing assistants and technicians. The number of academic staff had doubled over the period with a corresponding increase in research personnel. The balance of research activity had shifted markedly, but this was mostly associated with the dramatic increase in the general area of condensed matter physics, rather than with closing down activities. Mott was proud of the fact that his work on electrical and optical properties of non-crystalline materials spanned research in the Liquid Metals, Surface Physics, Electron Microscopy, Solid State Theory and Low Temperature Physics groups. If Bragg's Cavendish was full by 1950, the overcrowding in 1970 was unsustainable and the move to West Cambridge was to provide relief for the next decade or so.

It is intriguing that, while Mott accreted major activities to the department such as the Physics and Chemistry of Solids Group and started new activity in the Solid State Theory Group in the 1950s, the growth during the 1960s was largely associated with the expansion of the resources available for research and teaching in the universities. This was the period when there was a major increase in government investment in the university sector. Harold Wilson's speech of October 1963 on the 'white heat' of scientific and technological change was more than just political rhetoric, and was followed through with serious increases in funding for the universities. The Cavendish Laboratory rode the wave of these new opportunities.

The doubling of the scale of activity in the Laboratory coincided with the advent of funding for 'Big Science' in the UK. A good impression of the scale of the increase in funding for scientific research in the UK through the Mott era is provided by the article by Bernard Lovell which appeared in the 25th anniversary edition of the *New Scientist* magazine (Lovell, 1981). Lovell associated the development of 'Big Science' in the UK with his initiative in 1952 to fund the 250-foot (Lovell) radio telescope at Jodrell Bank. The DSIR and the Nuffield Foundation agreed to share the cost of £333,000 to build the radio telescope. In the same way, Ryle's success in funding his large aperture synthesis telescopes at Cambridge can be seen as part of UK investment in 'Big Science' (Chapter 14).

The background against which these development took place can be appreciated from the following extracts from Lovell's 1981 article (Lovell, 1981):

> Apart from a special annual grant of £250,000 for nuclear physics, in the five years beginning 1954–55, the DSIR Advisory Council could distribute only modest sums – £45,000 rising to £320,000 – for grants to Universities and technical colleges. As inflation in those years was only in the region of 2–4%, this represented significant growth.
>
> In the last year of its existence (1964–65) DSIR distributed £4.2 million for grants in addition to the special allocation for nuclear physics. . . . In 1980 the Science Research Council (SRC) dispensed about £180 million – a change of great significance. . . . When the SRC assumed responsibility for the research grants on the decease of DSIR, the new council embraced various activities additional to those previously supported by DSIR – especially the National Institute for Research in Nuclear Science, the Royal Observatories at Greenwich and Edinburgh, the Radio Research Station at Slough, the grants programme for the British Space Programme, and the budgets for the European Space Research Organisation (ESRO) and CERN, the European Centre for Nuclear Research. The SRC budget in that first year was £27 million.

Later, Lovell goes on:

> In the first meetings of the SRC in 1965 we were arguing that council could operate effectively only if the budget increased by 10–15 per cent per annum in real terms. The optimism of the previous decade still existed and indeed this part of the science budget continued to grow significantly. By 1970 the SRC was disposing of about £50 million per annum and the growth was still a real 5 per cent. Ten years later the chairman of the SRC complained that 'in 1979–80 the total gross expenditure of the Council was £176 million but inflation over the decade reduced its value, in terms of the amount of science we can support, to roughly the 1969–70 figure.'

Mott was undoubtedly lucky that he was Head of the Laboratory during a period when the resources for research were increasing by 5% per year in real terms essentially throughout his tenure of the Cavendish Chair. As Lovell makes clear, these halcyon days were not to last beyond about 1970.

13.3 Mott and education

When Mott returned to Cambridge he was convinced of the need for educational reforms, both of the Natural Sciences Tripos and of science education in the country as a whole. His main contribution to the reform of the Natural Sciences Tripos was to expand the amount of physics teaching which would be on offer to physics students during Part I of the Tripos.[2] When Mott arrived, physics was one of three experimental subjects which had to be taken in the second year. In Mott's view, this put Cambridge students at a disadvantage compared with other universities since by the time they arrived at the third and final year of the course, they were ill-equipped to cover the advanced topics to be included in that year. His view was strongly support by Alex Todd, the future Lord Todd and Professor of Organic Chemistry. They eventually succeeded in creating separate Natural Sciences and Medical Sciences Triposes. In addition, they proposed a solution to the need to increase the physics and chemistry content of the Tripos by the introduction of new courses entitled *Advanced Physics* and *Advanced Chemistry* in the second year. This meant that, for example, physics students in the second year could read physics, advanced physics and one other subject. Despite the objections of the biologists, who feared that this would decrease the numbers of students taking biological subjects, this reform was approved after three years of wrangling. This provided a much improved course in physics, enabling a smoother transition to the advanced topics in the third year. A similar pattern was to continue until the 1990s when physics courses throughout the UK increased from three to four years.

Another important contribution was Mott's involvement in strengthening the role of the physical science departments in the advisory and management structure of the University. Mott was convinced of the need for these departments, which required large resources to maintain teaching and research laboratories, to have a collective and coherent means of supporting these activities within the University. Following the Annan Report on the decision-making processes within the University, the Council of the School of the Physical

Sciences was given a renewed role in looking after the interests of the physical sciences. This was eventually to lead to the present arrangements in which the academic activity of the University is managed through six schools which have delegated financial authority for their programmes.

The other major area of Mott's interest in education concerned school teaching. He was a member of numerous committees and boards designed to improve the level and content of school physics and the sciences in general. To give some impression of his commitment, he was involved in the Institute of Physics Eduction Committee, the Nuffield Advisory Committee, the Committee on Education 16–18, the Ministry of Education's Committee on the Training and Supply of Teachers and the Education Committee of the Royal Society, and was Chairman of the National Extension College, the precursor of the Open University. This was a large commitment on his part, partly the result of his believing that his best years for research might be over. Nothing could have been further from the truth, but he did not know that when he became involved in these activities.

13.4 The evolving group structure of the Laboratory

With the changing direction of the Laboratory's research programme, it is often difficult to keep track of the evolution of the group structure. Figure 13.2 gives some impression of the changes as recorded in the booklets *Current Research in the Department of Physics* which were produced in the years 1962, 1965, 1967 and 1970 to replace Bragg's formal reports to the University. The group structure in 1962 corresponded closely to that bequeathed to Mott by Bragg, with the conspicuous additions of the Physics and Chemistry of Solids and Solid State Theory groups. Beneath each group heading, the main subgroup activities in each report are listed, and these give a good impression of the scope of the research within the Laboratory, as well as its evolution over the decade of the 1960s. The blue lines represent the continuity of the groups from one report to the next, while the red lines indicate significant changes in the group structure.

The changing pattern of the research groups reflects the natural evolution of the various sub-disciplines of physics, some subgroups splitting off to become groups in their own right. In other cases, the names of the subgroups evolved, corresponding to changes of scientific direction and to successes in winning funding for new areas of research. It is evidently impossible to do justice to all the sub-areas of research indicated in Figure 13.2 and so selection is necessary. Major research highlights of the Mott era are described in Chapters 14 and 15.

13.5 Planning the move to West Cambridge

Despite the departure of the molecular biologists to their new premises on the Addenbrooke's site, the pressures on space in the Laboratory were intolerable and became worse

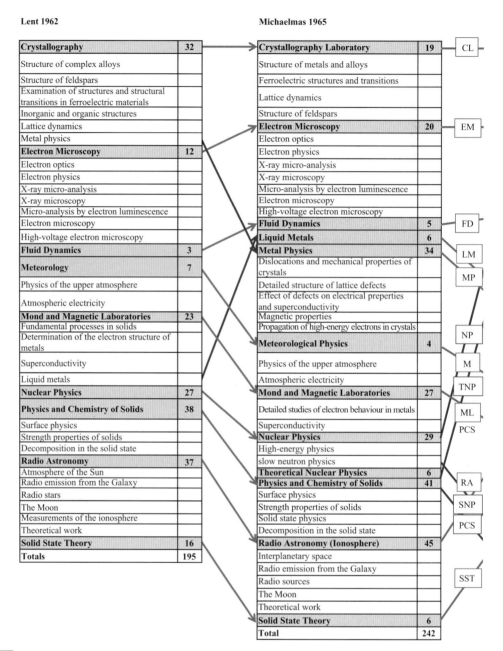

Fig. 13.2 The evolving group structure from 1962 to 1970. These data are taken from the booklets entitled *Current Research in the Department of Physics* which were produced in the years 1962, 1965, 1967 and 1970. The coloured boxes give the numbers of persons in each research group, the 1970 figures corresponding to those recorded in Table 13.1. Beneath each group heading are the principal research areas itemised in each report. Blue lines show the group headings from year to year. Red lines show significant changes in the group structure.

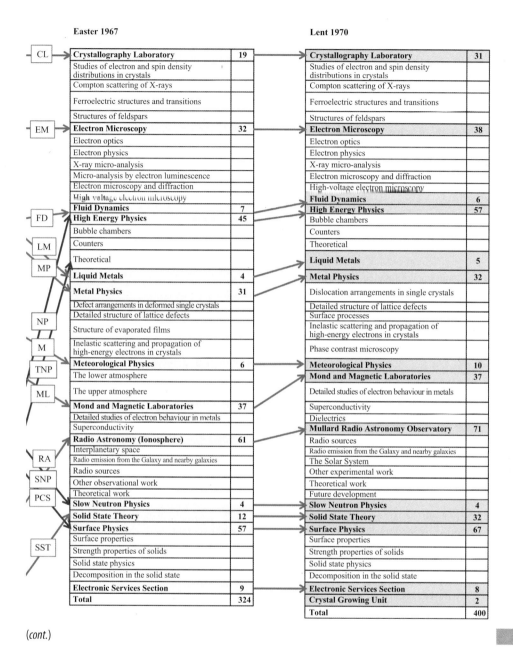

Easter 1967

CL	Crystallography Laboratory	19
	Studies of electron and spin density distributions in crystals	
	Compton scattering of X-rays	
	Ferroelectric structures and transitions	
	Structures of feldspars	
EM	**Electron Microscopy**	**32**
	Electron optics	
	Electron physics	
	X-ray micro-analysis	
	Micro-analysis by electron luminescence	
	Electron microscopy and diffraction	
	High voltage electron microscopy	
	Fluid Dynamics	**7**
FD	**High Energy Physics**	**45**
	Bubble chambers	
	Counters	
LM	Theoretical	
MP	**Liquid Metals**	**4**
	Metal Physics	**31**
	Defect arrangements in deformed single crystals	
	Detailed structure of lattice defects	
NP	Structure of evaporated films	
M	Inelastic scattering and propagation of high-energy electrons in crystals	
TNP	**Meteorological Physics**	**6**
	The lower atmosphere	
ML	The upper atmosphere	
	Mond and Magnetic Laboratories	**37**
	Detailed studies of electron behaviour in metals	
	Superconductivity	
	Radio Astronomy (Ionosphere)	**61**
	Interplanetary space	
RA	Radio emission from the Galaxy and nearby galaxies	
	Radio sources	
SNP	Other observational work	
PCS	Theoretical work	
	Slow Neutron Physics	**4**
	Solid State Theory	**12**
	Surface Physics	**57**
	Surface properties	
SST	Strength properties of solids	
	Solid state physics	
	Decomposition in the solid state	
	Electronic Services Section	**9**
	Total	**324**

Lent 1970

Crystallography Laboratory	31
Studies of electron and spin density distributions in crystals	
Compton scattering of X-rays	
Ferroelectric structures and transitions	
Structures of feldspars	
Electron Microscopy	**38**
Electron optics	
Electron physics	
X-ray micro-analysis	
Electron microscopy and diffraction	
High-voltage electron microscopy	
Fluid Dynamics	**6**
High Energy Physics	**57**
Bubble chambers	
Counters	
Theoretical	
Liquid Metals	**5**
Metal Physics	**32**
Dislocation arrangements in single crystals	
Detailed structure of lattice defects	
Surface processes	
Inelastic scattering and propagation of high-energy electrons in crystals	
Phase contrast microscopy	
Meteorological Physics	**10**
Mond and Magnetic Laboratories	**37**
Detailed studies of electron behaviour in metals	
Superconductivity	
Dielectrics	
Mullard Radio Astronomy Observatory	**71**
Radio sources	
Radio emission from the Galaxy and nearby galaxies	
The Solar System	
Other experimental work	
Theoretical work	
Future development	
Slow Neutron Physics	**4**
Solid State Theory	**32**
Surface Physics	**67**
Surface properties	
Strength properties of solids	
Solid state physics	
Decomposition in the solid state	
Electronic Services Section	**8**
Crystal Growing Unit	**2**
Total	**400**

(cont.)

Fig. 13.2

as the scale of the physics programme expanded. The need for remedial action was recognised by the University, who drew up plans for the redevelopment of the New Museums site with a number of multi-storey buildings. These proposals were strongly opposed by Mott and his colleagues in the Laboratory since the disruption to the research and teaching programmes would be unacceptable, and high-rise buildings were not appropriate for many of the highly sensitive experiments which required very stable foundations. This had been fully appreciated by Maxwell in his design of the Laboratory, but now the threat was very much worse. Although a move to the old Addenbrooke's Hospital site in Trumpington Street would have been a possibility, with the advantage of its proximity to the Chemistry and Engineering departments, this was also a constrained site and would lead to a repeat of the overcrowding in due course. Mott and Pippard's preferred solution was to move to a greenfield site at West Cambridge. After an impassioned speech in the Senate House by Pippard, the scheme proposed for the New Museums site was dropped.[3]

In 1965, the Deer Committee, chaired by Alex Deer, Professor of Mineralogy and Petrology and future Master of Trinity Hall, was set up to study the future development of accommodation for the physical sciences. Its conclusions were summarised by Mott in a fund-raising document of June 1966 entitled *The Cavendish Laboratory: The Need for a New Building*:

> In the view of this committee [the Deer Committee] it is not possible to allow a healthy redevelopment and expansion of Cambridge science unless one major department moves to an unencumbered site on the periphery of the City. The report suggests that this should be the Cavendish and the Cavendish welcomes this recommendation. It is not possible to rebuild on the present site without stopping the work of the Department. We believe that formal teaching will not suffer from the move, that practical teaching will benefit and that without this move research will be stifled. (Mott, 1966)

Mott left the detailed planning of the new Laboratory to Pippard, who was ably supported by the Secretary of the Department, Ian Nicol, who had succeeded Dibden. In turn, Nicol was to be succeeded by John Deakin in 1966. Pippard was mindful of the need for economy in designing the new Cavendish Laboratory. Mott had emphasised that 'a scientific laboratory should be plain and decent and [we] are anxious to co-operate with the University Grants Committee and the architects...in setting a standard of economy and utility which may be copied by other universities' (Mott, 1966). The approach is elaborated by Pippard in his essay on the move of the Laboratory to West Cambridge:

> Even in the late Sixties when the New Cavendish was being planned there were moments of grave doubt about the adequacy of the building grant, and this is reflected in two aspects of the buildings that eventually came into being: the cost was very low in comparison with earlier physics buildings elsewhere, and the hesitation that delayed the early stages of planning resulted in much more thorough discussion of the design than is usual with University buildings. This second point is worth stressing. The architects, Robert Matthew, Johnson-Marshall and Partners, were chosen for their notable work in the planning of York University, and by the time they arrived on the scene a building committee made up of members of the Department had already come to some conclusions about the desirable features of a physics laboratory. Even so, when a number of senior partners in the firm took up residence for a few days and talked to staff and students, they came back to the building committee with a battery of searching questions that the committee had not thought to ask themselves. As a result a clear view evolved about the organisation of a laboratory, especially the social organisation – who must have access to whom, who must be allowed to hide from public view, is it desirable for undergraduates to have free access to research laboratories etc. – the overall layout of the buildings reflects the outcome of these preliminary talks. The siting of the library, administration, stores and

The New Cavendish Laboratory on the West Cambridge site in 1974. From left to right, the buildings are the three-storey Mott Building, mostly for condensed matter physics, the long Bragg Building for teaching, management and administration and the two-storey Rutherford Building for high-energy physics and radio astronomy. Between the Bragg and Rutherford Buildings are the Main Workshops and the Link Building which was to house the mainframe computer.

Fig. 13.3

common room at the cross-roads to encourage chance meetings of people from different buildings; the embedding of Part II practical classes in the research areas; the open lay-out of the Mott Building, with research groups clustered on two or three floors around a hall and staircase, visible to the passer-by yet not wholly exposed since the individual rooms that lead off the hall have their own privacy; these are all examples of the benefits of leisurely pondering on the issues, and they have resulted in a laboratory that has most notably achieved what was intended – scope for individual initiative, and incentive for workers in different groups to cooperate. (Pippard, 1974)

Pippard's objective was to create the maximum amount of useable space for the available resource and he succeeded in completing the building within a single quinquennium during which the funds were guaranteed (Figure 13.3). The following description of the construction by John Payne, who was Secretary of the Buildings Committee throughout the construction phase, indicates how economies were achieved:

There is no basement; the building rests on a reinforced concrete ground slab and consists of a light steel frame, upper floor desks of precast reinforced concrete slabs, very light timber roof decks, and precast concrete external cladding with an 'aggregate' finish. Internal partitions are on light-weight 'Lignacite' block-work (a compressed concrete and wood-shavings mixture). At all levels there are false ceilings which conceal the steel structure and some of the services.

The design was strongly influenced by the Consortium of Local Authorities Special Programme (CLASP) prefabricated-design approach which had been adopted at York University. The cost of the building at June 1966 prices was estimated at £3,805,000 and the construction was to be carried out over the period 1969 to 1974. The building was successfully completed by 1974 and has what Payne described as 'a utilitarian external aspect . . . offset by extensive and imaginative tree planting'. The construction was well underway by the time of Mott's retirement in 1971 and his succession as Cavendish Professor by Pippard. Pippard had undoubtedly achieved his aim of providing the Laboratory with the necessary space for the existing and future programmes of teaching and research. What the building lacked in architectural distinction was made up for by the flexibility of its internal construction, which allowed the structure to be modified in the light of changing research priorities. At the time of writing, every laboratory area has been redeveloped three or four times in response to the requirements of the research programmes.

At the same time, the economical design was storing up problems for future Heads of Department. The CLASP design used large amounts of asbestos, which was not a banned substance at the time the building was constructed. Energy was cheap and so the large heat loss, for example through the light ceilings and single glazing, was not a problem. The thin flat roofs were subject to leakage and required regular maintenance. The design turned out to exceed the space norms for University buildings which would eventually lead to large space-charge costs. But Pippard undoubtedly achieved his aim of providing flexible, functional space with the potential for expansion on the neighbouring greenfield site.

The Mott era: radio astronomy and high-energy physics

The period of Mott's tenure of the Cavendish Chair from 1954 to 1971 was a revolutionary period in astronomy, astrophysics and cosmology. The nature of these disciplines changed in fundamental ways, highlights including:

- the emergence of high-energy astrophysics, in which high-energy particles and cosmic magnetic fields played a key role
- the discovery of the cosmological evolution of the radio source population
- the discovery of the cosmic microwave background radiation
- the discovery of pulsars, which were shown to be magnetised, rotating neutron stars.

The Radio Astronomy Group was at the centre of these events. At the same time, the Nuclear Physics Group changed direction from the earlier period when it was feasible for a University group to construct its own particle accelerators. The group was renamed the High Energy Physics Group in the mid 1960s, and successfully developed instrumentation and data analysis facilities in support of major international projects.

14.1 The growth of the Radio Astronomy Group

The story of radio astronomy in Cambridge up to 1953–54 was described in Section 12.7.2, the pivotal years when the discipline was about to develop into a 'Big Science' discipline. Ryle was the driving force behind these developments. As he wrote to me just before his death, he thought of himself primarily as an electrical engineer with the ability to make complex radio receiving systems a reality. In addition, however, he had remarkable physical intuition, which guided all his research activities in radio astronomy and its technology. His great contribution was the practical implementation of the concept of aperture synthesis, the technique by which images of the radio sky are created by combining interferometric observations made with modest-sized radio telescopes located at different interferometer spacings. Ryle's ambitious programme was to determine both the amplitudes and the phases of the incoming radio signals so that, by Fourier inversion, the detailed brightness distribution of the radio emission could be reconstructed. Although understood in principle, the key technical issues concerned whether or not these concepts could be realised in practice, given the problems of receiver sensitivity, the need to preserve phase coherence over long periods and the stability of the overall system performance.

Over a 25-year period starting in about 1950, Ryle and his colleagues developed a series of radio interferometers of increasing complexity and ingenuity that made it possible to

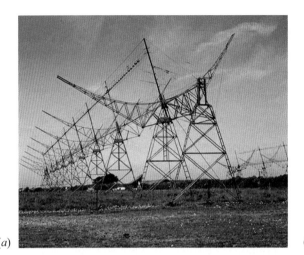
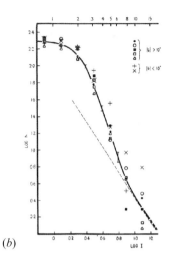

(a)   (b)

Fig. 14.1 (a) Two of the antennae of the 2C interferometer. (b) The integral number counts of radio sources from the 2C survey of radio sources. N is the number of radio sources brighter than flux density I, the units of I being 10^{-25} W m^{-2} sr^{-1}. The dashed line shows the plot expected for a Euclidean model of the source distribution (Box 14.1). The observations show a very large excess of faint radio sources relative to the expectations of the Euclidean model (Ryle, 1955).

carry out surveys of the sky and unravel the structures and nature of the radio sources. This programme involved a great deal of innovative electronics, such as the phase-switching interferometer, which enabled complete amplitude and phase information to be recovered. The practical development of aperture synthesis was a virtuoso technical achievement that involved a considerable team of researchers and support staff. Ryle put together a tightly knit group of physicists, as well as a strong support team of technicians and research students, many of whom were later to become leaders in radio astrophysics.[1]

14.1.1 The 2C survey and the controversy over the number counts of radio sources

Once Ryle was converted to the idea that the 'radio stars' observed in directions away from the plane of our Galaxy were in fact distant extragalactic objects, he and Hewish designed and constructed a large four-element interferometer to carry out a new survey of the northern sky at 81.5 MHz which would be sensitive to small-angular-diameter sources. Sufficient funds were made available by the Department of Science and Industrial Research to construct 50,000 square feet of collecting area. The four cylindrical radio antennae were constructed by the local firm of Donald Mackay on the rifle range site, behind Ryle's house and next to the University's rugby pitch (Figure 14.1(a)). The second Cambridge (2C) survey of radio sources was completed in 1954 and the first results published the following year (Shakeshaft *et al.*, 1955). Ryle and his colleagues found the startling results that small-diameter radio sources were uniformly distributed over the sky and that the number of sources increased enormously as the survey extended to fainter and fainter flux densities.

In any uniform Euclidean model, the number of sources brighter than a given limiting flux density S is expected to follow the relation $N(\geq S) \propto S^{-3/2}$ (Box 14.1). In contrast,

Ryle found a huge excess of faint radio sources, the slope of the source counts between 20 and 60 Jy being described by $N(\geq S) \propto S^{-3}$ (Figure 14.1(*b*)).[2] He concluded that the only reasonable interpretation of these data was that the sources were extragalactic, that they were objects similar in luminosity to the radio galaxy Cygnus A and that there was a much greater probability of finding radio sources at large distances than expected of a uniform distribution.[3] As Ryle expressed it in his Halley Lecture in Oxford in 1955 (Ryle, 1955),

> This is a most remarkable and important result, but if we accept the conclusion that most of the radio stars are external to the Galaxy, and this conclusion seems hard to avoid, then there seems no way in which the observations can be explained in terms of a steady-state theory.

These remarkable conclusions came as a surprise to the astronomical community. There was enthusiasm and also some scepticism that such profound conclusions could be drawn from the counts of radio sources, particularly when their physical nature was not understood and only the brightest 20 or so objects had been associated with relatively nearby galaxies. A rather acrimonious dispute arose between the proponents of steady-state cosmology and the standard Friedman world models, the former having been promoted enthusiastically by Fred Hoyle, Thomas Gold and Hermann Bondi. In the standard Friedman models, it is assumed that, as the Universe expands, the number of objects is conserved so that the galaxies continue to disperse with time. In the steady-state picture, it is assumed that, although the Universe is expanding, matter is created out of the vacuum of space to replace the matter which is dispersing – the average density of matter in the Universe is unchanging with time, despite the expansion. In the nearby Universe, the number counts of galaxies, or radio sources, would be expected to follow the same number count, $N(\geq S) \propto S^{-3/2}$, for both the Friedman and steady-state pictures (Box 14.1).

The Sydney group led by Bernard Mills was carrying out similar radio surveys of the southern sky with the Mills Cross Radio Telescope at about the same time, and they found that the source counts could be represented by the relation $N(\geq S) \propto S^{-1.65}$, which they argued was not significantly different from the expectation of uniform world models. In 1957 Mills and Bruce Slee stated:

> We therefore conclude that discrepancies, in the main, reflect errors in the Cambridge catalogue, and accordingly deductions of cosmological interest derived from its analysis are without foundation. An analysis of our results shows that there is no clear evidence for any effect of cosmological importance in the source counts. (Mills and Slee, 1957)

Thus, the controversy involved not only the proponents of the standard Big Bang and steady-state cosmologies, but also the Cambridge and Sydney radio astronomers.

The problem with the Cambridge number counts was that they extended to surface densities of radio sources such that the flux densities of faint sources were overestimated because of the presence of fainter sources in the beam of the telescope, a phenomenon known as *confusion*. Scheuer, who was Ryle's research student from 1951 to 1954, devised a statistical procedure for deriving the number counts of sources from the survey records themselves without the need to identify individual sources (Scheuer, 1957). Using what he called the $P(D)$ technique, which has since been adopted in many other astronomical

| Box 14.1 | Radio source counts in Friedman and steady-state cosmologies |

Suppose the sources have a luminosity function $N(L)\,dL$ and that they are uniformly distributed in Euclidean space. The number of sources with flux densities S greater than a particular limiting value in the solid angle Ω on the sky is denoted $N(\geq S)$. Consider first sources with luminosities in the range L to $L + dL$. In a survey to a limiting flux density S, these sources can be observed out to some limiting distance r, given by the inverse-square law, $r = (L/4\pi S)^{1/2}$. The number of sources brighter than S is therefore that within distance r in the solid angle Ω:

$$N(\geq S, L)\,dL = \frac{\Omega}{3} r^3 N(L)\,dL. \tag{14.1}$$

Substituting for r, the number of sources brighter than S is

$$N(\geq S, L)\,dL = \frac{\Omega}{3} \left(\frac{L}{4\pi S} \right)^{3/2} N(L)\,dL. \tag{14.2}$$

Integrating over the luminosity function of the sources,

$$N(\geq S) = \frac{\Omega}{3(4\pi)^{3/2}} S^{-3/2} \int L^{3/2} N(L)\,dL, \tag{14.3}$$

that is, $N(\geq S) \propto S^{-3/2}$, independent of the luminosity function $N(L)$. The result $N(\geq S) \propto S^{-3/2}$ is known as the *integral Euclidean source counts* for any class of extragalactic object. This is the relation which was used in the arguments about whether or not the observed counts of radio sources were consistent with this Euclidean prediction. In fact, it is better to work in terms of the *differential number counts*, in which the numbers of sources in each independent interval of flux density dS are counted. In this case $\Delta N(S) = n_0(S)\,dS \propto S^{-5/2}\,dS$.

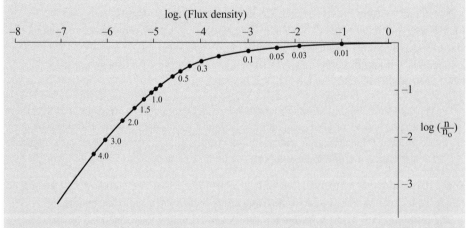

Once it was realised that the source population extended to large redshifts[4] at which the Euclidean calculation could no longer be used, the discrepancy with the observed number counts was very much greater. As an example, the above diagram shows the differential counts n of sources of a single luminosity class with spectral index $\alpha = 0.75$, defined by $S_\nu \propto \nu^{-\alpha}$, normalised to the Euclidean prediction n_0 for the reference Einstein–de Sitter world model. The redshifts at which the sources are observed are shown along the relation (Longair, 1974). The divergence from the Euclidean prediction is even greater in the steady-state model. Convolutions of the counts for a single luminosity class with the luminosity function $N(L)$ are shown in Figure 14.6.

contexts, he showed that the slope of the source counts from the Cambridge 2C survey records was actually -1.8. Ironically, this result, which we now know to be exactly the correct answer, was not trusted, partly because the mathematical techniques used by Scheuer were somewhat forbidding and also because his result differed from the prejudices of both Ryle and Mills. The dispute reached its climax at the Paris Symposium on Radio Astronomy in 1958 and the conflicting positions were not resolved (Bracewell, 1959). The strong feelings aroused by the contrasting views of the proponents of the evolutionary and steady-state models are vividly described in Kragh's book *Cosmology and Controversy: the Historical Development of Two Theories of the Universe* (Kragh, 1996).

14.1.2 The opening of the Lord's Bridge Observatory and the 3C and 4C surveys

Despite the ongoing controversy, it was clear that radio astronomy had the potential to provide new types of astrophysical and cosmological information. Ryle already had plans for the next-generation survey telescope, namely, radio interferometers with higher angular resolution and sensitivity which were less affected by source confusion. The space available at the rifle range was too small for the necessarily longer baselines and so, in 1956, the radio observatory moved to a disused wartime Air Ministry bomb store at Lord's Bridge, about 10 km to the south-west of Cambridge. This site also had the advantage that it was much better suited for radio astronomy because of the lower level of man-made radio interference. Funds had to be raised for the rental of the new site as well as the costs of the new telescope system, which would become the 4C radio interferometer, and its operational costs. In 1955 Mott and Ratcliffe persuaded the Mullard Company to make an endowment of £100,000 for the new telescope and associated facilities. This was supplemented by a grant of £40,000 from the Department of Scientific and Industrial Research (DSIR) and £40,000 from the University Development Fund to cover the costs of construction and operation for ten years. In acknowledgement of this major gift, the observatory was named the Mullard Radio Astronomy Observatory, and it was opened by Edward Appleton on 27 July 1957. In 1959, Ryle was appointed Professor of Radio Astronomy.

Once it was realised that confusion had resulted in far too many sources being included in the 2C catalogue, the 2C interferometer was upgraded with receivers for the higher frequency of 159 MHz, thus doubling the angular resolution and so decreasing the importance of confusion. The resulting Third Cambridge (3C) Catalogue contained 471 radio sources with flux densities greater than 10×10^{-26} W m^{-2} Hz^{-1} (Edge *et al.*, 1959).

The first truly two-dimensional maps of the sky employing the principles of aperture synthesis were carried out by John Blythe, who combined a fixed array of dipoles in the east–west direction with a moveable antenna along a north–south axis (Blythe, 1957a,b). The experiment was a success and proof of concept for the design of the very much larger and ambitious 4C interferometer system.

The 4C radio interferometer came into operation in the late 1950s, using the principles of aperture synthesis to create a much higher angular resolution and higher sensitivity survey of the northern sky. The very large cylindrical paraboloid was aligned east–west (Figure 14.2(*a*)) and a smaller cylindrical paraboloid was located on a north–south track located about 1 km to the east of the large antenna (Figure 14.2(*b*)). The telescope system

(a)

(b)

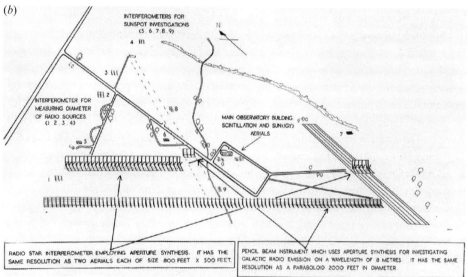

INTERFEROMETERS FOR
SUNSPOT INVESTIGATIONS
(5.6.7.8.9)

N

4 |||

3 |||

INTERFEROMETER FOR
MEASURING DIAMETER
OF RADIO SOURCES
(1.2.3.4)

|| 2

8

MAIN OBSERVATORY BUILDING
SCINTILLATION AND SUN(IGY)
AERIALS

7

5

9

| RADIO STAR INTERFEROMETER EMPLOYING APERTURE SYNTHESIS. IT HAS THE SAME RESOLUTION AS TWO AERIALS EACH OF SIZE 800 FEET X 500 FEET. | PENCIL BEAM INSTRUMENT WHICH USES APERTURE SYNTHESIS FOR INVESTIGATING GALACTIC RADIO EMISSION ON A WAVELENGTH OF 8 METRES. IT HAS THE SAME RESOLUTION AS A PARABOLOID 2000 FEET IN DIAMETER |

Fig. 14.2 (a) The long cylindrical paraboloid antenna of the 4C interferometer at the Lord's Bridge Observatory. (b) The layout of the various telescopes constructed on the Lord's Bridge site up to the early 1960s. The arrows attached to the lower left box show the long and moveable elements of the 4C array. The lower right box indicates the 38 MHz array, which also carried out a survey of the northern sky. Combined with the 178 MHz observations, the continuum spectra of the radio sources could be determined.

operated by setting both the fixed and moving antennae to a particular declination and observing the sky for 24 hours. Then the moveable antenna was moved along the track to obtain the amplitude and phase information for a different spatial combination of the long and moveable antennae. This procedure was repeated 24 times for a particular declination until observations at all positions along the north–south track had been observed. A map of a strip of sky 5° wide at a fixed declination with angular resolution 0.5 arcmin in right ascension and 3 arcmin in declination could then be synthesised from these data.

The Fourth Cambridge (4C) Catalogue of radio sources, containing almost 5000 radio sources with flux densities greater than 2×10^{-26} W m^{-2} Hz^{-1} at 178 MHz was published in two parts in 1965 and 1967 (Pilkington and Scott, 1965; Gower et al., 1967). The radio source counts derived from the 4C catalogues again showed an excess over the expectations of Euclidean world models, the slope of the integral counts being -1.8 (Gower, 1966), exactly the result found by Scheuer (1957). These observations showed that Ryle's conclusions of 1955 were basically correct but that the magnitude of the excess had been considerably overestimated. In fact, the discrepancies with both the uniform Friedman and steady-state models were much greater than this simple comparison suggested because, as indicated in Box 14.1, the predicted radio source counts converge rapidly as soon as the source populations extend to significant redshifts (Longair, 1971; Scheuer, 1975). By the mid 1960s, the evidence was compelling that there was indeed a very large excess of sources at large redshifts and this was at variance with the expectations of the steady-state theory.[5]

14.2 The 3CR catalogue and the discovery of quasars

The identification of extragalactic radio sources with very distant galaxies required radio positions more accurate than 1 arcmin, and this could now be achieved for the brighter sources with the elements of the 4C interferometer. In addition, these observations provided more precise flux densities of the sources already listed in the 3C catalogue. The results of these efforts was the compilation of the revised 3C (3CR) catalogue (Bennett, 1962), which contained 328 radio sources with flux densities greater than 9×10^{-26} W m^{-2} Hz^{-1} at 178 MHz and became the standard low-frequency catalogue of radio sources in the northern sky.[6]

The radio astronomical discoveries of the 1950s stimulated a great deal of astrophysical interest and led to major investments in the construction of radio telescope systems. Other radio observatories undertook systematic surveys of the sky in order to understand in more detail both the astrophysics of these intense sources of radio waves and their use as cosmological probes. Among the extragalactic radio sources which could be securely identified, most were found to be associated with some of the most massive galaxies known, which are very luminous and so could be observed to large redshifts. By 1960, the largest redshift known for any galaxy was that of the radio galaxy 3C 295 at $z = 0.46$, meaning that the galaxy emitted its light when the galaxies were closer together by a factor of 0.68 as compared with their present separations (Minkowski, 1960).

By 1962, Thomas Matthews and Allan Sandage had identified three of the brightest radio sources, 3C 48, 3C 196 and 3C 286, with 'stars' of an unknown type with strange optical spectra (Matthews and Sandage, 1963). The breakthrough came in 1962 when Cyril Hazard measured very precisely the position of the radio source 3C 273 by the method of radio lunar occultation using the recently commissioned Parkes 210-foot (64-metre) radio telescope in New South Wales, Australia (Hazard et al., 1963). These observations enabled

Fig. 14.3   The Hubble Space Telescope image of the quasar 3C 273, showing the optical jet ejected from the quasar nucleus. The faint smudges to the south of the quasar are now known to be galaxies at the same distance as the quasar (courtesy of NASA, ESA and the Space Telescope Science Institute).

3C 273 to be associated with what appeared to be a 13th magnitude star (Figure 14.3). The identification was rapidly relayed to Maarten Schmidt, who used the Palomar 200-inch telescope to obtain its optical spectrum. The spectrum contained prominent emission lines, but at unexpected wavelengths. The clue lay in the familiar pattern of lines, which Schmidt realised was the Balmer series of hydrogen but shifted to longer wavelengths with redshift $z = \Delta\lambda/\lambda_0 = 0.158$; at this large cosmological distance, this was certainly no ordinary star (Schmidt, 1963). 3C 273 was the first, and brightest, of this class of hyperactive galactic nuclei to be discovered. Its optical luminosity is about 1000 times greater than the luminosity of a galaxy such as our own. To make matters even more intriguing, searches through the Harvard plate archives revealed that the enormous optical luminosity of 3C 273 varied on a time-scale of years (Smith and Hoffleit, 1963). Nothing like this had been observed in astronomy before. 3C 48 was found to be associated with a star-like object with a red-shift of 0.3675 (Greenstein and Matthews, 1963), while 3C 47 and 3C 147 had redshifts of $z = 0.425$ and $z = 0.545$, respectively (Schmidt and Matthews, 1964). These sources were termed *quasi-stellar radio sources* and within a year this term had been contracted to the word *quasar*.

With the improved positions available from early observations with the 4C array, it became possible to search for the counterparts of many more of the 3CR radio sources. Those located away from the galactic plane were either very faint galaxies or quasars (Ryle and Sandage, 1964; Longair, 1965). By 1965 there were sufficient data to make improved estimates of the radio luminosity function of the sources and demonstrate clearly the very large evolutionary changes needed to account for the steepness of the radio source counts (Longair, 1966).

14.3 The development of earth rotation aperture synthesis

The 2C and 4C radio telescopes were transit survey instruments in which the antennae remained at fixed positions pointing at a particular declination and the sky drifted through the beam of the interferometer. The next step was to develop the technique so that a particular region of the sky could be tracked. In the early 1960s, Ryle and Ann Neville carried out the most ambitious experiment to date – the use of the rotation of the Earth to carry telescopes at fixed points on the Earth about each other as observed from a point on the celestial sphere (Figure 14.4(a) and (b)). The germ of this idea had already appeared in Ryle's notebooks in 1954. By 1959, digital computers were fast enough to cope with the data analysis requirements of this form of synthesis mapping (Ryle and Hewish, 1960). In a classic set of observations, Ryle and Neville created an Earth-rotation aperture synthesis map of a region of sky about the North Celestial Pole (Ryle and Neville, 1962). To achieve this, they used 45-foot segments of the long and moveable elements of the 4C interferometer, as well as a portable intermediate 45-foot element, to fill in all 75 interferometer pairs along the 3362-foot maximum baseline (Figure 14.4(c)). The reason for this was that the primary beam of each individual element determines the size of the area of sky which can be surveyed. In this case, the elements were all pointed in the direction of the North Celestial Pole and resulted in a synthesised map about the pole of radius 7.5° (Figure 14.4(d)). The angular resolution of the survey was 4.5 arcmin and the sensitivity eight times greater than that of the original antenna system. This experiment demonstrated convincingly the remarkable possibilities opened up by the technique of Earth-rotation aperture synthesis and was influential in promoting the case for future large aperture synthesis telescopes such as the Westerbork Synthesis Radio Telescope (WSRT) in the Netherlands and the Very Large Array (VLA) in the USA.

The success of this project pointed the way to the future. The succeeding generations of aperture synthesis arrays employed fully steerable antennae so that specific regions of the sky could be tracked and high angular resolution images of individual sources obtained. In order to 'flatten the wavefront' of the incoming radio signals, path compensation cables had to be continuously inserted into the signals coming from the elements of the interferometer so that the cross-correlation of signals arriving from different interferometer spacings could be obtained with the same phase. DSIR support for the project to the tune of £380,000 was sought and, after some delay, was forthcoming. As described in Section 13.2, Ryle was fortunate in bidding for the resources for the One-Mile Telescope, and subsequently for the Five-Kilometre Telescope, during an era when the resources available for pure scientific research were increasing in real terms. The One-Mile Telescope was completed in 1964 and consisted of three 120-ton, 18-metre diameter antennae (Figure 14.5(a)). Two of them were fixed, while the third could be moved along an 800 m rail track to 60 different stations. The telescope operated simultaneously at 408 MHz (75 cm) with angular resolution 80 arcsec and at 1.4 GHz (21 cm) with resolution 23 arcsec.[7]

The first radio maps made by the One-Mile Telescope revealed the power of the technique of Earth-rotation aperture synthesis for revealing the structures of galactic and extragalactic radio sources (Ryle et al., 1965) and led to new astrophysical challenges

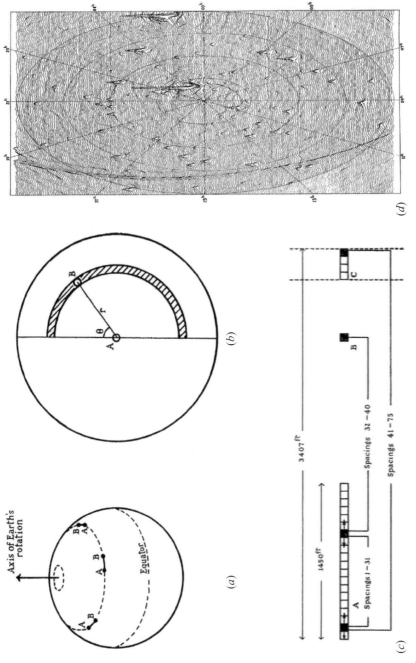

Fig. 14.4 (a) Illustrating how the Earth's rotation causes the baseline of an east–west interferometer to rotate about its centre as viewed from the sky, in this case as viewed from the North Celestial Pole. (b) Illustrating how a 12-hour observation with a two-element interferometer fills in an annulus of a large telescope. (c) Showing the location of the elements of the 4C fixed (A) and moveable (C) telescopes, as well as the supplementary telescope (B) which filled in the intermediate spacings (Ryle and Neville, 1962). (d) The radio survey of the region about the North Celestial Pole carried out by Ryle and Neville using the principles of Earth-rotation aperture synthesis (Ryle and Neville, 1962).

(a)

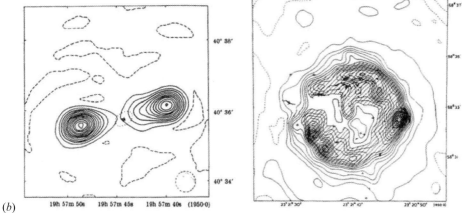

(b)

(a) The Cambridge One-Mile Radio Telescope. To the right are the two fixed telescopes, separated by half a mile in the east–west direction, and in the foreground the moveable antenna on its half-mile rail track. In the background, the long element of the 4C antenna can be seen. (b) The first radio aperture synthesis maps of the bright radio sources Cygnus A (left panel) and Cassiopaeia A (right panel), made with the One-Mile Radio Telescope at 1.4 GHz. Cygnus A is associated with a distant massive galaxy, indicated by the dashed ellipse between the radio lobes. Cassiopaeia A is a supernova remnant within our Galaxy. In both cases, the radio emission is the synchrotron radiation of ultra-high-energy electrons gyrating in magnetic fields (Ryle *et al.*, 1965).

Fig. 14.5

concerning the origin of the enormous fluxes of relativistic electrons and magnetic fields present in these sources (Figure 14.5(*b*)). In the case of the source Cygnus A, the double structure was clearly resolved, the location of the associated galaxy being indicated by the dashed ellipse in the left-hand panel. The energy requirements of such sources were

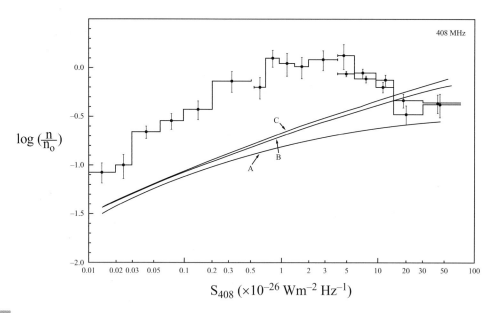

Fig. 14.6 Comparison of the expected differential counts of sources at 408 MHz with the observed source count. The curve (A) shows the expected count on the basis of galaxies with measured redshifts in the sample. Curve (B) shows the expected count using estimates of the redshifts where necessary. (C) is the same as (B) but assuming that all the unidentified sources had redshifts $z = 0.25$, a very conservative assumption (Longair, 1974).

enormous, corresponding to the conversion of the rest-mass energy of about 10^5 solar masses of material into ultra-high-energy electrons and magnetic fields. In the case of the supernova remnant Cassiopaeia A, the radio observations revealed a ring of radio emission, more or less coincident with the faint filaments observed on optical images, but forming a much more complete shell (right-hand panel). It was soon established that the high-energy electrons had to be accelerated in the radio shell itself. Supernova remnants such as these are the likely origin of the fluxes of cosmic rays observed at the top of the Earth's atmosphere.[8]

Some of the most important programmes of the One-Mile Telescope were the deep surveys of small regions of sky to extend the number counts of radio sources to very much lower limiting flux densities than the 4C surveys. By using observations made at all 64 separations along the rail track, deep maps of areas of sky about $3°$ diameter at 408 MHz could be obtained. The results of the 5C2 survey were published by Ryle and Guy Pooley in 1968 and extended to limiting flux densities of 10 mJy (Ryle and Pooley, 1968). These observations showed clearly the convergence of the number counts of radio sources at low flux densities. Figure 14.6 shows a slightly later compilation of radio source counts presented in differential, normalised form with examples of the expectations of uniform world models (Longair, 1974). By the time that review was undertaken, many more 3CR quasars and radio galaxies had been identified and the numbers of redshifts for them were growing steadily, despite the fact that these observations were at the very limit of capability of even the largest optical telescopes. It was clear that the discrepancy between the predictions of uniform world models and the number counts of radio sources was very large indeed. To account for this, it had to be assumed that the radio source population had been much greater in the past than it is at the present time. The same result was later found for the radio quiet quasars discovered by Sandage in 1965 (Sandage, 1965) and for X-ray sources.[9] From these data, it was shown that the most luminous members of the radio source population had maximum comoving space density at a redshift $z \sim 2$ and decreased at later cosmological epochs.

With the controversy over the number counts of radio sources resolved, the emphasis shifted to the study of the astrophysics of the radio sources. The success of the One-Mile Telescope encouraged Ryle to plan for a yet more powerful synthesis radio telescope, the Five-Kilometre Telescope, subsequently named the Ryle Telescope (Ryle, 1972). The Lord's Bridge site was not large enough for a baseline of this length, but the Cambridge to Bedford railway line had been closed and it so happened that it ran more or less exactly east–west along the northern boundary of the Lord's Bridge site. With a grant of £2.1 million from the Science Research Council,[10] Ryle built an eight-element interferometer operating at 15 GHz (2 cm wavelength) which resulted in an angular resolution of about 2 arcsec. The telescopes were 13-metre equatorial Cassegrain antennae, four of them being moveable on a 1.2 km rail track while the others were located at 1.2 km intervals to the west of the moveable antennae (Figure 14.7(a) and (b)).

The images of the radio sources and quasars observed by the Five-Kilometre Telescope were of extraordinary quality. These observations enabled strong constraints to be placed upon the high-energy astrophysics of relativistic plasmas and magnetic fields in cosmic environments. For example, they established beyond question the need for jets of relativistic material to be ejected from the nuclei of active galaxies and quasars to power the external radio lobes (Longair et al., 1973; Scheuer, 1974). This story is taken up in Section 17.1.2.

Both the One-Mile and Five-Kilometre radio telescopes were far ahead of the capabilities of any other radio telescope system in the world when they came into operation. Some measure of Ryle's achievement is the fact that, over a 25-year period, the sensitivity of radio astronomical observations increased by a factor of about one million and the imaging capability of the telescope system improved from several degrees to a few arcseconds, comparable to that of ground-based optical telescopes (Ryle, 1975). Ryle was personally involved in every aspect of these very complex telescope systems (see the quotation by Scheuer in Section 11.6.1).

Intellectually, Ryle relied almost completely upon his well-honed physical intuition as the way to solve any problem, be it in engineering, astrophysics or cosmology – indeed, he believed this was the only way research should be conducted. Radio astronomy played a crucial role in the realisation that high-energy astrophysical activity involving super massive black holes and general relativity are part of the large-scale fabric of our Universe. From the perspective of UK astronomy, these great breakthroughs in observational astronomy and the innovative science produced by them propelled the UK from a second-class role relative to the USA to the front rank of astrophysical and cosmological research. This greatly facilitated the remarkable growth of astronomy, astrophysics and cosmology in the UK from the 1970s onwards.

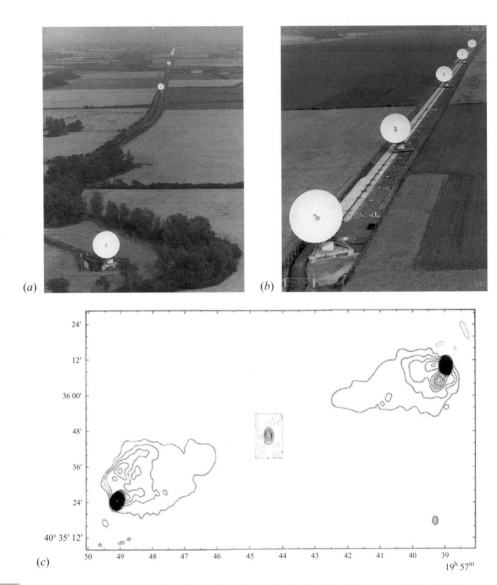

Fig. 14.7 (*a*) A view of the Five-Kilometre (Ryle) Telescope from the air, showing the eight telescopes on the line of the former Cambridge to Bedford railway line. (*b*) The most easterly of the fixed telescopes of the Five-Kilometre Telescope on the left and the rail track with the four moveable aerials as seen from the air. (*c*) The first map of the radio source Cygnus A observed with an angular resolution of 2.5 arcsec by the Five-Kilometre Telescope (Hargrave and Ryle, 1974). This image can be compared with the left-hand panel of Figure 14.5(*b*). A weak compact source was observed in the nucleus of the associated massive galaxy.

Ryle was knighted in 1966, appointed Astronomer Royal in 1972 and jointly awarded the Nobel Prize in Physics with Hewish in 1974 'for their pioneering research in radio astrophysics: Ryle for his observations and inventions, in particular of the aperture synthesis technique, and Hewish for his decisive role in the discovery of pulsars'. The latter discovery is the subject of the next section.

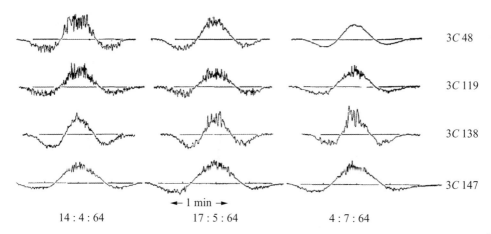

Illustrating the scintillation of compact radio sources as observed at different values of solar elongation, meaning the angle between the direction of the Sun and the source in the plane of the ecliptic (Hewish *et al.*, 1964). The smooth intensity profile at the top right was observed when the source 3C 48 was far from the Sun. The profiles to the left were observed much closer to the Sun and show the intense and rapid fluctuations caused by density irregularities in the plasma flowing out from the Sun.

Fig. 14.8

14.4 Interplanetary scintillation (IPS) and the discovery of pulsars

Hewish's pioneering studies of the use of the scintillations of the flux densities of bright radio sources to understand the properties of the ionosphere and the solar corona were described in Section 12.7.2. In 1954, Hewish had remarked in his notebooks that, if the angular sizes of the radio sources were small enough, they would illuminate the solar corona with a coherent radio signal and so give rise to rapid time variations in their intensities. This idea was forgotten until about 1962, when Margaret Clarke showed that two of the compact 3CR radio sources ($\theta \lesssim 2$ arcsec) were varying very rapidly in intensity.

By 1964, a number of radio quasars were known and some of these radio sources had small angular sizes. With Paul Scott and Derek Wills, Hewish showed that the radio scintillations were due to the scattering of the radio waves by inhomogeneities in the ionised plasma flowing out from the Sun, what is known as the *solar wind* (Figure 14.8). This wind had been predicted by Eugene Parker in 1958 and observed by the Soviet Luna satellites in 1959 and by the U.S. Mariner-2 satellite in 1962. The paper by Hewish, Scott and Wills showed how radio source scintillations could be used to map the outflowing solar wind (Hewish *et al.*, 1964).

Hewish realised that a large, low-frequency array dedicated to the measurement of the scintillations of compact radio sources would provide a new approach to the study of three important astronomical areas: (*a*) it would enable many more quasars to be discovered; (*b*) their angular sizes could be estimated; and (*c*) the structure and velocity of the solar wind could be determined. In 1965, he designed a large array to undertake these studies and was awarded a grant of £17,286 by the UK Department of Scientific and Industrial Research to construct it, as well as outstations for measuring the velocity of the solar wind. To obtain

Fig. 14.9   Hewish's 4.5-acre array for studies of the scintillations of compact radio sources. On the far right, the moving element of the One-Mile Telescope can be seen.

adequate sensitivity at the low observing frequency of 81.5 MHz (3.7 m wavelength), the array had to be very large, 4.5 acres (1.8 hectares) in area, in order to record the rapidly fluctuating intensities of bright radio sources on time-scales as short as 0.1 s (Figure 14.9).

Jocelyn Bell(-Burnell) joined the 4.5-acre array project as a graduate student in October 1965. The telescope was commissioned during July 1967 with the objective of mapping the whole sky once a week so that the variation of the scintillations of the sources with solar elongation could be studied. A key aspect of the array was that it had to be possible to measure the amplitudes of the scintillations in real time.

While the array was being constructed, Leslie Little and Hewish carried out a theoretical investigation of the strength of the scintillations as a function of heliospheric coordinates (Little and Hewish, 1966). They demonstrated how the angular sizes of the sources could be estimated from measurements of the amplitudes of the scintillations when sources were observed at different solar elongations. The key point is that the scintillations decrease to very small values when observed at large angles from the Sun, as they demonstrated in a plot of the scintillation index as a function of heliocentric coordinates.

The commissioning of the 4.5-acre array proceeded through the summer of 1967. Hewish suggested that Bell create sky charts for each strip of the sky each day, noting all the scintillating sources. This was a very demanding task requiring great persistence, patience and attention to detail on Bell's part, since she had to keep up with the high rate at which the charts were being produced by the telescope.

The discovery of the rapidly pulsating radio source, or pulsar, CP 1919 was made by Bell on 6 August 1967 (Figure 14.10(*a*)). The remarkable feature was that the source

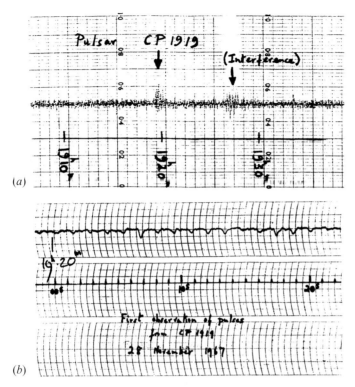

(*a*) The discovery record of CP 1919 taken on 6 August 1967. (*b*) The trace of the signal from the pulsar CP 1919 taken on 28 November 1967 with a short time constant in the recorder. This was the first time the individual pulses of the pulsar were distinguished (courtesy of A. Hewish and the Churchill College Archives).

Fig. 14.10

scintillated at roughly the 100% level in the opposite direction to the Sun, quite contrary to the expectations of the scintillation models of Little and Hewish. Furthermore, the source was highly variable and not always present. The source was not observed again until 28 November 1967, this time with a much shorter time constant in the recording system. With this improved time resolution, radio pulses were detected separately for the first time (Figure 14.10(*b*)). The signal consisted entirely of a sequence of pulses with repetition period 1.33 s. This period was found to be stable to better than 1 part in 10^6. The paper, *Observation of a rapidly pulsating radio source*, was published on 24 February 1968; the discovery of three other pulsars was also reported (Hewish *et al.*, 1968). One of these pulsars, CP 0950, turned out to have a pulse period of only 0.25 s, which was so short that it excluded the possibility that a white dwarf star could be the parent body of that pulsar (Pilkington *et al.*, 1968).

Within a few months, Thomas Gold convincingly associated the pulsars with the theoretical expectations of the properties of magnetised, rotating neutron stars (Gold, 1968). The radio pulses are caused by beams of very high-energy particles escaping from the magnetic poles of a rapidly rotating neutron star. When observations were made along the magnetic poles, an intense burst of radio emission was observed. The pulsars were of the greatest astrophysical importance as the last stable stars before collapse to a black hole ensues. The neutron stars represent matter in bulk at nuclear densities and offered many challenges for physicists and astrophysicists.

14.5 The cosmic microwave background radiation

In 1965, the cosmic microwave background radiation was discovered more or less by accident by Arno Penzias and Robert Wilson. In the late 1940s, Gamow and his colleagues Ralph Alpher, Robert Herman and James Follin had predicted that background radiation at a temperature of about 5 K should be observable today as the cooled remnant of the hot early stages of the Universe according to the Big Bang model (Alpher *et al.*, 1948, 1953). This prediction was largely forgotten when Gamow's model of the hot early Universe was unsuccessful in predicting the origin of chemical elements.[11]

Penzias and Wilson had joined the Bell Telephone Laboratories in the early 1960s with the intention of using the 20-foot horn reflector, which had been built to test telecommunications with the Echo satellite, for radio astronomical observations. Penzias and Wilson had the responsibility of calibrating the antenna for use at these frequencies, for which they had access to a 7.35 cm cooled maser receiver. Wherever they pointed the telescope on the sky, they found an excess antenna temperature, which could not be accounted for by noise sources in the telescope or receiver system.[12] Having carefully calibrated all parts of the telescope and receiver system, they found that there remained an excess noise contribution, corresponding to a black-body temperature of 3.5 ± 1 K (Penzias and Wilson, 1965).

At almost exactly the same time, Robert Dicke's group in Princeton were preparing exactly the same type of experiment to detect the cooled remnant of the Big Bang. Discussions with the Princeton group ensued and it became apparent that Penzias and Wilson had discovered the diffuse cosmic microwave background radiation, exactly what the Princeton physicists were searching for. Within a few months, the Princeton group had measured a background temperature of 3.0 ± 0.5 K at a wavelength of 3.2 cm, confirming the black-body nature of the background in the Raleigh–Jeans region of the spectrum (Roll and Wilkinson, 1966).

The Cavendish radio astronomers had had a long involvement with studies of the diffuse background radiation, but unlike the Bell Laboratories and Princeton experiments, theirs were at low radio frequencies and concerned the determination of the spectrum and intensity of the radio emission of the Galaxy and the integrated background radio emission due to discrete radio sources (Baldwin, 1967; Bridle, 1967). John Shakeshaft and Timothy Howell took up the challenge of attempting to measure the low-frequency, long-wavelength tail of the apparently thermal emission of the cosmic microwave background radiation. The problem was that, at frequencies less than 1.4 GHz, the contribution of the galactic radio emission and the integrated emission of discrete radio sources dominated the total sky emission.

Howell and Shakeshaft first made estimates of the diffuse background radiation at 1.407 MHz by placing their horn antenna inside one of the 7.5-metre Wurzburg telescopes

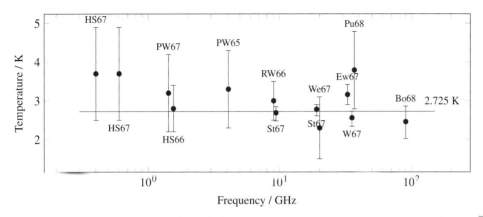

The spectrum of the cosmic microwave background radiation in terms of its thermodynamic temperature as a function of frequency, from measurements published in the years 1965 to 1968. These data are taken from the careful compilation by Lyman Page and his colleagues (Peebles *et al.*, 2009; see their Table A.2). The observations made by Howell and Shakeshaft are labelled HS66 (1.55 GHz) and HS67 (0.6 and 0.4 GHz). The red line at 2.725 K is the present best estimate of the thermodynamic temperature of the cosmic microwave background radiation from observations made by the COBE satellite (Mather *et al.*, 1990).

Fig. 14.11

which acted as a ground shield. After making allowance for the emission from the sky and other sources of emission, they concluded that the background radiation had a radiation temperature of 2.8 ± 0.6 K (Howell and Shakeshaft, 1966). They followed this up with even more challenging measurements at 610 and 408 MHz, with the result that their measurements were consistent with the presence of a background component of emission with radiation temperature 3.7 ± 1.2 K between 408 and 610 MHz (Howell and Shakeshaft, 1967). These observations confirmed that the cosmic microwave background radiation had a spectrum consistent with a pure black-body spectrum with radiation temperature about 3 K (Figure 14.11). This conclusion was subsequently spectacularly confirmed by the spectral measurements made by the COBE satellite (Mather *et al.*, 1990). These observations enabled constraints to be made to distortions of the thermal spectrum because of early heating of the intergalactic gas.

The Cavendish Radio Astronomy Group had little further involvement with the cosmic microwave background radiation until determined efforts were made to measure temperature fluctuations in the spatial distribution of the radiation and the detection of decrements in its radiation temperature due to the Sunyaev–Zeldovich effect in the directions of clusters of galaxies (Section 21.2.2).

14.6 High-energy physics

Mott's decision to stop the development of a linear accelerator in 1954 (Section 13.2) brought to an end the Laboratory's endeavours to construct its own particle accelerator.

Despite this setback, the nuclear physicists had in the meantime continued to develop advanced instrumentation for use with particle accelerators. In Frisch's own words,

> When I took up my appointment at the Cavendish Laboratory in 1947 as successor to Sir John Cockroft, it was still pioneering days for nuclear physics. Scintillation counters were just coming into general use, whereby gamma rays could be recorded with much greater efficiency and more precise timing. The primitive pulse height analysers ('kick sorters') which I had developed during the War were soon enormously improved by George Hutchinson, Denys Wilkinson and others, and so were many other techniques, created during the work on the atom bomb.
>
> Through these techniques it became possible to explore the angular distribution of the particles emitted from atomic nuclei under bombardment, and also the angular correlation of two particles emitted in succession by the same nucleus. A great deal was learned in that way about the excited states (for example, their spins and parities) of light nuclei, with Sam Devons as the driving spirit. Tony French and Denys Wilkinson pursued similar work, and it is sad to record that all three have since left Cambridge. (Frisch, 1974)

Instead, the Cavendish particle physicists used accelerators at other institutions, in particular the 28 GeV proton synchrotron at CERN and the Nimrod 7 GeV proton synchrotron at the Rutherford High Energy Laboratory. But new opportunities, very much to Frisch's taste, became a reality with the recently invented bubble chamber. As he wrote,

> All my life I have been interested in the design of scientific devices, even more than in the results which I or others might obtain with their help. (Frisch, 1980)[13]

14.6.1 Bubble chambers and Sweepnik

In 1952 the bubble chamber was invented by Donald Glaser, working at the University of Michigan, at Ann Arbor in the United States. Its principles of operation were similar to the cloud chamber but now the ionising events produce tracks consisting of strings of tiny bubbles in a superheated liquid, often liquid hydrogen but also liquid deuterium or propane. Bubble chambers can be built much larger than cloud chambers, and have many other advantages. They produce sharper tracks, as a liquid medium is more stable than the gas in a cloud chamber and also serves as a denser target material. Furthermore, cloud chambers needed minutes to recover after each expansion, whereas bubble chambers could operate at one photograph per second. This capability was well matched to the accelerators which at that time could produce a burst of 1 GeV fast particles every few seconds.

In 1955, at a conference at Ann Arbor Frisch met Glaser, who told him that his invention was now becoming a practical proposition and that huge numbers of photographs of high-energy particle collisions could be taken. Glaser stressed that there would be a huge bottleneck unless the particle track analysis could be carried out semi-automatically. Frisch took up the challenge, with the result that he and his group gained access to what he called the 'brotherhood of bubble-chamber physicists', as well as films containing the particle

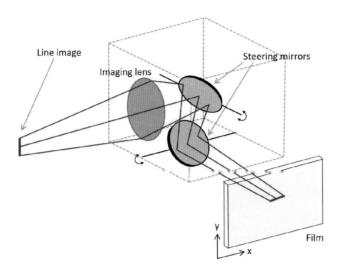

The Sweepnik probe is steered to any point on the film with an accuracy of about 1 μm by two mirrors (in grey) which are controlled by the computer. They are placed immediately behind the second imaging lens.

data. Frisch spent several years designing and building new machines for measuring bubble chamber tracks. Some worked better than others – and some didn't work at all. He explained: 'if an expert is one who has made every conceivable mistake, I became very expert in this field'.

In 1964 he came up with the concept that the measuring machine should not need an operator; instead, a beam of light would be used to follow the tracks automatically. The light beam could be steered by two mirrors, one to move it horizontally and the other vertically across the image on the bubble chamber film (Figure 14.12). Frisch used a line image which could be rotated in a circular sweep until it most closely fitted the track, instructing the machine about the direction in which a track was travelling.

These developments were aided by a number of fortunate circumstances. First, lasers became available to produce intense luminous light beams. Second, Frisch was provided by Mott with a small computer dedicated to the analysis of the scanned data. Third, the programme attracted Julian Davies, who wrote the necessary computer program, and Graham Street, who contributed greatly to the technical design and performance of the what became known as Sweepnik because of its ability to sweep up information like a sputnik (Figure 14.13).

In 1969, Frisch, John Rushbrooke and Street founded the company Laser Scan Limited, to manufacture the Sweepnik system. Originally established to commercialise Sweepnik, Laser Scan evolved into 1Spatial in 2006, and from a small group of academics into an industry-leading geospatial software and solutions provider. Its core business has been based on mapping and charting for organisations such as the Ordnance Survey of Great Britain and the UK Hydrographic Office.

Bubble chambers were used at CERN in the 1960s and 1970s for experiments which led to the discovery of the W and Z bosons in 1983. Sweepnik continued the analysis of bubble chamber photographs until 1987 (Figure 14.14).

Fig. 14.13   Sweepnik in operation identifying particle tracks (copyright High Energy Physics Group, Cavendish Laboratory).

14.6.2 Detector physics

At the same time, the Nuclear Physics Group maintained its expertise in counters and detectors, which was to prove vital for its future involvement in large-scale international projects, principally at CERN. The development of counter and spark-chamber techniques was carried out mainly at the Rutherford Laboratory using the Nimrod accelerator. The group members spent part of their time at the Rutherford Laboratory working on arrays of scintillator detectors, Cherenkov counters, spark-chambers and electronic logic circuits which operated down to the nanosecond region. The result was the ability to count events at rates in excess of a million particles per second. These techniques were used to study the decay of neutral K-mesons and determine the selection rules involved in weak decays.

14.6.3 Theoretical particle physics

Led by Eden, the Theoretical Nuclear Physics Group joined with the High Energy Physics Group. They carried out joint programmes with the experimentalists and also contributed to general theoretical developments in the study of elementary particles. The group attracted

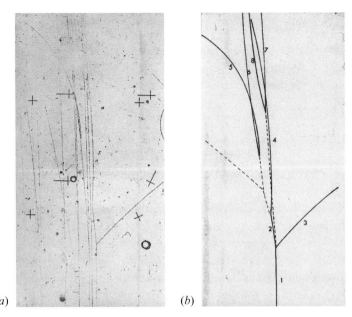

(a)   (b)

An example of particle track recognition from a bubble chamber photograph: (a) the bubble chamber photograph; (b) the Sweepnik reconstruction of the particle tracks (copyright High Energy Physics Group, Cavendish Laboratory).

Fig. 14.14

a number of outstanding young theorists including Jeffrey Goldstone and Michael Green. By the time of Mott's final report, the group's activities were listed as:

1. High-energy collisions of strongly interacting particles, including the study of resonances and asymptotic behaviour
2. General developments in Regge theory using complex angular momenta, which is closely related to topic 1
3. Analytic properties of collision amplitudes and S-matrix theory
4. Self-consistency methods for relating resonances and elementary particle interactions
5. Symmetry and group theoretic properties of elementary particles
6. The electromagnetic and weak interactions and their combination with strong interactions.

Eden was one of the authors of *The Analytic S-Matrix*, co-authored with Peter Landshoff, David Olive and John Polkinghorne, which has remained the definitive text on the subject (Eden *et al.*, 1966).

This arrangement of theorists supporting experimentalists whilst at the same time carrying out their own theoretical research has been a feature of the group since that time and has been particularly valuable in extracting the maximum science from the various experimental endeavours.

The Mott era: the growth of condensed matter physics

15.1 Low-temperature physics

During Mott's tenure of the Cavendish Chair, the low-temperature physics research carried out in the Mond Laboratory placed it at the forefront of the field internationally. The achievements during the Bragg era were to be reinforced by Pippard's and Shoenberg's ingenious experiments and insights into the nature of the Fermi surface, by Vinen and Hall's discovery of vortex quantisation in liquid helium and by Josephson's discovery of quantum tunnelling of Cooper pairs. The vision was to carry out experiments with relatively simple apparatus but great experimental skill, very much in the Rutherford tradition.

15.1.1 Determination of Fermi surfaces

The Fermi surface is a theoretical three-dimensional boundary in momentum space within which the conduction electrons of a metal are contained at absolute zero. For the free-electron model this surface is a sphere, but for a real metal the momentum states are determined by *band theory*, and the Fermi surface often has a remarkably complex shape, having crystal symmetry within the Brillouin zone structure of the atomic lattice, which is important in determining many of the properties of the metal (Box 15.1; see also Section 12.9.2). In the early 1950s there was no experimental determination of the shape of this surface for any metal, and band-structure calculations were not good enough to predict it reliably.

In 1954, however, Pippard showed that, in the extreme anomalous limit of the skin effect, the surface resistance provides a measure of one component of the curvature of the Fermi surface, averaged around the zone on which the electrons are moving parallel to the sample surface, and therefore 'effective'. He concluded that it might be possible, by making sufficient observations of the surface resistance in different orientations, to determine the geometry of the Fermi surface, at least in the fairly simple case of a metal such as copper.

He carried out this programme during the academic year 1955–56, which he spent on sabbatical leave at the University of Chicago, where Morrel Cohen had arranged for the Institute for the Study of Materials to grow a very large single crystal of copper. The success of the experiment depended on cutting extremely smooth surfaces through the crystal at different angles, and this was expertly performed at the institute. Pippard designed a new experimental arrangement to acquire the data, which were analysed on his

Box 15.1

The Fermi surface of a metallic crystal

As an example, a copper atom is composed of a positively charged nucleus surrounded by 29 electrons. The structure of the electrons is essentially that of an argon atom core plus filled 3d states and the one 4s electron; it is sometimes written as Ar $3d^{10}$ $4s^1$. The single 'outer shell' 4s electron is responsible for many of copper's physical properties, including its high electrical conductivity, its chemical stability and its reddish colour. In metallic crystalline copper, as with other metals, the 4s electron does not remain associated with any particular atom but becomes part of the electron cloud that pervades the crystal lattice.

Brillouin zones A free electron outside a crystal has kinetic energy $E = mv^2/2 = \hbar^2 k^2/2m$, which increases steadily with wave number k. Within a three-dimensional crystal, however, the energies of the freely moving valence electrons depend on the *wave vector* $\mathbf{k}$ in a more complicated way, which turns out to have lattice symmetry in k-space; the unit cell in this space is known as the *Brillouin zone*. For the FCC crystal structure of copper (*a*) it takes the form of the polyhedron shown in (*b*). Electrons lying on the surface of this polyhedron turn out to have wavelengths that satisfy the Bragg condition for reflection by the principal planes of the crystal, $\lambda = 2d \sin\theta$.

The Fermi surface is a three-dimensional boundary in momentum, or $\mathbf{k}$, space within which the conduction electrons of a metal are contained at absolute zero. In the absence of the lattice, the Fermi surface would be a perfect sphere. The presence of the atoms distorts the shape of the Fermi surface from a perfect sphere. The distorted surface must have lattice symmetry in k-space. The form it takes inside the unit cell is shown in (*c*). The model shows Pippard's determination of the Fermi surface of copper. The shape of the Fermi surface is important in determining many of the properties of the metal, for example, its de Haas–van Alphen and cyclotron resonance frequencies, and less directly its transport properties, such as electric and thermal conductivity, Hall coefficient and magnetoresistance.

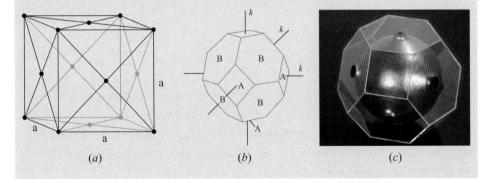

(a) (b) (c)

return to Cambridge (Box 15.2). From these data he was able by a process of geometric model-fitting to determine the curvatures of the Fermi surface in different crystal orientations (Box 15.3) and so map out the Fermi surface of copper, shown in the photograph of his model in Box 15.1(*c*), a widely recognised *tour de force* (Pippard, 1957). This was the first time that any Fermi surface had been measured experimentally.

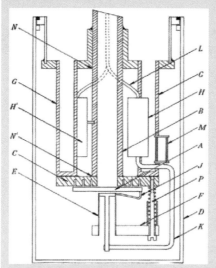

A number of thin discs in different orientations were cut from a high-purity single crystal of copper and carefully electropolished. Then they were in succession mounted at A below the end of a rectangular waveguide, B. The apparatus was placed into a cryostat containing liquid helium at a temperature of about 2.1 K. Microwaves produced by the klystron at the top of the apparatus passed down the waveguide to the sample where a small fraction of the microwave power, proportional to the surface resistance of the sample, was absorbed. The temperature of the sample was thereby raised, and the difference in temperature determined using a sensitive gas thermometer. By measuring the change in temperature, the corresponding surface resistance could be calculated.

Pippard could then calculate the radius of curvature of the Fermi surface for each angle of the crystal (Pippard, 1957).

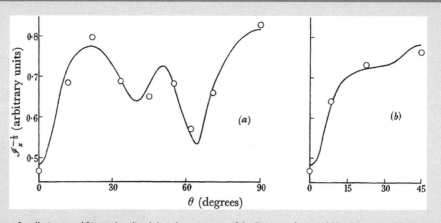

Ernst Sondheimer and Pippard realised that the curvature of the Fermi surface could be determined from studies of the surface resistance of a single crystal. Employing x, y and z coordinates in momentum space, where z is the normal to the sample surface, they showed that the surface resistance R could be decomposed into components in the x- and y-directions as $R = R_x \cos^2 \phi + R_y \sin^2 \phi$, where ϕ is the angle between the direction of current flow and the x-axis. The surface resistance in the x- and y-directions were given by

$$ R_x = \frac{\sqrt{3}}{2} \left\{ \frac{\pi \omega^2 h^3}{e^2 \int |\rho_y| \, dy} \right\}^{1/2} \quad \text{and} \quad R_y = \frac{\sqrt{3}}{2} \left\{ \frac{\pi \omega^2 h^3}{e^2 \int |\rho_x| \, dx} \right\}^{1/2}, $$

where ρ_x and ρ_y are the radii of curvature of the Fermi surface in the planes normal to the x- and y-axes, the integrals being taken over the effective zone and e being expressed in electromagnetic units (emu), the units then in use. Pippard used a process of successive approximation to find corrections to the spherical Fermi surface which would match the resistivity at different values of θ, the angle between the polished surface of the specimen and the planes of the lattice. An example of the success of the model fitting is shown in the figure for two series of measurements (Pippard, 1957).

In 1952, Lars Onsager had shown that the period in $1/B$ of the de Haas–van Alphen effect oscillations could be used to determine the maxima and minima of the cross-sectional area of a Fermi surface in planes normal to the magnetic field (Onsager, 1952), providing an alternative route to Fermi surface reconstruction. In principle, this had the advantage over Pippard's method that reconstruction of the surface was much easier, and indeed that two or more extremal areas arising from different pieces of the surface could be determined simultaneously by frequency analysis of the oscillations.[1]

In spite of his expertise in the relevant measurements, Shoenberg could not immediately deploy this approach for the most interesting noble metals, copper, silver and gold, which required much higher magnetic field strengths than he had available. He eventually solved the problem, elegantly and ingeniously, by discharging a bank of capacitors through a magnet coil to produce a very large field that rose rapidly to a peak and then fell again to zero. This field produced a time-dependent oscillatory magnetisation in the sample, detected by a search coil wound around it. In 1958 he at last succeeded in detecting the de Haas–van Alphen effect in copper. After 1962, commercial superconducting magnets became available, providing very strong and stable direct-current fields, and Shoenberg changed technique again, obtaining de Haas–van Alphen measurements that were even more precise and detailed. Thereafter, measurements of the effect became universally the preferred method of determining Fermi surfaces.

15.1.2 The dynamics of conduction electrons: Pippard and the Magnet Laboratory

Once measurements of the Fermi surface had been established, Pippard turned his attention to various aspects of the dynamics and scattering of conduction electrons, especially those orbiting in an applied magnetic field. In 1962 he published an important note showing that, in the de Haas–van Alphen effect, the field acting on the electrons is the magnetic induction and not the applied field, which explained why the magnetic moment was in some circumstances multi-valued, leading to the development of domains of different magnetisation (Pippard, 1963). He also introduced the novel idea of *magnetic breakdown*, in which the classical picture of electrons orbiting around the Fermi surface was abandoned and the possibility was allowed that they could pass by quantum mechanical tunnelling through momentum space to a nearby piece of the surface (Pippard, 1962, 1964). This idea made sense of the data on the de Haas–van Alphen effect that were threatening to become incomprehensible. He also developed important ideas in acoustic–magnetic resonance (Pippard, 1960). In 1965, his book *The Dynamics of Conduction Electrons* established his reputation as an omni-competent expert on Fermi surface physics (Pippard, 1965). His interests in these areas were matched by those of Ziman in the Solid State Theory Group, and they discussed these topics a great deal. Ziman's books *Electrons and Phonons* (1960), *Electrons in Metals* (1963) and *Principles of the Theory of Solids* (1972) covered much of the same ground.

Many of these investigations required strong magnetic fields, and until 1959 the Mond Laboratory had investigated such phenomena using strong pulsed fields in the Kapitsa tradition (see Section 10.3). By that time so many new phenomena were under investigation that Pippard decided to design a facility that would provide steady fields of 10 T over a two-inch bore. He was awarded a research grant of £43,000 and the new Magnet Laboratory, tucked into a corner of the old Cyclotron Laboratory, was opened in 1961. Its design was workmanlike and brute-force, using up to 2 MW of power, a large transformer and a bank of silicon rectifiers delivering 27 kA at 75 V. The magnet windings were flat coils of 75 mm × 1.5 mm copper strip through which cooling water was driven by powerful pumps from a large reservoir in the basement (Adkins, 1961).

In the Magnet Laboratory Pippard and his students conducted some remarkable and difficult experiments, particularly on magnetoresistance and helicons – low-frequency electromagnetic waves that can propagate in a magnetic field in solid conductors when the skin depth becomes large. The very complex subject of magnetoresistance remained for him a major preoccupation, culminating in his definitive text, *Magnetoresistance in Metals* (1989), which influenced research in cuprate superconductors. Although much was accomplished in the Magnet Laboratory, it was a short-lived enterprise because commercial superconducting magnets became available, providing greater steady magnetic field strengths, and no attempt was made to reproduce the facility when the group moved to West Cambridge in 1972.

15.1.3 Type II superconductors and quantised magnetic flux lines

It had long been realised that alloy superconductors often behave very differently and confusingly in magnetic fields from the traditional type I superconductors (Section 12.9.3). Clarification began in 1957 when Abrikosov first showed that, when the coherence length ξ is shorter than the penetration depth λ and the normal–superconducting interface energy becomes negative, a superconductor in the presence of a magnetic field breaks up into a *mixed state*, with much finer divisions than the intermediate state. Applying Ginzburg–Landau theory, Abrikosov showed that the mixed state should contain *quantised magnetic flux lines*, miniature flux vortices each containing a quantum h/e of flux. Under ideal conditions these flux lines would be arranged on a lattice. Materials that behave in this way are called *type II superconductors*. Because alloying shortens ξ and lengthens λ, most alloys are type II superconductors, but a few pure elements, such as niobium, belong to this type.

Abrikosov worked out magnetisation curves for the type II state and, for cases in which $\xi \ll \lambda$, the system could retain superconducting order to much higher fields. This immediately excited great interest because, provided some means could be found to pin the flux lines so that they could resist the magnetic forces acting on them, type II materials should be able to carry large currents in high magnetic fields, making high-field superconducting magnets a real possibility. These hopes were quickly realised. In 1962 Nb–Ti wire became commercially available and magnets generating 10 T could soon be bought from General Electric and other companies. They are now widely used, for example, in hospital NMR machines, in levitated trains and in the Large Hadron Collider at CERN.

The Mond Laboratory was not actively involved in the development of magnets because in 1962 David Dew-Hughes set up a superconductivity group in the Department of Materials Science, with the specific aim of developing new materials. This group quickly made many contributions to the fundamental understanding of type II materials. In 1966 Jan Evetts took over the leadership of the group and in 1972 published with Archibald (Archie) Campbell the influential *Critical Currents in Superconductors*, covering the whole field (Campbell and Evetts, 1972).

15.1.4 Second sound and quantised vortices in superfluid helium

The discoveries of superconductivity and superfluidity resulted in a burst of theoretical activity which continued after the war.[2] One such development was the prediction by Tisza and Landau of *second sound waves* – pure temperature waves in which heat takes the place of pressure. Such waves had been detected by Vasilii Peshkov in 1944, using a heater and a thermometer (Peshkov, 1944). In 1950 their velocity was measured in the Mond Laboratory as a function of temperature by Kenneth Atkins and Donald Osborne (Atkins and Osborne, 1950).

In another development, Landau had noted in 1941 that a consequence of the existence of an effective superfluid wave function in helium was that the velocity field of the superfluid should be zero – liquid helium II should not be able to rotate, and consequently, when a vessel containing helium II is rotated, the liquid surface should not take a parabolic form, as in the case of a normal fluid. To everyone's surprise, in 1950 the first experiments of this type showed that, even at very low temperatures, helium II behaved like any other liquid.

The solution, proposed by Onsager (1949) and elaborated by Feynman (1955), was that vortices appeared around singularities in the superflow, for which the circulation $\kappa = \oint \mathbf{v}_s \cdot d\mathbf{r}$ was quantised in units of h/m, where m was the mass of the helium atom. Quoting Leggett (1995),

> The explanation was that the liquid acquires vortex lines on which the density of the superfluid is zero, and around which the superfluid flows with quantised circulation... When the container is rotated, the vortices arrange themselves so that the macroscopically averaged velocity is the same as in an ordinary liquid and the macroscopically averaged surface... is correspondingly identical.

In the Mond Laboratory, Vinen and Hall carried out experiments to test this model by determining the mutual friction between the superfluid and normal components (Hall and Vinen, 1956a). If the vorticity were confined to a number of vortex lines, it would be expected that the ordered structure of the superfluid would act as scattering centres for the excitations which constitute the normal fluid and so give rise to mutual friction between the two fluids when they are in relative motion. This dissipative process was associated with the second sound waves, the velocity of which Atkins and Osborne had recently measured. Vinen and Hall showed that, according to the accepted models, there should be attenuation associated with the second sound which was proportional to the angular velocity of the uniformly rotating helium.

Using a resonance technique, they found that the dependence of the mutual frictional force followed exactly that predicted by the Feynman model (Figure 15.1). They concluded that

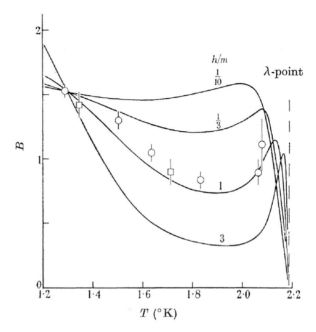

Fig. 15.1 The effect of changing the circulation κ about the vortex lines on the theoretical temperature dependence of the coefficient B in the mutual friction–angular velocity difference relation. The curves are calculated for the values of κ normalised to the value $B = 1.52$ at 1.30 K (Hall and Vinen, 1956b). Theoretical lines are shown for different quantised values of h/m.

> if the mutual friction is in fact due to collisions between normal fluid excitations and lines situated in the superfluid, then the number of lines per unit area cannot differ appreciably from the value $2\omega m/h$ given by Feynman's theory. (Hall and Vinen, 1956b)

In 1957 Hall went on to demonstrate that, even at low angular velocities, the superfluid component in a can of rotating helium II is set into rotation because of the angular momentum each quantum vortex carries (Box 15.4).

Hall and Vinen's experiments provided indirect evidence for quantisation of the circulation as predicted by Onsager and Feynman, but in a later experiment Vinen aimed to make a direct demonstration of the quantisation of a single vortex line (Vinen, 1961). Liquid helium was contained in a metal cylindrical tube down the centre of which a fine wire was stretched – the circulation was to be measured round the wire. Vibration of the wire in the absence of the rotating superfluid would result in two degenerate normal modes of vibration at right angles to each other but, in the presence of the circulation of the superfluid about the wire, the Magnus effect[3] lifted the degeneracy and the two resulting circularly polarised normal modes were separated in frequency by $\rho_s \kappa / 2\pi w$, where ρ_s is the density of the circulating fluid and w is the sum of the mass per unit length of the wire and half the mass of the fluid displaced by unit length of wire.

In his classic paper, Vinen describes the difficulties of observing the effects due to a single vortex line, but this was achieved by violently disturbing the system by a

A schematic diagram of the cryostat used by Hall in his demonstration of the accelerated motion of liquid He II (Hall, 1957).

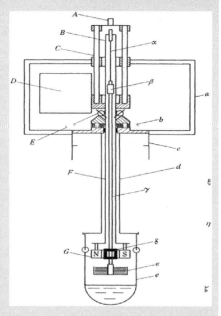

The helium is contained in a can filled with closely spaced discs (see figure). The can, initially at rest, is suddenly brought into rotation with a uniform angular velocity in the range 0.1 to 1 rad^{-1}. As the liquid gradually accelerates a torque must be exerted on the can in order to maintain its uniform motion. The experiment consists in measuring this torque as a function of time, and also the corresponding torque when the can is suddenly brought to rest, after the liquid has attained rotational equilibrium. By integrating the observed torque–time curves, the angular momentum of the liquid corresponding to each value of torque during the accelerated motion can be found. The measurements were made at a temperature of 1.27 K, at which the normal fluid is tightly coupled to the disc system by its viscosity, and the observed torque can be entirely attributed to the acceleration of the superfluid.

large-amplitude vibration which had the effect of detaching partially-attached vortices. The result was that the circulation decreased to exactly the value h/m predicted by theory.

In 1962, Vinen was appointed to the Chair of Physics at Birmingham University, where he set up a major low-temperature group, and then to the Poynting Chair in 1973. Hall moved to Manchester University in 1958 and was promoted to a professorship in 1961.

15.1.5 The BCS theory and the Mond Laboratory

The discovery of the Bardeen, Cooper and Schreiffer (BCS) theory of superconductivity (Box 15.5) had a powerful effect on research in the Mond Laboratory. For instance, when Andrew (Andy) Phillips joined the Laboratory as Pippard's research student in 1961, he investigated ultrasonic attenuation in superconductors. The lattice displacements in a sound wave create electric fields which act on the normal electrons, causing dissipation, and the BCS coherence factors for this process are different from those in microwave absorption, as Phillips confirmed. It was already known that the attenuation decreased in the superconducting state according to the Fermi function $f(\Delta/kT)$, where Δ is the energy gap, but Phillips showed that the scattering processes in an anisotropic superconductor were more complicated than had been thought, and that the simple BCS theory had to be modified

Box 15.5   **The route to the BCS theory of superconductivity**

Although Fritz London's ideas and the success of the semi-empirical Ginzburg–Landau theory had shown that superconductors appeared to have an effective superfluid wave function, there was no microscopic theory of how this came about, nor of what interaction caused condensation into the superconducting state, despite serious attempts by Heisenberg, Born and Cheng and others.

During the 1950s, some helpful clues emerged. Measurements by A. Brown, Mark Zemansky and Henry Boorse (1952) at Columbia University on the specific heat of superconducting niobium below $4.5°$ K indicated that at low temperatures the number of normal electrons decreased more rapidly than expected in the Gorter–Casimir two-fluid model. Then in 1956 Michael Garfunkel and his colleagues at Pittsburg showed, using millimetre microwaves, that there is an *energy gap* at the Fermi surface for the normal electrons in superconducting aluminium (Biondi *et al.*, 1956), a conclusion soon confirmed for other superconductors.

In 1950 Herbert Frölich at the University of Liverpool made the key suggestion that, for electrons sufficiently near the Fermi surface, an effective attraction might be mediated by exchange of a virtual phonon (Fröhlich, 1950), just as in quantum electrodynamics the Coulomb repulsion between electrons is mediated by a virtual photon. The idea that phonons were important was supported by the discovery, also in 1950, of the *isotope effect*. When the atomic mass M was changed by using a different isotope, the critical temperature for superconductivity varied as $M^{-1/2}$. This change was not expected to change the electronic wave functions themselves, but would alter the phonon frequencies, and hence the strength of the Frölich attraction.

Armed with these facts and concepts, John Bardeen, Leon Cooper and Robert Schreiffer of the University of Illinois formulated what became known as the BCS theory of superconductivity during the period July 1956 to July 1957 (Bardeen *et al.*, 1957).[4] The key step was the formulation of an approximate ground state in which the Fermi sea of electrons is replaced by

$$|\text{BCS}\rangle = \prod_{\mathbf{k}} \left(u_k + e^{i\theta} v_k c_{\mathbf{k}\uparrow}^{\dagger} c_{-\mathbf{k}\downarrow}^{\dagger} \right),$$

where the $c^{\dagger}$ are second-quantised creation operators for electrons, $u_k^2 + v_k^2 = 1$, and u_k increases smoothly from 0 below the Fermi surface to 1 above it. This expression can be thought of as representing the wave function for *Cooper pairs*, pairs of electrons bound together by the electron–phonon–electron interaction in a state somewhat similar to the Bose–Einstein condensation with a definite phase θ. But this simple picture neglects the fact that the Cooper pairs overlap very strongly with each other – the scale of the pairing turned out to be essentially Pippard's coherence length. Furthermore, the binding is a cooperative phenomenon – as each pair forms, it helps the condensation of the next.

The many successes of the theory were immediately apparent. For instance, it explained the magnitude and temperature dependence of the energy gap, and the temperature dependence of the free energy and consequently other thermodynamic parameters such as the heat capacity and the critical magnetic field. The BCS expression for the supercurrent reproduced almost perfectly Pippard's non-local expression, as the authors were at pains to point out, and it explained the order of magnitude of superconducting transition temperatures.

The theory also showed a fascinating but completely unexpected feature: the excitations corresponding to the normal fluid were a strange mixture of electrons and holes, changing character from pure-electron, well outside the Fermi surface, to pure-hole, deep inside. This proved to have many consequences, one of which was that all scattering matrix elements were drastically modified, by what BCS termed *coherence factors*.

(Phillips, 1969). This was an early move towards understanding non-equilibrium superconductivity.

Pippard continued his interest in the surface impedance of superconductors, and his students investigated a range of topics in different limits of the BCS theory. David Williams tackled aluminium, which has a particularly long coherence length and therefore fell into the Pippard limit, analogous to the extreme anomalous limit in normal metals. John Adkins, who joined the Laboratory as a postdoctoral fellow in 1959, pushed the group's surface impedance work up in frequency by a factor of 15, with the aim of directly measuring the anisotropy of the BCS energy gap in tin by microwave absorption. This was a decidedly difficult task with the microwave equipment then available, because it involved both stages of frequency doubling and the in-house construction of 2 mm waveguide by electro-deposition (Adkins, 1962). He also introduced the rest of the group to the transistor age by designing a simple temperature controller, which was soon adopted with enthusiasm, and relief, by the whole laboratory.

John Waldram joined the Mond Laboratory as Pippard's graduate student in 1959 and extended and developed Pippard's work on Sn–In alloys. The full implications of BCS for surface impedance had by now been worked out by Mattis and Bardeen (1958), and Waldram was able to show in detail how successful this theory was, in particular, how crucial the BCS coherence factors were to its success. He was also able to make one of the first determinations of anisotropy in the energy gap in tin. Having become an enthusiastic advocate for the BCS theory, with perhaps more enthusiasm than his supervisor, he reviewed all the new results on surface impedance in the light of the BCS theory (Waldram, 1964). He was appointed a University Demonstrator in 1963.

Another important development of 1960 concerned Ivar Giaever's first results on tunnelling using superconductors, carried out at the General Electric laboratory in Schenectady. When two metals are separated by a thin insulating oxide barrier, electrons can pass through the barrier by quantum mechanical tunnelling, directly from a quantum state on one side into a state at the same energy on the other side. Giaever reasoned that if one had a normal metal with a constant density of states on one side, and a superconductor showing a gap and a BCS density of states on the other, then if one measured the junction I–V curve at a low enough temperature, its derivative dI/dV should be proportional to the BCS density of states. This would allow the gap to be measured in a very simple way, and this was confirmed by his beautiful results (Giaever, 1960a,b).

Metal evaporators allowing the controlled deposition of metal films onto glass slides were by then commercially available, and Adkins, appointed a university demonstrator in 1960, soon became the local tunnelling expert. He and his graduate students investigated the energy gap in tin, higher-order processes in which several particles appeared to be tunnelling at once, and used tunnelling to investigate the approach to equilibrium in superconductors and so determine various scattering lifetimes (Gray et al., 1969).

A further development of BCS theory was the understanding of the *proximity effect*, the process whereby superconductivity can leak into an adjacent normal metal. New theories of the proximity effect were developed by Gor'kov, de Gennes and others, which involved a new form of BCS theory in which the self-consistency of the order parameter became non-local (de Gennes, 1966).

These developments were taken up in several ways in the Mond Laboratory. Adkins and his colleagues investigated the proximity effect by preparing a superconducting–normal sandwich and then tunnelling into its normal side to observe the local density of states. By varying the thickness of the normal layer the extent of the proximity effect and its effect on the density of states was observed. Waldram observed the microwave surface impedance of a thin layer of tin deposited by electroplating onto a lead rod. The proximity effect of a strong superconductor, lead, enhanced the order parameter of a weaker superconductor, tin. The proximity effect was also involved in studies of superconducting–normal–superconducting (SNS) junctions and of tin in the mixed state, by John Clarke and others (Section 18.4.3).

15.1.6 The Josephson effect

Brian Josephson arrived as an undergraduate in Cambridge to read mathematics in 1957, the same year the BCS theory of superconductivity was published. He switched to physics in his final (Part II) year, during which he wrote a ground-breaking paper on the Mossbauer effect (Josephson, 1960). Despite his remarkable ability and insight in theoretical physics, he decided to carry out research in experimental low-temperature physics in the Mond Laboratory under the supervision of Pippard. He performed high-frequency experiments and wrote his PhD dissertation on the magnetic field dependence of the superconducting penetration depth.

In 1962 Philip Anderson visited the Cavendish Laboratory for a year, on leave from the Bell Laboratories in New Jersey, and gave advanced lectures on theoretical condensed matter physics. Anderson's presence in Cambridge was crucial for Josephson. In particular, Anderson elucidated in his lectures the formal microscopic description of the superconducting order parameter, which behaves like a one-particle wave function having a well-defined local phase and describes the motion of the superfluid. Josephson discussed the theory of superconductivity intensively with Anderson.

It had been a puzzle why Giaever's simple interpretation of his tunnelling experiments on superconductors had worked as well as it did, because it made no allowance for the BCS coherence factors which applied to the normal electrons. In February 1962, however, Morrel Cohen, Leo Falicov and Bill Phillips validated Giaever's interpretation, treating the tunnelling according to BCS theory, and employing a simple tunnelling Hamiltonian (Cohen *et al.*, 1962). Josephson had been wondering whether there was some way in which tunnelling could be used to make the phase of the order parameter observable. The new paper provided a tractable formalism with which to tackle the issue.

Josephson soon realised that certain terms in the tunnelling current which Cohen and his colleagues had assumed averaged to zero might, with a different interpretation, be physically significant. He was initially seeking a phase-dependent term in the normal tunnelling, and did indeed find one, but in addition found another term of similar magnitude, the famous term describing the *superconductive tunnelling of Cooper pairs*, which takes the simple form

$$I_s = I_J \sin \phi, \tag{15.1}$$

where $\phi = \theta_2 - \theta_1$, the difference in superfluid phase between the two sides of the junction. He also deduced that the time dependence of the phase difference should be given by $\hbar(\partial\phi/\partial t) = 2eV$, where V is the voltage across the junction (Josephson, 1962). With help from Adkins in sample preparation, he promptly set out to check the effect himself experimentally, but failed to find it.

His predictions were immediately controversial, because superelectrons must tunnel as Cooper pairs, and the tunnelling rate for any pair process would be expected to be of higher order in the tunnelling Hamiltonian than that for single particles, and therefore unobservable. In July 1962, Bardeen published a magisterial rebuttal of Josephson's theory along these lines, and in September of the same year a famous, but very gentlemanly, confrontation took place at the Low Temperature Conference at Queen Mary College, London. Bardeen explained gently why he thought Josephson's results were wrong, but in this David-and-Goliath confrontation, Josephson stood up and explained very politely why he thought they were right.

Josephson was correct – because superpair tunnelling is a coherent phase-dependent process, it is *first* order in the appropriate matrix element, whereas normal particle tunnelling is an incoherent process, and like other scattering rates is *second*-order in its matrix element. Since supercurrent tunnelling involved two particles, this made the Josephson current of the *same* order as the normal current. The following January, Anderson and John Rowell at the Bell Laboratories conclusively showed experimentally that the Josephson effect was a reality (Anderson and Rowell, 1963). In 1973 Josephson was awarded the Nobel Prize in Physics, jointly with Leo Esaki and Ivar Giaever, for his discovery of the quantum mechanical tunnelling of Cooper pairs. The richness of the physical phenomena associated with the Josephson effect and its applications has turned out to be extraordinary.

Many of its applications are connected with the fact that when a magnetic field is introduced into a superconducting ring containing the junction, it affects the phase difference ϕ. The year after the first observation of the Josephson effect, Robert Jaklevic, John Lambe, Arnold Silver and James Mercereau at the Scientific Laboratory of the Ford Motor Corporation at Dearborn, Michigan demonstrated *quantum interference* in a superconducting ring containing two Josephson junctions when the magnetic field applied to the ring was varied (Jaklevic *et al.*, 1964). Subsequently, the first practical devices emerged and included point-contact SQUIDs, the acronym standing for superconducting quantum interference device, built by James Zimmerman and Silver (Zimmerman and Silver, 1964). Essentially they are exceedingly sensitive magnetometers which can detect magnetic fields as weak as 5×10^{-18} T and are widely used, for instance, in geophysical surveying and biomedicine. They can also be used to construct a direct-current voltmeter with sensitivity 10^{-16} V and a time constant of 1 s, which revolutionised many aspects of low-temperature experimental research.

In 1964 John Clarke became a research student in the Mond Laboratory under the supervision of Pippard, who encouraged him to develop a Cambridge version of the SQUID. This version came to be known as a superconducting low-inductance galvanometer, or SLUG (Clarke, 1966) (Box 15.6). Clarke has written a delightful account of how this came about (Clarke, 2011). In his words:

Box 15.6 **The superconducting low-inductance undulatory galvanometer (SLUG)**

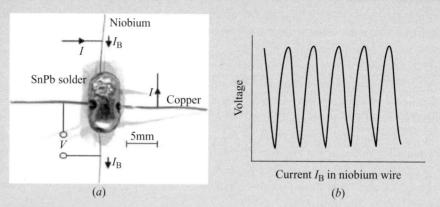

(*a*) Photograph of a SLUG showing the attachments of current (I), voltage (V) and flux bias (I_B) leads. (*b*) Voltage V versus I_B (Clarke, 2011).

The SLUG is a very sensitive picoammeter which makes use of the Josephson effect. A static current I is applied, as well as a sinusoidal current I_B through the SLUG, with the result that voltage pulses appear across the voltage leads. The area under these pulses depends on the critical current of the SLUG. This enabled Clarke to determine the critical current, and hence the current in the niobium wire, to an accuracy of about ± 1 µA. Clarke made the SLUG into a voltmeter by connecting the copper block in series with a manganin wire and the niobium wire in the SLUG. He used the SLUG as a voltmeter, mostly to investigate SNS Josephson junctions.

An important part of one's life at the Cavendish was – and still is – coffee at 11 am and tea at 4 pm. The Mond students and staff always sat at the same table. One tea-time, I discussed how I was looking for a Josephson junction technology that did not require thin films yet was mechanically stable. Paul Wraight, with whom I shared my lab, suddenly looked at me and said something like 'How about a blob of solder on a piece of niobium wire? Solder is a superconductor, and you keep telling me that niobium had a surface oxide layer.' We rushed back to our lab, where . . . I made two devices consisting of a blob of lead-tin solder melted onto a short length of niobium wire, attached some leads and lowered the devices into the helium bath. They both worked! . . . Paul and I were thrilled. The next morning Brian Pippard wandered into our lab to see how things were going, and Paul and I proudly showed him one of our new gadgets. Brian contemplated it thoughtfully for a while, and then – with a smile – said 'It looks as though a slug crawled through the window overnight and expired on your desk! (Clarke, 2011)

There are myriads of other applications of the Josephson effect. There are Josephson detectors, mixers and parametric amplifiers. The *Shapiro steps*, which appear in the *I–V* characteristic of a superconducting tunnel junction at the voltages $V_n = n\hbar\omega/2e$ when microwaves of angular frequency ω are applied to the junction, have proved so sharp and reproducible that Josephson junctions have long been employed in standards laboratories throughout the world for voltage calibration. Since 1988, the international volt has been defined using Josephson junctions.

15.2 Electron microscopy

During Mott's tenure of the Cavendish Chair, Cosslett continued his unremitting drive to exploit electron microscope techniques in a very wide range of scientific research areas. He devoted his management and entrepreneurial skills to developing advanced facilities and techniques for the many different ways in which focussed electron beams could be exploited, as illustrated by Figure 12.14. The discipline attracted many of the very best researchers from other groups in the Laboratory and other departments of the University. There was a strong link with the Engineering Department through William Nixon, with whom he had developed the X-ray projection microscope (Section 12.8) and with whom he wrote the first textbook on *X-Ray Microscopy* (Cosslett and Nixon, 1950). These pioneering instruments opened up the huge field of electron-probe techniques, which were soon to dominate the developing area of microstructural science.

A good example is Cosslett and Duncumb's development of the *scanning X-ray micro-analyser*. In 1951, Raimond Castaing of the Institute for Aeronautical Research of the University of Paris had constructed a static electron-probe X-ray microanalyser with a probe about 1 μm in diameter – he is regarded as the inventor of electron-probe techniques. By measuring the X-ray emission spectrum, chemical analyses could be carried out on the μm scale (Castaing and Descamps, 1955). As recorded by Mulvey,

> Cosslett decided that it might be more appropriate to use a modified scanning electron microscope for viewing the specimen and taking the spectrum . . . No industrial support was forthcoming, but Cosslett managed to find the funds needed to make an experimental instrument in the Cavendish workshop. [Peter] Duncumb took on the project and the scanning X-ray microanalyser was created from scratch, with minimal resources. The Cavendish RCA electron microscope column was 'cannibalised' to provide the electron probe and a new final lens and X-ray spectrometer fitted underneath. Complete success was swiftly obtained, and scepticism about the instrument rapidly evaporated. (Mulvey, 1994)

Whereas Castaing's electron beam had been focussed using an electrostatic lens, Cosslett and Duncumb scanned the specimen using a magnetic lens and then displayed the X-ray image of the surface by transferring the signal to a cathode-ray tube synchronised with the scanning beam. Using a proportional counter, they were able to select the X-ray lines of particular elements and so create pictures of the distribution of different elements over the surface of the specimen (Cosslett and Duncumb, 1956). With the collaboration of D.A. Melford of the Tube Investments Research Laboratory, a prototype commercial version of the microscope was developed, and subsequently marketed by the Cambridge Instrument Company.

A similar commercialisation took place with the development of the 'Geoscan' instrument by J.V. Long, who joined Cosslett's group in 1954 from the Chemical Research Laboratory at Teddington. His interest was in the use of electron microscopy for the analysis of geological specimens and, using all the technical refinements which the group had developed, the Geoscan electron probe analyser was commercialised, again by the Cambridge Instrument Company, in 1965.

15.2.1 The physics of dislocations

While Cosslett provided the electron microscopes, the pioneering science was carried out under the leadership of Peter Hirsch and his colleagues, particularly Michael Whelan and Archie Howie. The major breakthrough was the discovery that dislocations and other defects in crystalline solids could be observed in thin films by transmission electron microscopy. In Hirsch's pioneering theoretical analysis of diffraction effects in electron microscopic imaging, carried out jointly with Whelan and Howie, they demonstrated why dislocations and stacking faults appear as they do in electron microscope images (Hirsch et al., 1960). Most of the achievements of the group rest upon the application of these techniques, including those of bright- and dark-field imaging (Box 15.7). The pioneering calculations of Whelan and Howie, in developing the dynamical theory of electron diffraction to explain the details of the diffraction contrast images of dislocations and similar phenomena, depended crucially upon the computing power provided by EDSAC 2 (Howie and Whelan, 1961, 1962).

Soon after Mott arrived in Cambridge, a major breakthrough in these studies took place with the development by Hirsch, Robert Horne and Whelan of the technique of diffraction contrast transmission electron microscopy, which followed from Bragg's microbeam diffraction project.[5] Hirsch had previously been examining the concentrations of dislocations at sub-grain boundaries in metal foils using X-ray diffraction techniques. In 1956, Hirsch and his colleagues used the Siemens and Halske 'Elmiskop', acquired in 1955, to obtain high-resolution electron-optical images of very thin aluminium foils (Hirsch et al., 1956). The foils were beaten to a thickness of 0.5 μm and were found to be transparent to electrons over large areas. High-resolution electron micrographs were obtained by bright- and dark-field imaging from preselected areas of the foils. The techniques involved are described in Box 15.7 and some of the images obtained shown in Box 15.8.

In their pioneering experiments and interpretation of the electron microscopic images, they demonstrated that the spot and line features seen in the images in Box 15.8 are indeed dislocations in the crystal structure of aluminium, although the crystal planes themselves could not be resolved. Furthermore they observed the dislocations moving under stress in the interior of the metal, vindicating Taylor's insights of 20 years earlier. Most of the dislocations observed lay in the sub-boundaries of crystalline grains and relatively few inside the grains themselves. While similar structures had been observed in inorganic crystals on the scale of 1 μm, the spacing of the dislocations in the metal sample was on the scale of only 10 nm. These experiments opened up the quantitative study of the role of dislocations in metals. Mott recalled vividly the day in 1956 when Hirsch and his colleagues excitedly brought him the transmission electron microscope images of the movement of

Diffraction contrast and structure imaging

Box 15.7

For electrons traversing a crystal and making a small angle θ with the crystal planes of spacing d, Bragg diffraction occurs with a scattering angle 2θ, provided $\sin\theta = l/d$. The diffraction pattern, showing the different Bragg spots g, appears in the back focal plane A of the objective lens. The elastic strain field of a dislocation causes tilting of the crystal planes in its vicinity so that the Bragg conditions are different there from those in the perfect crystal far away. The diffracted beams still emerge in the same direction g but with different intensity near the dislocation.

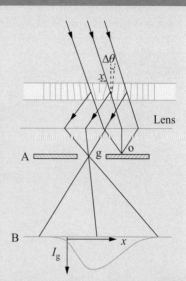

In *diffraction contrast imaging*, a dark-field image can be formed in the image plane B by selecting the spot g and inserting a screen with a suitable aperture in the plane A. This map of the Bragg reflection intensity shows a bright line near the disloca-tion if the strain field has tilted the crystal planes there closer to the Bragg condition. In simple cases the bright-field image, formed at the spot o in A, is complementary to the dark-field image. The image at B is fur-ther magnified by the later lenses in the microscope. These lenses can alternatively be focussed on the plane A to observe the diffraction pattern and the position of the aperture.

In the *structure imaging method* pioneered by James Menter, a larger aperture is used to enclose two or more Bragg spots. The crystal lattice can then be imaged directly if the resolution of the microscope is high enough. These and many other techniques are described in the book *Electron Microscopy of Thin Crystals* (Hirsch *et al.*, 1965).

dislocations within the aluminium foil. These experiments resulted in the discovery of new kinds of defects, which are vividly portrayed in the images in Box 15.8.

Other major achievements included the observation of stacking faults and the determination of their surface energies, the dependence of cross-slip on stacking fault energy and the observation of defects formed when vacancies condense on quenching a metal, for example, dislocation loops in aluminium and the discovery of the tetrahedral arrangement of defects in gold. The group also made the first observations of dislocation loops in work-hardened materials and as a result of radiation damage. The process of recrystallisation was shown to occur through the movement of existing grain boundaries, rather than by the nucleation of new ones. In analyses of cold-worked materials, they showed that the flow stress is inversely proportional to the distance between dislocations.

The distinctive feature of the remarkable achievements of the group was the combination of electron microscopy and electron diffraction with a thorough understanding of the principles of crystallography.

In 1956, James Menter made the first direct observation of crystal lattices by transmission electron microscopy. In most crystal structures, the separations between the planes of

Photographs of dislocations using the diffraction contrast technique

On an electron micrograph, dislocations show up as lines or spots, depending on their orientation.

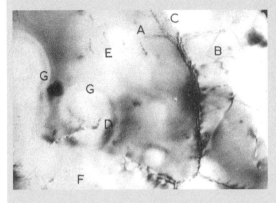

Dislocations in aluminium sheets (Hirsch *et al.*, 1956, Figure 5, Plate 24).

A: Dislocation node; the 'spotted' appearance of the dislocation lines is probably due to an interference effect

B: Crossing of two dislocations

C: Cross-grid of dislocations

D: Termination of boundary

E, F: Single dislocation lines

G: Extinction contour.

Magnification: $\times 100,000$.

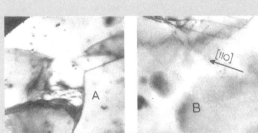

A, B: Square cross-grids of dislocations (Hirsch *et al.*, 1956, Figure 7, Plate 28). The dislocations in B are approximately parallel to [110] directions.

Magnification: $\times 100,000$.

Extended dislocations (stacking faults) in a thin foil of cobalt; the hexagonal phase has grown into the cubic phase on cooling.

Magnification: $\times 77,000$.

the crystal lattice are too small to be resolved by electron microscopy, but certain crystal materials are known from crystallographic studies to have large separations. In his study, Menter selected platinum phthalocyanine, for which the separation between the planes was about 1.2 nm, well within the capabilities of a Siemens Elmiscop 1 microscope operating at 80 kV. One of his classic images of a single edge dislocation in the crystal structure is

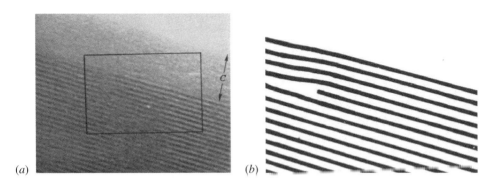

(a) Menter's 1956 image of a single dislocation in platinum phthalocyanine. The lattice plane spacing is 1.2 nm. (b) A sketch showing the lattice planes and a single edge dislocation within the box indicated in (a) (Menter, 1956).

Fig. 15.2

shown in Figure 15.2. His paper of 1956 had a profound impact on electron microscopy of crystals and laid the foundation for the modern high-resolution structure imaging method (Menter, 1956).

These studies opened up new possibilities for understanding the structure and properties of crystals, metals and other materials, and these were to be exploited by both physicists and materials scientists over the coming years. All aspects of electron microscopy were brought together in the published proceedings of the July 1963 Cambridge Summer School on advanced operational techniques (Hirsch *et al.*, 1965). *Electron Microscopy of Thin Crystals*, written by the pioneers of the discipline, Hirsch, Howie, Robin Nicholson, Donald Pashley and Whelan, provided comprehensive descriptions of all aspects of the techniques of electron microscopy.

The study of defects in the crystalline state now forms a substantial part of solid mechanics, treated in departments of engineering and central in the burgeoning discipline of materials science. Dislocation studies play an important role in the mechanisms at work in inorganic catalysts, as well as in semiconductor devices and minerals undergoing phase transitions. Following his promotion to a readership in 1963, Peter Hirsch was attracted to the Department of Metallurgy at Oxford in 1965, where he developed the department, familiarly known as 'Oxford Materials', to global prominence.

An important contribution to the reconstruction of images in the absence of phase information was carried out in the Electron Microscopy Group by R.W. Gerchberg and Owen Saxton (Gerchberg and Saxton, 1972). Their work depended upon the availability of high-speed computing facilities. Their procedure involved starting with the measured intensities in both the image and diffraction planes. They made initial guesses of the likely phases and then used fast Fourier transform (FFT) techniques to cycle between the image and diffraction plane information. The cycle converged to a sharp two-dimensional image. This work was not fully appreciated for some time, although it was used in secrecy by the US defence industry. The technique now underpins many lensless imaging techniques, including John Rodenburg's ptychography (Section 18.3). Similar approaches were to be developed independently by Baldwin and Warner in the Radio Astronomy Group (Baldwin and Warner, 1976, 1978) (Section 17.1.3).

15.2.2 Aberration correction

Because of the aberrations inherent in round electron lenses, electron microscopes only attained resolutions of about 0.3 nm, far less than the theoretical maximum imposed by the de Broglie wavelength of the electron. The most important aberration is spherical aberration (C_s). The history of attempts to improve the resolution of the electron microscope is told in some detail by Peter Hawkes (Hawkes, 2009), who was a member of Cosslett's group in the 1960s. The limits to the ability to correct for spherical and chromatic aberration had been treated in detail by Scherzer (1947), who noted that the first two terms in the expression for the potential of a cylindrical lens are those of a round lens and a quadrupole. In 1954, this led Geoffrey Archard of the AEI Research Establishment to propose that combined elements producing quadrupole and octopole fields simultaneously could be used to reduce the aberrations (Archard, 1956). Later, it was realised that hexapole correctors could also be employed to reduce the aberrations.

An early attempt by Hans Deltrap and Cosslett to develop a C_s corrector is illustrated in Box 15.9(a). Although it corrected C_s, it did not improve the resolution because of poor alignment of the corrector and insufficient power supply stability. Deltrap's original corrector consisted of four such eight-element units of the type illustrated in Box 15.9(b). The corrector was tested on an electron-optical bench, but not on an electron microscope. Although Deltrap demonstrated experimentally that C_s can be significantly reduced by a combined quadrupole/octopole lens, the system proved to be very difficult to align and adjust.

It was not until 1997, when Ondrej Krivanek, Mick Brown and their colleagues proposed a corrector that produces negative spherical aberration, that the resolution of the STEM

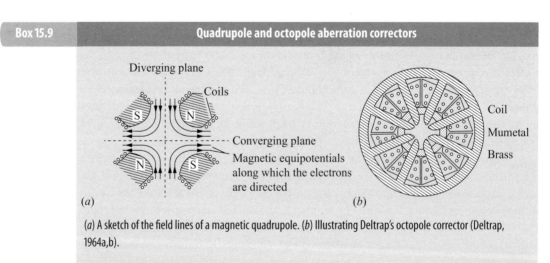

Box 15.9   **Quadrupole and octopole aberration correctors**

(a) A sketch of the field lines of a magnetic quadrupole. (b) Illustrating Deltrap's octopole corrector (Deltrap, 1964a,b).

A combination of quadrupole and octopole lenses provides a means of correcting spherical aberration. A single quadrupole lens has a converging and a diverging plane and this produces a line image of a point source. But combinations of quadrupoles can produce point images, and their spherical aberration can be cancelled using octopole lenses. In the centre is a magnetic quadrupole. On either side are units of eight elements, each carrying two coils. By suitably coupling the coils and shaping the ends of the poles, the unit provides both an octopole and a quadrupole field.

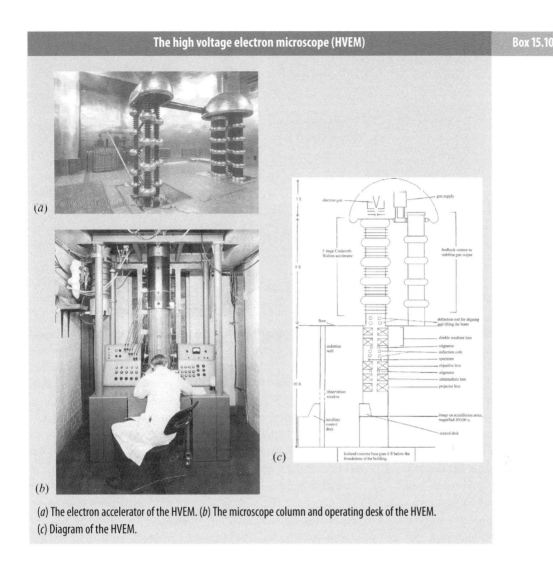

The high voltage electron microscope (HVEM)

Box 15.10

(*a*) The electron accelerator of the HVEM. (*b*) The microscope column and operating desk of the HVEM.
(*c*) Diagram of the HVEM.

improved to less than 0.1 nm, allowing individual atoms to he imaged and their spectra taken (Section 20.7). In the meantime, the only way of increasing the resolution of the electron microscope was the brute-force method of going to higher and higher voltages. This led to the Cambridge High Voltage Electron Microscope.

15.2.3 High voltage electron microscope

The design of the high voltage electron microscope (HVEM) was similar to that of a conventional TEM, with the addition of an electron accelerator between the electron gun and the lens system. X-ray radiation is a great hazard and so the microscope was heavily shielded and operated by remote control. The advantage of the HVEM was that much thicker specimens, up to about 3 μm thickness, could be examined, and so the bulk properties of metals and entire biological cells could be examined. The useable thickness tends to decrease with increasing density. As the voltage increases, chromatic aberration

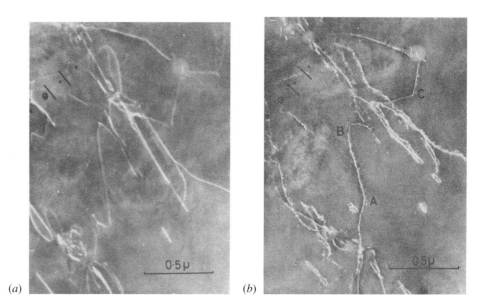

(a)   (b)

Fig. 15.3   (a) An HVEM dark-field image of dislocations prior to damage formation. (b) Clustering of damage around dislocation lines, following electron irradiation (Stobbs, 1973).

decreases so that, for a given resolution, the specimen thickness can be increased, roughly in proportion to the voltage.

The project was expensive and Cosslett became impatient with the delays in government funding. Instead, he applied for a grant from the Royal Society's Paul Instrument Fund to construct a 750 kV machine. Ken Smith joined the Electron Microscope Group from the Engineering Department and the design and construction were carried out in the old Maxwell wing of the Cavendish Laboratory with some assistance from the Engineering Department. The 750 kV generator (Box 15.10(a)) was purchased from Haefely in Switzerland. The HVEM, the first in Britain, was fully operational by 1966 and was far more flexible than any previously built. It could be operated at voltages from 75 to 750 kV. The design served as a prototype for the AEI EM7 million volt TEM, one of the most successful microscopes built by the company.

The main applications of the HVEM were in metallurgy. Inclusions in metals, the dynamics of moving dislocations, the domain structure of magnets and the nature of radiation damage were among the topics investigated (Figure 15.3). The microscope operated in Cambridge from 1966 to about 1990.

15.3 Crystallography

15.3.1 Molecular biology

After many years of effort, Kendrew and Perutz successfully applied the techniques of isomorphic substitution to determine the structures of the myoglobin and haemoglobin

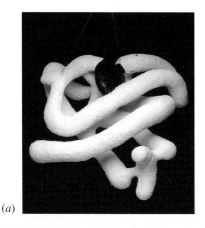

(a)   (b)

(a) Kendrew's 1958 model of the myoglobin molecule, with the haem group painted in red and the peptide backbone in white. (b) Max Perutz's balsa wood model of haemoglobin, built in September 1959. The segments represent peaks of high density in a three-dimensional map of the electron distribution obtained by X-ray crystallography. They outline the course of the four polypeptide chains, two shown in white and two in black, while the red discs represent the four haems. Myoglobin and haemoglobin were the first protein structures to be solved (courtesy of the MRC Laboratory of Molecular Biology).

Fig. 15.4

molecules (Section 12.6.1). In the case of myoglobin, the simpler molecule, Kendrew found four derivatives with heavy atoms attached at identifiable sites. The amount of computation involved to determine the electron density distribution scaled as the cube of the resolution, and the compromise was reached to aim for a resolution of 0.6 nm, at which it was expected that the structure of the protein would be revealed if, as was suspected, it was similar to an α-helix (see Section 12.6.2). Four hundred X-ray reflections were used and the necessary inversions were carried out in a reasonable time using the EDSAC computer.[6] The structure which emerged was a strange looped sausage structure (Figure 15.4(a)) which was interpreted as an α-helix winding round the density distribution of the haem group (Kendrew *et al.*, 1958). As they wrote in their paper,

> Perhaps the most remarkable features of the molecule are its complexity and lack of symmetry. The arrangement seems to be almost totally lacking in the kind of regularities which one instinctively anticipates, and it is more complicated than has been predicted by any theory of protein structure. (Kendrew *et al.*, 1958)

Then, in September 1959, Perutz used 40,000 measurements of haemoglobin crystals and six of its heavy-element derivatives to produce a three-dimensional structure of haemoglobin with 0.55 nm resolution (Perutz *et al.*, 1960). This resulted in his famous model for the electron distribution in the molecule, first created in balsa wood (Figure 15.4(b)) and then in thermoplastic. As Perutz remarked much later,

> [The structure] proved itself at first sight. The conformation of the two parts of chains in horse haemoglobin closely resembled that found by Kendrew. No conceivable combination of errors could have produced that striking similarity. (Perutz, 1997)

The success was achieved thanks to a continuing series of innovations in the techniques of X-ray crystallography. A far brighter fine-focus X-ray generator was developed by Tony

Broad, an automatic scanning densitometer designed by Peter Walker in Edinburgh was used to scan the spots on the X-ray plates, and new methods were developed by David Blow to determine the phase angles reliably. The continued development of the EDSAC computer by Wilkes and his colleagues was crucial for these studies. In addition to the two-dimensional Fourier transform programmes developed by Kendrew, the analysis involved a great deal of ingenuity to utilise to the full the tiny amount of memory available in EDSAC. In the following paper in *Nature*, Kendrew and his colleagues reported a preliminary structure of myoglobin with resolution 0.2 nm (Kendrew *et al.*, 1960). This analysis involved 200,000 measurements of the X-ray images and the maximum exploitation of the techniques developed within the MRC Unit. These achievements of Perutz and Kendrew were to be recognised by the joint award of the Nobel Prize in Chemistry in 1962.

As David Blow remarked, the MRC Unit had other major successes during the early Mott years (Blow, 2004), in addition to those described here and in Section 12.6. Hugh Huxley and Jean Hanson developed their sliding filament model for muscle contraction. Vernon Ingram, working with peptide fragments of human haemoglobin, determined the exact single amino acid substitution responsible for sickle-cell disease. Crick and Sydney Brenner demonstrated that DNA was a three-letter code, leading to Crick's *central dogma*, which resulted in rules for the flow of genetic information within a biological system at the molecular level.

The success of the MRC Unit and the dramatic expansion in both the scale and scope of its activity resulted in essentially insoluble accommodation problems at the New Museums site, as described in Section 13.2. Blow wrote:

> [The] Unit... took over a hut in the car park, erected for temporary wartime use (Figure 13.1). Happily, the X-ray generators remained in the basement of the Cavendish Laboratory's Austin Wing. In the following years, the group of scientists attracted to work with Perutz, Kendrew, Crick and Brenner became far too large to do experimental work in the hut. Every unused room in the vicinity was adopted for use. Kendrew built his model of myoglobin in the recently vacated cyclotron room at the Cavendish, while Brenner and his colleagues used Lord Rutherford's former stable for experiments in molecular genetics. (Blow, 2004)

The solution was the creation of the MRC Laboratory for Molecular Biology (LMB) on what became the new Addenbrooke's Hospital site, two miles south of the city centre. All the scientists associated with the MRC Unit transferred to the LMB and other distinguished molecular biologists soon joined them – Fred Sanger from Biochemistry and Aaron Klug from Birkbeck College, London. The LMB grew very rapidly and enjoyed extraordinary success. Since its foundation, nine Nobel Prizes, shared by 13 scientists, have been awarded for key discoveries made in Cambridge; these include, in addition to those mentioned earlier, Dorothy Hodgkin (Chemistry, 1964) and Aaron Klug (Chemistry, 1982).

15.3.2 Other crystallographic themes

Mott's almost biennial reports through the 1960s highlight a number of recurring themes which were pursued in the Crystallography Group under the leadership of Will Taylor. The major themes were the structure of complex alloys, the structure of feldspars, ferroelectric materials, lattice dynamics through neutron spectroscopy and metal physics. These investigations involved close collaborations with other groups in the Laboratory and also with other physical science departments, such as Metallurgy and Chemistry. The techniques of phase reconstruction pioneered in the studies of biological molecules were also applied to metals and alloys. In all these studies, the availability of the EDSAC and its successor TITAN computers was essential.

A number of highlights of the metal physics activities and the role of dislocations have been discussed in Section 15.2. The techniques of crystallography were now being used as a service to other groups in the Laboratory and elsewhere. Experimental neutron diffraction and spectroscopy were carried out in collaboration with the Harwell Laboratory and at the Chalk River Laboratory in Canada. Neutrons interact strongly with the nuclei of atoms in the crystal lattice and so provide complementary information about the distribution of nuclei in the unit cell. In addition, magnetic scattering experiments with neutrons carried out at Harwell enabled the spin density distribution in metals and ionic crystals to be determined.

Jane Brown, one of the leading crystallographers in the group, was appointed an Assistant Director of Research in 1965, supported by the Science Research Council and by Harwell. By 1970, she was directing the work of 12 researchers in the Crystallography Laboratory, including two postdoctoral workers. These activities involved studies of the electron and spin distribution in alloys. The plan was to obtain detailed information about the transition from non-integer magnetic moments observed in metals to the fixed integer moments associated with ionic materials by examining a range of magnetic materials from metals through semi-metals and semiconductors to insulators. These neutron physics studies required access to powerful neutron sources which could only be provided at national centres. When Brown moved in 1974 to the Institut Laue–Langevin at Grenoble, one of the major centres for neutron scattering and diffraction research, much of the neutron diffraction activity moved with her.

15.4 Physics and chemistry of solids

Mott was particularly proud of the addition of the Physics and Chemistry of Solids (PCS) Group to the programme of the Laboratory. Central to this development was Philip Bowden, an outstanding experimental physicist who was to bring new dimensions to the Laboratory's activities. In turn, these would lead to the spin-off of new research areas which would become major activities in their own right.

15.4.1 Bowden in Australia and Cambridge

Bowden was educated in Tasmania, where he carried out research in electrochemistry. In 1927 he was the recipient of an 1851 Fellowship which enabled him to come to

Cambridge to study with Eric Rideal, who was then a lecturer in the Physical Chemistry Department, located next door to the Cavendish Laboratory in Free School Lane. Bowden progressed rapidly up the academic ladder, becoming a demonstrator in 1930 and in 1937 the Humphrey Owen Jones Lecturer, a position he would hold until the end of the Second World War.

During the 1930s his interests evolved from electrochemistry to various aspects of friction. There followed a series of researches on many different aspects of friction and lubrication: kinetic friction, frictional 'hot-spots', frictional heating and surface melting, adhesion of clean surfaces and the measurement of the real, as opposed to the geometric, area of contact between stationary and sliding surfaces. The last topic was carried out in collaboration with David Tabor, who joined him as a research student in 1939. In a pioneering paper, Bowden and Tabor showed, by measurements of the electrical conductivity between crossed cylinders under different loads, that the area of actual contact between the cylinders was very small indeed compared with the geometric contact area and that the change of conductivity with increasing load showed that the surface asperities were subject to plastic rather than elastic distortion (Bowden and Tabor, 1939). As a result, when the load is applied, plastic deformation of the surface asperities continues until the increased surface of real contact can support the load (Box 15.11). It was also noted that, even at relatively small rates of sliding friction, very high-temperature flashes may occur, possibly resulting in local melting, although the bulk of the material remains cool. This was to inform his subsequent studies of the role of hot-spots in detonation. These researches accounted for the familiar established laws of friction, namely that the force of friction is (a) independent of the apparent area of contact and (b) proportional to the load, and were the inspiration for much of Bowden's subsequent work on friction and the physics of lubrication.

The work on friction and lubrication attracted the attention of the oil companies and other industrial organisations in the UK, the Netherlands and the USA and resulted in

Box 15.11   **Bowden and Tabor's theory of friction**

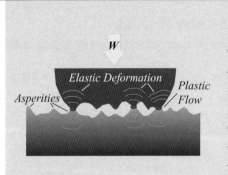

In Bowden and Tabor's theory of friction (Bowden and Tabor, 1939) all solid surfaces are rough compared with the length scale of molecular forces. Placing two surfaces in contact is like 'turning Switzerland upside down and placing it on Austria – the areas of intimate (real) contact are small compared with the apparent area of contact'. Further, the local pressure is high enough to deform the contact areas plastically, as shown in the figure, at the yield pressure of the material. Thus the real area of contact, at which friction is generated, is increased to support the compressive load and so is proportional to it.

the formation of a small research unit led by Bowden known as the Britannia Laboratory, dedicated to studies of wear and lubrication. With the approach of war, the Air Ministry, the Fuel Research Board and the War Office showed increasing interest in Bowden's activities; these would result in valuable contracts for the support of Bowden's research programme.

Bowden was in Australia when war broke out in 1939 and he decided to remain there, supporting Australian industry in its war efforts. In November 1939, the Council for Scientific and Industrial Research (CSIR), the forerunner of the Commonwealth Scientific and Industrial Research Organisation (CSIRO), set up a section in Melbourne entitled Lubricants and Bearings, with Bowden as its head.[7] Tabor joined the section from England in 1940. The subsequent researches were to foreshadow many of the topics which Bowden and Tabor would continue after the war on their return to Cambridge. These included special lubricants for machine tools and for aircraft, casting techniques for the production of aircraft bearings and flame-throwing fuels. New topics included the penetration of metal sheets by bullets, which led to the development of high-speed photography by the gifted Tasmanian engineer Jeofry Courtney-Pratt in order to measure the speeds of the bullets. Another new activity was the initiation of explosions and the detonation of explosive materials, in particular, the role of 'hot-spots'. Abe Yoffe took part in this activity when he joined the group in 1942.

After the war, Bowden resigned and returned to Cambridge. He had already written a prophetic memorandum entitled *A study of the physical and chemical phenomena associated with rubbing and with the impact of solids* in March 1944, in which he laid out his vision for his post-war activities in Cambridge. He proposed the continuation of the research carried out in Australia, but in addition added:

> War experience has shown clearly that, in pre-War days the academic and practical research were too widely separated and suggests that, in future, contact between the two should be maintained. The action of joint committees and frequent meetings and discussion can do something towards this. It is obvious that one method of making this contact more real is by the interchange of personnel, and it is suggested that in this case arrangements be made for certain members of the Service and Industry Research establishments to come to Cambridge and work for a time, normally for not less than a year, and conversely for members of the Cambridge laboratory to go into the Research establishments. (Tabor, 1969)

On his return to Cambridge in 1945, he set about implementing this programme from his base in the Physical Chemistry Department; he was promoted to a readership in 1946. He was joined by a number of his colleagues from the CSIR laboratory, including Courtney-Pratt, Yoffe and Tabor. Another of his CSIR colleagues from Melbourne, Robert Honeycombe, came to Cambridge on an ICI fellowship at the same time and joined the Metal Physics Group under Orowan. Honeycombe subsequently became Goldsmith's Professor of Metallurgy at Cambridge. In the aftermath of the war, there was little support for Bowden's programme from the Physical Chemistry Department and the University, but he obtained the resources to start his programme with funds from the Ministry of Supply and also the Department of Scientific and Industrial Research, where Appleton was Executive Secretary. The group was to rely upon industrial and defence funding, a pattern which was maintained throughout the numerous subsequent changes of name of the group. Through Bowden's many industrial and defence contacts, the scope of the programme expanded dramatically and Bowden took over whatever space was vacated on the New Museums site. He named his group the Physics and Chemistry of Rubbing Solids (PCRS) Group.

The turning point came in 1956 when the Chemistry Department began the move into its new buildings in Lensfield Road. There had already been strong collaborations between the group and members of the Cavendish Laboratory, for example, in the exploitation of electron microscopes to study dislocations and the properties of materials. In 1957 Bowden transferred his readership to the Department of Physics, and the group itself joined the Cavendish Laboratory as one of its sub-departments. The title of the group was changed to the Physics and Chemistry of Solids (PCS) Group and it brought to the Laboratory all the distinctive areas of research which had been fostered by Bowden and his colleagues. Bowden, Tabor and Yoffe became Reader, Lecturer and Assistant Director of Research respectively in the Physics and Chemistry of Solids Group of the Cavendish Laboratory. In 1966, just two years before his death, Bowden was promoted to an *ad hominem* Professorship in Surface Physics and the group changed its name to the Surface Physics Group. Following Bowden's death, Tabor became Head of the Physics and Chemistry of Solids/Surface Physics Group, and remained in that role until 1981 when he retired.

15.4.2 New scientific directions

The group continued its traditional interests in friction, lubrication, the strength of materials, explosive detonation and high speed-impacts. For the remainder of the Mott era, the research programme had four main themes.

Surface physics

The summation of Bowden and Tabor's pioneering research on friction and lubrication is contained in their classic texts *The Friction and Lubrication of Solids, Parts I and II* (Bowden and Tabor, 1950, 1964) and *Friction and Lubrication* (Bowden and Tabor, 1956). An example of the skill and ingenuity involved in experiments on surface physics was Tabor and Winterton's measurements of the strength and radial dependence of van der Waals forces between atomically flat surfaces (Tabor and Winterton, 1969). Generally, because of the roughness of the asperities on apparently flat surfaces, the previous experiments were limited to spatial separations greater than 100 nm. By using sheets of molecularly smooth mica surfaces, they were able to reach separations of only a few nanometers (Box 15.12).

The van der Waals forces between neutral molecules with no intrinsic dipole moment are attractive – the fluctuating dipole moment of one molecule induces a dipole in another nearby molecule, resulting in a force of attraction. At short distances, the interaction energy between the atoms or molecules is expected to depend upon separation as r^{-6}. At large

---

Tabor and Winterton's measurement of van der Waals forces between mica sheets   Box 15.12

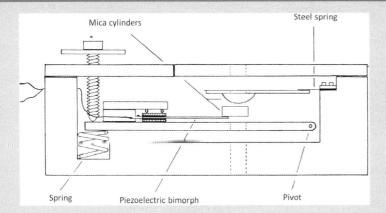

Van der Waals forces are attractive forces which act between all atoms including non-polar molecules. They are largely responsible for adhesion between non-metals. Tabor and Winterton carried out the first experimental investigation of van der Waals forces at short range (Tabor and Winterton, 1969). Theory predicted that the attraction between two crossed cylinders should be proportional to $1/x^2$ at short range (about 10 nm) and $1/x^3$ at long range (about 100 nm), where x is the separation between the cylinders. Previous workers had only been able to measure the long-range forces as they used surfaces which had asperities of about 100 nm amplitude. Experiments in the Laboratory had shown that mica can be cleaved to give surfaces that are molecularly smooth over relatively large areas.

With the apparatus shown in the diagram Tabor and Winterton measured the strength of the van der Waals forces over a distance range of 4.5 to 31 nm. Two sheets of mica are glued to crossed, cylindrical glass plates. The lower plate is attached to a moveable base and the upper to a cantilever spring. The lower plate is moved towards the upper one, and at some point the van der Waals attraction between the mica sheets overcomes the stiffness of the spring and the two plates jump together. A graph of 'jump distance' against 'spring stiffness' gives the van der Waals force law. The separation of the mica sheets is measured by multiple-beam interferometry. A new spring was used for each different 'spring stiffness'.

distances, however, the dipole moment of the first molecule changes before the second molecule feels the effect of the induced electric field, what is known as the *retarded van der Waals force* – the dependence of the potential energy changes to r^{-7} in this case. When the total attractive force between a sphere and a flat sheet is evaluated, the expected force laws become $1/x^2$ at short range and $1/x^3$ at long range, where x is the minimum separation between the sheet and the sphere.[8] In their experiments, outlined in Box 15.12, Tabor and Winterton found that the normal van der Waals force predominates for distances less than about 10 nm and the 'retarded' force for separations greater than 20 nm.

These experiments were refined by Jacob Israelachvili and Tabor, who confirmed these conclusions over the distance range 1.5 to 130 nm (Israelachvili and Tabor, 1972). Tabor's

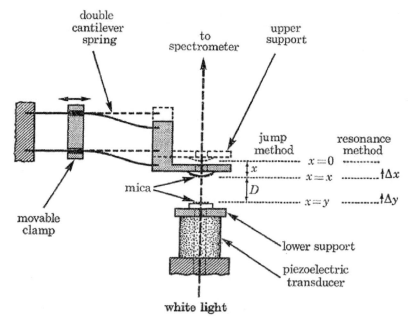

Fig. 15.5 A schematic drawing of the main parts of Israelachvili's apparatus (Israelachvili and Tabor, 1972). In the jump experiments, a double cantilever spring was used and the experiment carried out as described in Box 15.12. In resonance experiments a single cantilever 'bimorph' spring was used. The natural frequency f_∞ of the upper support depends on the spring stiffness and the inertia of the support. If the two surfaces approach each other, the increasing van der Waals attraction between them causes the natural frequency to decrease until at the jump separation it falls to zero. The principle of the resonant method is to measure the resonant frequency f_D of the upper support as a function of gap distance D and from this to deduce the law force in the range of distances measured.

interest in what he termed 'unconventional techniques of force measurement' is exemplified by the measurement of van der Waals forces at very short distances using his surface force apparatus, shown in Figure 15.5. In their experiment, the forces were measured between two curved mica surfaces by the jump method described in Box 15.12 (see Tabor, 1991) and by the resonance method described in Figure 15.5. In both approaches, they confirmed the earlier results that the direct force, proportional to r^{-6}, is dominant for separations r less than about 10 nm while the retarded force, proportional to r^{-7}, dominates at larger separations.

Another example of Tabor's interest in the practical application of investigations into friction is his work on hysteresis loss in rolling friction. This can be demonstrated by rolling two identical steel spheres down grooves in different types of rubber (Box 15.13). At one extreme, a high-modulus rubber, with a small hysteresis loss, produces little friction and this ball moves faster. In contrast, a low-modulus rubber with large hysteresis loss causes the sphere to move slowly (Eldredge and Tabor, 1955; Tabor, 1955; Greenwood *et al.*, 1960). Hysteresis loss can be measured by loading a rubber band with weights and then unloading it as illustrated in Box 15.13. These studies led to a patent for highly hysteretic rubber tyre treads to reduce skidding on wet roads.

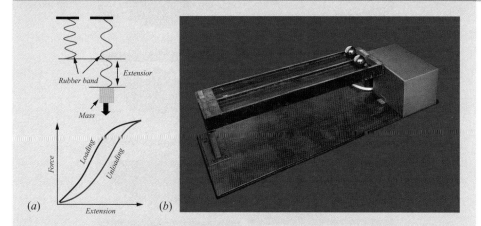

Box 15.13

Rubber band

Extensior

Mass

Force

Loading

Unloading

(*a*)

Extension

(*b*)

Measurement of hysteresis loss in rubber. (*a*) If the force versus extension follows the same curve for loading and unloading then the response is elastic and there is no hysteresis loss. If the loading or unloading curves are not identical, then the area between the curves is a measure of the hysteresis loss. A major application of these studies is in the choice of rubber for the manufacture car tyres. (*b*) A simple illustration of the differences in hysteresis loss for different types of rubber. The channels are lined with different rubbers, resulting in different speeds of rolling downhill when the tracks are tilted downwards.

Strength properties of materials

This heading includes a wide range of topics of importance for industry, including the properties of materials which melt at temperatures above 2000°C, rolling friction and elastic hysteresis and the use of high-speed cameras to study hyper-velocity impacts and deformation by liquid impact.

On his arrival in Cambridge, Courtney-Pratt continued his development of high-speed image converter cameras and completed his Cambridge dissertation on this topic in 1947. In this system, the images of events are focussed on the photocathode of an image tube, resulting in a beam of electrons reproducing this information. The electron beam can then be deflected electronically to different parts of the output screen by a shutter and shift plates. The result is a time-series of images on the output screen which can be photographed by a conventional camera. This scheme could operate at extremely high frame rates, up to 6×10^8 frames per second, and could typically record 6 to 20 images. These cameras were fully electronic and so the images could be enhanced by coupling to an image intensifier. Surveys of this and other techniques of high-speed photography were reviewed by Courtney-Pratt in 1957 and by John Field in 1983 (Courtney-Pratt, 1957; Field, 1983).

An alternative approach, which was fully exploited by the group, was to use a rotating mirror camera. In 1958, in recognition of the group's contributions to research, the American DuPont Company presented Bowden with a commercial version of one of these

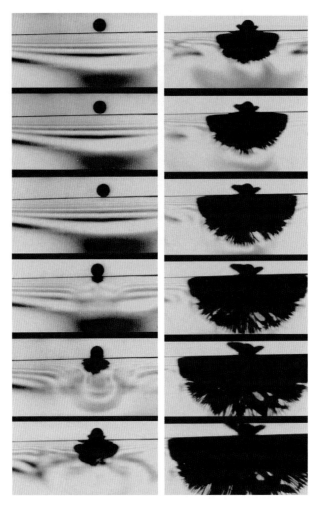

Fig. 15.6 The catastrophic failure of thermally toughened glass caused by small-particle impact. The images were taken with the Beckman and Whitely 189 rotating mirror high-speed camera. The projectile was a 2 mm diameter tungsten carbide ball. The inter-frame time was 1 µs and the impact velocity 150 m^{-1} (Chaudhri and Chen, 1986).

cameras, a Beckman and Whitely 189. This camera could image 25 frames at framing rates of up to about 5×10^6 frames per second with superior quality as compared to that of the image converter camera. The event was viewed through an objective lens which focussed the image onto a mirror driven by a turbine rotating at up to about 17,000 revolutions per second. After reflection from the mirror, the light passed through 25 separate sets of lenses and slits onto stationary photographic film. Pulses from the turbine were used to trigger the light source, which had to have a duration of less than one rotation time of the mirror. The camera was used by Field and Munawar Chaudhri from 1958 until 2008 to study fracture, impact, explosive ignition and propagation, laser damage, the visualisation of shock waves and more. These studies included the high-speed impact of liquid drops onto metals and other substances, which were important in understanding the effects of erosion. As an example of the capabilities of the technique, Figure 15.6 shows a sequence of photographic images of the impact of a 2 mm diameter tungsten carbide ball onto toughened glass at a speed of 150 m s^{-1}. The stress pattern was visualised using a photoelastic arrangement for observing isochromatics (Chaudhri and Chen, 1986).

Decomposition in the solid state and solid state physics

Bowden and Yoffe's pioneering research on explosives and detonation were summarised in their classic texts *Initiation and Growth of Explosions in Liquids and Solids* (1952) and *Fast Reactions in Solids* (1958). The scope of the activity expanded to include the oxidation of metals and the optical and electrical properties of insulators, semiconductors and ionic solids. Other new areas included the physical properties of solids under high pressure, electrical breakdown and radiation damage in solids. Gradually Yoffe changed the direction of his research to solid state physics and this became the source of many new areas of research, including the study of amorphous materials, an area very close to Mott's heart.

15.5 Solid state theory

By the time of Mott's arrival in Cambridge, experimental solid state physics at the Laboratory was a flourishing industry. The major developments initiated in low-temperature physics by Shoenberg and Pippard in mapping the Fermi surface of metals and Pippard's introduction of the coherence length in superconductivity, Bowden and Tabor's work on surface physics, the new opportunities opened up by Cosslett and Hirsch in electron microscopy and Crick and Watson's discoveries in molecular biology would benefit enormously from the support of a strong group in theoretical solid state physics. With the appointment of John Ziman in 1954, the Solid State Theory Group was set up. There was little opposition from those theorists who joined the Department of Applied Mathematics and Theoretical Physics when it was created in 1957. The division of interest occurred quite naturally, as has been splendidly described by Volker Heine in his essay, *TCM Group History*:[9]

> Physics (and Physics research) divides broadly into (*a*) the basic laws of nature and (*b*) the physical phenomena of nature. In so far as one tries to understand the phenomena in relation to the basic laws, it is not incorrect to describe (*b*) as application of the basic laws. However over the last 50 years, and one still meets it today, there have been overtones of status difference, with (*a*) being 'fundamental' and (*b*) 'applied', with all sorts of implications for job, teaching and resource allocation. In this view the BCS theory of superconductivity, the Mott transition between metallic and insulating behaviour, Anderson localisation due to disorder, super-exchange as the origin of antiferromagnetism in oxides, ... are merely applied quantum mechanics. And similar judgements apply to

understanding the existence of rubber or reptation in polymer flow [and so on] in relation to statistical mechanics.

Of course, it is nonsense. There is no way that armed with the Schrödinger equation anyone could have deduced superconductivity or the other things. It required just as much of an insight and a similar imagination to understand these phenomena as to develop the theory of quarks. The point has been well expressed by Phil Anderson . . . by saying 'many is different' (Heine, 2015).

A quite different mind-set and technical approach were needed to tackle what can be broadly classified as *many-body theory* and *strongly correlated electron physics*.

15.5.1 The development of the Solid State Theory Group

On his appointment as a University lecturer, Ziman set about creating a theoretical group which would cover much of the ground which now falls within the sphere of the theory of the solid state. His previous work in Bristol had largely been on magnetism, but now he started again from scratch. These studies resulted in his discovery of new research themes as he went along, among the more important of which were his studies of electron transport in disordered media. The result was the publication of his monumental work *Electrons and Phonons: The Theory of Transport in Solids* (Ziman, 1960).

With his prestige in condensed matter theory, Mott was successful in inviting a number of world-leading many-body theorists to come to Cambridge as visitors. The most important in the early days of the group were Hans Bethe from 1955 to 1956 and Philippe Nozières and Philip (Phil) Anderson for the academic year 1961–62. In addition, a number of outstanding graduate students, all from outside the UK, came to study with Mott and Ziman. Volker Heine arrived from New Zealand in 1954, becoming a research fellow in 1957 and a demonstrator in 1958. Lu Sham came from Hong Kong, Neil Ashcroft from New Zealand, Maurice Rice from the Irish Republic, Leo Falicov from Argentina and Federico Garcia-Molnar from Spain. They have all become distinguished physicists and hold highly respected positions in the academic community. As recounted earlier, space was a problem which was only relieved after 1959 with the construction of the additional floor on the Austin Wing, which was to house the theorists. Before that time, the theorists were largely located in the experimental groups.

In 1964 Ziman left to become Professor of Theoretical Physics at Bristol, where he wrote his books *Elements Of Advanced Quantum Theory* (1969) and *Principles of the Theory of Solids* (1972). These important books were based upon the graduate courses on solid state theory which he had introduced while he was Head of the Solid State Theory Group and which were the first systematic presentations of the discipline in the Cavendish Laboratory – they expounded the elements of quantum field theory as applied to condensed matter physics. Following Ziman's departure, one of the most important developments for theoretical solid state physics was the creation of a new professorship, formally entitled *Professorship of Physics 1966*, thanks to Mott's and Pippard's efforts. Anderson was appointed to the chair on a half-time basis, the other half of his time being spent at the Bell Laboratories, New Jersey; he also became Head of the Solid State

Theory Group. This was an inspired appointment since his achievements spanned huge ranges of solid state physics, including the theories of localisation, antiferromagnetism, symmetry breaking and superconductivity. Heine was by now a University lecturer and, as second-in-command to Anderson, ran the administrative side of the group's activities. With Anderson's arrival, the number of UK graduate students began to increase. In Anderson's words,

> From the mid-1960s we began to get really first-rate UK students including Richard Palmer, Alan Bishop, John Armytage, Mike Cross, Duncan Haldane, Roger Bowley and John Inkson on the many-body side, with Dave Bullett, John Inglesfield, John Pendry, Denis Weaire on 'one-electron' theory.

It was always the intention that the Solid State Theory Group should support the experimental work in the Laboratory. Thus, from the beginning the approach was very much to tackle problems of value in supporting the experimental activities. This is well illustrated by Heine's introduction to Mott's style of research when he came to the Laboratory as a research student:

> In fact, when Volker [Heine] arrived in 1954 to do his PhD with Mott, he did not realise that Nevill did not really like having research students, preferring to have his own ideas in conjunction with more senior experimentalists producing the latest results. Nevill tried to palm him off onto others around the Cavendish, but after thinking about it for a day Volker felt he had not come half way around the world to be put off as easily as that. He therefore promised to be no problem as long as he could have access when he needed it, and so Nevill gave him his project which was 'Why don't you go down to the Low Temperature Group and see if you can make yourself useful.' That was Nevill's concept of a theoretician, and it accorded with Volker's New Zealand culture.

Until the 1980s, the Solid State Theory Group was often closely involved with experiments done elsewhere, particularly in the USA. This is perhaps not so surprising because of the important influence of Anderson and his continuing association with the Bell Laboratories. In addition, in the early days there were no other strong condensed matter theory groups in the UK, and so Heine worked regularly with collaborators in the USA, very much to the benefit of the group. As part of these collaborations, a key element was the winning of travel grants to enable distinguished theorists from abroad to come to the Laboratory.

Another key feature of the work of the group from the very beginning was the application of state-of-the-art computing to the elucidation of theoretical problems in solid state physics. Heine had begun this trend as soon as he arrived in the Laboratory and was one of the pioneers in using the EDSAC computer in theoretical solid state physics. Computational condensed matter physics became one of the great strengths of the group, making pioneering contributions which have succeeded in making such studies an integral part of condensed matter physics research. As a result, the experimental groups benefitted greatly from the local expertise in condensed matter theory.

The impact of the theoretical activities of the Condensed Matter Theory Group on the work of the Laboratory as a whole are well expressed in a note to me by John Waldram:

[Many] theoretical themes . . . had very wide application for many groups, such as Fourier analysis and k-space; the development of band theory and Fermi surface ideas [following on from] Hartree's approach to atomic orbitals; lattice dynamics and phonons; and many-body quantum mechanics with its applications to particle theory, nuclear theory, atomic and metallic electrons and the idea of the quantum fluid. [There was a] general sense that theory was in some cases driving experiment rather than the other way around [as well as] the importance for the Cavendish of the general theory climate. There was a lot of important embedded theoretical work outside the formal theory groups, for instance in electron dynamics in solids, in crystallography, in lattice dynamics under Bill Cochran, and so on.

A good example of this is Howie's multi-pronged attack on computing the scattering of conduction electrons by stacking faults (Howie, 1960). In his analysis, the Fourier components of the pseudopotentials were derived from Shoenberg and Pippard's determination of the Fermi surface of copper and gold. These were then used to compute the scattering of conduction electrons by stacking faults, resulting in good agreement with the measured conductivities of these materials.

The biennial reports on *Current research in the Department of Physics* give a good impression of the range of topics studied and how they developed through the 1960s. Highlighted topics and the interactions with the experimental groups included:

- *The electric and magnetic properties of metals.* The objective was to work out theoretically the shape of the Fermi surface of metals, starting with the simpler cases of silver and gold and then proceeding to more complex metals (Pippard and Shoenberg).
- *The scattering of electrons by impurities* involved studies of wave propagation in anisotropic media, which was of wider application than just to electrons in metals.
- *Electron–phonon interactions.* These calculations led to corrections to the basic band structure and were significantly advanced by the introduction of the concept of the *pseudopotential*, which moves bodily with the ion in the lattice vibration. These calculations began with the EDSAC 2 computer and were then extended using its successor, TITAN. These calculations were also important in advancing band theory and supported the work of Pippard and Cochran in understanding transport properties.
- *First-principles calculations of the microscopic potential inside a metal.* This area was to develop into a major area of study and depended crucially upon the availability of high-speed computers.
- *Liquid metal calculations.* These studies necessarily led to the determination of the energy levels in disordered structures. In these cases, the individual electron eigenstates could not be studied in the usual way but rather statistical averages had to be determined for the liquid as a whole (Faber).
- *Electron energy levels at surfaces.* The calculations were of particular importance for the study of the work function of semiconductors (Pepper).

A notable feature of the Laboratory during the Mott era was that many members of the department who might be classed as experimentalists made important contributions to theory. The above example of Howie's analysis of the electrical conductivity of copper and gold were made whilst he was a member of the Crystallography Group and used the expertise of the Low Temperature and Solid State Physics groups. I recall being told by colleagues outside Cambridge when I was a demonstrator in the Laboratory that many of the staff regarded as experimentalists within the Laboratory were thought of as theorists by the rest of the UK.

15.5.2 Mott's contribution

Although Mott maintained his overarching interest in all aspects of solid state physics during his tenure of the Cavendish Chair, his original publications in the major journals principally concerned metal–non-metal transitions, particularly in amorphous solids and liquids, metal–ammonia solutions and supercritical mercury. These interests had their origin during the 1930s while he was at Bristol. In 1937, J.H. de Boer and E.J.W. Verwey reported that nickel oxide NiO, which according to the band theory of solids as understood at the time should have been metallic, was in fact an insulator (de Boer and Verwey, 1937). After the war, Mott returned to this topic and wrote his famous paper on the *Mott transition*, described in Box 15.14 (Mott, 1949). This was a somewhat controversial area of theory and it was made more complex by Anderson's equally important paper in which he demonstrated that large numbers of random scatterings, treated quantum mechanically, could also lead to localisation of the electron distribution, in what became known as *Anderson localisation* (Anderson, 1958) (Box 15.14). These papers proved to be a remarkable stimulus to research in solid state physics.

Between 1954 and 1971 the Mott transition was a recurrent theme in the numerous papers Mott published, while at the same time he dealt with his many University and national preoccupations. By the end of 1971, his series of papers entitled *Conduction in non-crystalline systems* had reached number VIII, often in collaboration with Edward (Ted) Davis, and the series was to continue after his retirement up to paper XI. In 1971, he published *Electronic Processes in Non-Crystalline Materials* with Davis (Mott and Davis, 1971). In addition, papers continued to appear in the major journals on a variety of topics including the strength of materials and the photographic process. But this was only a prelude to a golden summer after he retired. In 1977 he, Anderson and Anderson's supervisor John van Vleck were awarded the Nobel Prize in Physics, 'for their fundamental theoretical investigations of the electronic structure of magnetic and disordered systems'.

But he had also consolidated and expanded the scope of experimental and theoretical solid state physics in the Laboratory so that by 1971 the combined activities in these areas was the largest research field within the Laboratory.

15.6 The teaching of theoretical physics

While experimental physics remained the core of the Laboratory's activities, the dramatic increase in theoretical physics within the department and the support it provided to the experimental programme was not reflected in the undergraduate teaching programme. The

Box 15.14	Some terminology

The pseudopotential

A *pseudopotential* is an approximation for the potential experienced by a valence electron in the presence of the many 'core' electrons which surround the nucleus (see Heine 1970). This modified effective potential is included in Schrödinger's equation in place of a pure Coulomb potential and results in pseudo-wave functions which can be described by far fewer Fourier terms. Pseudopotentials were successfully applied to solid state physics by Walter Harrison to determine the Fermi surface of aluminium (Harrison, 1958) and by James Phillips to the covalent energy gaps of silicon and germanium (Phillips, 1958). Phillips and his colleagues extended this approach to many other semiconductors, in what they called *semi-empirical pseudopotentials*.

The Mott transition

A Mott transition is the transition of a material from an insulator to a conductor because of screening of the electrostatic potential of the electrons by other electrons in the material. As this pseudopotential becomes more localised, this can be thought of as a narrowing of the potential well, and there is a corresponding increase in the energy of the finite number of bound states. The highest-energy state moves out of the well until, as the density of electrons increases, there are no more bound states for the outer electrons above a certain electron density and the material becomes a metal. The critical electron density corresponds to a critical screening wave number k_s for which the last electron state is no longer bound: $k_s < 1.19/a_0$, where a_0 is the Bohr radius. The electron density may be changed by temperature, pressure, external fields or by the doping levels in semiconductors. Mott's theoretical researches opened up the field of amorphous materials, which were studied experimentally by John Adkins (Section 20.6.2).

Anderson localisation

As Philip Anderson wrote in his 1977 Nobel Prize lecture:

Very few believed [localisation] at the time, and even fewer saw its importance; among those who failed to fully understand it at first was certainly its author. It has yet to receive adequate mathematical treatment, and one has to resort to the indignity of numerical simulations to settle even the simplest questions about it. Only now, and through primarily Sir Nevill Mott's efforts, is it beginning to gain general acceptance. (Anderson, 1992)

According to the quantum mechanical picture of electron conduction, the wave properties of an electron in a crystal lattice cause it to be diffracted, the resistance of the material only appearing when electrons scatter from imperfections in the crystal. The more the impurities, the smaller the mean free path and the lower the conductivity. Anderson addressed the question of what happens when the number density of impurities becomes very large, and showed that, beyond a critical impurity scattering level, the diffusive motion of the electron is not just reduced but is reduced to zero. The electron becomes trapped and the conductivity vanishes. The effect is associated with the way in which the amplitudes of the wave functions between any two points in the disordered medium are summed and squared. Thus, above a certain critical impurity level, the material changes from a conductor to an insulator, a *metal–insulator transition*. These concepts now find application in a very wide range of disciplines, including classical waves, light, microwaves, acoustics and photonic band-gap materials (see Lagendijk *et al.*, 2009).

quality of the three-year physics course had been enhanced with the introduction of advanced physics, but there remained the problem that mathematics teaching was in the hands of the Department of Applied Mathematics and Theoretical Physics. As Heine has expressed it,

> All 'proper' theoretical physics teaching was organised by the Department of Applied Mathematics and Theoretical Physics in the Mathematical Tripos, but it wasn't really theoretical physics: it was the mathematics of physics. (Heine, 2015)

At stake was much more than just nomenclature. The theoretical physicists needed applicable mathematics to place the results of experiment in a more formal and widely applicable context. Those of us who have taught the mathematisation of physics know from experience that putting the behaviour of matter and radiation in the real world together with the necessary mathematical constructs is perhaps the most difficult part of the art and craft of theoretical physics.

To help remedy the situation, Heine introduced a voluntary set of examples classes in theoretical physics during the second year of the physics course, for those who wished to concentrate on more theoretical topics in the later parts of the course. The objective was to treat the topics the students were learning in a more theoretically satisfactory manner, using what they had been taught in the formal mathematics courses. The course became well established and popular with students of a more theoretical disposition and helped them greatly with their physics education in general.[10]

As recounted in Section 12.10.4, Hartree had introduced a course in theoretical physics as part of Bragg's reforms, and after he died in 1958 the course was taken over by Richard Eden. He and Heine realised that, with only a few modest changes, they could devise a much more satisfactory course in theoretical physics. In place of experimental work, the students could take a number of theoretical physics courses in the third and final year and obtain an honours degree in Physics and Theoretical Physics. With the increased effort in solid state theory, there was no lack of teaching effort to support these courses. This option proved to be popular with the students, particularly for those of a more mathematical bent.

15.7 Mott's legacy

If Bragg had begun major changes in the direction of the Laboratory after the war, from 1954 onwards Mott consolidated these changes and strengthened the role of the Laboratory in solid state physics. The Laboratory doubled in size and, with Mott's skill in operating at a high level within the University, was very effective in improving the Laboratory's infrastructure and buildings, culminating in the successful bid to move the activities to West Cambridge. The pre-war dominance of nuclear physics was a thing of the past, and was replaced by spectacular advances in radio astronomy and molecular biology and the dramatic strengthening of experimental and theoretical solid state physics. At the same time, the needs of industry were strengthened, particularly with the incorporation of the

Physics and Chemistry of Solids Group into the Laboratory and the advances in electron microscopy.

In 1971 Mott bequeathed to his successor Brian Pippard a large, fully functioning operation with a brand-new Laboratory which Pippard had largely designed. But times were to get tougher. As has been emphasised, Mott had a good deal of luck in being Head of the Laboratory at a time when the resources allocated to scientific research were increasing at about 5% in real terms per year so that it was not difficult to implement his vision of the necessary changes. These heady days were not to return.

PART VIII

1971 to 1982

Alfred Brian Pippard (1920–2008)
Oil painting by Paul Gopal-Chowdhury

The Pippard era: a new Laboratory and a new vision

16.1 Pippard as Cavendish Professor

Unlike Bragg, Rutherford and Mott, Brain Pippard was an 'internal' candidate for the Cavendish Chair. He had spent most of his academic career in the Cavendish Laboratory where he had carried out outstanding research in low-temperature physics (Sections 12.9.2 and 15.1.1). As recounted in Section 13.5, he had been deeply involved in planning the move of the department to West Cambridge and in the detailed design of the new buildings. In addition, in 1965 he became the first president of the new graduate college, Clare Hall, where again he was fully involved in the planning and construction phases of the splendid new buildings designed by Ralph Erskine and formally opened in 1969. Mott regarded Pippard as his natural successor as Cavendish Professor and this duly came about in 1971 on Mott's retirement.

During the 11 years of Pippard's tenure of the Cavendish Chair, there were numerous changes in the way the Laboratory was managed in the light of changing circumstances. With the doubling in size of the research and teaching activity under Mott, the Head of Department could no longer assume the commanding role of a Rutherford, but rather acted as the chairman of a 'company', the main role being to keep the whole operation on the road. In addition, Pippard was temperamentally quite different from his predecessors. He was very much a hands-on experimental physicist who prided himself on not having held a Research Council grant and asserted that he had never done any experiment which could not be built in the Laboratory workshops. These attitudes were very much in the experimental physics tradition established by J.J. Thomson and Rutherford. Pippard often expressed provocative views, as is illustrated by his banquet speech at an IBM international conference on superconductivity in 1961. John Waldram and I wrote in his biographical memoir for the Royal Society:

> He took as his title 'The cat and the cream', and startled his audience by asserting that physics, apart from fundamental particle physics, was largely worked out. Perhaps it was tongue in cheek, or perhaps he felt it so in a personal sense, for his extraordinary research productivity did decline somewhat thereafter. But this must not be exaggerated. He was always interested, his views were eagerly sought, and he continued to contribute important ideas. (Longair and Waldram, 2009)

Whatever his intention, the timing was unfortunate, as new techniques and advances were being made in condensed matter physics and radio astronomy. Indeed, during his occupancy of the chair, condensed matter physics became the largest subject area in the Laboratory.

It was characteristic of Pippard that he gave his inaugural lecture, *Reconciling physics with reality*, to the undergraduate Cambridge Physics Society on 7 October 1971, almost exactly a century after Maxwell's inaugural address (Pippard, 1971). It included a set of physics demonstrations, the outcomes of which were to be predicted by a show of hands. As he intended, most of the audience got the answers wrong. The experiments were designed to illustrate how confused even a professionally educated physics audience could be about the physics questions he posed. Several of these experiments foreshadowed his later interest in the physics of non-linear systems and chaos. These fruitful areas of research are also adumbrated in his books on *The Physics of Vibration* (Pippard, 1978a,b) and *Response and Stability* (Pippard, 1985), although he did not make the leap to the universality of chaotic phenomena.

His serious message was the need to maintain a balance between fundamental physics and the aspects of physics which are of benefit to society. The lecture was delivered at a time when the public had lost faith in the sciences, and in his view it was largely the physicists' fault for concentrating too much on physics for its own sake – the claims of 'fundamental physics' were a particular bugbear. Some flavour of his message can be inferred from the following remarks towards the close of his lecture:

> At this stage, you may begin to think that I am proposing to dismiss fundamental physics as not worthy of serious attention, but this is not so. I am, rather, concerned to examine the reasons why it is worthy of serious attention, which are not the reasons given by fundamental physicists when they are seeking government approval for spending another hundred million pounds. They argue most eloquently, and they are eloquent folk, that fundamental physics is important in a special way; it is, no less, seeking out the Basic Truth of the Universe. If you have followed my argument to this point, I think that you may find so bold a claim slightly pretentious. There is indeed a certain importance to any branch of study which throws light on, or encourages critical appreciation of, the general condition in which we find ourselves; thus, in so far as the fundamental study of physics has led to the construction of a cosmology which is demonstrably better than the legendary cosmologies of primitive religions, to that extent fundamental physics has played an important role in the life of man. But in so far as it has failed to show how to proceed from fundamental laws to the behaviour of other systems, to that extent it has no other importance to mankind in general.

John Waldram and I offered further insight into his individual approach to physics and theoretical physics when we wrote:

> In trying to assess Brian as a thinker, it must be said that he certainly disliked mathematical formalism, feeling that it often obscured the essence of things. Yet despite this professed antipathy, he was expert at using relatively elementary mathematics to solve problems that he chose to study. He loved unexpected solutions, such as phase-locking in nonlinear systems – not least in Josephson junctions. As a teacher he had irrational horrors of the grand canonical ensemble and four-vectors in special relativity. He always preferred to think of quantum problems in terms of semi-classical wave packets, never eigenfunctions and eigenvalues; and he refused to learn second quantization, which is necessary for a full appreciation of BCS and Josephson theory, just as he refused to learn to drive. His strength was his unerring instinct for the right, often subtle, illuminating

idea. His concepts of ineffectiveness and coherence length, for instance, although qualitative and heuristic, were essentially correct, and were immediately grasped by a whole generation of physicists and applied to a wide range of new discoveries. (Longair and Waldram, 2009)

As can be imagined, Pippard's views were not necessarily welcomed by many members of the department. The leaders of the research groups went ahead with considerable success in implementing their own visions of the most productive areas for future research in physics. Those of us involved in the Cavendish Teaching Committee (Section 16.3) recall with relish the splendid confrontations between Pippard and most of the other members of the committee regarding the teaching of physics. In the end, we had little difficulty in maintaining within the core syllabus four-vectors, quantum operators, grand canonical ensembles and so on, all topics which the students needed, and wanted, to appreciate in order to understand the contemporary literature. But he did succeed in introducing a 'physical insight' paper into the final year of the physics examination, which terrified the staff as much as the examination candidates.

As Cavendish Professor, Pippard automatically assumed the role of Head of Department, but he changed that rule in the late 1970s when he decided to split the posts. In 1978 he stood down as Head of Department and Alan Cook, appointed Jacksonian Professor of Natural Philosophy in succession to Frisch in 1972, took over as Head of Department from 1979 to 1984.

16.2 The new Cavendish Laboratory

The development of concepts for the new Cavendish Laboratory was described in Pippard's words in Section 13.5. As he made clear, towards the end of the 1960s the money for capital investment in buildings was drying up, and economies had to be made to provide the desired space for the Laboratory within a single funding quinquennium by the University Grants Committee.[1] There can be no doubt that Pippard was successful in ensuring that the building was as close as was feasible to his intentions and fulfilled his scientific and academic goals (Figure 13.3). Figures 16.1, 16.2 and 16.3 show the overall layout of the new Laboratory, the ground floor of the Mott Building and its second and third floors, respectively (Crowther, 1974).

The overall useable floor space amounted to 16,000 m^2 and was distributed among the three major buildings as follows:

- The *Mott Building*, labelled 'Phase I' in Figure 16.1, had a useable floor area of 5,800 m^2 on three storeys and contained research laboratories, workshops and offices, third-year teaching laboratories and seminar rooms. The activities in the Mott Building were largely concerned with experimental and theoretical condensed matter physics. The ground floor accounted for half the total floor area in the building, and the upper floors had progressively smaller areas, what has been referred to as a 'ziggurat' design. The space on the

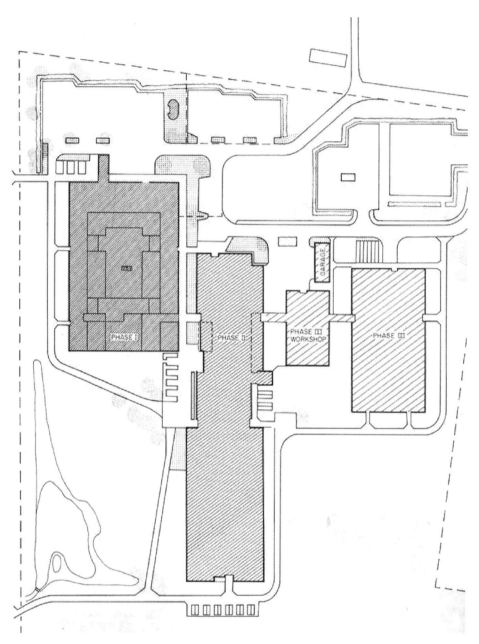

Fig. 16.1   The plan of the new Cavendish Laboratory on the West Cambridge site in 1974 (Crowther, 1974).

ground floor slab gave maximum flexibility for experimental work and allowed the coexistence of vibration-producing machines and vibration-sensitive equipment. The whole building was pierced by four vertical halls, labelled 'voids' in Figure 16.2, which acted as focal points for researchers to meet and also provided the main vertical circulation routes. The condensed matter theorists were allocated offices on the top floor, which contained

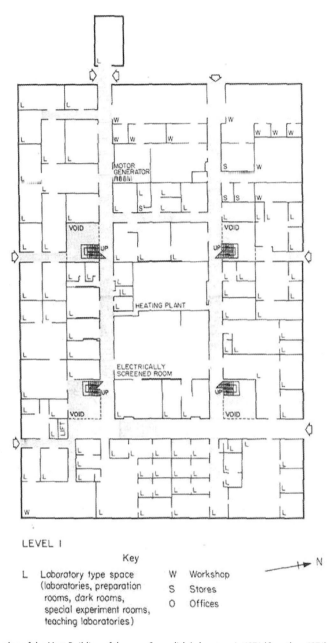

LEVEL I

Key

L   Laboratory type space   W   Workshop
    (laboratories, preparation   S   Stores
    rooms, dark rooms,   O   Offices
    special experiment rooms,
    teaching laboratories)

→ N

The ground floor plan of the Mott Building of the new Cavendish Laboratory in 1974 (Crowther, 1974).

Fig. 16.2

two seminar rooms (Figure 16.3). Some third-year undergraduate teaching laboratories were located on the second floor. All the laboratory spaces were serviced with circulatory cooling water, and a helium recovery system was installed in those laboratories involved in low-temperature physics experiments.

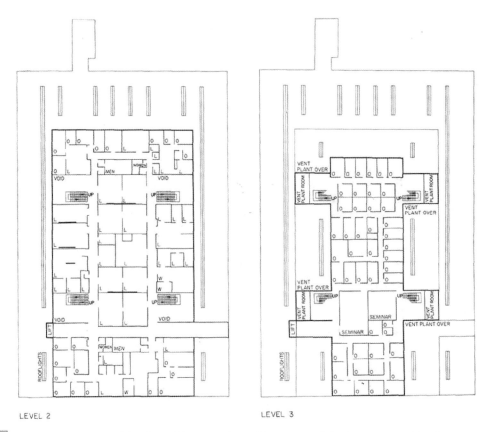

LEVEL 2   LEVEL 3

Fig. 16.3   The plans of the second and third floors of the Mott Building of the new Cavendish Laboratory in 1974 (Crowther, 1974).

- The *Bragg Building*, labelled 'Phase II' in Figure 16.1, had a useable floor area of 5,700 m^2 on two storeys and housed the central library and stores, common rooms, lecture theatres, first- and second-year undergraduate teaching laboratories and the administrative hub of the Laboratory. The west end of the building was on two storeys and housed administration and central services, including the Rayleigh Library and the main stores. The east block included two lecture theatres, now called the Pippard and Small lecture theatres, with capacities of 450 and 200 people, respectively. Further east is the long single-storey block for first- and second-year undergraduate teaching in practical physics.
- The *Rutherford Building* as well as the *Main Workshop* and *Computer Centre*, labelled on Figure 16.1 as 'Phase III', had a useable floor area of 4,600 m^2. The Rutherford Building contained research laboratories, workshops and offices, third-year teaching laboratories and seminar rooms, primarily for the Radio Astronomy and High Energy Physics groups. As compared with the Mott Building, it was a much simpler rectangular two-storey building and consisted mainly of research laboratories and offices. The main workshop was located on the ground floor of what was called the Link Building, between the Bragg and Rutherford buildings. The upper floor of the Link Building contained a remote job entry

A view of the Mott Building across Payne's Pond, so named by Pippard to acknowledge the major contributions of John Payne in effecting the successful move of the Laboratory from the New Museums site to West Cambridge. The overall layout of the site as it was in 1974 is shown in Figure 13.3.

station to access the Computer Laboratory's mainframe IBM 370/165 computer, which was located on the New Museums site and which was heavily used, particularly by the Radio Astronomy, High Energy Physics and Theory of Condensed Matter groups.

The appearance of the buildings was utilitarian and somewhat industrial in style, not out of keeping with other buildings of that era (Figure 16.4). Whatever reservations there might have been on aesthetic grounds were more than made up for by the much-needed space for teaching, research and offices for all classes of staff; teaching officers now had their own offices. But, most importantly, the fragmentation of the research and teaching areas at the New Museums site, illustrated in Figure 5.3, was at last resolved. Fears that the Laboratory would be isolated from the rest of the University were largely allayed by the scheduling of first- and second-year morning lectures in the centre of Cambridge, enabling the students to make a quick transit to other lecture locations in the vicinity. The third-year lectures were all delivered at West Cambridge. The complex matter of moving the Laboratory and all its equipment from central to west Cambridge was managed successfully by John Payne. Only the electron microscopes constructed by Cosslett and his colleagues remained at the New Museums site.

Pippard was well pleased with the way in which the new Laboratory encouraged collaboration between research groups, which had previously been housed in separate buildings. Although he was not an unbiased observer, no one would disagree with his assessment of how careful architectural design can enhance interdisciplinary research programmes:

The work on disordered materials in the Mott Building exemplifies this point. Although the research groups that had developed separate, even competitive, identities in the old buildings still hold to their primal loyalties, with their own assistants, secretaries, workshops and darkrooms, there is much more exchange of ideas than hitherto between those in the Physics and Chemistry of Solids group whose primary interest is in the electrical behaviour of amorphous semiconductors, and those in (then) Metal Physics and Low Temperature Physics who are more concerned with the mechanical and thermal properties of glasses, both dielectric and metallic. One unfortunate consequence of shortage of funds, that the theoreticians who require less heavily serviced accommodation have to occupy the top floor instead of being mixed up among the experimenters, has not prevented valuable collaborations developing. The move away from well ordered crystalline solids towards the more difficult problems of disorder finds its parallels upstairs in the development of analytical and computational techniques for handling atomic groupings without the simplifications introduced by periodic structure. And catalysing the whole process, from a vantage point deeply embedded in the experimental areas, is the father-figure of disorder, as of so many other major advances in solid-state physics, Professor Mott himself. (Pippard, 1974)

It is fair to say that, although many of the developments in physics in the Laboratory were not necessarily to his taste, Pippard's worst fears about the end of physics and its role in society proved to be unfounded. The physicists found many new and important challenges which spanned the range from the 'fundamental' to the 'applied' ends of the spectrum. In particular, many of the problems of complexity became more tractable as computing power increased exponentially. New fields of research developed in semiconductor and low-temperature physics, the programme expanded under his successors as Head of Department into the physics of materials, biology and medicine. Indeed, Pippard himself advocated what many of us regard as the motto of the Laboratory, 'Physics is what Physicists do'.

New activities developed remarkably quickly. With the appointment of Alan Cook as Jacksonian Professor of Natural Philosophy, initiatives in laboratory astrophysics were undertaken, reflecting his interest in precise measurement. Within the High Energy Physics Group, Eden developed his interest in energy issues and set up the Energy Research Group. Within Radio Astronomy, Ryle combined his concerns about the impending energy crisis with his interest in sailing to create a new subgroup of the Radio Astronomy Group dedicated to wind energy. In the area of condensed matter physics, every effort was made to encourage cooperation between groups; that story is taken up in the next chapter.

16.3 Teaching

Pippard was deeply involved in the development of the teaching of physics within the department. In 1968, Mott had set up the Cavendish Teaching Committee and Pippard became its first chairman. The secretaries of the committee had a key role in implementing as far as was feasible its recommendations for the teaching of physics, the operation and content of the practical classes and the fair distribution of teaching and managerial responsibilities among the staff. Tom Faber was the first secretary of the committee, and was succeeded by John Waldram. In practice, the secretaries ran the teaching programme and many innovations were made. These included student questionnaires about the quality and delivery of the lecture courses and practical classes, feedback to the lecturers and the introduction of a Staff–Student Consultative Committee, of which I was chair for a number of years.

The challenges continued to increase with the large numbers of students to be taught and the changing preparation of students for the physics course. The first-year course was delivered in parallel by two lecturers, one for those students who had strong preparation for the course, including a good grasp of the necessary mathematics, the second for those with weaker preparation and who might well not continue with physics in the later years of the Natural Sciences Tripos.

Another important issue was the increasing scope of physics and the desire of both the staff and students that provision be made for exposure to a wide range of these new topics, and those in cognate disciplines such as medical physics. The organisation of the course became an increasing challenge, and this was to continue to be an issue until the introduction of the three/four-year course structure in the mid 1990s.

An important development during the 1970s was the introduction of project work for final-year students. Interestingly, the initiative began with the students themselves – the Staff–Student Consultative Committee requested that the final-year students be allowed to replace part of the final examination requirements by independent project work, supervised by a member of staff. At the time, this was quite a challenging request because of the rigour of the examination requirements. I remember vividly, however, that the prospective final-year students desperately wanted to do projects and, however complex and difficult the requirements, they were determined to do them. A solution was found and the students did extremely well, setting a precedent for what was to become the norm within a few years.

Pippard and sympathetic colleagues made a valiant attempt to enhance the teaching of practical physics by devising various original experiments, designed to improve the students' perception of what was involved in carrying out any experiment in physics. In many ways, his objectives were exactly those advocated by Maxwell in his inaugural lecture concerning the development of students' powers of observation and the dexterity of their handling of equipment (Section 3.4). These experiments were perhaps of limited success: the better students may have appreciated what they were expected to learn, but the others could not understand why they were being exposed to somewhat primitive experiments with equipment inferior to what they had been used to at school.

There were also significant enhancements in the teaching of theoretical physics. The system was made somewhat more flexible so that students could combine experimental and theoretical options before they entered the final year of the course. We reviewed the essential elements of the teaching of theoretical physics for the Long Vacation and second-year examples classes. For me personally, one of the most significant innovations was a non-examinable course which I gave during the four-week Long Vacation course for those students entering their final year. It was an attempt to bring together the various strands of theoretical physics to provide a more coherent picture of the subject. To my delight, although given at 9.00 am on Mondays, Wednesdays and Fridays during the most glorious summer days in Cambridge, I had as large a turn-out of both theorists and experimentalists as for the examinable courses – a real tribute to the enthusiasm and dedication of the physics undergraduates. This course evolved into my book *Theoretical Concepts in Physics* (Longair, 2003).

A significant aspect of the teaching programme was the incentive it offered for attracting the best students into the research areas of the groups. Those groups with a prominent profile in the undergraduate programme, particularly in the final year of the course, had a significantly better harvest of the best undergraduates than the other groups. High Energy Physics, Theory of Condensed Matter, Low Temperature Physics and Radio Astronomy were particularly successful in this regard. This eased the problem of the Teaching Committee in encouraging academic staff members to play a full role in undergraduate teaching, very much to the benefit of both the teaching and the research programme.

16.4 Implementing Pippard's vision for condensed matter physics

Pippard held to his vision that the Mott Building offered an ideal opportunity for encouraging closer collaboration between the many different strands of condensed matter physics, which had developed in separate locations at the New Museums site. Throughout his tenure of the Cavendish Chair, the annual booklets entitled *Current Research in the Cavendish Laboratory*[2] repeat the same vision:

> Almost all the work in the Mott Building is concerned with experimental and theoretical studies of solids (and liquids). There are a number of research groups which have developed independently over the years and still retain a large measure of individuality. In recent years, however, problems have come to the fore which are of interest to members of more than one group and, moreover, are of such complexity as to demand the pooling of all available expertise. Since the groups have evolved through the study of different types of problem, and are still recognizably different in their attitudes and skills, the combined attack can prove very powerful.

Identical words were used in the 1975 to 1983 editions of the booklet, aside from the dropping of the words 'and liquids' in later editions. There then followed brief summaries of the areas of research in condensed matter physics. The continuously evolving programme of research can be appreciated from the summaries of activities for the years 1975 and 1983, shown in Table 16.1. The group names, acronyms and academic staff members are shown in Table 16.2. There were significant changes of emphasis during this period as well as the seeds of the next generation of developments in Condensed Matter Physics. Table 16.1 reflects a number of features of the research in each of the groups:

- The scope of activities in the Physics and Chemistry of Solids Group continued to expand to encompass new activities which would lead to the creation of new groups in optoelectronics, semiconducting polymers and soft condensed matter physics in the 1980s. Note also the increasing use of national facilities such as the Daresbury Synchrotron Radiation Source and the Spallation Neutron Source at the Rutherford–Appleton Laboratory.

Table 16.1 Research group activities in condensed matter physics in the Cavendish Laboratory in 1975 and 1983

	Activities 1975	Activities 1983
PCS	Unconventional techniques of force measurement, high-speed photography, electron diffraction and microscopy, Auger spectroscopy, study of band structures by electron and optical spectroscopy. New work involves laser Raman spectroscopy, the study of one-dimensional systems, synchrotron radiation spectroscopy, electron energy loss, diffusion in metals and in polymers.	Unconventional techniques of force measurement, high-speed photography, electron diffraction and microscopy, electron spectroscopy, surface studies, study of band structure by optical spectroscopy and high-pressure experiments, study of lattice dynamics and phase transitions by laser Raman spectroscopy, X-ray diffraction of solids under high pressure and at low temperatures, mass spectroscopy, TG and DSC studies of decomposition kinetics. New work involves the study of low-dimensional systems, synchrotron radiation and photoelectron spectroscopy, EXAFs, electron energy loss, surface vibrational spectroscopy, surface plasmon–polariton spectroscopy; diffusion in polymers; preparation of amorphous thin films of metals and semiconductors and glasses and their study by electrical and optical techniques
MP	Deformation studies with particular emphasis on electron microscopy for the examination of fine details of dislocations and other defects. This includes measurements of the energy loss of fast electrons and stereo-imaging of structures.	Development and application of high-resolution scanning transmission electron microscopy and microanalysis. Study of local structure, composition, electronic, mechanical and optical properties of a wide range of materials including metals, semiconductors, glasses and catalyst structures. Lattice defects and fatigue deformation.
LTP	Mainly electrical and magnetic measurements at temperatures below the boiling point of helium (4.2 K).	Measurements at temperatures down to 50 mK and in magnetic fields up to 10 T in metals, crystalline and amorphous semiconductors, thin films and glasses. The measurements explore a wide range of properties – electrical, magnetic, thermal, acoustic and superconducting, and the dependence of some of these on the carrier concentration when it can be varied by applied electric fields. A recent development is the use of point contacts and tunnelling spectroscopy to probe phonons and magnons in glasses and crystals.
CSN	X-ray and neutron diffraction, electrical and optical measurements associated with structural phase transitions. Inelastic neutron scattering in magnetic studies and lattice dynamics (neutron measurements are carried out at AERE Harwell and at the Institut Laue–Langevin, Grenoble).	
TNP		Scattering of thermal neutrons to study the structure and conformation of non-linear polymers. Other interests are spectrometers for use on the Spallation Neutron Source at the Rutherford–Appleton Laboratory, and precision measurements of nuclear parameters for reactors.
TCM	Applications of quantum mechanics to single-particle and many-body problems in solids, liquids, neutron stars etc.	Applications of quantum mechanics to single-particle and many-body problems in solids, liquids, glasses, rubbers etc.
LSP	Classical experimental techniques applied to liquid crystals.	Experimental and theoretical work on the properties of nematic liquid crystals.

PCS = Physics and Chemistry of Solids; MP = Metal/Microstructural Physics; LTP = Low Temperature Physics; CSN = Crystallography and Slow Neutron Physics; TNP = Thermal Neutron Physics; TCM = Theory of Condensed Matter; LSP = Liquid State Physics.

Table 16.2 Research groups and academic staff in condensed matter physics in 1975 and 1983 (new staff are indicated in red font)

Research group		Staff members 1975[a]	Staff members 1983[a]
Physics and Chemistry of Solids	(PCS)	Tabor (P), Mott (PE), Yoffe (R), Field (L), Davis (ADR),[b] Liang (D), Briscoe (SAIR), Chaudhri (SAIR), Lee (SAIR)	Tabor (PE), Mott (PE), Yoffe (R), Field (L), Hughes (L), Liang (L), Willis (L), Allison (D), Friend (D), Klein (D), Donald (RSF), Chaudhri (SAIR)
Metal Physics (Microstructural Physics)	(MP)	Brown, L.M. (L), Howie (L), Metherell (L)	Brown, L.M. (R), Howie (R), Winter (L)[c]
Low Temperature Physics	(LTP)	Pippard (P), Shoenberg (P), Waldram (L). Adkins (L), Phillips (D)	Pippard (PE), Shoenberg (PE), Adkins (L), Lonzarich (L), Phillips (L), Waldram (L), Pepper (RSF)[d]
Crystallography and Slow Neutron Physics	(CSN)[e]	Squires (L), Glazer (SAIR)	Squires (L)
Theory of Condensed Matter	(TCM)	Anderson (P), Edwards (P), Josephson (P), Heine (R), Pendry (SAIR)	Edwards (P), Heine (P), Josephson (P), Inkson (L), Ball (D), Needs (D)[f]
Liquid State Physics	(LSP)	Faber (L)	Faber (L)
Electron Microscopy	(EM)[g]	Cosslett (R)	Cosslett (RE)

[a] Notation: P = Professor, PE = Professor Emeritus, R = Reader, RE = Reader Emeritus, L = Lecturer, ADR = Assistant Director of Research, D = Demonstrator (Assistant Lecturer), RSF = Royal Society Fellow, SAIR = Senior Assistant in Research.
[b] Promoted to Lecturer in 1976.
[c] Dennis McMullan was an SAIR in the Metal/Microstructural Physics Group from 1979 to 1982.
[d] Pepper also had a close association with Mott who chose to take up residence in the Physics and Chemistry of Solids Group on his retirement.
[e] Became Thermal Neutron Physics (TNP) in 1976.
[f] From 1977 to 1982, Roger Haydock was a University Demonstrator in the Theory of Condensed Matter Group.
[g] Remained at the New Museums site.

- The Metal Physics Group continued the development of innovative techniques for the understanding of the microstructure of many different types of material, through the use of the electron microscopes constructed by the Electron Microscopy Group and the development of commercial instruments.
- The Low Temperature Physics Group was about to open up new ranges of studies in strongly correlated electron physics by extending measurements to extremely low temperatures, high pressures and very strong magnetic fields.
- The crystallography and neutron diffraction activities gradually declined with the retirement of Will Taylor in 1971, the formation of the Cambridge Crystallographic Data Centre in the early 1970s under Olga Kennard and the move of Jane Brown to Grenoble. Gordon Squires remained the only staff member whose primary interest was in neutron physics, an area to which he had made important original contributions earlier in his career.[3]
- The Theory of Condensed Matter Group continued its leading roles in this discipline, an important new research dimension being added by the appointment of Sam Edwards as John Humphrey Plummer Professor of Theoretical Physics. His ground-breaking contributions in theoretical soft condensed matter would lead to the establishment of research groups in biological and soft condensed matter physics in the 1980s.
- The liquid crystal activity was led by Tom Faber, who had begun research in low-temperature physics, then changed to the study of liquid metals, research which was completed in 1972 with the publication of his major book *Introduction to the Physics of Liquid Metals* (Faber, 1972).[4]
- Ellis Cosslett and the Electron Microscopy Group remained at the New Museums site, where space was found for the construction and operation of the new 600 keV high resolution electron microscope (HREM). At the same time, a new VG HB5 scanning transmission electron microscope (STEM) was installed in the Mott Building at the West Cambridge site.

The list of staff members associated with each research activity shows the natural evolution of the composition of the staff with retirements and the move of staff to other universities. It is noteworthy that the post of demonstrator remained a non-tenured position and not all those who held these posts were promoted to lectureships after their fixed five-year terms as demonstrators.

Following the dramatic expansion of the scope of the Laboratory's programme during the Mott era, the Pippard era was one of consolidation. This is illustrated by the data in Figure 16.5(*a*) showing the total numbers of members of the Laboratory for the period 1975 to 2001. The staff numbers remained more or less flat through the 1970s and early 1980s, reflecting the levelling off of support for the sciences by government. These numbers are also reflected in the academic staff employed in the area of solid state physics over the same period. Table 16.2 shows that the non-emeritus staff were 25 in 1975 and 27 in 1983. The corresponding figures for other staff members in the area were 38 postdoctoral workers, 77 research students and 27 technical assistants in 1975, compared with 31, 77 and 22 respectively in 1983.

Figure 16.5(*b*) shows the dramatic increase in research grant income over the same period. This increase reflected a number of important changes in the way research in the UK

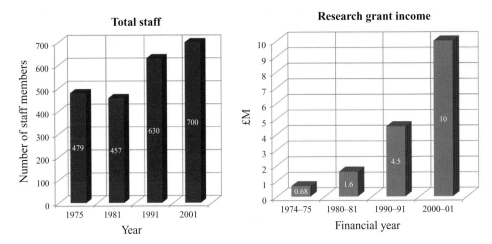

Fig. 16.5 (*a*) Total staff, 1975–2001. These numbers include all academic and support staff, postdoctoral researchers and research students. (*b*) Research grant income, 1975–2001.

was funded. Most important was the fact that the funding for university research was to pass from the University Grants Committee (UGC) to the Research Councils. Whereas previously the resources for research had largely been part of the UGC block grant to the universities, now research funding depended upon winning research grants competitively from the Research Councils through the peer-review process. As a result, funding became more selective and unproductive areas ceased to be funded. Furthermore, equipment and instruments were becoming more expensive and more staff to support these had to be funded by Research Council grants. The scale of the major experimental projects meant that Research Council funding had to be the principal route for obtaining the large resources required. As can be seen from Figure 16.5(*b*), this trend was to continue dramatically through the succeeding years. By 2001, essentially all the Laboratory's income for research came from external research grants, principally from the Research Councils but also from charities, industrial collaborations and philanthropic gifts.

In the 1975 edition of the *Current Research* booklet, Pippard summarised his vision of a concerted attack on the major challenges of condensed matter physics under four headings: electronic structure, defects, amorphous materials and surfaces. Additional research areas included liquid crystals, superconductivity, critical phenomena, inelastic neutron scattering and other theoretical problems. There were some successful collaborations between groups, but there was a tendency to revert to loyalty to established groups. There were good reasons for the type of group structure established by Bragg, but group projects became all the more significant during the 1970s and 1980s when large research grants were needed to support projects at the cutting edge of physics. In addition, experience and practice has shown that the optimum size of a research group is about 50 to 100 persons. On that scale it is generally possible to provide effective coverage of a major area of physics, have a wide-ranging research group seminar programme and not be so large that individuals no longer know each other. The new regime of research grant funding by the Research Councils also encouraged the successful subgroups to seek independence from less successful, and consequently less well-endowed, groups. Nonetheless, there were numerous successful inter-group collaborations, resulting in new areas of research.

By 1983, the list of shared problems in the *Current Research* booklet included low-dimensional conductors, defects, amorphous materials, surfaces, macromolecular physics, superconductivity, magnetic fluctuations in metals, liquid crystals, electron microscopy, semiconductor physics, high-speed photography, point contact and electron tunnelling spectroscopy and theoretical problems – topics which bear more than a passing resemblance to the future research group structure of the Laboratory. As remarked by Mick Brown,

> It is notable that if it is successful, the Cavendish style of research into condensed matter, which emphasises pioneering physical studies of topical interdisciplinary problems, spawns whole new cognate subjects with their own line of development. (Mick Brown, personal communication, 2015)

The Pippard era: radio astronomy, high-energy physics and laboratory astrophysics

17.1 The Mullard Radio Astronomy Observatory

17.1.1 The extragalactic radio sources

The construction of the One-Mile and Five-Kilometre (Ryle) telescopes opened up the study of the astrophysics of galactic and extragalactic radio sources. The radio emission of these sources had been established as the synchrotron radiation of ultra-relativistic electrons gyrating in the magnetic fields within the radio sources. The One-Mile Telescope carried out deep surveys of small regions of sky which demonstrated the convergence of the number counts of radio sources at low flux densities (Section 14.3), a key result for astrophysical cosmology. In addition, over the three years from 1965, complete samples of radio sources were imaged with an angular resolution of about 23 arcsec at 1.4 GHz, using the telescope as a grating synthesis instrument. The positions and structures of over 200 of the brightest extragalactic sources in the northern sky were determined, the results of this large programme being published in three papers in the late 1960s (Macdonald *et al.*, 1968; Mackay, 1969; Elsmore and Mackay, 1969). With these data, it was possible to begin the astrophysical analysis of the population of powerful extragalactic radio sources (Longair and Macdonald, 1969). The vast majority of them were double-lobed sources, associated both with radio galaxies and with radio quasars, but the details of their structures were not resolved.

With the success of the mapping of the sources at 0.408 GHz and 1.4 GHz, the next challenge was to develop receivers for the One-Mile Telescope for the higher frequencies of 2.7 GHz and 5 GHz, providing three times greater angular resolution at the higher frequency. As always, Cygnus A was the first target because of its enormous flux density compared with all other powerful extragalactic radio sources in the northern sky.[1] Simon Mitton and Martin Ryle (1969) published radio maps of Cygnus A at these higher frequencies; the 5 GHz map is shown in Figure 17.1. These observations showed convincingly that there are regions of very high energy density in relativistic electrons and magnetic fields in the compact radio structures observed towards the 'leading edge' of the radio structures. There was clearly much to be understood from observations with yet higher angular resolution and, this was to be provided by the Five-Kilometre Telescope, which was designed to have outstanding performance at 5 GHz.

The radio image of Cygnus A observed by the Five-Kilometre Telescope (see Figure 14.7(*b*)) clearly showed the high surface brightness 'hot-spots' at the leading edges

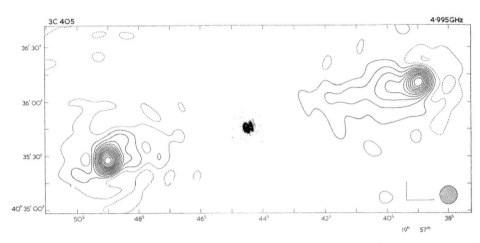

The 5 GHz map of the radio source Cygnus A, showing the 'hot-spots' at the advancing edges of the radio lobes (Mitton and Ryle, 1969). Note that the image is in telescope coordinates so that the beam of the telescope is round. In fact, the angular resolution in right ascension and declination is elliptical, as indicated by the 'L' in the bottom right of the map. The location of the parent galaxy and its structure are sketched midway between the radio lobes. **Fig. 17.1**

of the radio lobes, as well as a compact radio source associated with the nucleus of the radio galaxy (Hargrave and Ryle, 1974). This characteristic structure was soon shown to be very common among the most luminous extragalactic radio sources. There were many consequences of these observations for the future of high-energy astrophysics and cosmology. In 1974, not all the extragalactic sources in the 3CR catalogue had been associated with distant galaxies – often the source positions were not well enough determined, but now, with the understanding that the associated galaxy was to be found between the double radio lobes and often associated with a compact nuclear radio source, the search area was much refined and many more radio sources could be identified. Over the next few years, superb maps of many of the bright 3CR radio sources with angular resolution about 2 arcsec were published by Guy Pooley, Julia Riley and Chris Jenkins (Riley and Pooley, 1975; Jenkins *et al.*, 1977).

In 1977, I became a member of the Science Working Group for what was to become the Hubble Space Telescope, at a time when the first prototypes of charged-coupled device (CCD) cameras for astronomy were being developed as part of that project. By good fortune, James Gunn was interested in finding the optical counterparts of the unidentified radio sources and we used the prototype CCD camera, called PHUEI,[2] with the Palomar 200-inch telescope to find the galaxies associated with many of them. With the accurate positions and structures from the Five-Kilometre Telescope and the deep CCD-camera images, we were able to identify most of the previously unidentified radio sources (Gunn *et al.*, 1981). It was a further piece of good fortune that these galaxies were very luminous indeed, as luminous optically as the brightest galaxies in clusters of galaxies, and so observable at large redshifts with the new generation of CCD cameras. Furthermore, the optical spectra of the faint galaxies had strong emission lines, which enabled Hyron Spinrad and his colleagues to measure redshifts for most of them (Smith and Spinrad, 1980).

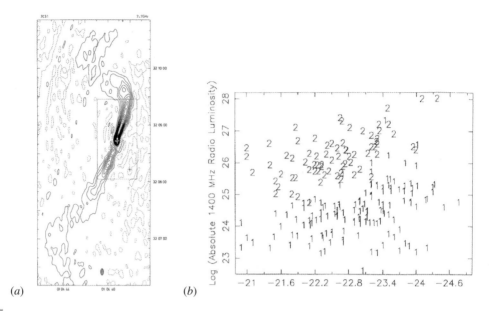

(a)   (b)

Fig. 17.2   (a) The 2.7 GHz map of the radio source 3C 31, showing the bright jet originating in the nucleus of the associated galaxy NGC 383, the location of which is indicated by the cross at the base of the radio jet (Burch, 1977). (b) Illustrating the Fanaroff–Riley correlation between radio luminosity and the location of the highest brightness radio emission in the radio structure (Fanaroff and Riley, 1974; Owen and Ledlow, 1994). The FR1 and FR2 sources are labelled 1 and 2 respectively in the diagram. The abscissa is the optical absolute magnitude of the galaxy or quasar of the associated optical object.

Many of the faint radio galaxies had redshifts of about 1 and greater, spanning essentially the same range of redshifts as the much more optically luminous 3CR radio quasars.[3] This was the most distant sample of large redshift galaxies for many years.

While many of the most luminous radio sources had hot-spots at the leading edges of the radio lobes, in the less luminous sources there are no hot-spots, but there are jets in the inner regions of the radio structures. A good example is the radio source 3C 31 shown in Figure 17.2(a), which is in contrast to the structures of sources such as Cygnus A (Figure 17.1). Bernard Fanaroff and Julia Riley showed that there is a remarkable correlation between radio luminosity and the structural features of the radio sources. Those sources with hot-spots at the advancing edge of the radio structures were referred to as Fanaroff–Riley Class 2 sources (FR2), while those in which the highest brightness regions occurred less than halfway to the outer radio contours were called Fanaroff–Riley Class 1 sources (FR1) (Fanaroff and Riley, 1974). The diagram by Fraser Owen and Michael Ledlow (1994) illustrates vividly this structural dichotomy between the two classes (Figure 17.2(b)).

17.1.2 The theory of extragalactic radio sources

The high-resolution maps provided important constraints upon the theory of extragalactic radio sources, in particular, the very high energy densities in relativistic electrons and magnetic fields which had to be present in the hot-spots of the most luminous sources.[4] In an important contribution, David De Young and Ian Axford pointed out the similarity between the structures of the hot-spots with their trailing lobes and that of the magnetosphere about the Earth (De Young and Axford, 1967). This provided a mechanism for confining the large energy densities of relativistic material within the hot-spots. If the hot-spots were blobs of relativistic plasma ejected at high speed from the nucleus of the associated galaxy, the ram pressure of the surrounding intergalactic gas would provide the confining pressure. The problem with this picture was that the relativistic material would inevitably suffer adiabatic losses as the blob expanded from a compact source in the nucleus to a component kiloparsecs in size. Martin Rees (1971) proposed a solution to this problem in which the source components were continuously supplied with low-frequency electromagnetic waves from a source in the nucleus, by analogy with the proposal of Gunn and Ostriker that such low-frequency waves could be emitted by magnetised rotating neutron stars (Gunn and Ostriker, 1969).

Peter Scheuer, Ryle and I analysed the implications of the high energy densities in the hot-spots and considered a wide range of models which would ensure that such hot-spots could exist at distances far from their primary source of energy. We demonstrated that what was needed was a continuous supply of energy from the nucleus to replenish the hot-spots with relativistic material (Longair *et al.*, 1973). This had to be supplied by some form of relativistic jet. These concepts were elaborated by Scheuer into a physical model in which the hot-spots were continuously supplied with energy from the nucleus and high-energy electrons accelerated in the hot-spots. The accelerated electrons then escaped from the hot-spots, leaving behind in their wakes what were identified as the lobes of the radio source (Scheuer, 1974). This model, illustrated in Figure 17.3(*a*) and developed in more detail by Alexander (2006), has become the standard picture of the astrophysics of double radio sources. The FR1 sources were associated with less powerful jets of relativistic material which were decelerated in the vicinity of the parent galaxy.

Another important insight was that of Stephen Gull and Kevin Northover, who realised that the extended radio lobes, once inflated by the injection of relativistic material, constituted a bubble of light gas embedded in the denser intergalactic gas. Consequently, the radio lobes would be buoyant in the presence of a gravitational potential gradient and move up that gradient (Gull and Northover, 1973). This is particularly important for radio sources in clusters of galaxies in which the hot intracluster gas forms a more or less isothermal atmosphere within the deep gravitational potential well formed by the dark matter in the cluster. Bubbles of radio-emitting plasma have now been observed in a number of clusters of galaxies in which the intracluster gas is a strong X-ray emitter, the radio-emitting clouds forming 'bubbles' in the distribution of hot X-ray emitting gas (Fabian *et al.*, 2000) (Figure 17.3(*b*)).

In 1978, Anthony Bell, one of Scheuer's graduate students, published two important papers on the acceleration of charged particles in strong shock waves (Bell, 1978a,b). Other investigators had independently come to similar conclusions, but Bell's papers gave clear insights into what is now called the *first-order Fermi acceleration* mechanism. By simple physical arguments, he showed that high-energy particles bounce back and forth across a strong shock front if they are randomly scattered in the plasma on either side of the shock

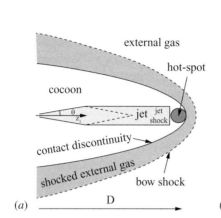

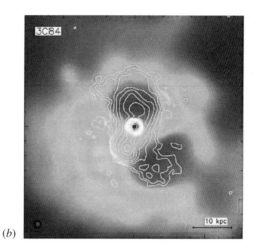

Fig. 17.3 (*a*) A model for one side of an FR2 double radio source according to the model of Scheuer, as elaborated by Alexander (Scheuer, 1974; Alexander, 2006). The jet feeding the hot-spots is self-collimated by the surrounding cocoon of relativistic material and hot gas. Electrons are accelerated in the shock waves and then escape backwards to form the cocoon. Alexander showed that the structure is self-similar, accounting for the similarity in appearance of the compact and extended double structures. (*b*) 'Radio bubbles' in the core of the Perseus cluster of galaxies. The radio emission is shown as contour lines and the diffuse X-ray emission of the hot intergalactic gas in the cluster is shown in colour (Fabian *et al.*, 2000).

front, resulting in an increase in the energy of the particles at each crossing. A simple calculation also shows that the particles should be accelerated with a power-law energy spectrum, as observed, with more or less the correct spectral index. The great merit of this model was that it resulted in the acceleration of charged particles where they are needed, namely at the shock fronts in supernova remnants and in the vicinity of those inferred to be present in double radio sources. This type of acceleration is now generally accepted as the means by which cosmic ray electrons, protons and nuclei are accelerated to high energies in astronomical environments.

17.1.3 New directions

The great success of the One-Mile and Five-Kilometre Telescopes gave the Cavendish Laboratory a world-leading position in radio astronomy and it was to be some years before other radio observatories were able to compete with and eventually surpass the capabilities of these telescopes. This began to take place in the late 1970s with the commissioning of the Very Large Array in New Mexico in the USA. Consisting of 27 25-metre radio antennae in a Y-shaped configuration, it had superior performance to the Five-Kilometre Telescope, which continued to carry out excellent science but was no longer unique and had to become a more specialised instrument to continue carrying out front-line research.

An important development, which was to prove of considerable significance for the future of synthesis imaging not only in the radio but also in the optical and infrared wavebands, was the publication of two papers by John Baldwin and Peter Warner on *phase-less aperture synthesis* (Baldwin and Warner, 1976, 1978). In the standard procedures at that time, the signals measured by two separated antennae were cross-correlated, resulting in amplitude and phase information for points in the aperture, or Fourier, plane. In their first paper, Baldwin and Warner demonstrated how aperture synthesis images could be reconstructed from the cross-correlation functions between pairs of telescopes without using the phase information, provided the flux density of one of the point sources far exceeded that of the sum of all the fainter sources in the field. This procedure is similar to that used in the determination of the structures of complex molecules by the heavy-atom method in X-ray crystallography, or in the use of a reference beam in the construction of holograms. The experimental validation of the technique was performed using observations of a region close to the North Celestial Pole with the Half-Mile Telescope.[5] The technique was exploited by Riley and Pooley (1978) to map the radio source 3C 123 at 15 GHz with large dynamic range with the Five-Kilometre Telescope.

The procedure was generalised in the second paper by Baldwin and Warner to the case in which there was no dominant single source or in which the sources had complex structures. They validated the procedures using observations of a field from the seventh Cambridge (7C) survey. These were important steps in improving the techniques for reconstruction of images from interferometric data without explicitly using the phase data. The further development of these procedures led to closure phase and self-calibration techniques, which have become standard data reduction techniques in radio astronomy and were also to be central to the development of optical–infrared interferometry in the 1980s and 1990s.

Another important development for the analysis of interferometric data obtained by radio and all other types of telescope was the application of maximum-entropy statistical techniques to the observations (Gull and Daniell, 1978; Skilling and Bryan, 1984). These procedures extracted the maximum amount of information from the data, producing maps smoothed to the resolution of the telescope system, subject to very general constraints such as that the resulting maps should not result in negative intensities. These techniques were successfully commercialised by Stephen Gull and John Skilling. They were also to lead to the application of Bayesian techniques in which general prior assumptions about the final images can be incorporated in the analysis procedures – these are now commonly used in essentially all branches of astronomy.

There were other reasons for changing the direction of the research programme. Whereas radio astronomy had been a somewhat separate discipline from the mainstream of astronomical research up to about 1970, it gradually became a much more integral part of the astronomy scene. For example, to understand the physics of the radio sources, access to the largest optical telescopes with the most sensitive spectrographs was essential. The work we carried out on the identification of extremely distant radio galaxies illustrated some of the astrophysical directions in which the subject was to develop.

In addition, advances in other wavebands could not be ignored. The role of hot intergalactic gas in providing the ram pressure needed to confine the radio source components indicated the importance of observations in the X-ray waveband. With a great struggle, Peter Willmore and I were able to use the tiny X-ray detector on board the Copernicus satellite to detect the X-ray emission from Cygnus A, which, as we suspected, turned out to originate in the hot intergalactic gas in the surrounding cluster of galaxies, of which the Cygnus A galaxy was by far the brightest member (Longair and Willmore, 1974).

In the same way, infrared astronomy in the wavebands to the long-wavelength end of the optical spectrum was becoming possible with the inauguration of the UK Infrared Telescope (UKIRT) and the NASA Infrared Telescope Facility (IRTF) on Mauna Kea in Hawaii in the late 1970s. For the identification and study of the most distant radio galaxies, UKIRT had the great advantage that the large redshifts of these sources meant that they were relatively easier to detect in the infrared as compared with the optical waveband. In fact, using single-element photometry in the K-waveband at 2.2 μm at UKIRT, we were able to complete the identification of the 3CR radio sources for a sample away from the galactic plane (Lilly and Longair, 1984).

Another burgeoning discipline was millimetre and submillimetre astronomy. The UK community had strong backing for as large a telescope as could be built to explore these wavebands, and the radio observatories at Cambridge and Jodrell Bank were the obvious organisations to take a leading role. The Jodrell Bank programme was fully occupied with the development of what was to become the MERLIN long-baseline interferometer and so Ryle and his colleagues were faced with the decision of whether or not to take up the challenges of the quite different radio astronomy of molecules and dust. Richard Hills, a leading expert in millimetre and submillimetre techniques, returned from Germany and the USA as the project scientist for a 15-metre millimetre/submillimetre telescope, the construction of which would be carried out at the Rutherford–Appleton Laboratory. The group had to decide what directions they would pursue after the pioneering days of the Five-Kilometre Telescope were over.

Ryle was ambivalent about the way ahead. His entire career had been devoted to the mastery of interferometry and aperture synthesis, culminating in the award of the Nobel Prize in Physics in 1974. In the end, he gave his positive support to millimetre astronomy as well as to the construction of large optical telescopes for extragalactic research, but it was clear that his heart was not as engaged as it had been in the heroic days of the development of aperture synthesis. He regarded the Five-Kilometre Telescope as more or less the end of that particular line of activity in Cambridge. His health had been deteriorating, partly because of the strain of constructing new radio telescope systems for the previous 30 years, and in 1974 he underwent major heart surgery. In 1977, he had a further operation for lung cancer. From that time, he shifted his interests to alternative energy sources and to his abhorrence of nuclear weapons and nuclear power.

17.2 Wind power

The energy crisis of the early 1970s spurred two initiatives in the Laboratory, Richard Eden's Energy Research Group (Section 17.4) and Martin Ryle's Wind Power Group. Ryle had been a passionate opponent of nuclear weapons, believing that the UK's activities at the Atomic Energy Establishment at Harwell and the Atomic Weapons Establishment

Fig. 17.4

Ryle's experimental wind turbine at the Lord's Bridge Radio Observatory. In the background are one of the fixed elements of the Five-Kilometre Telescope, the control building for the Five-Kilometre Telescope and the Lord's Bridge railway station.

at Aldermaston, although nominally intended to create fuel for domestic nuclear energy generation, were in fact a cover for the production of weapons-grade plutonium. With the oil crisis of the early 1970s, he became an advocate of wind power as a viable alternative source of renewable energy. This matched another of his passions, sailing and the design of sailing boats – he had already built innovative trimarans with his own hands.

His case is best described in his own words:

> There is an urgent social need to utilise the energy available with recyclable fuels such as wind power and waves; they are clean; they do not pose pollution problems; they will be there when oil, coal and nuclear fuels are exhausted. The United Kingdom is particularly well-endowed with a wind regime that is not only the equal of any other European country but also superior to that of the large continental land masses, such as the Americas, India and Australia. Although wind power has been harnessed to do useful work for at least 5000 years, when the first sailing boats plied their trade on the Nile, and later allowed the development of N. America by pumping irrigation water with a total power equal to about 5 nuclear power stations, only very recently has it been possible, with the aid of aerodynamics, to design high performance wind turbines. (Ryle, 1982)

Ryle was joined in his experimental programme by Paul Scott and Donald Wilson from the Radio Astronomy Group and by Allen Metherell from the Laboratory Astrophysics Group, as well as by a number of research students and postdocs. The main thrust of the research was to achieve efficiencies as close as possible to the theoretical maximum of 59% and to understand all the technical and materials issues which went along with the new concepts (Figure 17.4). The innovation was to use aerodynamics to increase the power output of the turbines. The interaction between neighbouring turbines was also studied. The programme was prescient and successful, well ahead of the time when wind turbines became commercially viable. With Ryle's death in 1984, the initiative came to an end, but his advocacy of wind power has been vindicated by the large-scale development in the UK of wind farms, which are making a significant contribution to the UK's national energy budget (MacKay, 2008).

17.3 High-energy physics

17.3.1 Particle physics 1955 to 1983: the growth of CERN

After the war, the leadership role in high-energy physics passed to the USA where the physicists benefitted from the large investments ploughed into fundamental research by the US governmental and military organisations. The Manhattan Project had vastly increased the capability of the USA in high-energy physics and demonstrated the role of fundamental physics in the country's strategic goals. More and more powerful accelerators were constructed and by the 1950s these machines, with their controlled high-intensity beams of particles, replaced cosmic rays as the preferred source of high-energy particles. These experimental capabilities were supported by a matching investment in theory. Table 17.1 lists a selection of major advances in experimental and theoretical particle physics from 1955 to 1983. A number of inferences can be drawn from this table.

- The dominance of the USA in experimental particle physics continued until the CERN experiments came into operation in the 1970s.
- The combined efforts of the theorists and experimentalists led to fundamental insights into the basic structure of matter, the pace of discovery continuing at a dramatic rate through the 1950s, 1960s and 1970s. The standard model of particle physics was consolidated in the early 1970s, providing new challenges for both experimentalists and theorists.
- The European physicists could only compete with the USA if they collaborated internationally. This led to the formation in 1952 of the provisional Conseil Européen pour la Recherche Nucléaire (CERN), or European Council for Nuclear Research, with the mandate of establishing a world-class fundamental physics research organisation in Europe. The convention establishing CERN was ratified in 1954 by 12 countries in western Europe, with the title Organisation Européenne pour la Recherche Nucléaire,

Table 17.1 Selected discoveries and advances in high-energy physics, 1955–1983	
Year	Events
1955	Antiproton discovered by Owen Chamberlain, Emilio Segrè, Clyde Wiegand and Thomas Ypsilantis at Berkeley, California
1956	Electron antineutrino detected by Frederick Reines and Clyde Cowan at the Hanford and Savannah River nuclear reactors in the USA
1962	Muon neutrino shown to be distinct from the electron neutrino by Leon Lederman, Melvin Schwartz and Jack Steinberger at the AGS accelator at the US Brookhaven National Laboratory
1964	Quarks predicted independently by Murray Gell-Mann and George Zweig, both at California Institute of Technology
1964	Higgs boson proposed: François Englert and Robert Brout; Peter Higgs; Gerald Guralnik, Carl Hagen and Tom Kibble
1964	Charm quark proposed by James Bjorken and Sheldon Glashow, also by Sheldon Glashow, John Iliopoulos and Luciano Maiani in 1970
1964	Xi baryon discovered at the Brookhaven National Laboratory in the USA
1968	Unification of weak and electromagnetic interactions and the predictions of neutral currents and the W and Z bosons by Sheldon Glashow, Abdus Salam and Steven Weinberg
1969	Partons observed in deep inelastic scattering experiments between protons and electrons at SLAC (Stanford Linear Accelerator Center); partons became identified with quarks and gluons, leading to the discovery of the up quark, down quark and strange quark
1971	Standard model of elementary particles
1973	Neutral weak currents caused by Z boson exchange discovered at CERN in the Gargamelle experiment
1973	Asymptotic freedom and quantum chromodynamics
1973	Bottom quark proposed by Kobayashi and Maskawa to explain CP violation
1974	J/ψ-meson discovered by groups headed by Burton Richter at SLAC and Samuel Ting at Brookhaven, demonstrating the existence of the charm quark
1974	Supersymmetry proposed
1975	Tau lepton discovered by a group headed by Martin Perl at the SLAC and Berkeley laboratories
1977	Bottom quark discovered in the US Fermilab E288 experiment led by Leon Lederman
1979	Gluon observed indirectly in three-jet events at DESY in Germany
1980	Minimal supersymmetric standard model
1981–86	String and superstring theory
1982	Detection of hadronic jets in the UA2 experiment at CERN
1983	W and Z bosons discovered by Carlo Rubbia, Simon van der Meer and the UA1 and UA2 collaborations at CERN

although the acronym CERN was maintained.[6] Although the formal title of the new organisation suggested that nuclear physics was at the heart of the research programme, the core activities were in fundamental particle physics and the basic constituents of matter, as was reflected in the later name for the organisation, the European Laboratory for Particle Physics. The laboratory is based in the suburbs of Geneva on the Franco-Swiss border. Twenty European states, as well as Israel, are now full members of CERN, in addition to which there are several associate member countries.

It is helpful to list the major projects which were successively undertaken at CERN and then review how the High Energy Physics Group exploited these opportunities.

- The *600 MeV Synchrocyclotron*, CERN's first accelerator, started operations in 1957 for experiments in nuclear and particle physics. After 1964, the machine concentrated upon nuclear physics studies.
- The *Proton Synchrotron* began operations in 1959 and was CERN's flagship accelerator for a number of years. When the next generation of accelerators was constructed in the 1970s, the Proton Synchrotron's principal role was to provide the initial stage of acceleration of particles for the higher-energy machines. The circumference of the synchrotron was 628 m, enabling the accelerator to operate up to 25 GeV. In addition to protons, it could accelerate α-particles, oxygen and sulphur nuclei, electrons, positrons and antiprotons. The first major particle physics discovery at CERN was the detection in 1973 of *neutral currents*, which had been predicted by the electro-weak theory of Glashow, Salam and Weinberg. This was achieved by the French-led Gargamelle experiment, a giant bubble chamber detector designed mainly for the detection of neutrino interactions – the muon neutrino beam was produced by the CERN Proton Synchrotron.
- During the 1960s, physicists fully appreciated that colliding two particle beams head-on would give much higher collision energies in the centre of momentum (laboratory) frame than a single beam of particles colliding with a fixed target. The concept became a reality with the construction of the *Intersecting Storage Rings* (ISR), which consisted of two intersecting rings each with diameter 150 m. Protons circulated in opposite directions and collided with a maximum centre-of-momentum energy of 62 GeV. The proton beams were initially accelerated by the Proton Synchrotron and then injected into the ISR. The great challenge was to produce narrow, high-intensity beams of protons which would result in large number of collisions. This was achieved under the leadership of Simon van der Meer, who first demonstrated the technique of *stochastic cooling*, by which both the transverse dimensions of the beam and the spread in the energies of the particles were greatly reduced. The machine came into operation in 1971 and recorded the first-ever proton–proton and proton–antiproton collisions. This technique was to be used with great success in proton–antiproton collisions in the *Super Proton Synchrotron* (SPS). For the next 13 years the ISR provided a unique tool for particle physics and enabled European physicists, including the members of the Cavendish High Energy Physics Group, to gain valuable knowledge and expertise for subsequent collider projects, leading ultimately to the Large Hadron Collider (LHC).
- In 1968, George Charpak invented the multiwire proportional counter, a detector which enabled the tracks of particles through the target region of the accelerator to be determined entirely electronically, rather than by the use of the photographs obtained in hydrogen bubble chambers. The important feature of these detectors was that they could operate at vastly greater detection rates than bubble chambers. This development would lead to a wide range of new types of detector for particle physics, including spark chambers, calorimeters and so on which had the advantage of directly measuring the energies and momenta of the products of particle collisions.

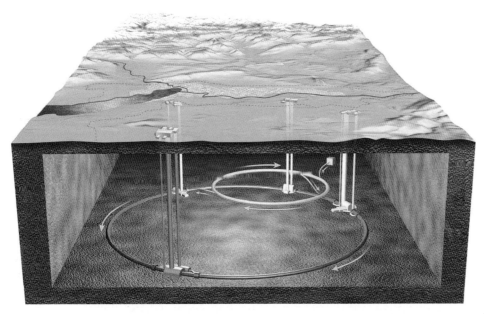

Fig. 17.5

A schematic diagram of the layout of the SPS (smaller grey ring) and the LHC (large blue ring) in the underground laboratory at Geneva. The SPS was the workhorse of CERN from its opening in 1976 until the LEP ring, which was to became the LHC tunnel, was constructed. The circumferences of the SPS and LHC rings were 7.5 km and 27 km respectively, on the border between Switzerland and France. The depth of the LEP-LHC tunnel ranges from 50 m to 175 m underground. The SPS became the injector accelerator for the LHC when the latter became operational in 2008 (copyright CERN).

- The CERN Council approved the construction of the Super Proton Synchrotron in 1971 with the intention of building a world-leading particle accelerator, originally specified as a 300 GeV machine but in fact capable of greater energies. The accelerator measured nearly 7 km in circumference (Figure 17.5). Particles were injected from the Proton Synchrotron and then accelerated to energies of up to 450 GeV. The SPS could accelerate many different kinds of particles – sulphur and oxygen nuclei, electrons, positrons, protons and antiprotons. By the official commissioning date of the SPS on 17 June 1976, an operating energy of 400 GeV had been achieved, but this had already been exceeded by the Main Ring Accelerator at Fermilab, which had attained an energy of 500 GeV on 14 May of that year. The SPS became the workhorse of CERN's particle physics programme when it came into operation in 1976.
- By 1976, a major objective of the particle physics community was the creation and detection of the W and Z bosons predicted by the standard model of particle physics. The electroweak theory had already had a major success with the prediction and subsequent detection of neutral currents in the Gargamelle experiment. The ultimate objective was, however, the production and detection of the W and Z gauge bosons, which were predicted to have masses of roughly $m_W \sim 65$ GeV/c^2 and $m_Z \sim 80$ GeV/c^2. Out of these considerations came the proposal for the Large Electron–Positron Collider (LEP), the report being completed late in 1976 following the success of the SPS in achieving and then

exceeding its design goals. But the LEP experiment was to be a huge long-term project, requiring the construction of a new tunnel 27 km in circumference (Figure 17.5). A faster route to the discovery of the W and Z bosons was desirable in the face of competition from the USA.

- In the same year, 1976, Carlo Rubbia came up with his imaginative solution to convert the newly inaugurated SPS into a proton–antiproton collider. Many accelerator physicists believed the project was not feasible, but Rubbia persisted and, in collaboration with van der Meer, came up with the solution. The project was approved in 1977 without jeopardising the future construction of the LEP collider. The project was very ambitious, each element of the design requiring innovative thinking, but the feasibility of obtaining adequately collimated beams of particles and antiparticles was demonstrated by the *Initial Cooling Experiment (ICE)* and the successful development of the *Antiproton Accumulator* design. At the same time, the UA1 experiment and, six months later, the UA2 experiment were approved as the key initial experiments for use with the proton–antiproton collider, which was often referred to as the $Sp\bar{p}S$ experiment.

17.3.2 The regeneration of the High Energy Physics Group

UK policy in particle physics was to exploit the opportunities offered by CERN and other international collaborations, supported by grants provided by the Science Research Council (SRC) and its successor the Science and Engineering Research Council (SERC). Through the UK's annual subscription to CERN, the UK particle physicists were guaranteed involvement in CERN projects but, to maintain serious capabilities in experimental particle physics, grants had to be won from the SRC and SERC to provide the instrumentation for the particle accelerators. Because of the increasing magnitude of the experimental projects, most of the UK instrumental contributions had to involve collaborations between University particle physics groups and their European colleagues.

Within the Cavendish Laboratory, the pattern of research in high-energy physics established during the Mott era continued under Pippard. Frisch retired at the end of the academic year 1970–71 and the Jacksonian Chair was not refilled in high-energy physics, but in laboratory astrophysics. With the foundation of the Energy Research Group by Richard Eden in 1974, he and Ken Riley shifted their interest to various aspects of energy research. The activity in high-energy physics reached a rather discouraging low ebb in that academic year, with only John Rushbrooke as the Head of Group, Bill Neale, Bryan Webber and Janet Carter holding academic posts. It is ironic that, just as the major advances in experimental and theoretical particle physics were coming thick and fast throughout the Mott and Pippard eras (Table 17.1), the Cavendish High Energy Physics Group was fighting for survival.

Despite the diminished scale of the High Energy Physics Group, the three main lines of research – the analysis of bubble chamber photographs, instrumentation for high-energy physics experiments, and theory – were maintained. The appointment of Webber, a former student of Eden's, as a University Demonstrator ensured that theoretical particle physics studies flourished, and in the following years new appointments were made which enabled the group to play a serious role in particle physics experiments in Europe and the USA – new staff members included Tom White (1975), Richard Ansorge (1975) and David Ward (1978). The technical expertise of the group remained focussed on experimental particle physics and this was supported by a number of research associates and a large number of personnel who were variously described as data assistants or scanners to support the Sweepnik machine.

Frisch's strategy of developing world-leading film-scanning techniques for use with hydrogen bubble chamber photographs had paid off handsomely with the successful construction of the Sweepnik machine (Section 14.6.1). During the 1970s, the group members participated in the selection, preparation and operation of experiments at CERN and gained access to data from major accelerators in the USA. At CERN, this involved studies with the Proton Synchrotron, the Intersecting Storage Rings and the SPS, and at Fermilab in the USA with the 500 GeV proton synchrotron. In addition, the Cavendish high-energy physicists had access to bubble chamber images from the 12 GeV proton synchrotron at the Argonne National Laboratory in the USA. A particular interest of the high-energy physicists at the Cavendish Laboratory was the study of 'resonances' at 6 and 21 GeV and understanding the quantum selection rules governing the interactions of K-particles.

The higher-energy experiments using 500 GeV protons were of importance in understanding proton–antiproton interactions above 100 GeV. The bubble chamber images had to be supplemented by measurements with other detectors, including Cherenkov detectors, multiwire proportional counters and spark chambers to obtain a complete characterisation of the disintegration products of the particle collisions. These developments resulted in new instrumentation challenges for the High Energy Physics Group, which would point the way to the future. The experimental team, particularly Rushbrooke and Carter, took part in the development phase of the Intersecting Storage Rings and SPS projects which were to be important for their future involvement in the UA2 and UA5 projects of the $Sp\bar{p}S$ experiment. At the same time, use was made of the hybrid spectrometer at the Stanford Linear Accelerator Center (SLAC) to study proton–antiproton interactions at the lower energies of about 9 GeV. Webber played a major role in understanding the total proton–antiproton interaction cross-sections for particle collisions at the highest accessible particle energies.

17.3.3 The UA1, UA2 and UA5 experiments[7]

The High Energy Physics Group joined the UA2 experiment at the beginning of an upgrade programme to enable the W and Z masses and their widths to be measured precisely. Electron identification was one of the great strengths of the UA2 experiment and was the focus of the group's participation in the experiment. The W and Z bosons decay very quickly after they are produced but the decays leave characteristic energy deposits in the detectors. In the case of the decay of the lower-mass W particle, the signature consisted of a single high-energy electron and an undetectable neutrino, whose presence could be inferred from the 'missing energy' in the calorimetry of the event. In the case of the higher-mass Z particle, the signature was decay into a high-energy electron–positron pair.

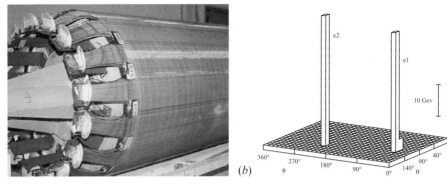

Fig. 17.6 (*a*) The scintillating fibre detector for the UA2 experiment. (*b*) The pattern of energy deposition seen in the UA2 calorimeter, interpreted as the decay of Z bosons into electron–positron pairs (copyright High Energy Physics Group, Cavendish Laboratory).

The international UA2 project involved the design, installation and operation of a new scintillating fibre detector (SFD) that improved the tracking and electron identification performance of the experiment (Figure 17.6(*a*)). The detector involved a combination of several novel technologies: scintillating fibres as the active element, custom image intensifiers with integrated CCD readout to collect the signals and custom digitiser modules designed and built in the High Energy Physics Group to process the CCD images.[8]

The light produced by the passage of charged particles through a fibre travelled to its end where it was collected and amplified by image intensifiers. Many individual fibres were grouped into a bundle and imaged together by a single image intensifier and CCD detector. By associating the detected light with the correct fibre, those hit by the particle could be reconstructed and then the position and direction of the particle track determined in three dimensions. The SFD also incorporated a layer of lead that allowed the energy deposited by photons to be more precisely localised than was possible using calorimetry alone. An example of the energy distribution due to Z bosons decaying into electron–positron pairs recorded in the UA2 experiment is shown in Figure 17.6(*b*).

The Sp$\overline{\text{p}}$S proton–antiproton collider began taking data in 1981 and soon the machine was operating at the highest intensities ever produced in collider experiments. In the 1982 run, among about one thousand million collisions, six W particles were identified at the UA1 experiment and four at UA2, the mass of the particle being about 80 GeV/c^2. The rarer Z particle decay required even greater beam intensities, and this was achieved by April 1983. By the end of the 1983 run, the UA1 and UA2 experiments had identified about a dozen Z particles with average mass 93/c^2 GeV and about 100 W particles with mass about 81/c^2 GeV. The co-discovery of the W and Z bosons by the UA1 and UA2 experiments was spectacular validation of the electroweak theory. The standard practice in high-energy physics is that the best estimate followed by the statistical error, the internal systematic error and the external systematic error are given. The final UA2 results for the masses of the electroweak particles were then:

- W boson mass, $m_W = 80.84 \pm 0.22 \pm 0.17 \pm 0.81 \, \text{GeV}/c^2$
- Z boson mass, $m_Z = 91.74 \pm 0.28 \pm 0.12 \pm 0.92 \, \text{GeV}/c^2$
- W boson width, $\Gamma_W = 2.10 \pm 0.14 \pm 0.08 \, \text{GeV}/c^2$
- $\sin^2 \theta_W = 0.2234 \pm 0.0072$.

The width of the W boson resonance Γ_W is crucial for determining the number of modes of decay of the particle, while $\sin^2 \theta_W$ involves θ_W, the mixing angle. Rubbia and van der Meer were awarded the 1984 Nobel Prize in Physics for their major contributions to the discovery of the W and Z particles.

In addition to the properties of the W and Z bosons, the UA2 experiment made important measurements in quantum chromodynamics and in searches for new particles. The measurements included:

- the strong coupling constant, $\alpha_s(M_W^2) = 0.123 \pm 0.018 \pm 0.017$
- jet cross-sections and angular distributions
- direct photon production
- the mass of the top quark, $m_t = 160 \pm 60 \, \text{GeV}/c^2$ (for $m_H = 100 \, \text{GeV}/c^2$).

Later, the Cambridge group worked on the development of a new silicon pad detector for a subsequent upgrade of the experiment. Silicon detector development continued to be an important area of expertise within the group, and was to be exploited both in the OPAL experiment at LEP and the ATLAS experiment at the LHC. The UA2 experiment ran from 1981 to 1990.

According to the standard protocol for large international particle physics experiments, a key role was played by the *spokesperson* for each experiment, essentially playing the role of the leader of the project with the approval of all the international partners. John Rushbrooke was selected as the spokesperson for the UA5 and UA5/2 experiments with the SPS, with the titles *Investigation of proton–antiproton events at 540 GeV c.m. energy with a streamer chamber detection system* and *An exploratory investigation of antiproton–proton interactions at 800–900 GeV c.m. energy at the SPS collider*.

The basic detector of the UA5 experiment (Figure 17.7(a)) consisted of two large (6-metre) streamer chambers, triggered by hodoscopes[9] at either end, and viewed by cameras with associated image intensifiers. Charged tracks could be observed down to $0.75°$, and hence over most of the pseudorapidity[10] range in which they were produced; photons were observed over the same pseudorapidity range. Neutral and charged particle decays were also identified in the apparatus.

Later, a calorimeter 4 m long was installed to identify neutral particles in the central region, to give a rough energy measurement for charged and neutral particles, and to provide a trigger on their transverse energies. A small wire chamber closely fitted to the vacuum chamber permitted triggering selectively for high-multiplicity events. Together with a beryllium vacuum chamber introduced to reduce secondary interactions, the upgraded UA5 detector was used to continue hadron physics studies at the highest energy available.

The UA5 collaboration investigated many features of the physics of 540 GeV proton–antiproton collisions, including

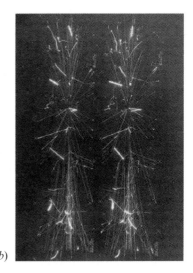

(a)   (b)

Fig. 17.7   (a) The UA5 experiment. (b) Collision events detected by the two 6-metre streamer chambers (copyright CERN).

- charged particle production; pseudorapidity and multiplicity distributions
- photon production; pseudorapidity distributions
- charged-charged and charged-neutral particle correlations
- neutral and charged strange particle production and their p_T-distributions
- a study of high-multiplicity events; search for Centauro events.

The experiment ran from 1979 to 1989.

17.4 The Energy Research Group

Two events stimulated Richard Eden's interest in what became known as the field of energy research, meaning the study of energy usage domestically, nationally and globally and its future supply. The first was the publication of the influential book *The Limits to Growth* (Meadows *et al.*, 1972), which concerned computer simulations of the consequences of the predicted exponential economic and population growth given the finite resources available on the planet. The second was the 1973 oil crisis, which resulted when members of the Organization of Arab Petroleum Exporting Countries (OAPEC) instituted an oil embargo in response to the outcome of the Arab–Israeli conflict. By the end of the embargo, the price of oil had risen from \$3 per barrel to nearly \$12. There was a global oil shortage which focussed interest on the fragility of the world's energy resources.

From the time of his first interest in these topics, Eden had made contact with economists, large commercial energy providers in the coal, gas and nuclear industries, and politicians; this encouraged him to set up the Energy Research Group in 1974. In the first instance, the group consisted of Eden and Ken Riley, a lecturer in the High Energy Physics Group, as well as three research students who were funded by the Science Research Council and a postdoctoral fellow supported by Shell International. Intentionally, the Energy Research Group was highly interdisciplinary, involving staff members from the Economics and Engineering departments as well as strong engagement with the relevant industries. The oil crisis had enormous impact upon the perception of energy use world-wide. Eden was successful in attracting excellent research students to work in these areas, although energy research scarcely existed as an academic discipline at the time.

Eden was fortunate that Sam Edwards had been appointed to the John Humphrey Plummer Chair of Physics in 1972 and held that chair in the Cavendish Laboratory. Almost immediately, Edwards was appointed Chairman of the Science Research Council (SRC), which entailed four years' leave of absence from the Laboratory. Edwards was strongly supportive of Eden's vision of an interdisciplinary energy research programme and took initiatives within the Research Council to encourage such activities within the UK as a whole. As a result, by a combination of Research Council grants and industry support, the Energy Research Group expanded to ten graduate students and ten postdoctoral workers by 1981.

Throughout this period, the group carried out a wide range of research projects, grouped under three headings.

- *Industrial energy*. This area concerned the analysis of the technology and economics of energy in manufacturing industries, including the study of energy conservation and interfuel substitution. Specific case studies were carried out in the paper and aluminium industries, as well as comparative studies of the UK, German and Italian industries.
- *United Kingdom energy studies*. The topics selected for study concerned policy issues for the UK as a whole, with an emphasis upon energy conservation, the impact of the 1973 energy crisis, modelling of energy use in various sectors of the UK economy and cost–benefit analyses for new energy technologies.
- *International energy studies*. The group contributed to two major international studies of energy prospects for the world as a whole: the Workshop on Alternative Energy Strategies (WAES), involving 15 national groups with a horizon of 2000, and the World Energy Conference (WEC), which focussed upon scenarios for world energy demand in 2020.

These activities were supported by the development of systems and information processing tools and some modest experimental studies of heat loss in buildings and in industry. Many more details of the group's activities are contained in Eden's autobiography (Eden, 2012). He was promoted to a personal Professorship in Energy Studies in 1983. He and his group continued to be active internationally until his retirement in 1989 when the group was wound up. There can be no doubt that Eden was ahead of his time in advocating the importance of physics-based interdisciplinary research for the benefit of society.

17.5 Laboratory astrophysics

Alan Cook had a distinguished career at the National Physical Laboratory, becoming Superintendent of the Quantum Metrology Division in 1966. In 1969, he moved to Edinburgh University to take up the newly created Professorship of Geophysics. There he was charged with setting up a new Department of Geophysics, which was successfully achieved, but after only three years he moved to the Cavendish Laboratory as Jacksonian Professor of Natural Philosophy in succession to Frisch. His scientific interests were very wide, but the subjects closest to his heart were those in the areas of precise measurement, particularly in the fields of gravity, masers and lasers and their application in geophysics and astronomy, and the setting of absolute standards of measurement (Quinn, 2005). These interests were reflected in a number of well-received books which he wrote during his period as Jacksonian Professor: *Physics of the Earth and Planets* (Cook, 1973), *Celestial Masers* (Cook, 1977), *Interiors of the Planets* (Cook, 1980), *The Motion of the Moon* (Cook, 1988) and *Gravitational Experiments in the Laboratory* (Chen and Cook, 1993).

In 1975 Cook was joined by Allen Metherell, who transferred from the Electron Microscopy Group to work in the field of experimental gravity. In 1976, Bob Butcher was appointed as a University Demonstrator to start up activities in millimetre and infrared laser spectroscopy. Cook was clear that he saw a role for laboratory astrophysics in the Cavendish Laboratory in support of the activities in radio and millimetre astronomy. As he stated in his goals for the group,

> [The] aim of the Laboratory Astrophysics Group is to pursue studies in physics suggested by astrophysical problems . . .

Through the 1970s, three main themes were developed: experiments on gravitation, molecular masers and laser spectroscopy.

17.5.1 Experiments on gravitation

Experimental gravity is a notoriously difficult area of laboratory physics because of the extreme weakness of the gravitational force. Cook's interests inspired two areas of research. The first was the redetermination of the value of the gravitational constant G. The most ambitious experiment involved a collaboration with Antonio Marussi at the University of Trieste and with the National Physical Laboratory. It was to involve the construction of a scaled-up version of the traditional torsion-balance technique of the original Cavendish experiment. It was to be located in the Grotta Gigante, a huge cave on the Italian side of the Trieste Karst, in which Marussi had already set up gravitational tiltmeters for geophysical studies. The length of the tiltmeter suspension was 95 m. The concept was to use a similarly long torsion fibre and large masses to produce a relatively large gravitational signal. Although the masses were prepared and measured at the National Physical Laboratory, sadly the early death of Marussi put an end to that project.

Experiments to measure the gravitational constant were continued by Cook's graduate student Clive Speake using enhanced measurement techniques, but, as in so many of these experiments, the results were subject to subtle systematic errors.[11]

The most successful of the experimental gravitation experiments were those carried out by Y.T. Chen, Cook and Metherell to test the inverse-square law of gravity at small distances (Chen *et al.*, 1984). Their experiment was inspired by the proposal that, according to the theory of Yasunori Fujii (1971, 1972) which attempted to unify the gravitational force with the other forces of nature, the gravitational potential between two point masses should be modified to

$$V = -G\left(\frac{M_1 M_2}{r}\right)(1 + \alpha\, e^{-\mu r}), \qquad\qquad (17.1)$$

where μ^{-1} is the Compton wavelength of the dilaton, a hypothetical particle introduced to make the unified theory scale-invariant; α is the coupling constant, which Fujii estimated should have the value $\frac{1}{3}$. The experiment involved meticulous attention to the details of all aspects of the experiment, including the torsion balance, the optical lever, the amplifier and servo-systems and the recording system. As part of the analysis, they used the closed analytic solution for the attraction of a finite cylinder at an arbitrary point which Cook and Chen had discovered a couple of years earlier (Cook and Chen, 1982). Their results were inconsistent with any fractional deviation from the inverse-square law of gravity greater than about 0.0001 at distances of the order of 0.1 m. An alternative way of expressing their result was that the Compton wavelength of the exchange particle must be greater than 4.9 m.

17.5.2 Celestial masers

Cook had had an interest in astronomical masers since the discovery in 1965 of the maser emission of the hydroxyl molecule in compact regions of star formation (Weaver *et al.*, 1965; Weinreb *et al.*, 1965). He carried out his theoretical studies while a Visiting Fellow at the Institute for Laboratory Astrophysics at Boulder, Colorado (Cook, 1966). In this paper and in subsequent publications he made pioneering contributions to the understanding of the conditions under which extreme observed brightness temperatures of 10^9 K and greater could be produced in the four lines of the hydroxyl molecule, the polarisation properties of the lines, and the regions within the sources in which the velocity dispersion along the line of sight would be narrow enough to result in beamed maser emission.

Not long after he took up the Jacksonian Chair, the opportunity to test his model for hydroxyl maser sources arose with Five-Kilometre Telescope observations of the compact region of ionised hydrogen W3(OH) in the nebula W3 by Baldwin, Harris and Ryle of the Radio Astronomy Group (Baldwin *et al.*, 1973; Cook, 1975). The positions of the maser sources determined by long-baseline interferometry lay at the periphery of the region of ionised hydrogen, almost certainly associated with the ionisation fronts about the region of ionised hydrogen and consistent with Cook's preferred model for the source. Cook's involvement with astronomical masers culminated in his book *Celestial Masers* (1977).

He continued to advocate the importance of laboratory measurements of the properties of the millimetre, submillimetre and infrared lines which are observed in astronomical spectra, for example, in his 1978 Presidential Lecture as President of the Royal Astronomical Society (Cook, 1978) and at the 1981 Discussion at the Royal Society on *Molecules in Interstellar Space* (Cook, 1981). His graduate student Michael Foale made a theoretical investigation of acoustic laboratory techniques for determining the basic parameters of transitions of relevance for molecular line astronomy. Cook was particularly proud of the fact that Foale, who had joint US–UK nationality, went on to become the UK's first astronaut, and he attended the launch of STS-45, Foale's first shuttle mission, in March 1992.

17.5.3 Laser spectroscopy

Bob Butcher was a laser physicist who was not only gifted experimentally, but also an outstanding teacher of experimental physics. Appointed as a University Demonstrator in 1975, he quickly built up world-leading laser facilities for a wide range of laser spectroscopic techniques, many of these matching Cook's perceived requirements for laboratory astrophysics. Butcher's emphasis was upon using these highly monochromatic and intense light sources in novel spectroscopic experiments. It is simplest to quote from the description of the work of Butcher and his colleagues from 1982 Cavendish Laboratory *Current Research* booklet:

> Current topics are:
>
> - stable fixed frequency carbon dioxide lasers combined with electric field tuning of molecular energy levels,
> - waveguide mode tunable lasers,
> - optically pumped submillimetre lasers,
> - measurements of very weak transitions by laser–radio frequency double resonance techniques,
> - tunable infrared generation by laser modulation and by laser mixing.
>
> The first and second areas allow a spectral resolution of 10^8 to be obtained using the techniques of saturated absorption. This is good enough to observe, for example, the coupling of nuclear spin to molecular vibration. The submillimetre work is intrinsically significant in that it employs high levels of the rotational manifold of the lasing molecule, and in addition there is a lack of powerful conventional sources in this region. The modulation and mixing experiments allow wide ranging access to large parts of the infra-red spectrum.

These topics were successfully addressed in a series of papers throughout the 1980s by Butcher, his graduate students and postdocs.

17.5.4 Cook's contributions to management and administration

Besides the activities associated with being Head of the Laboratory Astrophysics Group, Cook devoted a great deal of effort to the management of the Laboratory and the University and to national affairs. To mention only a few of his roles, he was Chairman of the National Committee for Geodesy and Geophysics (1972–78), Chairman of the Joint SERC/NERC Committee on Climatology and Chairman of the SERC Astronomy, Space and Radio Board (1984–88); within the University he was Head of the Cavendish Laboratory (1979–84), member of the General Board of the University (1981–84), Chairman of the Council of the School of Physical Sciences and Chair of the Syndics of Cambridge University Press, as well as being a member of the Wass Committee on the governance of the University in 1986. At the same time, he was Master of Selwyn College from 1983 to 1993, a position he greatly enjoyed and in which he carried out his responsibilities with dignity and quiet good humour.

Given this considerable load of responsibilities, it is perhaps not surprising that his personal involvement in the work of the group remained at a modest but effective level. When Metherell left for the USA in the mid 1980s and Cook retired in 1990, Butcher was left as the only staff member in the Laboratory Astrophysics Group. Although by the 1980s and 1990s the Radio Astronomy Group was deeply involved in the construction and science of the UK's 15-metre James Clerk Maxwell Telescope for millimetre and submillimetre astronomy in Hawaii, there was not the intellectual overlap of interests which might have been expected. Butcher carried on making pioneering investigations in laser spectroscopy, but his collaborators were mostly at the CNRS Laboratoire de Physique des Lasers in Paris, where he was a regular visitor.

The Pippard era: condensed matter physics

18.1 Physics and chemistry of solids

The Physics and Chemistry of Solids (PCS) Group continued to expand its areas of interest, as indicated in Table 16.1. While Tabor continued as Head of Group until his retirement, Yoffe played a key role in broadening the experimental activities of the group. As he states in his memoir of Nevill Mott,

> [Mott] was to have a great influence on my research activities and career . . . . I did not collaborate directly with him on any specific topic, since this would not have worked out, given our approaches. But through contact with him, I moved into areas such as semi-conductors, low-dimensional solids, amorphous solids and metal-insulator transitions. (Yoffe, 1998)

Much of the research was at the interface between physics and chemistry. Yoffe's research had originally been concerned with the physics of explosive materials, but now he acted as the leader for the activities which would inspire important initiatives in the Laboratory, many of them the result of Mott's advocacy. The PCS Group had grown to be the largest in the Laboratory and proved to be a magnet for many of the brightest young researchers – Yoffe's graduate students included Richard Friend, Yao Liang and John Wilson, while among John Field's students was the successful entrepreneur Hermann Hauser.[1]

18.1.1 Mott and amorphous materials

Undoubted highlights of the PCS programme were the studies of amorphous thin films of metals and semiconductors and of glasses using a variety of electrical and optical techniques. This work was greatly stimulated by Mott's pioneering activities on the electronic structure of magnetic and disordered systems, especially amorphous semiconductors, which, following his 'retirement', he carried out in the PCS Group. He was awarded the 1977 Nobel Prize in Physics for his work on amorphous materials.

In crystalline silicon, the crystal structure is very regular. Looking down any of the bond directions, the atoms display tetragonal symmetry and this structure is repeated throughout the crystal, with each atom being bonded to four others. The bonds are all of the same length and at the same angle to one another, forming rings of six atoms (Figure 18.1(*a*)). In amorphous materials such as glasses or amorphous silicon, the structure is irregular and there is no such symmetry. Each atom is still bound to four other atoms and the bond lengths are the same as in the crystal, but the bond angles are slightly distorted, the atomic

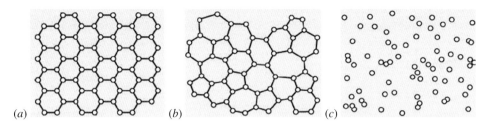

(a) In a crystalline solid, the atoms are arranged in regular patterns that repeat themselves. There is complete periodicity, or order, in the atomic structure. (b) In glass, an example of an amorphous solid, this regular pattern is lacking. There is short-range order about each atom, but no long-range periodicity as is present in the repeating units of crystals. (c) The differences between the crystalline and the non-crystalline structures of crystals, amorphous solids and gases are compared in two dimensions in the illustrations (a), (b) and (c) above.

Fig. 18.1

rings containing five, six or seven members (Figure 18.1(b)). This type of structure is referred to as a continuous random network.

Glasses also have disordered structures and yet they are transparent. Mott wrote in his Nobel Lecture:

> The striking fact about glass is that it is transparent and one does not have to use particularly pure materials to make it so. (Mott, 1992)

The material must therefore possess a large energy gap, as in the case of crystals (Box 18.1). The problem was that, according to the Bloch theory of conduction in metals, the regular structure of perfect crystals meant that the lattice excitations could propagate through the crystal with a dispersion relation which was only real for certain wavenumbers, leading to the formation of a band gap, as illustrated in Box 18.1(a). But if the crystals lacked this regular structure, it was not obvious how the band gap could arise. The problem was solved by Mott, who applied the concept of *Anderson localisation* to quantum mechanical tunnelling between localised states. In particular, the concept of *variable-range hopping* between these states resulted in basically the same type of band structure as in the crystalline state.

The key realisation was that the broad features of the band structure are determined by the immediate environment of the atoms, and this tends to be the same in both amorphous and crystalline solids. The lack of long-range order in the glass smoothes out the details of the band but does not destroy the gap. Disorder and the presence of defects, such as dangling bonds or impurity atoms, can introduce discrete energy levels in the gap (Box 18.1(b)). Although these do not greatly change the visible transparency of the material, they lead to electrical and optical properties which are the basis of technological applications, for example in amorphous silicon for the semiconductor industry.

The whole of silicon technology depends upon the fact that if, for example, a phosphorus atom with five electrons is added, four of the electrons form bonds but the fifth is very loosely bound (Figure 18.2(b)). An example of the type of sputtering chamber used in the production of amorphous thin films is shown in Box 18.2. This pioneering work, described in the editions of the book *Electronic Processes in Non-crystalline Materials* by

| Box 18.1 | The energy gaps in band structure of glasses and amorphous materials |

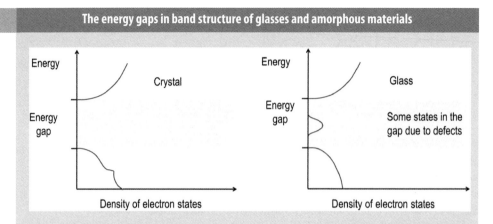

The broad features of the band structure are determined by the immediate environment of the atoms, and this tends to be the same in both amorphous and crystalline solids. The lack of long-range order in the glass smoothes out the details of the band but does not destroy the gap. But the disorder and the presence of defects, such as dangling bonds or impurity atoms, can introduce discrete energy levels in the gap. These, although they do not affect much the visible transparency of the material, lead to electrical and optical properties which can be the basis of technological applications.

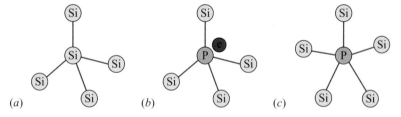

Fig. 18.2 (*a*) Tetrahedral bonding on a pure silicon crystal. (*b*) If a phosphorus atom is introduced, the tetrahedral bond is maintained but there is a loosely bound electron which affects the electrical properties of the crystal. (*c*) If a phosphorus atom is introduced into a glass, all electrons are used in bonding, but the tetrahedral symmetry is then broken.

Mott and Davis (1971, 1979), was to result in the remarkable development of the technologies of amorphous semiconductors for a myriad of applications. Amorphous selenium plates were used in the first Xerox machines. Nowadays, the photoreceptive material is more likely to be amorphous silicon, and it is coated onto the drum of the Xerox copying machine. Other applications of amorphous silicon include solar cells, threshold switches, memory devices and flat-screen displays.

Solar cells convert sunlight into electrical energy. Layers of semiconducting films are separated by insulating films so that the electrons and holes formed when light strikes the material are separated. The discovery by Walter Spear and Peter LeComber (1975) at Dundee University that amorphous silicon could be doped opened up a huge range of new possibilities, including its use in practical and economically viable solar cells. The

Preparation of glasses and amorphous materials

Box 18.2

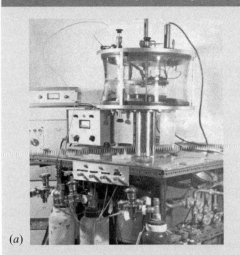

(a)

(b)

(a) A sputtering chamber for the production of amorphous thin films. (b) The chamber in operation.

Most methods of preparation, except for radiation damage, depend upon cooling a liquid or gas so quickly that the atoms do not have time to arrange themselves into a regular crystalline lattice. Quenching from the liquid is the only method of preparing large samples. Vacuum evaporation, sputtering and glow discharge decomposition can only produce thin films.

significance of Spear's work can be appreciated from this quotation from his *Biographical Memoir of Fellows of the Royal Society* (Adams, 2009):

> Perhaps the most important breakthrough in the field was achieved in 1975 when, contrary to prevailing opinion, the Dundee group was able to demonstrate that plasma-deposited a-Si (and a-Ge) could be doped very effectively from the gas phase during deposition. As described in detail in the Philosophical Magazine paper by Spear and LeComber (1975), the addition of small but accurately measured volumes of the gaseous hydrides phosphine (PH_3) or diborane (B_2H_6) to the silene flowing through the reactor could control specimen conductivities over 12 orders of magnitude.

With doping, the energy conversion efficiency is significantly enhanced as the wavelength response can be better matched to the solar spectrum.

The properties of amorphous materials have been important theoretically for the light they can throw on the physics of other types of disordered systems. The physics of disordered systems brought together activities in the Physics and Chemistry of Solids, Theory of Condensed Matter, Low Temperature Physics and Microstructural Physics groups, with the continued encouragement of Mott. He and his colleagues tackled a variety of problems in non-crystalline systems: electrical conduction in heavily doped crystalline semiconductors, photogeneration, structure, transport and optical properties of thin films, glasses and liquids, exciton effects, hopping conduction and more. The strength of the activity was due to active interdisciplinary and inter-group interactions, as well as strong connections with industry, including Xerox, Pilkington, CERL and many others. The resulting research contributions were on a par with the best in the community.

18.2 Pepper and the quantum Hall effect

When Mott retired from the Cavendish Chair in 1971, he became a consultant to the electronics company Plessey. This was an important time in the development of the semiconductor industry as it moved towards the use of silicon in the manufacture of very large-scale integrated circuits with very high transistor packing densities. At the same time, the technology of III-V semiconductors such as gallium arsenide GaAs and indium antimonide InSb was advancing rapidly. Mott was interested in these materials as he sought examples of localisation in a wide range of disordered systems. Michael Pepper was working as a research physicist at Plessey and drew Mott's attention to a number of key features of conduction in silicon–silicon dioxide materials. These included conduction in the two-dimensional electron gas in the very narrow silicon conducting layer in a metal–oxide–semiconductor (MOS) transistor. Mott recognised that this provided a model system for investigating Anderson localisation, but the facilities at Plessey were inadequate to study the temperature dependence of the various effects. It was agreed that Pepper should spend a sabbatical year starting in 1973 working with John Adkins of the Low Temperature Physics Group, where suitable facilities were available (Pepper *et al.*, 1974a,b; Pollitt *et al.*, 1976). The arrangement was extended by various grants, until Pepper was awarded a Warren Research Fellowship of the Royal Society, in association with the Plessey Company in 1978.

This was a very productive period for Pepper and his colleagues. As he wrote in his contribution to *Nevill Mott: Reminiscences and Appreciations* (Pepper, 1998),

> During the period 1973 to the early 1980s the work using the two-dimensional electron gas in the inversion layer of the silicon device allowed us to explore variable-range hopping, the nature of the wave function decay and numerous consequences of disorder, a programme which was extended to gallium arsenide by exploiting impurity band conduction. We also investigated localisation in the presence of a quantising magnetic field, which led to the collaborative project that found the quantum Hall effect. It was a very exciting time as the entire field of semiconductor structures for the pursuit of basic physics was developing rapidly.

The collaboration to which Pepper refers was with Klaus von Klitzing, who was working at the high magnetic field laboratory in Grenoble. The silicon-based samples used in the experiments were developed by Pepper in Cambridge and by Gerhard Dorda, who worked as a physicist in the MOS Silicon Semiconductor Division at the Siemens Research Laboratories in Munich. The *integral quantum Hall effect* was discovered in 1980 by Klitzing, Dorda and Pepper (1980) – at temperatures of about 4 K, the Hall resistance of a two-dimensional electron system was found to display plateaus with values of exactly $R = (h/e^2\nu)$, where $\nu = 1, 2, 3, \ldots$; h is Planck's constant and e is the electron charge (Box 18.3). In the ranges

Box 18.3

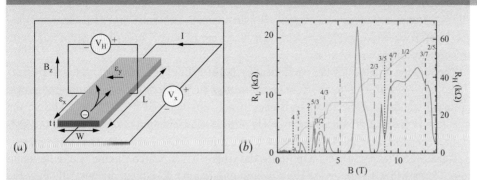

(*a*) Illustrating the circuit arrangement for measurements of the Hall effect. (*b*) The variation of the Hall resistance as a function of the magnetic flux density B. The Hall resistance is shown by the blue line, flat steps corresponding to the integral and fractional Hall effect, the values of ν being indicated above each plateau. The red line shows the variation of the direct current along the specimen as a function of magnetic flux density (courtesy of the Cavendish Semiconductor Physics Group).

The classical Hall effect is the potential difference V_H and current produced when a strong magnetic field acts perpendicularly to the current in, for example, the current sheet shown in (*a*). The $v \times B$ force acting on the electrons (or holes) in the material causes a potential difference normal to the directions of both v and B. In this case, the Hall resistance is proportional to the magnetic flux density.

The quantum Hall effect was discovered experimentally in 1983 in a set of experiments by von Klitzing, Dorda and Pepper (Klitzing *et al.*, 1980). They discovered that the resistance is quantised in the two-dimensional semiconductor structures they investigated. The effect is associated with the quantisation of the magnetic energy levels and can be understood in terms of the Landau levels. Writing the effect in terms of the conductance σ of the material, $\sigma = I/V_H = \nu(e^2/h)$, where $\nu = 1, 2, 3, \ldots$. The quantum theory of Landau levels can account for this integer quantum Hall effect.

In the **fractional quantum Hall effect**, the quantised energy levels are given by fractions such as $\nu = 1/3, 2/5, 3/7, 2/3, 3/5, 1/5, 2/9, 3/13, 5/2, 12/5, \ldots$, rather than by integers. Unlike the normal quantum Hall effect, the origin of these fractional values is not understood, the likely explanation being a manifestation of the effects of strongly correlated electron physics within the semiconductor materials.

of magnetic flux density where the resistance plateaus occur, the resistance measured along the direction of the current flow drops to negligible values. The quantum theory of Landau levels can account for the normal quantum Hall effect. Klitzing was awarded the 1985 Nobel Prize in Physics for this discovery; Pepper played a full role in the experiments and their interpretation.

Two years later, Horst Störmer, Daniel Tsui and Arthur Gossard (Tsui *et al.*, 1982) discovered the *fractional quantum Hall effect*: when cooled below about 2 K, the Hall resistance of the two-dimensional electron system also shows plateaus but now with fractional values of ν, for example, $\nu = 1/3, 2/5, 3/7, 2/3, 3/5, 1/5, 2/9, 3/13, 5/2, 12/5, \ldots$, rather than integral values. Robert Laughlin, Störmer and Tsui were awarded the 1998 Nobel Prize in Physics 'for their discovery of a new form of quantum fluid with fractionally charged excitations'.

The measured values of the coefficients ν in the Hall conductance are integral or fractional multiples of e^2/h to an accuracy of nearly one part in a billion, resulting in a new practical standard of electrical resistance which was adopted in 1990. In addition, the quantum Hall effect provides an extremely precise determination of the fine structure constant $\alpha = (1/4\pi\epsilon_0)(e^2/\hbar c)$.

Pepper left Plessey in 1982 and joined the GEC Hirst Research Centre, where he set up joint Cavendish–GEC projects. His research programme developed dramatically over the coming years, culminating in the formation of the Semiconductor Physics Group in the Laboratory in 1985 and his promotion to the post of Professor of Physics in 1987.

18.3 The HREM, STEM and metal physics

In 1970, Ellis Cosslett had begun discussions with the SRC about the next electron microscope project in Cambridge and came up with the concept of the high-resolution electron microscope (HREM) with which to achieve atomic resolution. It was to be a joint project between the Laboratory and the Department of Engineering. There was no space in Engineering, and the Laboratory was about to move to West Cambridge, but somehow space was found in the old Cavendish Laboratory. Despite many technical and administrative difficulties, Cosslett put together a collaboration involving AEI Instrumentation Division, Haefely Switzerland, the Department of Trade and Industry and Cambridge University to secure funding for the project. The HREM project was approved in 1972 and, after Cosslett retired in 1975, it was managed by William Nixon of the Engineering Department and brought to fruition with the appointment of the gifted Australian electron microscopist David Smith.

To attain atomic resolution the accelerating voltage of the electron microscope had to be about 600 kV or more but, in order to reach the theoretical resolving power, very stringent requirements were placed on the specification of the microscope (Figure 18.3). The electron beam had to be highly monochromatic, the high voltage being stable to one part per million, the electron lenses had to be free of machining defects to better than 0.5 μm and the specimen stage had to provide accurate positioning of the specimen to within 0.5 milliradians, in the case of crystals. Furthermore, all external sources of magnetic, electrical, thermal and mechanical disturbance were kept to a minimum. In order to isolate the microscope, weighing about seven tonnes, from the influence of site vibrations, it was suspended from the walls of the microscope room by a special three-point suspension system, resting on cylinders of compressed air (Figure 18.4(a)).

Construction of the 600 kV HREM started in early 1973 and, after a prolonged development phase when many technical issues had to be resolved, the first images were

The HREM, showing the vertical column of magnetic electron lenses, control desks and part of the mechanical suspension system.

Fig. 18.3

obtained in September 1977. Atomic resolution of close to 0.2 nm was reached by the time of its official opening in 1979. An example of a micrograph of a small gold particle is shown in Figure 18.4(*b*). The HREM was devoted to the study of many different materials: metals, ceramics, minerals, catalysts, organic crystals and biological proteins. It operated successfully for 13 years following its inauguration and for several years was the leading high-resolution electron microscope worldwide.

The HREM was a brute-force method of attaining atomic resolution but it had the disadvantage that the electrons were of such high energy that the beam caused radiation damage of the specimens under study. An alternative route was adopted by Howie, Brown and their colleagues, who realised the great potential of the scanning transmission electron microscope (STEM), the concept for which had been developed by Albert Crewe at the Argonne Laboratory and the University of Chicago. Rather than wide-field imaging, the STEM required a high-intensity beam of electrons with energies typically in the range 100 to 300 keV. To achieve the necessary high-intensity electron beam, Crewe and his colleagues developed the field emission gun, which produced an intense beam of electrons

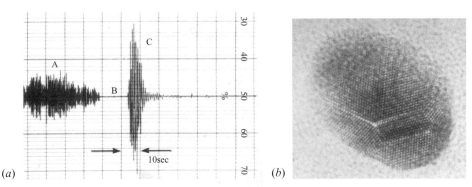

Fig. 18.4 (*a*) Trace of the attenuation of site vibrations. The trace demonstrates the attenuation obtained after lifting the microscope free of the mechanical wall support (A) onto its cushion of air (B). The damping effect after a sudden short impulse of energy is shown at (C). (*b*) Micrograph of a small gold particle, recorded at the optimum defocus, with an accelerating voltage of 575 kV, showing individual rows of atoms, viewed end-on (Cosslett *et al.*, 1979).

from a tiny tip. In addition, they pioneered the use of *electron energy loss spectroscopy* (EELS) as a means of obtaining the energy spectrum of the electrons ejected by the electron beam. An advantage of this technique was that the scattered electron energy spectrum included the signatures of the chemical nature of the atoms in the specimen. By the early 1970s, STEMs were being sold commercially by the Hitachi and Vacuum Generator (VG) companies. A VG HB5 STEM was acquired by the Cavendish Laboratory in 1974 and was at the centre of technical development of many facets of electron microscopy in the Laboratory for the next 25 years (Figure 18.5).

Crewe also pioneered the *high-angle dark-field imaging method*. The annular dark-field detector (ADF) is a doughnut-shaped scintillator which collects electrons scattered incoherently well away from the axis of the incident beam. With it, Crewe was able for the first time to image single heavy atoms. When Crewe's methods were applied to crystalline materials, especially to industrial catalysts, the annular dark-field detector also collected the coherent (Bragg) beams which interfered with one another, producing fringes which obscured the image and required deconvolution techniques to interpret the data. Howie (1979) recognised that if the doughnut-shaped ADF were made with a central hole large enough to exclude the coherent Bragg beams, a direct image could be obtained from the incoherently scattered electrons. This technique resulted in the development of the *high-angle annular dark-field detector* (HAADF) which is now universally used in the STEM.

By scanning the beam after the specimen, the STEM enables diffraction patterns to be recorded from every point at which the probe is focussed, resulting in *convergent-beam electron diffraction* (CBED) patterns, also known as *microdiffraction* or *nanodiffraction*. The patterns record information on spacings much smaller than the resolution limit of the microscope. The technique, known as *ptychography*, was developed by John Rodenburg for STEM measurements and is an important technique in X-ray and optical microscopy (Rodenburg, 1989). Using the convergent-beam diffraction information from the nanometre probe, it proved possible to distinguish the short-range order in localised domains in alloys.

(a)

(b)

(a) The VG HB5 scanning transmission electron microscope (STEM) as installed in 1974. The instrument was to undergo many upgrades over the succeeding 25 years. (b) The VG HB5 in the early 1980s. Many enhancements and additional detectors were added to the STEM, including an X-ray detector.

Fig. 18.5

18.4 Low-temperature physics

In 1972 the Mond Laboratory Group was the first to move to the new Cavendish Laboratory at West Cambridge, and was renamed the Low Temperature Physics (LTP) Group.

John Ashmead had designed and installed a completely new helium recovery system, and a commercial Collins liquefier was purchased with the capacity to supply not only the Laboratory but also low-temperature research in the Materials Science and Engineering departments. All high-field magnets were now superconducting and two screened rooms were constructed for work involving Josephson devices.

18.4.1 Non-metallic conduction

Between 1971 and 1974 John Adkins and his group, using the film deposition equipment that he had built up to study superconductive tunnelling, published a series of papers on problems in non-metallic conduction in thin films, including amorphous carbon and heavily doped germanium, and also began to consider theoretical models of hopping conduction between local donor sites.

As recounted in Section 18.2, in 1973 Mott invited Michael Pepper to visit the Laboratory and use Adkins' facilities to examine threshold conduction in field-effect transistors. This led to a series of projects in which Adkins collaborated with Pepper, and also with Mott and Anderson. The first of these was concerned with hopping conduction in silicon inversion layers (Pepper *et al.*, 1974a,b). Mott had published an ingeniously simple analysis of variable-range hopping through a random array of donor centres, which predicted a temperature dependence following his celebrated $T^{1/4}$ law. Intrinsic to this model was the idea of a *mobility edge*, an energy above which carriers could propagate as free particles, while carriers below this energy had to tunnel between discrete localised states. Most of the group's early work in this area concentrated on the temperature dependence of conductivity.

Mott's mobility-edge model failed completely, however, to describe the observed Hall effect in field-effect inversion layers. According to his model, the number of carriers activated should be variable and their mobility constant, but experiments in the group and other laboratories from 1978 showed the exact reverse to be the case. Adkins was the first to demonstrate that this could be explained by the dominance of carrier–carrier interaction energies.

The question of non-metallic conduction often depends on understanding the point at which the electron states change from being extended in space, as in the band theory of metals, to localised. Adkins' group carried out experiments related to Anderson localisation (see Box 15.14) in silicon inversion layers which is created by an excessive density of scattering centres. The group published an important series of papers on the effect with Pepper and Mott (Mott *et al.*, 1975; Pepper *et al.*, 1975). It had become apparent that tunnel junctions could be remarkably fast phonon detectors, and, with his student Andrew Long, Adkins put this to good use in a study of ballistic phonon propagation along a sapphire rod (Long and Adkins, 1973).

18.4.2 Amorphous solids

In 1973 Andrew (Andy) Phillips returned to Cambridge and established a group working on the low-temperature properties of amorphous solids, building on his model of local atomic states that he had developed in the USA. This model had its origins in work by Pippard

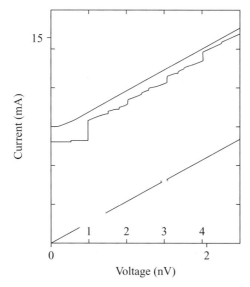

The $I-V$ characteristic of a wide SNS junction as computed from (18.1). The top curve is for the unexcited junction: note the supercurrent step at zero voltage. The curve below it is for a junction excited by radio-frequency current at frequency ω: note the Shapiro steps at voltages $V_n = n\hbar\omega/2e$, and smaller submultiple steps. The lowest plot is for purely resistive conduction with no supercurrent. The calculations fitted Clarke's observations well (Waldram *et al.*, 1970).

Fig. 18.6

in the late 1960s on the dielectric loss of plastics, in response to industrial interest in the possibility of high-power superconducting transmission lines. Part of the observed loss could be identified as due to the quantum tunnelling of atoms between sites, and this model, when extended to take account of random variations in the local environment in glasses and other disordered solids, successfully explained the universally observed term proportional to temperature in the heat capacity, and also the unexpected quadratic dependence of the thermal conductivity on temperature (Phillips, 1972, 1987). The Phillips model predicted a wide range of phenomena which were observed in amorphous solids, including anomalous thermal expansion, saturable ultrasonic attenuation and dielectric loss.

18.4.3 Further developments in superconductivity

Josephson's theory of tunnel junctions was originally developed for small junctions and assumed a direct-current voltage source, but it was quickly realised that this could be generalised in many ways. An intriguing example was provided by John Clarke's superconducting normal superconducting (SNS) junctions, which were often sufficiently large for quantum interference to occur within a single junction. When radio-frequency currents were applied to them they developed a current–voltage (I–V) characteristic of the type shown in Figure 18.6 in which, in addition to the supercurrent step at zero voltage, Shapiro steps appear at the multiple voltages given by the quantum relation $2eV_n = n\hbar\omega$, with submultiple steps also visible. It turns out that the Josephson phase ϕ obeys the equation

$$\frac{\hbar C_0}{2e}\frac{\mathrm{d}^2\phi}{\mathrm{d}t^2} + \frac{\hbar}{2eR_0}\frac{\mathrm{d}\phi}{\mathrm{d}t} + J_0\sin\phi = \frac{\hbar}{2e\mu_0 d}\frac{\mathrm{d}^2\phi}{\mathrm{d}x^2}, \tag{18.1}$$

where C_0, R_0 and J_0 are the capacitance, resistance and Josephson current per unit area, d is the magnetic thickness and x is distance along the junction; the right-hand side describes the effects of quantum interference within it. Time-dependent solutions of this equation were computed, which fitted Clarke's results perfectly (Waldram *et al.*, 1970).

From Clarke's work on SNS junctions it became apparent that when the normal thickness d became too great to allow the supercurrent to cross the barrier, interesting effects appeared at the point where the supercurrent (S) changed to the normal current (N). A strange consequence of BCS theory is that, because of the energy gap, at low temperatures an electron approaching S from N cannot penetrate S, and is reflected not as an electron but as a hole below the Fermi surface, the phenomenon of *Andreev reflection*. As Pippard, John Shepherd and David Tindall demonstrated, this has the effect that energy is reflected, leading to a large thermal boundary resistance – but charge is *not* reflected, since a pair of electrons enters S as a Cooper pair, and there is no corresponding electrical boundary resistance. The theory of these effects was developed by Pippard and improved by Waldram (Pippard *et al.*, 1971; Waldram, 1975).

At higher temperatures electrons may have enough energy to surmount the BCS gap, in which case they can enter S. This leads to the strange effect of *charge imbalance*, in which the superconductor contains more electrons above the Fermi surface than holes below (Waldram and Battersby, 1992).

The Josephson effect was now seen not only in tunnel junctions and SNS junctions but also in other varieties of *weak link* between superconductors, including point contacts and microbridges, for which there was no exact theory. It became of interest to see how precise the Josephson relation (18.1) was in such cases, and Waldram devised a scheme for measuring the Josephson relation directly using quantum interference in an evaporated closed circuit; in 1975 he and John Lumley published the first such measurement (Waldram and Lumley, 1975). Waldram's student Robert Brady also realised that it was possible to use quantum interference to measure absolute rotation with respect to an inertial frame. When he tried to patent this idea, the Ministry of Defence clamped a secrecy order on it, and his work had to be completed at the Admiralty Compass Laboratory, with some success. Eventually, the technique was judged insufficiently precise for the navigation of ballistic missiles and was published in 1980.

18.4.4 Itinerant electrons

When Shoenberg retired in 1978, Gilbert (Gil) Lonzarich, who had studied for his PhD with Shoenberg's former student Andrew Gold at the University of British Columbia, was appointed as a University Demonstrator to run the group working on quantum oscillations. Lonzarich's early work in Cambridge exploited and advanced the de Haas–van Alphen techniques that had been pioneered by Shoenberg. He set up a new measurement system with the sensitivity needed to detect quantum oscillations in complex materials, beyond the elemental metals being investigated in Cambridge at that time. He and his students also set up a novel high-purity crystal growth laboratory, employing ultra-high-vacuum methods, in which the samples were melted by radio-frequency induction heating on water-cooled copper hearths. This led to breakthroughs in materials science and the growth of crystals with record low levels of impurities

He also took the experiments in new directions. The magnetism of iron, cobalt and nickel is very strong – it appears at high temperature, and the magnetic effects become large and very non-linear at low temperatures. The resulting non-linearities are difficult to understand. Soon after arriving in Cambridge, Lonzarich had the important insight, which was to lead to many important advances over subsequent decades, that the much weaker magnetism in materials such as Ni_3Al, Ni_3Ga, $ZrZn_2$, $TiBe_2$ and MnSi – metals on the boundary of magnetic order at low temperatures – would be easier to understand. Also, in extreme cases, such materials might exhibit unconventional metallic behaviour and new forms of superconductivity mediated by magnetic interactions rather than by phonons as in the traditional BCS theory.

Working with his students Thorstein Sigfusson, Nicholas (Nick) Bernhoeft and Louis Taillefer, Lonzarich initially studied the archetypal weak itinerant ferromagnets Ni_3Al and MnSi. The results led him to develop a quantitative spin-fluctuation model of the coupling of the conduction electrons to magnetic fluctuations, which yielded, in terms of empirically determined microscopic parameters only, an excellent fit to the low-temperature magnetic equation of state. In particular the model explained the mechanism for the collapse of magnetic order at the Curie temperature in these materials (Lonzarich, 1984, 1986; Taillefer and Lonzarich, 1985).

Again with his students, in particular with Taillefer, he made the first direct observation of *heavy quasiparticles*, or *heavy fermions*, through quantum oscillatory effects in metals on the verge of magnetic order at low temperature in which the itinerant electrons are 'dressed' by a cloud of magnetic fluctuations that they drag around with them. This cloud gives them unusual properties: for example, they behave like electrons having a mass much larger than that of a bare electron. These measurements were carried out in a number of compounds, Ni_3Al, Ni_3Ga, Pr, MnSi and, in particular, UPt_3 in which they observed dressed electrons with masses up to a hundred times larger than the bare electron mass (Taillefer and Lonzarich, 1988). Their measurements made novel use of helium-3 and dilution refrigerators that allowed them to cool samples to the mK temperature range, as well as the latest high-field superconducting magnet from Oxford Instruments, which had a maximum field of 14 T. This work firmly established the existence of heavy fermions and the validity of the Landau theory of a Fermi liquid even in extreme cases where it was widely expected to break down.

In addition, with his graduate students Bernhoeft, Stephen Hayden and Sarah Law, Lonzarich used the neutron scattering facilities at Harwell and at the Institut Laue–Langevin in Grenoble in a novel way to observe slow, overdamped fluctuations in the local magnetisation of Ni_3Ga, UPt_3 and other materials (Bernhoeft *et al.*, 1983). These are the elusive paramagnons that played a central role in the phenomenological spin-fluctuation theory of itinerant-electron magnetism and – of particular importance to the group in later years – of magnetically mediated superconductivity.

The above experimental and theoretical studies were described in a unified way in Lonzarich's papers of this period (Lonzarich, 1984, 1987, 1988).

18.5 Theory of condensed matter

With Anderson's departure in 1975, Heine was promoted to the Professorship of Physics 1966, very much continuing his role of overseeing the many different directions in which theoretical condensed matter research was developing. There were areas in which collaborations between the solid state theorists and the experimenters in the Mott Building were established, but these were fewer than might have been anticipated. As Heine expresses it in his informal *TCM Group History*,[2]

> However it has often been a matter of regret that the TCM group up to the 1980s was mostly more closely allied to experiments being done elsewhere around the world than those at the Cavendish. (Heine, 2015)

There were a number of reasons for this. Heine gives the example of the development of the techniques of low-energy electron diffraction (LEED) which enabled the atomic structure of surfaces to be determined. Tabor in the PCS Group was a world authority on surface physics with a strong experimental group and with the capability of creating very clean surfaces in high vacuum, but to carry out the experiments would have required the development of new, complicated instruments and techniques. It turned out that John Pendry, a graduate student in the TCM Group, had just completed a brilliant analysis of the theory of the complexities of the LEED process, in which he carried out the exact calculation for forward scattering and then added in the back scattering by successive approximations as in a perturbation calculation (Pendry, 1969a,b,c,d). The experiment was subsequently carried out in collaboration with a Swedish group (Andersson and Pendry, 1972).

Another factor was that a great deal of Anderson's innovative theoretical activity was carried out in collaboration with his experimental colleagues at the Bell Laboratories in the USA. Furthermore, Heine was a regular visitor to the USA, where he was in demand as an expert in theoretical condensed matter physics.

Nonetheless, it was certainly the case that there were important collaborations between theorists such as Ernst Sondheimer, Leo Falicov, Phil Anderson, Jim Phillips, John Ziman and Heine himself with members of the Low Temperature Physics Group during the 1950s and 1960s, particularly after the breakthrough of the discovery of the Bardeen, Cooper and Schreiffer (BCS) theory of superconductivity (Bardeen *et al.*, 1957).

The theoretical activity in the Laboratory received a major boost in 1972 with the appointment of Sam Edwards to the John Humphrey Plummer Professorship of Physics, which he chose to hold in the Laboratory. He had carried out his PhD studies under Julian Schwinger at Harvard University on the structure of the electron, which involved using advanced techniques in quantum field theory. Seeking challenges outside particle physics, he realised that he could apply these techniques to complex problems in condensed matter physics and the theory of superconductivity. His seminal paper, entitled *A new method for the evaluation of electric conductivity in metals*, opened up a vast field of research in the quantum mechanics of electrons in random potentials (Edwards, 1958). As David Khmelnitskii has written,

> Ryogo Kubo and Rudolf Peierls were the first to ask whether conductivity was a subject of quantum mechanics. Sam Edwards, then at Birmingham, was introduced to the problem by Peierls and made the decisive step by considering the transport of electrons elastically scattered by a random potential. This step had dual importance: first of all, Edwards came up with a field theory, which corresponded to averaging over random potentials; secondly, he developed a very effective diagrammatic technique, which allowed him to calculate the Drude conductivity and gave to those who came after him an efficient tool for further research. (Khmelnitskii, 2004)

In the years prior to his arrival in Cambridge, Edwards published innovative papers in many diverse areas: *The statistical dynamics of homogeneous turbulence* (1964), *The statistical mechanics of polymers with excluded volume* (1965), *The theory of polymer solutions at intermediate concentration* (1966) and *Statistical mechanics with topological constraints, II* (1968).

Edwards came to Cambridge from Manchester University, where he had held the Professorship of Theoretical Physics since 1963. He brought quite new dimensions to both the theoretical and experimental work of the Laboratory. As stated in the book *Stealing the Gold: A Celebration of the Pioneering Physics of Sam Edwards* (Goldbart *et al.*, 2004),

> Over the course of nearly half a century, Sam Edwards has led the field of condensed matter physics into new directions, ranging from the electronic and statistical properties of disordered materials to the mechanical properties of granular materials. Along the way, he has provided seminal contributions to fluid mechanics, polymer science, surface science and statistical mechanics.

Among his major contributions was the expansion of the range of theoretical and experimental work in polymer science and statistical physics. In order to encompass this expansion of the scope of the Solid State Physics Group, it was renamed the Theory of Condensed Matter (TCM) Group, thus incorporating polymer and complex fluids into its portfolio of interests.

Almost immediately, however, on 1 October 1973, Edwards became Chairman of the Science Research Council, a position he was to hold for four years. Nonetheless, he continued to supervise his graduate students throughout this period, which saw some of his most original contributions to condensed matter theory. Of particular importance was the *theory of spin glasses* which he developed with Anderson (Edwards and Anderson, 1975). He employed the technique known as the *replica trick*, which he had already used in his study of polymers, to work out the ground state and properties of spin glasses (Box 18.4). In disordered systems, the ground state can be difficult to determine. For example, disorder in a spin system such as a spin glass, with different types of magnetic links between spin sites, results in many configurations having the same energy. The replica trick enables the ground state of the system to be found. The free energy $\ln Z[J_{ij}]$ is averaged over J_{ij}, which

Box 18.4   **Anderson and Edwards on spin glasses**

(a)

(b)

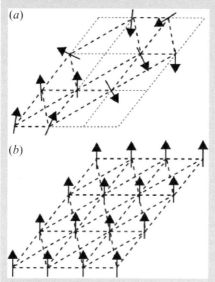

Schematic representation of the random spin structure of a spin glass (a) and the ordered spins of a ferromagnet (b).

Following is the abstract of the paper by Edwards and Anderson (1975).

A new theory of the class of dilute magnetic alloys, called the spin glasses, is proposed which offers a simple explanation of the cusp found experimentally in the [magnetic] susceptibility [as a function of temperature]. The argument is that because the interaction between the spins dissolved in the matrix oscillates in sign according to distance, there will be no mean ferro- or antiferromagnetism, but there will be a ground state with the spins aligned in definite directions, even if these directions appear to be at random. [As the temperature is lowered], at the critical temperature the existence of these preferred directions affects the orientation of the spins, leading to a cusp in the susceptibility. This cusp is smoothed by an external field. Although the behaviour at low T needs a quantum mechanical treatment, it is interesting to complete the classical calculations down to $T = 0$. Classically the susceptibility tends to a constant value at $T = 0$, and the specific heat to a constant value.

is often of Gaussian form. The trick involves using the formula

$$\lim_{n \to 0} \frac{Z^n - 1}{n} = \ln Z \qquad (18.2)$$

to replace an average over $\ln Z[J_{ij}]$ by one over a power series in Z, which results in a much simpler series of exponential integrals. Edwards and Anderson were able to define an order parameter, and hence could develop a mean field theory of the spin glass transition; in a second paper, they quantised the spin glass (Edwards and Anderson, 1976). As Heine expresses it, 'A whole industry on spin glasses and then neural networks developed'.

Edwards' second major contribution during this period was the theory of the dynamics of polymers, the process known as *reptation*, with Masao Doi (Box 18.5). Entangled long-chain molecules wiggle as if they were confined to a tube, the motion consisting of extending out one end of the tube and retracting at the other. The dynamics of reptation described by their theory proved to be very successful and now underpins the huge, industrially important field of rheology. The summation of their pioneering work was published in their influential book *The Theory of Polymer Dynamics* (Doi and Edwards, 1986).

What is remarkable is that Edwards made these fundamental contributions to theoretical physics while carrying out his responsibilities as Chairman of the Science Research Council in London. He would supervise research students on the train and work out multidimensional integrals during meetings, filling up successive 'little red books'.

Box 18.5

Doi and Edwards on reptation

Reptation is the thermal motion of very long, linear, entangled macromolecules in polymer melts or concentrated polymer solutions. The term is derived from the word *reptile*, suggesting the movement of entangled polymer chains, analogous to snakes slithering passed one another. The concept was introduced by Pierre-Gilles de Gennes in 1971 to explain the dependence of the mobility of a macromolecule on its length (de Gennes, 1971). Edwards and Doi later refined reptation theory in a number of articles in 1978–79 and subsequently a book on the subject in 1986 (Doi and Edwards, 1978a,b,c, 1979, 1986).

(Diagrams and text courtesy of Rae Anderson, University of San Diego)

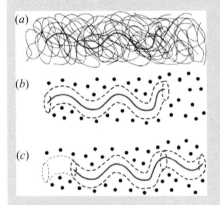

Sketches illustrating the process of reptation for entangled polymers. (*a*) An illustration of an entangled polymer solution, the polymer of interest being shown in red. (*b*) The 'tube' formed by the surrounding polymers, which restricts motion in directions transverse to the outline of the polymer. (*c*) As the polymer reptates out of its original 'tube', a new tube is formed as the old one disappears.

The ways in which Edwards brought new experimental activities to the Laboratory are described in more detail in Chapter 19, but already during the Pippard era, he brought new theoretical initiatives to the TCM Group which were to prove to be major growth areas in the future. In particular, he realised the need to make full use of his industrial connections to support theoretical physics activities. The Industrial CASE Award scheme[3] provided opportunities for some outstanding graduate students to study with Edwards. In general terms, these new initiatives were in soft condensed matter physics, an area which Edwards had made his own. Three future staff members capitalised upon Edwards' innovations in the statistical mechanics of complex polymers.

- Robin Ball was awarded a CASE studentship jointly with Unilever in 1977, and in the following years developed the *replica theory of polymer networks*. He was appointed a University Demonstrator in 1983.
- Edwards was the PhD supervisor of Mark Warner, who carried out pioneering work on the theory of liquid crystal elastomers, materials which, as Heine writes, 'can change their lengths by factors of 4 or 5 on slight heating or exposure to light, strange solids that change their shape without energy cost, can tell the difference between right and left, lase when lightly pumped and change emission colour on stretching, and so on'. After SRC and IBM research fellowships, as well as posts at the Rutherford–Appleton Laboratory and the Institut Laue–Langevin at Grenoble, Warner became a University lecturer in the Laboratory in 1986, initially supported by Unilever, thanks to the efforts of Edwards.

- Michael Cates held a CASE studentship in collaboration with the Esso Petroleum Company, in the area of the viscoelastic properties of polymer mixtures, which led to his interest in the statistical mechanics of complex polymers. Following the award of a Royal Society University Research Fellowship, he became a University Assistant Lecturer in 1989.

These initiatives foreshadowed the future development of experimental and theoretical soft condensed matter physics, which was eventually to extend into the physics of biology and medicine.

Meanwhile, Heine's pioneering use of computers to tackle frontier problems in condensed matter physics continued as more and more powerful computing facilities became available. Examples include the collaboration between Roger Haydock, who became a University Assistant Lecturer in 1977, and Chris Nex, who was the first Computer Officer dedicated to supporting computational activities in the TCM Group, both in the computational mathematics aspects of the activities and in overseeing hardware and software development. Haydock and Nex used Nex's recursion method to calculate the electronic structure and lattice vibrations in amorphous materials and also in iron with the local magnetic moments pointing randomly or non-aligned (Haydock and Nex, 1984). At the time, no other group was capable of carrying out such computations. This was an example of the use of computational physics in theoretical studies which were beyond analytic treatment.

A further enhancement of computational physics took place with the appointment of Richard Needs as an Assistant Lecturer in 1983, under the 'New Blood' scheme introduced by the government to regenerate the ageing population of university departments.[4] Needs had been working with Edwards in polymer simulations and changed the direction of his research to electronic structure calculations, which was to prove to be a growth area during subsequent decades.

18.6 The Pippard era concluded

Pippard's main achievement as Cavendish Professor was the move of the Laboratory to West Cambridge from its unsatisfactory accommodation at the New Museums site. This was a bold move in that it separated the Laboratory geographically from the academic centre of Cambridge, involving a ten-minute bicycle ride to the large lecture halls in the centre where the first- and second-year undergraduate lectures were delivered – all the practical laboratories were incorporated in the new Laboratory. It had been expected that other physical science and engineering departments would also move to the site, but this was not to take place to any significant degree until about 25 years later. Despite this separation from the historic centre, the huge advantages of having a consolidated Physics Department on a single site more than compensated for any minor inconveniences.

Pippard also believed strongly in the importance of good teaching, and this became a significant factor in the appointment of new staff. This was a further example of his long-sightedness as the government began to take a closer interest in the quality of university teaching, and it was to lead to the programmes of Teaching Quality Assessment in the late 1990s.

Pippard changed the rule that the Cavendish Professor was automatically Head of Department when he decided to split the posts. In October 1979, he stood down as Head of Department and Alan Cook took over that role for the next five years. When the University of Cambridge offered a very generous early retirement package in 1982, Pippard took that opportunity. His successor, Sam Edwards, was appointed Cavendish Professor in 1984 and assumed the role of Head of Department after Cook's five-year period of office.

Pippard and Cook consolidated the activities in the department under conditions of increasing financial stringency and competitiveness of physics research. Renewed expansion and regeneration of the physics programme was to take place under Edwards' stewardship.

PART IX

1984 to 1995

Samuel Frederick Edwards (1928–2015)
Oil painting by Paul Gopal-Chowdhury

The interregnum between Pippard's resignation from the Cavendish Chair in 1982 and Sam Edwards' assumption of the position in 1984 was covered by Alan Cook as Head of Department from 1979 to 1984. On his appointment, Edwards took on the role of Head of Department for the next five years. Unlike Pippard, Edwards had been deeply involved in national and international science politics for many years. He had served as a member of the Council of the European Physical Society from 1969 to 1971. He had been a member of various committees of the Science Research Council since 1968 and of the Council's Science Board since 1970. In 1971 he was appointed a member of the University Grants Committee and was then Chairman of the Science Research Council from 1973 to 1977. This was followed by his Chairmanship of the Defence Scientific Council from 1977 to 1980, and he was Chief Scientific Adviser to the Department of Energy from 1983 to 1988. He had also served as Vice-President of the Royal Society and of the Institute of Physics, and had been President of the Institute of Mathematics. Thus, he had a very wide range of contacts in government and industry and used that experience to begin a major expansion of the Laboratory's activities, to remarkable effect. He was famous for hosting dinners for senior figures in industry and government in his college, Gonville and Caius College, where he had accumulated a superb, and large, wine collection. When I took over as Head of the Laboratory in 1997, his only advice to me was: 'Have dinners!'

19.1 Expansion of the Laboratory's programme

During the Pippard era, the numbers of staff members remained roughly constant (see Figure 16.5(a)). New initiatives were needed and this was brought about largely through the vision of Edwards during his five-year term as Head of Department. The funding pressures on the University with the gradual erosion of support for research and the universities meant it was a major challenge to increase significantly the numbers of tenured academic posts, despite the 'New Blood' scheme initiated by the government to regenerate research and teaching activity in the universities.

Edwards fully appreciated the gravity of the situation from the inside through his various roles on the national scene and his many contacts. He realised that the government and the Research Councils could not be relied upon to provide the resources for new activities. Rather, the way to do new things was to become much more closely associated with the needs of industry and to enhance the support they could provide to the research programme. This was also attractive to the government, who were keen to promote

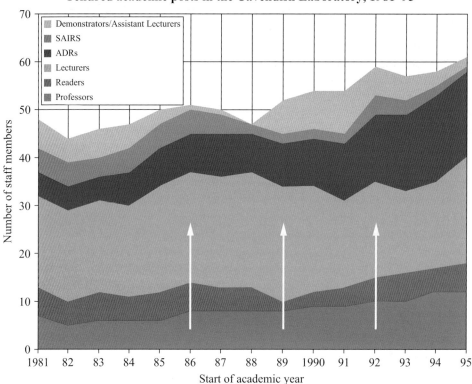

Tenured academic posts in the Cavendish Laboratory, 1981–95

Fig. 19.1 The distribution of academic staff members among academic grades. The data are taken from the Annual List of University Officers, published in the first week of each academic year in the University *Reporter*. The white arrows indicate the academic years in which Research Assessment Exercises took place.

research which would be of benefit to industry. Often matching funding from the Research Councils and government could be obtained, as well as studentships through a variety of incentive schemes. During Edwards' five-year period as Head of Department new groups were created: Microelectronics led by Haroon Ahmed (1983), Semiconductor Physics by Michael Pepper (1984), Optoelectronics by Richard Friend (1987), Polymers and Colloids by Athene Donald (1987) and the Interdisciplinary Centre for High Temperature Superconductivity, a collaborative effort between a number of departments (1987), and all of these new activities had strong industrial connections. These initiatives grew largely out of the activities of the Physics and Chemistry of Solids Group, much of the stimulus being provided by Yoffe in encouraging many of the best graduate students to exploit the opportunities for innovation in these new disciplines. The result was a significant decrease in staff numbers in the PCS Group.

Figure 16.5(*a*) showed the dramatic increase in total staff numbers from 457 in 1981 to 630 in 1990. How was this achieved against a background of decreasing resources from the government? The changes in the numbers of established academic staff posts are shown in Figure 19.1, using data from the annual information published in the University *Reporter*.

It can be seen that the numbers of professors, readers and lecturers remained rather static through this period, the main growth in staff numbers occurring in the employment of Assistant Directors of Research (ADRs). The long-standing ADR posts in radio astronomy and particle physics had long ago been converted into permanent staff posts supported by the University, but the new posts could be fixed-term appointments supported by research grants provided by industry and the Research Councils. This proved to be an effective means of supporting new activities. Eventually a number of these posts were converted into University Lectureships and Readerships as the burden of funding the ADRs fell on the departments rather than the University. At the beginning of the Edwards era there were five ADRs, which had increased to 16 by 1995.

The programme was also invigorated by the appointment of more University Assistant Lecturers, formerly known as University Demonstrators, many of whom were members of the new research groups. The creation of permanent posts depended primarily upon the retirement or resignations of tenured staff members. The various changes in emphasis of the programme in the period 1981 to 1995 are tracked in Tables 19.1 and 19.2. There was a tendency to 'store up' vacant posts, which were frozen when economies had to be made by the University, only for there to be a blitz of new appointments with the approach of the regular Research Assessment Exercises (RAE) which began in 1986, as can be seen from the white vertical arrows in Figure 19.1.

But this still does not account for the very large increase in total numbers shown in Figure 16.5(a). This was largely the result of the influx of large numbers of research students, research fellows and research assistants, most of the increase being associated with the new groups. The nature of this increase can be appreciated from Table 19.2, which shows the changing numbers of tenured staff associated with the groups over this 15-year period. The Energy Research, Fluid Dynamics, Liquid Metals, Neutron Physics and Radio groups had mostly been reduced to single senior staff members and, when they retired, their activity ceased. As can be seen, a total of 17 tenured academic posts were created in the new areas. Since each of these staff members would have numerous graduate students, research fellows and postdoctoral research assistants, it is no surprise that the total number of staff increased dramatically through the Edwards era.

19.2 Teaching: the three/four-year physics course

The three-year structure of the undergraduate physics course had remained essentially unchanged since the 1970s. There was, however, a serious concern about the first-year physics courses in that the school syllabus in physics was becoming progressively eroded, with the result that many students were not well prepared for the first-year course in Cambridge.[1] Consequently, more elementary physics had to be introduced into the first year of the course and more advanced topics transferred to later years. Soon after he took over as Head of the Laboratory in 1984, Edwards decided that the summer term courses, which had provided excellent preparation for the final year, could no longer be supported and that material had to be incorporated into the final year of the course. With many students wanting to carry out

Academic year	Staff member	Research group	Notes	Staff member[a]	Research group[b]	Notes[c]
	Departures			Arrivals		
1981–82	A.B. Pippard	LTP	Retirement	R.F. Willis	MP	UL
	M. Ryle	RA	Retirement	P.J. Warner	RA	SAIR
	K. Budden	Radio	Retirement			
	R. Haydock	TCM	To U. Oregon			
1982–83	D. McMullan	MP	Retirement			
1983–84	A.A. Townsend	FD	Retirement	W. Allison	PCS	UD
	J. Klein	PCS	To Oxford U.	R.C. Ball	TCM	UD
				R.J. Needs	TCM	UD
1984–85	J.C. Inkson	TCM	To U. Exeter	R.E. Hills	RA	ADR
				S.D. Berger	MP	SAIR
				A.N. Lasenby	RA	SAIR
1985–86				H. Ahmed	ME	Reader
				A.M. Donald	PC	Lecturer
				G.L.T. Fasol	PCS	Lecturer
				G.A.C. Jones	SP	ADR
1986–87	A.D. Yoffe	PCS	Retirement	S. Chandrasekhar[d]		
1987–88	W.W. Neale	HEP	To Rutherford Lab.	M. Pepper	SP	Professor
	S.D. Berger	MP	To Bell Labs, USA			
	P.V. Head	PCS	Retirement			
1988–89	R.J. Eden	Energy	Retirement			
	J.G. Rushbrooke	HEP	To Bond U., Australia			
	R.F. Willis	MP	To USA			
	A.J.F. Metherell	MP	To USA			
1989–90	A. Cook	LA	Retirement	M. Warner	PC	Lecturer
	A. Hewish	RA	Retirement	A.L. Bleloch	MP	SAIR
	G.L.T. Fasol	PCS	Dir. Hitachi Cam. Lab.	E.A. Marseglia	OE	SAIR
				P. Alexander	RA	UAL
				J.R. Batley	HEP	UAL
				J.A.C. Bland	PCS	UAL
				D.D.C. Bradley	OE	UAL
				M.E. Cates	TCM	UAL
				R.A.L. Jones	PC	UAL
				M.A. Parker	HEP	UAL
1990–91	W.A. Phillips	LTP	To GEC-Marconi	*R.E. Hills*	RA	Prof. Radio Astron.
	G.L. Squires	NP	Retirement	M.S. Longair	RA	Jacksonian Prof.
				J.R. Carter	HEP	ADR

Table 19.1 Arrivals and departures of academic university officers, 1982–1995

			Table 19.1 *(cont.)*			
Academic year	Staff member	Research group	Notes	Staff member[a]	Research group[b]	Notes[c]
	Departures			Arrivals		
1991–92				J.R.A. Cleaver	ME	ADR
				C.J.B. Ford	SP	UAL
				J.R. Cooper	HTS	ADR
				R.D.E. Saunders	RA	ADR
				M.C. Payne	TCM	UAL
1992–93	T.E. Faber	LM	Retirement	R. Padman	RA	ADR
	A.T. Winter	PCS	To School of Phys. Sci.	A. Rennie	PC	ADR
	D.D.C Bradley	OE	To Sheffield U.	D.A. Ritchie	SP	ADR
				R.T. Phillips	OE	SAIR
				J.M. Wheatley	HTS[c]	SAIR
1993–94	–	–	–	–	–	–
1994–95	S. Edwards	PC	Retirement	*W.Y. Liang*	HTS	Prof. Supercond.
	M.E. Cates	TCM	To U. Edinburgh	S. Withington	RA	ADR
				V. Gibson	HEP	UAL
				D.A. Green	RA/AP	UAL
				R. Padman	RA/AP	ADR
1995–96	S. Kenderdine	AP	Retirement	*R.H. Friend*	OE	Cavendish Prof.
				S.R. Julian	LTP	UL
				B.D. Simons	TCM	UL
				C.G. Smith	SP	UL

[a] Names in italic font are elections to named Professorships, the appointee already being a member of the Laboratory.

[b] HTS = Interdisciplinary Research Centre for High Temperature Superconductivity.

[c] Notation: Prof. = Professor, R = Reader, SUL = Lecturer, UL = Lecturer, ADR = Assistant Director of Research, UD = University Demonstrator, UAL = University Assistant Lecturer, SAIR = Senior Assistant in Research.

[d] Subrahmanyan Chandrasekhar was Nehru Visiting Professor and delivered the Scott Lectures in April–May 1987 on the topic *The Physics of Liquid Crystals*.

project work in the final year, the third-year course was vastly overstretched and very demanding. This was not only a problem for Cambridge physics but for physics departments throughout the country.

The Institute of Physics took the bold step of recommending that the standard physics course for students wishing to proceed to a postgraduate degree should be four years long rather than three, whilst keeping the option of leaving with a degree after three years if the student did not intend pursuing physics at the postgraduate level. Somewhat to the surprise of the physics community, the government agreed to this proposal, provided the cost to the government did not increase. In fact, the costs of the additional year did not fall on

	Table 19.2 Changes in numbers of academic staff group members, 1981–1995			
Research group	Departing staff	New staff	Net gain or loss[a]	Notes[b]
High Energy Physics	2	4	+2	
Laboratory Astrophysics	1	0	−1	
Low Temperature Physics	2	1	−1	
Microstructural Phys.	4	3	−1	
Physics and Chemistry of Solids	10	4	−6	
Radio Astronomy	3	8	+5	
Theory of Condensed Matter	3	5	+2	
Optoelectronics	1	6	+5	New Activity
Polymers and Colloids	1	4	+3	New Activity
High Temperature Superconductivity	0	3	+3	New Activity
Microelectronics	0	2	+2	New Activity
Semiconductor Physics	0	5	+5	New Activity
Energy Research Group	1	0	−1	End of Activity
Fluid Dynamics	1	0	−1	End of Activity
Liquid Metals	1	0	−1	End of Activity
Neutron Physics	1	0	−1	End of Activity
Radio Group	1	0	−1	End of Activity

[a] The data are compiled from the information in Table 19.1.
[b] New activities are shown in green and terminating activities in red.

the government since the funding of undergraduate students was the responsibility of the Local Educational Authorities.

I returned to Cambridge in 1991 and became Chair of the Cavendish Teaching Committee in the following year. The committee was faced with reconstructing the physics course, changing it from a three-year to a four-year structure. This was a major relief for the teaching programme. The level of the first year was moderated and the final year became much more a preparation for research, with courses spanning most of contemporary physics. All students now carried out research projects in the fourth year. The core of physics, which all students had to absorb, was contained in the first three years of the course, while the final year consisted of a series of Major Option courses of 24 lectures, spanning major areas of physics in the Michaelmas term, and a very wide range of Minor Options of 16 lectures each in the Lent term (Table 19.3). The students could select topics which matched their interests; these ranged from particle and condensed matter theory to astrophysics and medical physics. There was the option of students graduating after three years if they did not wish to carry on to a research career, and courses appropriate to this career choice were introduced. Typically, of about 120–140 third-year students, 10 to 15 would take the three-year option.

Physics was the first subject to adopt the three/four-year course structure in Cambridge and other subjects in the physical sciences soon followed suit. The implementation of the

Table 19.3 Part III lecture courses delivered in the academic year 1998–1999

MICHAELMAS 1998 Major options	LENT 1999 Minor options

DR W. ALLISON
  Solid State Physics. Tu. Th. S. 11
PROF. A.M. DONALD
  Structure and Properties of Condensed Matter.
    M. W. F. 9
PROF. A.C. FABIAN, DR A.N. LASENBY AND
    PROF. M.J. REES
  Gravitational Astrophysics and Cosmology.
    M. W. F. 12
DR M.A. PARKER AND DR J.R. BATLEY
  Particle Physics. M. W. F. 11
DR K.F. PRIESTLEY AND DR A.J. HAINES
  Physics of the Earth as a Planet. Tu. Th. S. 10
DR B.D. SIMONS
  Theoretical Concepts in Physics. Tu. Th. S. 12
	DR B.R. WEBBER
	  Gauge Field Theory. W. F. 12
	DR D.J.C. MACKAY
	  Information Theory, Pattern Recognition and
	    Neural Networks. M. W. 11
	DR R.F. CARSWELL
	  General Relativity. Tu. Th. 9
	DR J.A.C. BLAND
	  Low Dimensional Magnetism. M. W. 12
	DR B.D. SIMONS
	  Phase Transitions and Collective
	    Phenomena. Tu. Th. 12
	DR J.A. COOPER
	  Superconductivity. Tu. Th. 9
	DR C.H.W. BARNES
	  Quantum Properties of Electron Systems in
	    Semiconductors. Tu. F. 11
	DR G.A.C. JONES
	  Microelectronics and VLSI. M. W. 9
	DR N.C. GREENHAM AND DR D.R. RICHARDS
	  Optoelectronics. Tu. Th. 12
	PROF. J.E. FIELD AND DR N.K. BOURNE
	  Shock Waves and Explosives. M. W. 11
	DR J. MELROSE
	  Polymers and Colloids. M. W. 11
	DR A.N. LASENBY AND DR C.J.L. DORAN
	  Physical Applications of Geometric (Clifford)
	    Algebra. Tu. Th. 11
	PROF. R.E. HILLS
	  The Frontiers of Experimental
	    Astrophysics. Tu. Th. 11
	DR P.P. DENDY AND OTHERS
	  Medical Physics. Tu. Th. 10

Courses from Part III Mathematics	*Courses from Part III Mathematics*[a]

DR N.S. MANTON
  Quantum Field Theory. M. W. F. 10
PROF. D.O. GOUGH AND DR C.A. TOUT
  Structure and Evolution of Stars. M. W. F. 9
	PROF. P.V. LANDSHOFF
	  Advanced Quantum Field Theory.
	    M. W. F. 12
	DR A. BURGESS AND DR H.E. MASON
	  Atomic Astrophysics. M. W. F. 9

[a] Advanced Quantum Field Theory could not be taken together with Gauge Field Theory.

new system began with a new first-year course in the academic year 1993–94 and the first cohort of students, who were the ground-breakers for all four years of the new course, completed their studies four years later. There is no question but that this stretching out of the course was a tremendous relief for the teaching of physics, although it imposed an additional year of student teaching on the staff. Fortunately, much of the additional teaching was in the final, more specialist, year of the course and staff often found this an attractive option since the courses were closer to their research expertise. It also provided a splendid opportunity to attract enthusiastic undergraduates to become research students in their research areas.

Some impression of the range of topics to which the students were exposed in their fourth year is given by the lecture syllabus for the Michaelmas and Lent terms of 1998–99, when the new arrangements had been in place for a year (Table 19.3).[2] The 24-lecture Major Option courses in the Michaelmas term were substantial surveys of broad ranges of physics, using all the tools the students had acquired in the first three years of the course. Some of the courses were provided by cognate departments such as the Institute of Astronomy and the Department of Geophysics. It was also possible to take a limited number of courses from Part III of the Mathematical Tripos. In the Lent term, the 16-lecture Minor Options also involved other departments, including the Medical School. The Minor Option courses could change from year to year and significant new topics could be introduced. For example, a Minor Option in Entrepreneurship was introduced in collaboration with the Judge Business School. Later, a Minor Option in Physics School Teaching was introduced, the students spending part of their time teaching in schools.

The same basic structure has continued to the present day, although the preparation of many students remains a concern, including their appreciation of mathematics.

The Edwards era: new directions in condensed matter physics

20.1 Pepper and semiconductor physics

Michael Pepper's association with the Laboratory dated back to 1973 when his collaboration with Mott began and he established a joint research programme with John Adkins to study the low-temperature behaviour of two-dimensional electron gases. His research programme was supported for the period 1978 to 1986 by a Warren Research Fellowship of the Royal Society, the 1978 advertisement stating that the fellowship was

> for research in metallurgy, engineering, physics or chemistry, or for the use or application of such research or its results in or for industry and industrial development.

Pepper's involvement in the discovery of the quantum Hall effect during the tenure of the fellowship was an outstanding example of the pursuit of the aims of basic research and industrial application (Section 18.2). In 1982, he transferred his research activity as Principal Research Fellow to the GEC Hirst Research Centre.

The formation of the Semiconductor Physics Group took place formally in 1984 under Pepper's leadership. This marked the beginning of a major expansion of the activity and facilities for semiconductor physics, Pepper using his industrial contacts to support the rapid development of the group, which was to become one of the largest in the Laboratory. Following his pioneering work on Anderson localisation, variable-range hopping and related topics, inspired by his association with Nevill Mott, he led the construction of a series of facilities for the production of semiconductor devices using the techniques of molecular beam epitaxy (MBE) and electron beam lithography (EBL). MBE is an ultra-high-vacuum technique for the deposition of thin films of various semiconductors and insulators layer-by-layer from an evaporated beam of particles, what is known as *epitaxial growth*. This technique is used to grow very high-purity III-V semiconductors in which the thickness of layers can be controlled with sub-monolayer precision (Box 20.1(*a*)).[1] The resulting layered materials are referred to as heterostructures or heterojunctions. The substrates on which the patterns are laid down are created by EBL, an example of the current Leica Nanowriter facility being shown in Box 20.1(*b*). As well as being used for many projects within the Laboratory, samples were provided for many other universities and for industrial organisations.

From the early 1980s onwards, Pepper and his colleagues produced a flood of papers which exploited the developing suite of facilities to create semiconductor devices in which quantum mechanics played a central role in their operation. Among the many significant and influential advances were:

Box 20.1   **EBL and MBE facilities in the Semiconductor Physics Group**

(a)

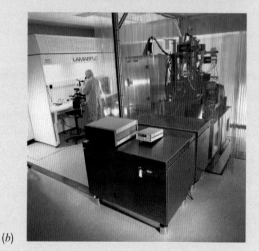

(b)

The VG V80H MBE growth chamber began operation in 1986 and since then has grown about 4000 wafers in the InGaAs-AlGaAs-GaAs material system. The chamber is attached to a surface decontamination chamber and a focussed ion beam system, both of which have been used for the fabrication of 3D semiconductor structures. In 1991, a Varian Gen II system was installed in the Laboratory for the growth of very high mobility 2D electron and hole gases in the GaAs-AlGaAs system.

An example of a state-of-the-art EBL system, which was installed at the Laboratory in 2002. In collaboration with the Leica Company, this Leica VB6 UHR electron beam lithography machine for the patterning of semiconductor structures uses an electron beam of diameter 4 nm and energy up to 100 kV. It is capable of patterning substrates of up to 200 mm diameter with a resolution as high as 10 nm. This instrument is used to produce samples for a wide variety of academic and industrial users.

- the first separation and identification of quantum interference, or weak localisation, and electron interaction mechanisms (Davies *et al.*, 1981)
- the use of local resistance as a means of measuring electron temperature and of exploring the low-temperature thermal properties of mesoscopic and two-dimensional electron gases
- extension of the use of semiconductor structures for studying the role of dimensionality transitions on quantum interference (weak localisation)
- the invention and development of the 'split-gate' method for producing one-dimensional behaviour in GaAs-AlGaAs heterojunctions and the general shaping of an electron gas by 'electrostatic squeezing'. This work led to rapid progress in mesoscopic physics, including topics such as quantisation of one-dimensional resistance
- the invention and demonstration of zero-dimensional quantum dots. These structures are of great significance, topics emerging from the application of these devices including Coulomb blockade, edge current interference and other effects.

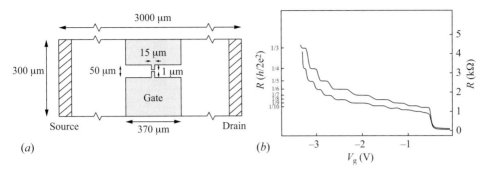

(a) The structure of a split-gate heterojunction (Berggren et al., 1986). (b) In the experiments carried out in 1988, the split-gate dimensions were reduced to 0.5 μm wide and 0.4 μm long. The channel resistance at $T = 0.1$ K is plotted as a function of gate voltage for two different carrier concentrations induced by illumination. The existence of a resistance quantised in units of $h/2ie^2$, where i is the number of occupied sub-bands, is illustrated in the diagram (Wharam et al., 1988).

Fig. 20.1

These researches resulted in the creation of new areas of low-temperature solid state physics. In 1987, the 1996 Professorship of Physics, which had been held in abeyance for a number of years, was advertised and Pepper was duly appointed to the chair. At the same time, he was appointed to a Professorial Fellowship of Trinity College.

The new technologies enabled the construction of semiconductor devices on very small scales such that the length L of the conductor became much smaller than the mean free path l. This form of transport is termed 'ballistic', meaning that the electrons are not scattered during the time it takes to travel through the conductor. The split-gate method, illustrated in Figure 20.1(a), provided the means of creating a one-dimensional structure in which the resistance is quantised, the values of the resistance being $h/2ie^2$, where i takes integral values. This prediction was dramatically confirmed by the experimental results shown in Figure 20.1(b) (Wharam et al., 1988).

The next step was to go further and create zero-dimensional structures by confining the one-dimensional quantum wire within narrow barriers, resulting in a quantum 'box' or quantum dot. The structure is illustrated in Figure 20.2. As Smith, Pepper and their collaborators wrote in their paper,

> In this experiment we have created a three-dimensional potential and by transport measurements we have probed the states defined within it. At low gate voltages we have observed resonant scattering of one-dimensional sub-bands, while with higher gate voltages resonant tunnelling is seen. By applying a source-drain voltage it is possible to measure the energy separation of the zero-dimensional states as 1.4 meV, which is in good agreement with what would be expected in a parabolic well of width 0.3 μm. (Smith et al., 1988)

These experiments marked the beginning of the detailed control of the wave function of an electron in an electrostatic potential of prescribed form.

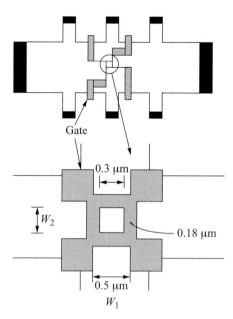

Gate

0.3 μm

W_2

0.18 μm

0.5 μm

W_1

Fig. 20.2 A schematic diagram of a zero-dimensional quantum dot, showing the general layout of a short split-gate device on a GaAs-AlGaAs heterostructure. The inset shows the detail of the gate pattern and its dimensions (Smith *et al.*, 1988).

The development of these facilities for cutting-edge semiconductor research was an expensive undertaking. It was well supported by a rolling grant from the EPSRC, but considerable capital investment was needed beyond what could be obtained from that source. Pepper used his contacts with the semiconductor industry to enhance support for these endeavours. The most important of these was the foundation in 1991 of the Toshiba Cambridge Research Centre, now known as the Cambridge Research Laboratory (CRL) of Toshiba Research Europe. Pepper became part-time Director of the Toshiba Cambridge Research Centre from the time of its foundation. The attraction of locating in Cambridge is described on Toshiba's website:

> The Cambridge Research Laboratory (CRL) was Toshiba's first overseas corporate-level R and D laboratory. . . . it was established in 1991 to undertake scientific studies which may lead to the semiconductor technology of the 21st Century. Cambridge was chosen for the opportunities to collaborate with the University of Cambridge, one of the world's top universities, and with the growing cluster of hi-tech companies based in the city . . . Close collaboration is undertaken with the Cavendish Laboratory and Engineering Department of the University of Cambridge, as well as other Toshiba R and D groups. We have achieved a number of world firsts, successfully transferred technology into Toshiba products, and spun-out TeraView Ltd in 2001 to develop terahertz (THz) technology.

This close collaboration with Toshiba meant that advances in basic physics could be rapidly developed into commercial products.

An excellent example of this collaboration was the development of terahertz (THz) technology for application in fields as diverse as astrophysics and medicine. The group specialised in developing new facilities and techniques for the construction of innovative semiconductor structures, and among the most successful was the quantum cascade laser, which provided high-intensity beams of coherent radiation in the far infrared and submillimetre wavebands. Until the development of this new technology, such sources were not available in the THz waveband. In collaboration with the Toshiba Cambridge Research Centre, Pepper and Edmund Linfield developed imaging and spectroscopic techniques for application in medicine and dentistry using these devices (Ciesla *et al.*, 2000; Townsend, 2013). Box 20.2 shows an example of the application of THz technology for the location of cancerous tissues. The advantages of THz radiation over X-rays are that the radiation is non-ionising and less hazardous to use, with power levels generally lower than everyday background THz radiation.

The company TeraView was spun off from Toshiba Research Europe in 2001 by its co-founders, Pepper and Don Arnone, to exploit the intellectual property and expertise developed in THz sources and detectors, using the innovative technologies developed jointly by the Semiconductor Physics Group and the Toshiba Cambridge Research Centre. The range of applications of the technology is very wide, including non-invasive medical and dental diagnostics, detection of hidden objects and weapons, drug discovery and formulation analysis of coatings and cores, characterisation of electron carriers and metamaterials, improvement of the performance and quality of solid-state properties of semiconductors and metamaterials, non-contact material imaging of coatings and composites and non-contact imaging for conservation of paintings, manuscripts and artefacts.

Among the more important discoveries of the collaboration were single-photon sources from electrically activated quantum dots, the photons having energy about 1.3942 eV (Yuan *et al.*, 2002) (see also Box 22.10(*b*)). As the authors write, these InAs self-organised quantum dots can be tailored to emit at the wavelength used for long-distance fibre optic communications, about 1.3 μm.

Within the Laboratory, a major collaboration developed with Stafford Withington of the Astrophysics Group in the development of array detectors for millimetre, submillimetre and far-infrared astronomy. These find application in both space- and ground-based astronomy facilities for these wavebands, examples including the James Clerk Maxwell Telescope in Hawaii, the ALMA array in Chile and space missions such as SPICA. Withington formed what became the Quantum Sensors Group in the early 2000s, with the capabilities described in Section 22.8.7.

20.2 Microelectronics

One of Pepper's collaborators during the remarkable early years of the Semiconductor Physics Group was Haroon Ahmed, who had been a demonstrator and lecturer in the Engineering Department. He began his researches into the processing of semiconductors by particle beams in 1973, supported by Standard Telephones and Cables Ltd and, from 1975 to 1978, by the SERC. The success of these researches resulted in an expansion of his research team and facilities which could not be accommodated within the Engineering

Box 20.2 THz imaging of cancerous tissues

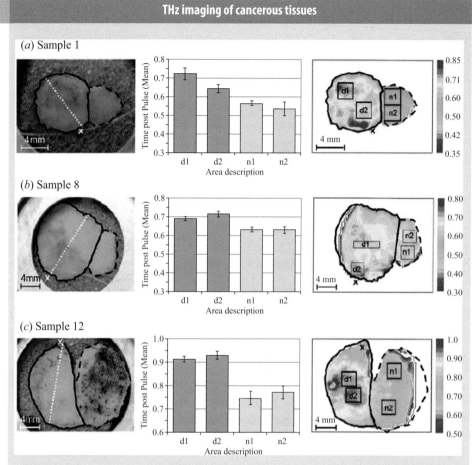

(*a*) Sample 1

(*b*) Sample 8

(*c*) Sample 12

A comparison between the visible images, on the left, and the THz images, on the right, of three samples of cancerous tissue. The diseased tissue, on the left of the visible image, is marked by a solid boundary, the normal tissue on the right by a dashed boundary. The histograms show the mean time post pulse (TPP) values and the error bars for the areas highlighted by the boxes in the THz images. The equal-size areas d1 and d2 were located on the diseased tissue, and n1 and n2 on the normal tissue.

Terahertz pulse imaging (TPI) in reflection geometry for the study of skin tissue and related cancers has been demonstrated both in vitro and in vivo (Woodward *et al.*, 2003). The sensitivity of terahertz radiation to polar molecules, such as water, makes TPI suitable for studying the hydration levels in the skin and the determination of the lateral spread of skin cancer pre-operatively. By studying the terahertz pulse shape in the time domain, diseased and normal tissue can be differentiated in the study of basal cell carcinoma (BCC).

Department. The solution was the formation of the Microelectronics Research Laboratory on the Cambridge Science Park, founded by Trinity College in 1970. Ongoing support from 1978 to 1984 was provided by a large rolling grant from the SERC, as well as GEC, British Telecom, other industrial companies and the Royal Society's Paul Instrument Fund.

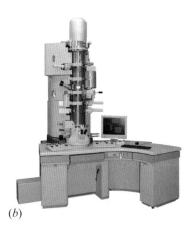

(*a*)   (*b*)

(*a*) The Microelectronics Building, opened in 1990. The photograph is more recent and shows extensions to the building to house the expanding research programme. (*b*) One of three electron microscopes presented to the University in 2009 to celebrate the 20th anniversary of the Hitachi Cambridge Laboratory and the 100th anniversary of the Hitachi Group.

Fig. 20.3

In 1984, Ahmed was promoted to a readership. According to the University's Statutes and Ordinances, such an *ad hominen* post could be held in the department of the post-holder's choice. Because of his close and fruitful collaboration with Pepper, he transferred his readership and the Microelectronics Research Laboratory to the Cavendish Laboratory, resulting in a major increase in capabilities in this rapidly developing field. By 1988 there were 32 research workers, including 15 research students, in the group, supported by grants from the SERC, the Ministry of Defence, the University, Trinity College and industry.

The range of studies complemented those of the Semiconductor Physics Group. In summary they included

- the use of electron beams for thermal annealing of ion-implanted semiconductors, including detailed investigations of diffusion and crystal growth
- the use of a double electron-beam system for the melting and re-crystallisation of polysilicon layers and for making silicon-on-insulator devices
- the construction and application of a focussed ion-beam system for direct ion implantation into selected areas of a semiconductor, for scanning ion microscopy and for microsectioning of devices and circuits
- the fabrication of nanometre-scale structures.

Ahmed had already been collaborating with the Hitachi Company and in 1989 he came to an agreement with them to establish a research laboratory embedded within the Cavendish Laboratory. At the same time, he raised a considerable amount of money for a new £850,000 building on the Cavendish site, with the result that the Microelectronics Research Laboratory could be transferred from the Cambridge Science Park to the new building (Figure 20.3(*a*)). Space was provided in the new building for the new Hitachi Cambridge Laboratory (HCL).

Research initially focussed on microelectronics and the development of transport techniques for the fabrication of devices on wafer surfaces. This rapidly resulted in the demonstration of the first single-electron memory (Nakazato *et al.*, 1993) and the first single electron logic circuit (Nakazato and Ahmed, 1995). This successful long-term collaboration has been of great value to both parties and has provided a model for academic–industrial partnerships. Hitachi provided both material support (Figure 20.3(*b*)) and studentships and fellowships, which have been notably successful.

The collaboration between the HCL and the Microelectronics Group was organised so that research was concentrated on the basic physics of new concepts and techniques. Once these projects led to potentially new commercial products, the research and development activities were transferred to Hitachi's Central Development Laboratories in Japan. An example of this is the invention of the PLEDM memory chip, the acronym standing for phase-state low electron (hole)-number drive memory.

On retirement from his University post in 2003, Ahmed arranged for the upgrading of his underlying post to an endowed professorship with a generous gift from the Hitachi Company. Henning Sirringhaus was appointed to the Hitachi Chair of Electron Device Physics in 2004 and brought new activities to Microelectronics Group and the Hitachi Cambridge Laboratory. Since then, the Cavendish–Hitachi collaboration has extended to other research groups on the Cavendish site, including the Optoelectronics, Semiconductor, Thin Film Magnetism and Atomic, Mesoscopic and Optical Physics groups, with the aim of creating new concepts in advanced electronic and optoelectronic devices.

20.3 Polymers and colloids

If Sam Edwards was the inspiration for the theory of polymers, colloids and complex materials, experimental work in these areas in the Laboratory was pioneered by Athene Donald. Her PhD had been in metal physics in the Cavendish Laboratory, but during a four-year period at Cornell University in the USA she changed discipline to the experimental study of polymers using electron microscope techniques, under the inspiring leadership of Edward (Ed) Kramer. Her work on glassy polymer crazing was an immediate success and proved to be very influential. This was followed by innovative studies of shear deformation in liquid crystal polymers. In 1981, she returned to Cambridge supported by a SERC fellowship, which she held in the Department of Metallurgy and Materials Science, to be followed by a Royal Society Research Fellowship in the Cavendish Laboratory in 1983.

In 1985, a lectureship in experimental polymer physics opened up in the Laboratory and she was duly appointed to that position.[2] She became Head of the newly formed Polymer and Colloids Group with the unstinting support of Edwards. This was to be a period of rapid development in the study of polymer and colloid physics, to which she brought all the advanced techniques of analysis of experimental solid state physics, including neutron and X-ray scattering, optical microscopy and infrared spectroscopy. The study of the properties of foodstuffs began in the late 1970s and was to be greatly enhanced by a large AFRC-funded project with the Institute of Food Research. As she wrote in 2015,

[Edwards'] close links with industry had major importance for me in that he was able to pull all the relevant parties together to create a linked grant from government and industry on the topic of colloids, at £3M a huge sum back in 1992 for the Cavendish to win. This was a grant I was then fingered to lead. His was the hard work that brought the grant to fruition, but I was one of the group who derived the benefit of scientific credit. This wasn't the first time he had done so for me either: he had previously brought Food Physics to the Cavendish with another major grant which I then took the scientific (and experimental) lead on. My scientific reputation, if you like, derives from his vision. It was demanding to fulfil that vision but I had a strong financial platform on which to build. (Donald, 2015)

This was a strikingly new initiative. The important role which physics can play in these areas is admirably proclaimed in the summary of her review, *Physics of foodstuffs*, in *Reports of Progress in Physics* (Donald, 1994):

The aim of this article is to demonstrate that foodstuffs can be usefully and excitingly studied within the framework of physics. Many of the same issues that exercise the mind of researchers in traditional areas of soft condensed matter can be found within these homely materials, issues such as percolation, the nature of the glass transition and mechanisms of phase separation. By applying the conventional tools of the physicist new insights can be obtained into the structures and responses of foodstuffs. In turn these insights may lead to improvements in the overall quality of the food to be consumed – by the knowledge contributing to improvements in processing, texture or storage for instance.

In the introduction to the article, she gives further examples:

cooking actually involves many processes and concepts familiar to physicists: diffusion (e.g. water into a potato as it is boiled), phase transitions (e.g. melting of chocolate), and van der Waals forces which play a crucial role in stability (has the salad dressing unfortunately phase separated before serving?) to name but a few. (Donald, 1994)

These issues are of real importance for the food industry, as can be appreciated from the acknowledgement of support from Nestlé, Dalgety-Spillers and Unilever, as well as from the Agriculture and Food Research Council.[3]

A particular area of innovation and expertise was in the development and application of the environmental scanning electron microscope (ESEM) to the study of the physics of soft condensed matter (Figure 20.4). The key feature of this variant of the techniques of electron microscopy is the ability to study samples in their natural state, for example, as wet samples, opening up a new range of possibilities for imaging dynamical phenomena, including live systems. The list of materials to be studied by this technique included cement, textiles, lacquers, detergent hydration, emulsions, aggregation of polymer colloids, foods, plant systems, biomedical materials, cells and tissues. There was a natural progression of the interests of the group from soft condensed matter to biological physics, including the study of proteins, polysaccharides and cellular biophysics. As noted in the citation for the award to her of the 2010 Faraday Medal of the Institute of Physics:

Donald's mid-career launch into biological physics followed naturally from this polymer work leading to the physics of food and thence to starch. The starch granule structure and its changes during different processing histories were brilliantly analysed using a novel

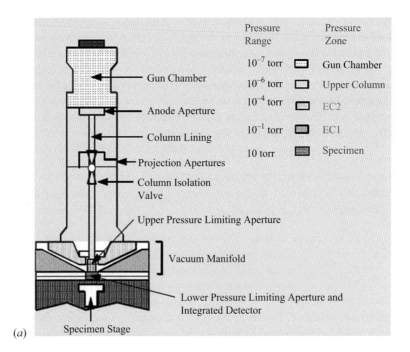

Pressure Range		Pressure Zone
10^{-7} torr		Gun Chamber
10^{-6} torr		Upper Column
10^{-4} torr		EC2
10^{-1} torr		EC1
10 torr		Specimen

(a)

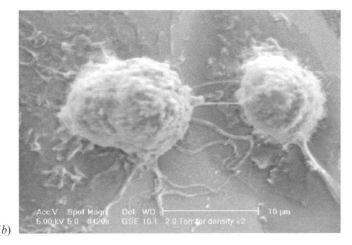

(b)

Fig. 20.4 Athene Donald and her colleagues developed and exploited the capabilities of the environmental scanning electron microscope (ESEM) for the study of soft matter. The advantage of the ESEM is that biological specimens can be kept 'wet' during the imaging. (a) The microscope involves a range of pressures in the different sections of the microscope so that, by the specimen stage, living matter can be imaged. This microscope has been used in a wide variety of projects in soft matter and food physics. (b) ESEM image of human macrophages, the types of white blood cell that engulf and digest cellular debris, foreign substances, microbes, cancer cells and so on.

X-ray scattering technique. Structural changes during cooking, with the amylopectin molecule imaginatively treated as a side chain liquid crystalline polymer, brought understanding to different processing treatments. The mis-folding of proteins forming amyloid fibrils is well recognized in the aetiology of many diseases, particularly those of old age. Donald's recent work has demonstrated that this important and challenging problem can be powerfully addressed by the approaches of polymer science and furthermore suggests an intriguing connection between the structures observed in both fields.

These activities were well funded by research grants from the Research Councils and industry, and in due course the number of staff members grew to support these activities. Mark Warner was appointed to a lectureship in 1989, the early years of his appointment being supported by Unilever, organised by Edwards, until a University post became vacant. Richard Jones joined the group in 1989 as an assistant lecturer and Adrian Rennie was appointed an ADR in 1992. When Jones moved to a Chair in Sheffield in 1997, his post was filled by Eugene Terentyev, who, although trained as a theorist, launched himself into the experimental aspects of this work with determination and enthusiasm. In addition, the appointments of Robin Ball in 1983 and Michael Cates in 1989 in the Theory of Condensed Matter Group greatly expanded the strength of the Laboratory in soft condensed matter physics. A significant area of the Bragg Building was refurbished to house these activities, the area being designated the Tabor Laboratory in recognition of David Tabor's pioneering researches in these areas.

Her leadership of the Polymers and Colloids Group led to Donald's promotion to a readership in 1993 and to a professorship in 1998, the first woman to be promoted to such a post in physics in the history of the University.

20.4 Optoelectronics

The remarkable expansion of research in optoelectronics and the formation of the Optoelectronics Group was driven by the dynamic leadership of Richard Friend. As a research student he had been supervised in the Physics and Chemistry of Solids Group by Abraham (Abe) Yoffe, who had encouraged his graduate students to investigate a broad range of topics which he considered ripe for exploration and exploitation in experimental condensed matter physics. Friend completed his PhD in 1979 on transport properties and lattice instabilities in one- and two-dimensional metals, a significant part of the work being carried out under the supervision of Denis Jérome at the Laboratoire de Physique des Solides of the Université Paris-Sud, Orsay. His interests in condensed matter physics spanned a very wide range of the topics and techniques being pursued in the PCS Group at the time, as can be appreciated from his list of publications. Among the early signs of things to come were his numerous papers on organic metals by the date of his appointment as a University Demonstrator in the Laboratory in 1980. As he remarked in an interview in 2008,

> I had been elected to a Research Fellowship at St John's in 1977 and that is when I switched from the world of inorganic materials to working with molecules, carbon-based

conductors, because that was the main research line in Paris; I have continued on that theme in various ways. (Friend, 2008)

As a result, much of his most fruitful work developed at the interface between physics and chemistry, a legacy of the inspiration of Yoffe, who remained Head of the PCS Group until his retirement in 1987.

Some impression of the vast range of Friend's activities can be appreciated from the citation associated with his election to the Royal Society of London in 1993:[4]

> Distinguished for his experimental study of the electronic properties of novel materials, principally organic materials, both semiconductors and metals, and inorganic materials with 'low-dimensional' electronic structure, including layer structure transition metal dichalcogenides and cuprate superconductors. He established the pressure/temperature phase diagrams for transition metal dichalcogenides, showing conditions for CDW (Charge Density Wave) phases and superconducting phases in TaS_2, band crossing transition in TiS_2 intercalation of transition metal dichalcogenides with various Lewis bases (alkali metals, amines), and use of controlled charge transfer to the host layer to fine-tune electronic structure in order to establish conditions for CDW superlattice formation, and mechanisms for charge transport.
>
> He has made a major contribution to understanding the conditions for metallic, superconducting, magnetic and insulating ground states in organic charge transfer salts. Established the pressure/temperature phase diagram for the incommensurate and commensurate Charge Density Wave phases of TTF-TCNQ. He made the first observations of de Haas–van Alphen oscillations in magnetic susceptibility in an organic metal. He and his group have developed polymer processing techniques for conjugated polymers, and demonstrated non-linear electronic excitations through electrical and optical measurements. First construction of MOSFET (Metal Insulator Semiconductor Field Effect Transistor) with polyacetylene as active semiconductor, and demonstration of novel mechanism of operation, with novel behaviour and made the first construction of efficient, large area, polymeric semi-conductor LED's (Light Emitting Diodes), based on polyphenylene-vinylene [PPV].

With Yoffe's retirement in 1987, the PCS Group began to fragment, and Friend set up the Optoelectronics Group, specialising in the physics and chemistry of organic polymers for use as semiconductor materials. Howard Hughes and Elizabeth Marseglia were founder members of the group; Donal Bradley became an Assistant Lecturer in 1989, but left for a readership at the University of Sheffield in 1993. Richard Phillips joined the group as a Senior Assistant in Research (SAIR) in 1992.

The first major breakthrough took place in the late 1980s when Friend and his collaborators constructed semiconductor devices with silicon-like properties made of organic conjugated polymer materials. In 1988 they succeeded in making the first polymer field-effect transistor (Burroughes *et al.*, 1988). As Friend stated,

> The original structures we made were actually made purely to explore; the transistor which we first got working in 1988 was a tool to understand how, when you start moving electrons around, unlike the case of silicon where they travel just by themselves and the lattice of silicon atoms remains in place, if you do that with a molecular system you disturb the positions of all the carbon atoms and that makes a big difference to how

Box 20.3

The physics of organic polymer light-emitting diodes

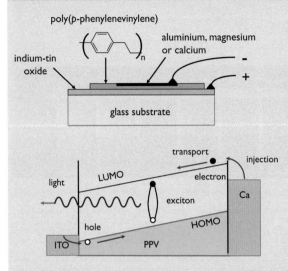

poly(*p*-phenylenevinylene)

aluminium, magnesium or calcium

indium-tin oxide

glass substrate

transport
light
LUMO
injection
electron
exciton
Ca
hole
HOMO
ITO
PPV

The diagram shows the structure of a typical polymer light-emitting diode (PLED). Although PPV-related polymers are still of interest elsewhere, much of the effort in Cambridge is now directed at polyfluorenes.

In a polymer LED, light is emitted when positive hole and negative electron charge carriers are injected into the organic semiconductor metallic contacts on opposite sides of a semiconducting polymer film. When electrons and holes come under the influence of their mutual Coulomb attraction inside the device, they recombine, emitting a photon. The wavelength of the emitted light depends on the band gap of the semiconducting polymer. Hence, the output colour can be altered by chemically tailoring the polymer.

things move; our transistors were absolutely useless for any practical application but they showed beautiful characteristics and we managed to get some very clean science out of that. (Friend, 2008)

The attraction of this approach to semiconductor technology was that the techniques used were relatively simple to implement. Inorganic transistors required massive vacuum systems and complex manufacturing processes, whereas organic polymer materials can be dissolved in organic solvents to create 'inks' that can be used to 'print' circuits under normal atmospheric conditions. This research laid the foundation for the production of low-cost transistors using flexible plastic materials.

The second major breakthrough took place in 1989, almost by accident. A piece of semiconductor material was sandwiched between two metal layers (Box 20.3). When a voltage was applied across the electrodes, green light was emitted from the polymer material. As Friend remarked, 'It was good fortune that the top electrode was thin enough – it was semi-transparent'. He and his colleagues realised the enormous potential of this discovery and Friend, Bradley and Jeremy Burroughes immediately filed a patent for their invention. The paper describing the discovery of electroluminescence from polymers was subsequently published in *Nature* (Burroughes *et al.*, 1990). This and the many follow-on publications made Friend the most cited physicist in the UK for the decade 1990–99, according to the Institute for Scientific Information. The colour emitted by the diode depends upon the band gap, and this can be altered by creating different conjugated polymers by chemical processing (Figure 20.5).

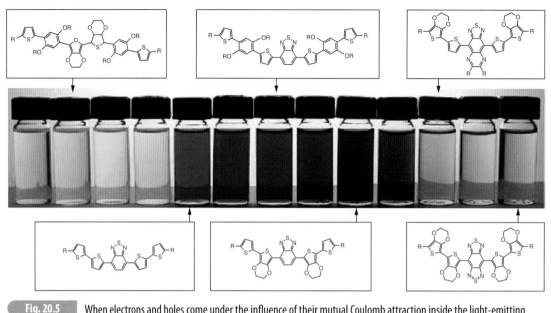

Fig. 20.5   When electrons and holes come under the influence of their mutual Coulomb attraction inside the light-emitting diode, they recombine, emitting a photon. The wavelength of the emitted light depends on the band gap of the semiconducting polymer and so can be altered by chemically tailoring the polymer, as illustrated in the diagram. The range of colours shown is a combination of two-photon absorption of normal light at short wavelengths and emission in the near-infrared wavebands (Ellinger *et al.*, 2011). (Diagram courtesy of John Reynolds, University of Georgia, USA. The image is a modified version of that appearing in the paper by Ellinger and his colleagues.)

The success of these activities and the industrial interest they generated resulted in an expansion of the activities of the group. Friend was elected to the Cavendish Professorship of Physics in 1995, in succession to Edwards. Neil Greenham was appointed an SAIR in 1996 and then became an Assistant Lecturer in 1998. Henning Sirringhaus joined the group in 1997 as a Senior Research Fellow and was appointed a University Lecturer in 2001. He was promoted to Reader in 2003 and then to the Hitachi Professorship of Electron Device Physics in 2004. Ullrich Steiner was appointed John Humphrey Plummer Professor of Physics in 2006 and had a loose association with the work of the group. The scale of the activity can be appreciated from the statistics for the academic year 2003–04: one professor, three readers, one lecturer, one ADR, three research fellows, 12 research associates and 30 graduate students.

Friend had a strong interest in commercialising these inventions and within a couple of years had secured venture capital funding for the formation of the company Cambridge Display Technologies (CDT). The highlights of the subsequent history of the company are presented in Box 20.4.

The discovery of electroluminescence from polymers and the earlier construction of organic transistors were turning points in the development of polymer electronics, and a huge industry followed up these pioneering efforts. Friend and his colleagues now set about understanding the complex physics behind these processes as well as tackling the

The commercialisation of polymer light-emitting diodes   Box 20.4

Richard Friend played a major role in setting up Cambridge Display Technologies to exploit the discovery of electroluminescence from polymers.

1992: CDT founded by Cambridge University and seed venture capital obtained

1996: CDT secures first licensees

2000: CDT and Seiko–Epson demonstrate the world's first full color active matrix inkjet-printed polymer light-emitting diode (PLED) display, measuring 2.5 square inches and 2 mm thick

2002: First high-profile commercial PLED product when Philips launches its innovative shaver with electronic display

2005: CDT shows 14-inch demonstrator with 1280 x 768 pixel resolution, produced using inkjet printing

2006: Seiko–Epson announces world's first polymer-organic light-emitting diode (P-OLED) printhead for use in printers, scanners etc. Toppan Printing and CDT show roll-printed display.

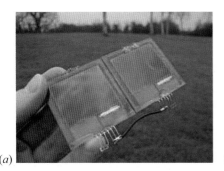

(a)   (b)

(a) A prototype organic solar cell; (b) image of an inkjet-printed organic light-emitting diode (OLED) display made by   Fig. 20.6
Cambridge Display Technologies (courtesy of Jeremy Burroughes and CDT).

physics and engineering problems of converting their inventions into viable commercial products. The vision was to create light-emitting diodes (LEDs), polymer electronics and solar cells by processes akin to inkjet printing and so produce huge arrays of these devices on flexible plastic substrates at low cost.

The potential for solar cells followed immediately from the discovery of photoluminescence. If light could be produced by the recombination of electrons and holes in polymer materials, then the absorption of light by such materials would lead to the formation of electron–hole pairs and hence to organic photovoltaics. When light is absorbed by a conjugated polymer, a bound electron–hole pair, a so-called 'exciton', is created. To generate a photocurrent these excitons need to be split up into free charges. The electrons and holes need to be able to move freely through the polymer film to the respective contacts to be collected. The collaboration between the Optoelectronics Group and CDT resulted in the first viable organic solar cells (Figure 20.6(a)).

One of the challenges was to develop techniques by which electronic circuits could be created using soluble polymers. In addition to his work on the fundamental physics of

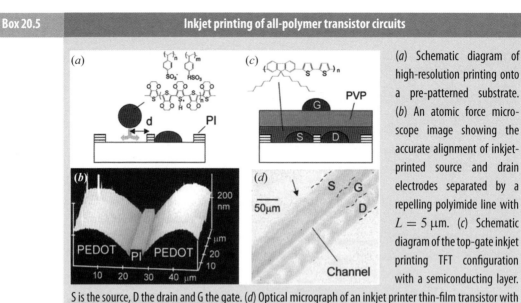

Box 20.5 — Inkjet printing of all-polymer transistor circuits

(*a*) Schematic diagram of high-resolution printing onto a pre-patterned substrate. (*b*) An atomic force microscope image showing the accurate alignment of inkjet-printed source and drain electrodes separated by a repelling polyimide line with $L = 5$ μm. (*c*) Schematic diagram of the top-gate inkjet printing TFT configuration with a semiconducting layer. S is the source, D the drain and G the gate. (*d*) Optical micrograph of an inkjet printer thin-film transistor with $L = 5$ μm (Sirringhaus *et al.*, 2000).

polymer semiconductors, Sirringhaus led the efforts to create polymer electronic circuits with narrow electrically conducting polymer 'wires' by inkjet printing (Box 20.5). As they describe it in their pioneering paper (Sirringhaus *et al.*, 2000),

> Our approach for overcoming this problem is to confine the spreading of waterbased conducting polymer ink droplets on a hydrophilic substrate with a pattern of narrow, repelling, hydrophobic surface regions that define the critical device dimensions.

In 2000, Friend and Sirringhaus set up the company Plastic Logic, which attracted a large amount of investment from major companies. The sum of these pioneering activities is that plastic electronics came of age. Using these techniques, a vast range of organic polymer devices have been developed, including light-emitting diodes for displays and lighting, photodiodes for light detection and solar energy conversion, field effect transistors, lasers, optical amplifiers and switches for data communications.

20.5 High-temperature superconductivity and the IRC

The first ceramic oxide superconductor to be discovered was in the Ba-La-Cu-O system in 1986, by IBM research workers Georg Bednorz and K. Alex Müller (1986).[5] Its transition temperature was $T_c = 30$ K, and similar materials with much higher values of T_c were soon discovered, including $YBa_2Cu_3O_{6+x}$, known as YBCO, at 93 K, and $Tl_2Ba_2Ca_2Cu_3O_{10+x}$, designated Tl-2223, at 125 K. The new materials were all similar: they contained planes

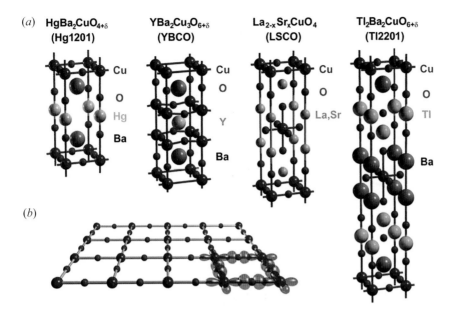

Examples of the crystal structures of four high-temperature cuprate superconductors (Barišić, 2013). (a) The unit cells **Fig. 20.7** with the total number of atoms. (b) The universal CuO_2 building block of the high-T_c cuprates; the most important electronic orbitals are shown in the bottom right of the diagram.

of CuO_2 molecules with the atoms arranged on a simple square or nearly square lattice, separated by planes of other oxides which served to dope the CuO_2 planes with holes so that they became conducting (Figure 20.7). It is primarily in these CuO_2 planes that superconductivity appears. Bednorz and Müller were jointly awarded the Nobel Prize in Physics in 1987.

The appearance of superconductivity at such high temperatures came as a complete surprise to the physics community. According to BCS theory, ordinary metallic superconductors have transition temperatures limited by the strength of the electron–phonon–electron interaction. By the 1980s, it was generally agreed that values of T_c greater than about 30 K were not to be expected.

The discovery of cuprate superconductors opened up the possibility of room-temperature superconductivity, which would be of enormous economic importance for loss-free transmission of electricity and many other technological applications. The UK Science and Engineering Research Council (SERC) under its Chairman William (Bill) Mitchell was already keen to develop a programme of Interdisciplinary Research Centres (IRCs) to improve cross fertilisation between pure science and technology. The discovery of high-T_c superconductivity encouraged the council to select this area for the first such centre. Cambridge was chosen because of the strong base of experimental and theoretical research in the Cavendish Laboratory, combined with more applied work on type II materials under Evetts in Materials Science and applications of superconducting magnets under Campbell in Engineering. The IRC was to be independent of the participating departments, but overseen by the School of the Physical Sciences. It was formally established in 1988, funded by

The Cambridge Interdisciplinary Research Centre in Superconductivity is the first IRC to be granted by SERC, and is now established as a new Department of the University. With funding available for six years, the Centre will concentrate on high temperature superconductivity with research in the fields of:

MATERIAL PREPARATION AND CHARACTERISATION

Ceramic processes, growth of single crystals and thin films, chemical synthesis, electron microscopy and analytic techniques.

ELECTRON, PHONON AND MAGNETIC PROPERTIES

Experimental and theoretical approach to understanding superconductivity and related phenomena.

FABRICATION, PROCESSING AND APPLICATIONS

Polycrystalline material, improvement of critical current, production of thin film electronic devices and of wires and monoliths for electrical machines.

Extensive work has already taken place in these areas in five Departments of the University – Chemistry, Earth Sciences, Engineering, Materials Science and Physics – and the task of the Centre is to build on this work and develop it as a coherent programme.

Fig. 20.8   Transcript of the advertisement for research posts in the IRC published in *New Scientist* on 7 July 1988.

the SERC for an initial six-year period, with Peter Duncumb from Tube Investments as Director to ensure the applied side was not neglected. The advertisement for research workers published in *New Scientist* (Figure 20.8) gives a clear indication of what was intended.

The IRC collaboration had five co-directors: Archie Campbell from Engineering, Jan Evetts from Materials Science, Ekhard Salje, who had a strong interest in the relevant theory, from Earth Sciences and Paul Attfield, an expert in the relevant synthesis techniques, from Chemistry, while W.Y. Liang was seconded from the Cavendish Laboratory as Co-Director for Physics.

Liang was a member of the Physics and Chemistry of Solids Group and an expert experimentalist in the electronic properties of solids. From 1968 to 1987, he had worked extensively on layered structure transition metal dichalcogenides, which have a wide variety of electrical, optical and vibrational properties. In particular, these studies included charge density wave formation, superconductivity, incipient ferroelectric phase transitions, intercalation and fast ion conduction. He made numerous experimental studies of these materials including angle-resolved photoemission spectroscopy, which was well suited to the study of layered structures. Within the IRC his research initially focussed on the electronic modelling of the superconductors and high-resolution specific heat measurements, with a particular aim of understanding their transport properties.

Part of the IRC's experimental work remained within the collaborating departments, but it needed a headquarters, which was first housed in the Bragg Building of the Laboratory. That building was already overcrowded, and conditions were not satisfactory for interdisciplinary research. It was agreed that new accommodation was needed on the West Cambridge site, where there was plenty of room for expansion and easy access to the

The building in the foreground housed the activities of the Interdisciplinary Research Centre in Superconductivity. **Fig. 20.9** Since the end of that programme it has housed a number of activities related to the Physics of Sustainability programme, and has been renamed the Kapitsa Building.

Cavendish Laboratory for liquid helium, stores, administrative support, postal services and the common room where meals were served. But the initial plans were rather ambitious, costing £2.55 million, more than could be afforded. After a considerable amount of to-ing and fro-ing, it was agreed that a more modest building should be constructed using the same CLASP design which had been used in the construction of the new Cavendish and the Microelectronics Research Centre (Figure 20.9). The SERC provided £800,000 and the University the remainder of the £1.55M total.

There was on-going concern about whether an IRC was the best way of advancing the subject. After initial enthusiasm, support waned for research carried out in this manner, and the situation was not helped when Duncumb resigned as Director in 1991, having reluctantly concluded that his personal expertise was too far removed from the physics of superconductivity; he was replaced by Liang. The new building was finally opened in early 1992. There was concern that it had taken five years from the date of the discovery of high-temperature superconductivity before the IRC was fully operational with its own building.

In 1991 Waldram was seconded as Co-Director to replace Liang, while continuing his own research in LTP. Examples of the physics activities he was overseeing in the IRC give an impression of the interdisciplinary nature of the research programme:

- Fundamental theory of high-temperature superconductivity with Joe Wheatley (IRC), Alexander (Sasha) Alexandrov (IRC), Mott, who took up residence in the IRC, and Salje from Earth Sciences
- Transport properties with John Cooper (IRC)
- Thermal properties of high-temperature superconductors with John Loram (IRC)
- High-magnetic-field experiments with Gil Lonzarich (Low Temperature Physics), Andrew MacKenzie (IRC) and Stephen Julian (Low Temperature Physics)

Box 20.6	The problems of high-temperature superconductivity

The problems of understanding high-temperature superconductivity have turned out to be much tougher than anticipated, chiefly because the *normal* state of the cuprate superconductors proved to be hard to understand. The undoped material is an antiferromagnetic insulator. The Née temperature, above which an antiferromagnetic material becomes paramagnetic, falls as the hole concentration rises, but in this region conduction is by hopping between localised states. At the doping level at which superconductivity first appears, the normal state has only just made a Mott transition into a metallic state and still shows strong antiferromagetic fluctuations. T_c is maximised at a doping of 1.15 holes per unit cell, and there is evidence that a Fermi liquid has by that stage appeared, with a nearly two-dimensional Fermi surface. Fully metallic behaviour only appears, however, at still higher dopings.

On the theoretical side, confusion about the normal state led to an extraordinary multiplicity of models to replace the BCS theory in cuprate superconductors. The IRC recruited as its lead theoretician Joe Wheatley, who had worked with Anderson. He developed Anderson's resonating valence-bond model and other theories based on the t–J Hamiltonian suggested by Zhang and Rice. Theories of this type assume that the interaction responsible for superconductivity is a magnetic interaction between spins, the same interaction that is responsible for antiferromagnetism. Sasha Alexandrov, who joined the IRC from Moscow after Wheatley left the Laboratory, worked on a less fashionable proposal, that the important interaction involved the exchange of bipolarons. But all this effort has been to no avail: no microscopic theory of the cuprate superconductors has so far proved successful.

- High-T_c junctions with Ed Tarte (IRC)
- Electron energy loss spectroscopy (EELS) with Mick Brown (Microstructural Physics)
- Infrared spectroscopy with Salje (Earth Sciences)
- STM and SFM microscopy with Adkins (Low Temperature Physics) and Mark Welland (Engineering).

Waldram organised lecture courses within the IRC, which also provided an annual winter school in high-T_c superconductivity for the UK community. In 1996 he published his influential book *Superconductivity of Metals and Cuprates* (1996), which, as well as providing a general introduction to superconductivity, also summarised what was known about high-temperature superconductivity by that date, including much of the physics research of the IRC.

It turned out that the problems of understanding high-temperature superconductivity were much tougher than anticipated (Box 20.6). Nonetheless, good progress was made on the experimental side, in both the physics and the engineering aspects of the discipline.

John Cooper joined the IRC from Zagreb, and brought with him a wide range of experimental expertise in the fundamental transport and thermodynamic properties of unusual compounds, such as single crystals of quasi-one-dimensional conductors, organic conductors and high-T_c oxide superconductors. He and his graduate students carried out systematic measurements of transport properties of the normal state such as electrical conductivity, Hall effect and thermopower as a function of temperature and doping, which were important in elucidating the transition from insulating to metallic behaviour (Figure 20.10). He also initiated measurements on the transport properties of single crystals in a magnetic field, magnetisation and magnetic susceptibility, and the superconducting penetration depth. These experiments helped demonstrate universal properties of all hole-doped cuprates, as well as the importance and ubiquity of the mysterious *pseudogap*, a sort of persistence of the BCS energy gap into the normal state.

John Loram made an important series of thermodynamic measurements on the cuprates, mainly using a difficult technique which allowed the difference in heat capacity between doped and undoped materials to be determined, which clearly brought out the electronic contribution. In the normal state these measurements followed the transition from insulator to full metallic behaviour, and again demonstrated the existence of the pseudogap. In the superconducting state his measurements showed the greatly enhanced fluctuations near T_c associated with the very short coherence length in the cuprates, and also supported the model of *d*-wave superconductivity. In other areas, the IRC made substantial advances in understanding the type-II behaviour of cuprates, and more practically in sample and thin-film preparation and magnet design.

The programme was reviewed after six years and the grant extended for a further three years. After that time, the SERC stopped direct funding of the Centre, the participants being invited instead to apply for standard collaborative research grants. The management arrangements also changed. The General Board of the University ruled that

> The Centre's Building in West Cambridge will continue to accommodate interdisciplinary research in superconductivity, and will be assigned for administrative purposes to the Department of Physics for an interim period of two years from 1 April 1998. Permanent support posts associated with the Centre will also be transferred to Department of Physics for the same period . . . It will be managed by a Research Committee which will include all the holders of relevant research grants and a number of co-opted members.

Research continued for a number of years, but eventually the decision was taken by the School of Physical Sciences that the activities of the participating departments should be relocated to their home bases. The building was refurbished and renamed the Kapitsa Building, acknowledging its low-temperature legacy. It now houses the Optoelectronics and Nanophotonics Groups. Low-temperature activities were to be reinvigorated with new appointments and the LTP Group was renamed the Shoenberg Centre for Quantum Matter (Section 22.8.4).

20.6 Low-temperature physics

Research in low-temperature physics was significantly strengthened with the appointment of Gilbert (Gil) Lonzarich as a demonstrator in 1978 (Section 18.4.4). He was promoted to a lectureship in 1981, to a readership in 1990 and to the post of Professor of Condensed

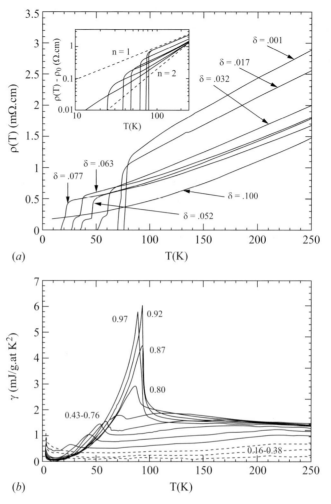

(a) Electrical resistivity of overdoped ($Tl_{0.95}Cu_{0.05}$)$Ba_2CuO_{6+\delta}$, showing abnormal normal state behaviour (Wade et al., 1994). The overdoping increases from top to bottom. Although the behaviour in the optimally doped top curve is superficially similar to what one would have observed in an ordinary metal such as copper, the corresponding electron concentration revealed by the group's Hall effect observations is not constant but varies roughly as T and so the observed resistivity implies a scattering rate roughly proportional to T^2, probably associated with electron–electron scattering rather than the phonon scattering that dominates in copper. (b) Electronic specific heat capacity of $YB_2Cu_3O_{6+x}$ measured by Loram and his colleagues, plotted as $\gamma(x, T) = C_{el}(T)/T$, which would have been, constant in a normal metal, for various levels x of hole doping shown on the diagram (Loram et al., 1993). The dashed lines correspond to the antiferromagnetic insulating state. It is clear that ordinary metallic behaviour above T_c only appears for the heaviest dopings. Note the cusp-like peak at T_c, associated with large critical fluctuations due to the short coherence length, which would not have been seen in a low-T_c superconductor such as tin.

Matter Physics in 1997. In parallel with activities in high-temperature superconductivity fostered in the IRC, Lonzarich focussed upon the physics of materials which displayed quite new phases of matter under extreme conditions of low temperature, high pressure and high magnetic fields. The necessary facilities for carrying out these experiments were built up during the Edwards era, Lonzarich benefitting from the support of an outstanding series of graduate students and postdocs, among them Stephen Julian from Toronto, who came as a postdoctoral research fellow in 1988 and then was appointed a University lecturer in 1995.

20.6.1 Itinerant-electron magnetism, quantum phase transitions and unconventional superconductivity

Lonzarich's research interests continued to expand during the Edwards era. His insights into how spin fluctuations, or paramagnons, affect the magnetic equation of state and increase the mass of quasiparticles led him to consider the possible breakdown of Fermi liquid theory and the emergence of novel types of ground states on the border of magnetic quantum phase transitions at low temperatures. The study of such phenomena required not only the purification of carefully chosen materials but also techniques for fine-tuning their properties without introducing unwanted disorder. This led to further development of the pressure-tuning technique already used in his early studies in iron and nickel.

In the mid 1980s, during a visit to Los Alamos, Lonzarich began using high pressure to tune materials as close as possible to magnetic instabilities. On his return he organised the development of apparatus in which pressures up to 30 kbar could be applied to systems at mK temperatures. This was achieved with very significant contributions by Stephen Julian.

The Laboratory facility currently used for these experiments is shown in Figure 20.11(a). The cryogenic system extends deep underground and is located on a mounting vibrationally isolated from the rest of the building, because of the necessity of maintaining a vibration-free environment. Figure 20.11(b) shows the construction of an artificial diamond cell for creating extreme pressures, as well as the external modulation coil and internal sensing microcoil. These facilities, in conjunction with the ovens and apparatus needed for creating ultra-pure samples of the materials studied, were to prove to be the workhorses for a very wide range of studies of matter under extreme conditions.

In parallel, Lonzarich and his theorist colleagues, in particular Greg McMullan and Philippe Monthoux, developed the theory of spin fluctuations, non-Fermi liquid behaviour and magnetically mediated superconductivity, which helped identify promising candidate materials for more detailed pressure-tuning investigations (Lonzarich, 1994, 1997; McMullan and Lonzarich, 1997; Monthoux and Lonzarich, 1999). These efforts led to a number of major advances during the 1990s, including the seminal discovery of unusual forms of superconductivity near the border of metallic antiferromagnetism or ferromagnetism in $CePd_2Si_2$, $CeIn_3$, $CeNi_2Ge_2$ and UGe_2 (Mathur *et al.*, 1998; Saxena *et al.*, 2000).

These and many related discoveries in the next two decades showed that new forms of superconductivity and other exotic states can arise on the border of magnetic instabilities in samples with sufficiently low levels of quenched disorder. This breakthrough challenged

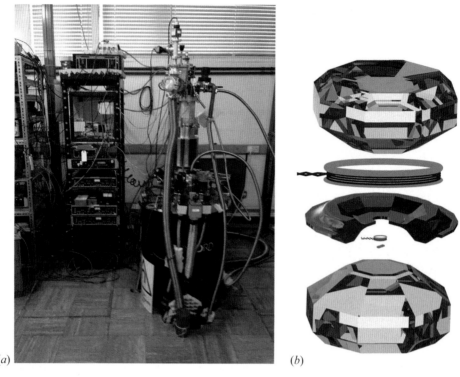

(a)   (b)

Fig. 20.11 (a) The cryogenic facility for the study of materials under extreme conditions of low temperature, high magnetic field and high pressure, developed by Lonzarich, Julian, Walker and their colleagues (see also Box 22.8). (b) The structure of a diamond cell for the study of materials under extremely high pressures.

traditional views on the incompatibility of magnetism and superconductivity and on the pair-breaking tendency of slow spin fluctuations near magnetic quantum critical points. The discoveries were, however, consistent with the pairing of fermion quasiparticles through the exchange of spin fluctuations, or more generally through magnetic interactions, rather than by the 'standard' exchange of phonons as in the BCS theory (Monthoux and Lonzarich, 1999). A simple introduction to some of these topics and the concept of quantum criticality is given in Box 20.7, based on the illuminating review by Piers Coleman and Andrew Schofield (Coleman and Schofield, 2005). Some examples of these exotic phases of matter are included in Box 20.8.

Also during this period, high-pressure measurements on MnSi by Lonzarich's student Christian Pfleiderer with the help of Julian and others, revealed a peculiarly robust breakdown of the conventional theory of metals, which became known as *strange metal behaviour* (Pfleiderer *et al.*, 1997), for which a new theoretical description was developed (Lonzarich, 1997; McMullan and Lonzarich, 1997). This marked the beginning of a new area of research, which culminated in the detection of skyrmions – magnetic vortices with promising technological applications.

Meanwhile techniques for the detection of magnetic quantum oscillations were also being advanced in the group. Lonzarich and Julian, together with Ross Walker and Sam

Despite their diversity, different phase transitions – for example, the crystallisation of water into snowflakes, the alignment of electron spins inside a ferromagnet, the emergence of superconductivity in a cooled metal – often share many fundamental characteristics. The specific heat when water turns to steam at a critical pressure has exactly the same power-law dependence on temperature as that of iron when it is demagnetised above the Curie temperature. Understanding such 'critical phenomena' was a triumph of twentieth-century physics. One of the key discoveries was that the onset of order at such continuous phase transitions is preceded by the formation of short-lived droplets of nascent order that grow as the system approaches the critical point. At the critical point, the material is spanned by droplets of all sizes. The melting of ice is, like most phase transitions, caused by the increase in random thermal motion of the molecules which occurs as the temperature is raised.

In quantum criticality, the randomness is not associated with thermal motion, but with the quantum fluctuations which must be present according to Heisenberg's uncertainty principle. These quantum fluctuations are called 'zero-point motion'. The thermal motion of particles ceases at absolute zero temperature but the atoms and molecules cannot be at rest because this would simultaneously fix their positions and velocities. Instead they adopt a state of constant agitation. If the zero-point motion becomes large enough, the material can 'melt', but in this case the melting takes place at absolute zero.

Among the best studied examples of quantum phase transitions is magnetism in metals. Iron is magnetised when all the spins align in parallel, but in other materials the spins form a staggered, alternating or antiferromagnetic arrangement, left-hand part of the image). These fragile types of order are susceptible to melting by zero-point fluctuations. Such 'quantum critical matter' offers the real prospect of new classes of universal electronic behaviour developing independently of the detailed material behaviour, once the material is driven close to a quantum critical point.

Figure (a) is a schematic phase diagram of a heavy fermion metal–antiferromagnet phase transition near a quantum critical point (Coleman and Schofield, 2005). The regions outside the V are stable phases, in this example antiferromagnetic in region A and a heavy fermion metal in region N. The boundary of the low-temperature phase can be 'tuned' by changing the chemical composition of the material or by the application of very high pressures or very high magnetic fields – collectively, the possible means of changing the boundary of criticality are described as *tuning parameters*.

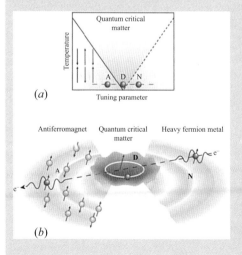

(a)

(b)

The presence of a quantum critical point distorts the structure of the phase diagram, creating a 'V-shaped' phase of quantum critical matter fanning to finite temperatures from the quantum critical point D.

(b) Physics inside the V-shaped region of the phase diagram probes the interior of the quantum critical points (D), whereas the physics in the heavy fermion metal (N) or antiferromagnet (A) reflects their exterior. Experiments that tune a material from the normal metal past a quantum critical point force electrons through the 'horizon' in the phase diagram, into the interior of the quantum critical matter, from which they ultimately re-emerge through a second horizon on the other side into a new universe of magnetically ordered matter.

Box 20.8   **Novel states of matter under extreme pressures, temperatures and magnetic fields**

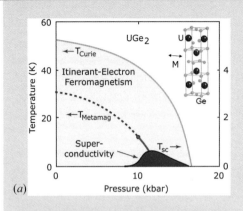

(a)

Exotic states of matter on the edge of magnetism

The ferromagnet UGe_2 is an example of a material which exhibits exotic states of matter. With the application of hydrostatic pressure in a pressure cell, the magnetic transition temperature T_c is suppressed towards zero and at the same time a superconducting state is found at low temperatures (Figure (a)). In this case, superconductivity is most naturally understood in terms of magnetic as opposed to lattice interactions, and by a spin-triplet pairing rather than the spin-singlet pairing normally associated with nearly antiferromagnetic metals (Saxena et al., 2000).

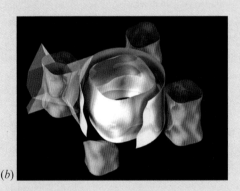

(b)

Electronic structure of correlated electron materials

The exact form of the de Haas–van Alphen oscillations and their dependence on field and temperature enable both the shape of the Fermi surface and the effective quasiparticle masses to be determined. An example of the complex Fermi surface of the oxide material Sr_2RuO_4 measured in the quantum physics laboratories is shown in Figure (b). For such materials, the results are often completely surprising and strongly deviate from theoretical predictions. Studies of heavy-fermion materials can yield effective mass values 100–1000 times greater than electron masses in vacuum which cannot be inferred from first principles. Such studies about the metallic state provide key information for theories of superconductivity and other Fermi liquid instabilities (Bergemann et al., 2000, 2003).

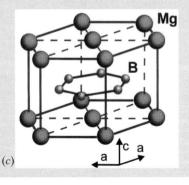

(c)

Novel superconductors

The superconductor MgB_2 can be thought of as magnesium layers sandwiched between hexagonal boron layers, as depicted in Figure (c). With a remarkably high T_c of 39 K, this compound has attracted considerable interest, from both a theoretical and applications perspective. The Quantum Physics Group has been involved in studies of its Fermi surface using piezo-resistive torque cantilevers to probe oscillations in the magnetisation.

Brown, developed a state-of-the-art cryomagnetic system featuring a dilution refrigerator capable of reaching 6 mK, a nuclear demagnetisation stage reaching below 1 mK and a superconducting magnet system reaching 18 T in the sample region and 6 T in the demagnetisation stage. Two early examples of the use of the system illustrate the richness of this field of low-temperature physics research:

- Andrew (Andy) MacKenzie, a member of the IRC in Superconductivity who worked with Julian, observed for the first time quantum oscillations in Sr_2RuO_4, one of the few perovskite superconductors not containing copper (MacKenzie et al., 1998b). This established the detailed nature of the normal state of this important and unusual superconductor. The exotic nature of the electron pairing state was strikingly demonstrated in one of MacKenzie's studies, which showed that the superconducting transition temperature is extremely sensitive to the presence of both magnetic and non-magnetic impurities (MacKenzie et al., 1998a). Sr_2RuO_4 is now believed to be a spin-triplet p-wave superconductor, which is of great interest in the search for a new type of emergent particle known as the Majorana fermion.
- Building on MacKenzie's collaboration with John Cooper of the IRC on understanding the copper oxide superconductors, MacKenzie and Julian also used the cryomagnetic system to investigate the upper critical field of the new cuprate superconductors $Tl_2Ba_2CuO_{6+}$ and $YBa_2Cu_3O_7$ down to temperatures of a few mK and showed that their behaviour was in sharp contrast to that predicted by BCS theory (MacKenzie et al., 1993). This led to the understanding of the important vortex liquid state, a chief distinguishing characteristic of the copper oxide superconductors and other exotic superconductors.

20.6.2 Physical processes in amorphous solids

Following the move to West Cambridge, Adkins invested in a dilution refrigerator, allowing tunnelling work to be pushed to temperatures below 1 K, where the energy resolution becomes sharper, yielding more precise measurements of superconducting energy spectra. In 1984 he realised that tunnelling experiments could be used to conduct inelastic tunnelling spectroscopy, and a long sequence of experimental and theoretical work followed, much of it in collaboration with Phillips. This started with studies of phonons in amorphous SiO_x films and moved on to explore surface phonon modes and phonon modes of adsorbed molecules (Payne et al., 1984; Adkins and Phillips, 1985).

Phillips and his students explored the links between two-level states and the vibrational and configurational behaviour of amorphous solids at higher temperatures, including the dynamics of glasses as they pass through the glass transition. Neutron scattering studies at the Institut Laue–Langevin in Grenoble led to the identification the microscopic nature of these states, and to an explanation of how they were a fundamental property of glasses (Buchenau et al., 1986, 1988; Phillips et al., 1989).

The continuing interest in this field is a reflection of the universal occurrence of two-level states in non-crystalline materials and of the recognition that they are the origin of $1/f$ noise. Phillips's model of two-level states has been extended to explain $1/f$ noise in a great

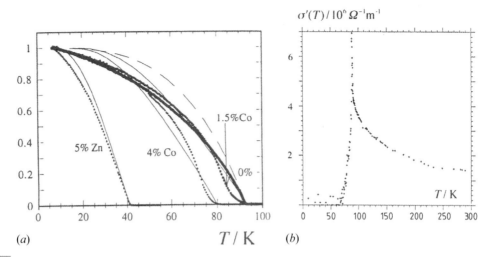

Fig. 20.12 Physics of cuprate superconductivity revealed by microwave measurements. (*a*) Temperature dependence of $n_S(T)$ for YBCO doped with gap-breaking zinc impurities and non-gap-breaking cobalt impurities. The solid and dashed lines show the BCS predictions for a conventional superconductor in the weak coupling and strong coupling limits. The quite different behaviour observed at low T is due to the presence of gap nodes on the cuprate Fermi surface (Porch *et al.*, 1993). (*b*) Normal electron conductivity $\sigma'(T)$ in YBCO, showing the effect of the very large critical point fluctuations near T_c (Waldram, 1996, after Porch *et al.*, 1990).

variety of electronic devices, ranging from semiconductor transistors to superconducting qubits.

20.6.3 Microwave properties of high-T_c superconductors

The discovery of high-T_c superconductors in 1986 generated much activity in the LTP Group, particularly those working with microwaves. Initially only small sintered discs were available, and it was decided that the best approach was to grind them up and return to Shoenberg's technique of measuring the susceptibility of finely divided powders. Because the materials were highly anisotropic, it was necessary to align the powders, but this was easily achieved using a superconducting magnet, and useful results soon followed (Cohen *et al.*, 1987). By 1991, the group had moved on to studying the microwave properties of good-quality thin films prepared in the Materials Science section of the IRC.

   The cuprate superconductors proved to have very short coherence lengths. This meant that Pippard non-locality could safely be ignored, which made the analysis of microwave results much simpler, and the group was able to report quickly on the magnetic penetration depth of YBCO for currents flowing parallel to the CuO_2 planes. It was immediately apparent that the superelectron density $n_S(T)$ rose much more slowly with decreasing temperature than BCS theory predicted (Figure 20.12(*a*)). This in turn suggested that the energy gap must be smaller than the BCS gap, at least on some parts of the Fermi surface. It turned out that the cuprate superconductors were *d-wave superconductors*, in which the gap parameter has a pair of nodes in the k_x–k_y plane. On these nodes the gap falls to zero, which explained the unusual temperature dependence of $n_S(T)$. The appearance of d-wave superconductivity suggested at once that, even if a ground state of the BCS form were present, the attractive interaction between the electrons in the cuprates was very unlikely to be the electron–phonon–electron interaction, and was most probably magnetic in origin.

The microwave measurements also provided values of the conductivity $\sigma'(T)$ due to the normal electrons, and by comparing results at different frequencies it was possible to determine the temperature dependences of both the density of normal electrons $n_N(T)$ and their scattering time $\tau(T)$, within the superconducting state. The behaviour of $n_N(T)$ confirmed the results of the penetration depth measurements: there were more normal electrons than in conventional BCS theory because of the reduced gap near the nodes of the gap parameter.

Near T_c the short coherence length meant that the cuprates, unlike conventional superconductors, showed conspicuous critical fluctuations. This led to a large coherence peak in the microwave resistive losses, which the group was able to explain quantitatively using an effective medium model (Figure 20.12(b)) (Waldram *et al.*, 1999).

When good single crystals of cuprate superconductors prepared in the IRC became available, Rodrigo Ormeno developed a sapphire-rod resonator in order to make accurate microwave measurements. This work was taken up and developed much further by David Morgan and Dave Broun, who greatly increased the precision, sensitivity and frequency spread of the measurements, making systematic studies particularly of the low-temperature excitations of various cuprates and the scattering that they undergo (Broun *et al.*, 1997; Waldram *et al.*, 1997).

20.7 Microstructural physics

The capabilities of the scanning transmission electron microscope (STEM) continued to advance through the Edwards era, the activity being supported by the appointment of Steven Berger, an expert in electron energy loss spectroscopy (EELS), as an SAIR from 1984 to 1988. Mick Brown had a long-standing interest in diamond as an ideal subject for study by EELS since it has a large band gap and relative stability under an intense beam of electrons. Natural diamond contains defects known as platelets, and Brown and Bruley were able to show that the nitrogen present is not in the platelets but in tiny bubbles, known as voidites, at very high pressure (Bruley and Brown, 1989). As a consequence, the diamonds must have formed rather deeper in the Earth's mantle than had been previously assumed.

The next major advance towards achieving sub-Å resolution was made by Ondrej Krivanek in collaboration with members of the Microstructural Physics Group. Krivanek, from the former Czechoslovakia, had completed his PhD studies in the Laboratory in 1975 under the supervision of Howie. In the late 1970s, having moved to the USA, he developed a series of electron energy loss spectrometers (EELS) which were commercially very successful. He co-authored the *EELS Atlas*, which became a standard reference for EELS

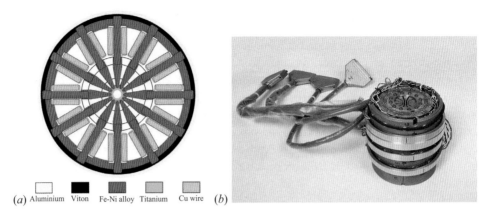

(a) Aluminium Viton Fe-Ni alloy Titanium Cu wire (b)

Fig. 20.13 Krivanek's corrector for the first aberration-corrected scanning transmission electron microscope (STEM). (a) The layout of one of the lenses of Krivanek's corrector. (b) Krivanek's automated spherical aberration corrector.

measurements in electron microscopy (Ahn and Krivanek, 1983). In the meantime, the development activities in electron microscopy continued with the appointment of Andrew Bleloch as an SAIR in 1989.

Krivanek returned to the problem of eliminating aberrations in electron microscope imaging, to which he had already made significant contributions in the 1970s. He had developed algorithms for tuning the performance of electron microscopes, but now planned to develop sub-Å imaging in working STEMs. Having failed to obtain funding in the USA, he, Brown and Bleloch applied successfully to the Royal Society for funding, and he was granted leave of absence from the Gatan Company to come to the Laboratory to develop the system.

As described in Section 15.2.2, Hans Deltrap had made pioneering advances in correcting the aberrations of the STEM, but alignment and adjustment problems prevented the practical implementation of his scheme. Krivanek's objective was to solve these problems and then correct the aberrations to third order, a key challenge for electron microscopy. During his period of leave in Cambridge, he developed a corrector with 84 independently adjustable optical elements and automated computer control. Twelve elements were used for the correction of spherical aberration (C_s). The rest served for alignment and for correcting 11 aberrations other than C_s (Figure 20.13). The system was controlled by two computers. One computer acquired images and measured all the important aberrations on-line, while the other nulled the aberrations by adjusting the system's 84 precision power supplies. Each horizontal section of the corrector contains one strong quadrupole and one strong octopole (Figure 20.13(a)). The complete corrector consists of a stack of six identical stages (Figure 20.13(b)). These are energised by windings distributed as appropriate over 12 poles and connected together. There is also a weak auxiliary winding on each pole. The computer controls these windings in order to produce weak multipole moments for 'fine tuning'. The team reported their progress towards the first working quadrupole–octopole probe C_s corrector in 1996 and 1997 (Krivanek *et al.*, 1996, 1997b).

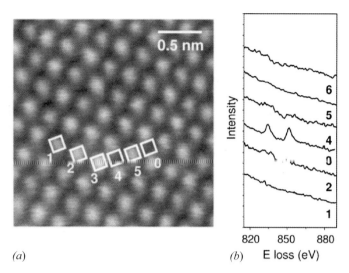

(a)   *(b)*   E loss (eV)

EELS observations of atoms in a crystal with subatomic resolution. (*a*) An image of a crystal consisting of a stack of CaTiO$_3$ and La$_x$Ca$_{1-x}$TiO$_3$ molecules with a small number of calcium atoms replaced by lanthanum (La) atoms, $x = 0.04$. The atom labelled 3 is lanthanum. (*b*) EELS traces showing spectroscopic identification of a single La atom at atomic spatial resolution, with the same beam used for imaging. The M4;5 lines of La are seen strongly in spectrum 3. The spectra from neighbouring columns show a much reduced or undetectable La signal (Varela *et al.*, 2004).

Fig. 20.14

Together with his colleagues in the Microstructural Physics Group and Niklas Dellby of the Gatan Company, Krivanek achieved the projected improvement in imaging by 1997. In the same year, Krivanek and Dellby set up the Nion Company, where a new corrector design was developed (Krivanek *et al.*, 1997a, 1999). By 2000, the first aberration correctors had been sold commercially, and these were to become the industry standard for sub-Å imaging. Subsequent versions of Krivanek's corrector achieved resolutions of 0.1 nm in 100 keV STEM measurements. Nion correctors delivered to the Oak Ridge National Laboratory produced the first sub-Å resolution electron microscope images of a crystal lattice of CaTiO$_3$ containing a small number of lanthanum atoms. Figure 20.14 shows the image of the crystal with the atoms clearly resolved and the EELS spectrum of a single atom of lanthanum in a bulk solid (Varela *et al.*, 2004). These technologies were to be used in the SuperSTEM project at the Daresbury Laboratory, the UK's EPSRC National Facility for Aberration-Corrected Scanning Transmission Electron Microscopy. Bleloch was appointed project scientist for this national project and moved to Liverpool University and the Daresbury Laboratory.

20.8 Physics and chemistry of solids

The figures in Table 19.2 show the exodus from the formerly dominant Physics and Chemistry of Solids Group of a number of its most creative physicists to form new groups in

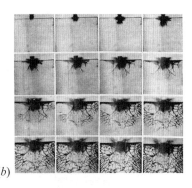

(a)   (b)

Fig. 20.15   (a) Munawar Chaudhri with the Beckman and Whitley model 189 ultra-high-speed framing camera, now in the Cavendish display of scientific instruments. (b) Fracture propagation in a $50 \times 50 \times 6.2$ mm sheet of toughened windscreen glass after impact by a lead airgun pellet (Field, 1971).

areas which had been nurtured in PCS but which now took on lives of their own. With the retirement of Yoffe, John Field took over the reins, looking after three main areas of research: (1) the traditional area of high-energy materials, explosives and shock waves, which was Bowden's legacy and Field's particular area of expertise; (2) surface physics, which benefitted from the appointment of Bill Allison as a demonstrator in 1983, and (3) a new area, thin-film magnetism, the area of expertise of Tony Bland, who was appointed to a University Assistant Lectureship in 1989.

20.8.1 Shock waves, explosives and high-energy materials

John Field had continued the PCS Group's involvement in energetic phenomena, in particular the physics of explosions, shock waves, the effects of high-speed impacts, erosion and so on, very much in the spirit of Bowden. This work was well supported by grants from government and military establishments, enabling a vibrant programme of research to be maintained. Through close association with Pilkington, British Aerospace, AWE Aldermaston, QinetiQ, Boeing and major governmental and industrial organisations, as well as the EPSRC, Field and his colleagues built up state-of-the-art facilities in areas such as ultra-high-speed photography, X-ray and optical techniques and ballistic impact, including liquids, solid particles and projectiles. The development of such facilities was clearly in the national interest and Field and his colleagues trained a large number of students in these techniques.

Examples of the facilities developed during the Edwards era and beyond include:

- high-speed photography facilities with cameras capable of 10^8 frames per second (Figure 20.15)
- optical and X-ray facilities allowing precise quantitative measurement of displacements during static or dynamic loading for processes involving fracture, impact or shock
- techniques using X-rays and digital cross-correlation speckle to study internal deformations in opaque solids to sub-micron accuracy

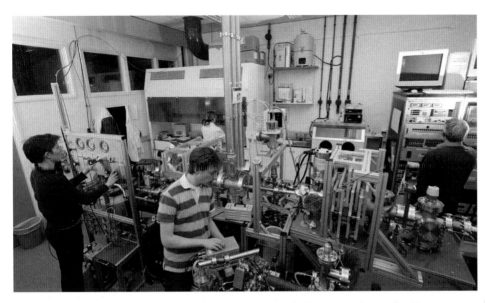

The helium microscope laboratory in 2008. A photograph of the spin-echo laboratory is shown in Box 22.14.   **Fig. 20.16**

- mechanical studies of the properties of thin solid films and bare surfaces using sophisticated nano-indentation techniques.

20.8.2 Surface physics

Surface physics had been a staple component of the Laboratory's research programme from Bowden's time. With Tabor's retirement in 1979, activity in this area had contracted, but it was taken in new directions with the appointment of Bill Allison as a University Demonstrator in 1983. Allison began a programme of high-resolution helium-atom scattering for surface studies as part of the activities of the PCS Group. He was joined by John Ellis in 2001 when the Physics Department at the University of Essex was closed. The techniques involved the scattering of low-energy helium atoms and measuring the velocities and directions of the atoms scattered from surfaces in a time-of-flight apparatus.

In the inelastic scattering measurements, after the beam leaves the source it is modulated by a mechanical chopper before interacting with the sample. Scattered atoms then traverse a fixed distance before reaching the detector, where their time of flight from the chopper is recorded. The quasi-elastic broadening of the velocity distribution is determined by comparing time-of-flight distributions before and after scattering. The helium microscope laboratory as it was in 2008 is shown in Figure 20.16.

The *spin-echo* approach overcomes the main limitation imposed by the velocity distribution of the beam of atoms. Now a polarised beam of ^{3}He atoms is used which have small but finite nuclear magnetic moments. The technique determines the energy change on scattering using NMR-like spin precession of the nuclei of the ^{3}He beam. Improvements in resolution of nearly four orders of magnitude relative to unpolarised atoms have

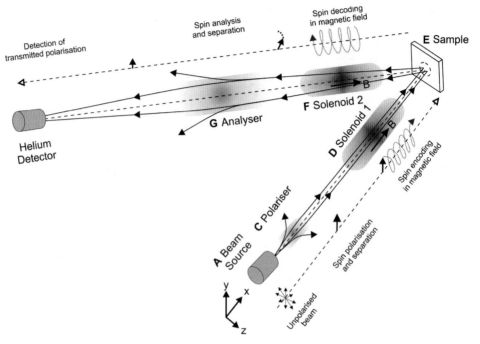

Fig. 20.17 Classical schematic representation of the spin-echo technique (Jardine *et al.*, 2009). An unpolarised ^{3}He beam is first produced in beam source A before passing through the nuclear spin polariser C. The polarised spins are transferred into the precession solenoid field D, before scattering from the surface E. Some of the scattered atoms pass through an identical, but reversed field F, before being spin-analysed at G and counted in the detector, to provide a beam-averaged polarisation measurement, following procedures described by Alexandrowicz and Jardine (2007).

been achieved. The technique is illustrated schematically in Figure 20.17. On the incident leg, between the source and sample, there is a beam polariser, which aligns the ^{3}He nuclear spins, followed by a spin-precession solenoid. On the outgoing leg, the scattered atoms pass through a further spin-precession solenoid before entering a spin analyser and detector. If the atoms are scattered perfectly elastically from the surface, the original polarisation is recovered. This recovery of the polarisation is called the 'echo'. The technique is much less sensitive to the initial velocity distribution than the time-of-flight method. Different magnetic field strengths correspond to measurements over different time-scales. Although the spin-echo spectrometer is primarily aimed at dynamical measurements, it is also uniquely suited to the measurement of helium–surface potentials through selective adsorption resonance. In these processes, selective adsorption resonances occur where a helium atom becomes transiently trapped in one of the energy levels of the helium–surface potential, instead of diffracting into a real, observable beam. These techniques can be applied to a wide range of surface structures and processes and demonstrate the real-time movement of atoms and molecules over the surface (Figure 20.18). The Surface Physics Group is unique among surface science groups in specialising in helium atom scattering, which is complemented by more traditional techniques.

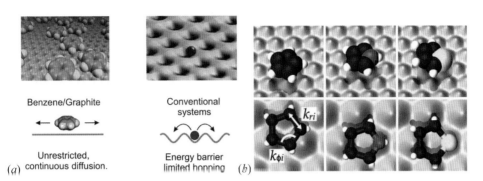

(a) Diagrammatic comparison of the interaction between the mobile species and the surface for the observed atomic-scale Brownian behaviour involving the unrestricted continuous diffusion of benzene on graphite (left), and for conventional hopping between sites on a surface (right) (Lechner *et al.*, 2013). (b) Atomic-scale friction of five-membered ring molecules on surfaces (Hedgeland *et al.*, 2009).

Fig. 20.18

20.8.3 Thin-film magnetism

Tony Bland carried out his PhD studies in surface particle scattering at the Laboratory under the supervision of Roy Willis. After a year as a research scientist at the Institute Laue–Langevin in Grenoble and three years as a Research Fellow at the Clarendon Laboratory at Oxford University, he returned to Cambridge in 1987. There he started a dynamic and ambitious programme of research into thin-film magnetism, a topic of immediate importance because of its application in data storage and in recording media. Almost immediately, he made a number of key contributions to the discipline. As his colleague Bretislav Heinrich wrote in 1994,

> His expertise in the polarised neutron diffraction technique (PNR) is unmatched anywhere in the world. His quantitative understanding of diffuse scattering due to interface roughness allowed him to determine the magnetic moments in Fe/non-ferromagnetic metal interfaces. In fact, Dr. Bland's spin polarised neutron diffraction results were the first clear experimental determination of enhanced magnetic moments at such interfaces.
>
> This enhancement is central to the understanding of magnetism in low dimensional systems. This effect was theoretically predicted at the beginning of the era of magnetic metallic molecular beam epitaxy. However, for many years, this crucial prediction escaped clear experimental verification. In this respect the polarised neutron reflection work carried out by Dr. Bland's group represents a truly important discovery.
>
> Recently they have added to their experimental repertoire a fully vectorial Magneto-Optical Kerr Effect (MOKE) apparatus including an imaging system. These are powerful tools for determining the modes of magnetisation reversal in the film specimens. These allowed them to answer many questions concerning the nucleation of magnetic domains, domain wall propagation and magnetisation rotation. This led to the publication of a book on these topics. (B. Heinrich, personal communication, 2015)

The fields of thin- and ultra-thin-film magnetism became feasible with the development of molecular beam epitaxy, but were now focussed upon the creation of magnetic

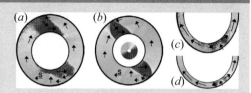

A key issue in nanomagnetism is to understand and control magnetic switching precisely. A geometry that fulfils these criteria is the ring geometry, in particular, narrow ferromagnetic rings. This geometry is well suited to the investigation of fundamental magnetic properties such as domain wall trapping and current-induced domain wall displacement. Magnetic measurements and micromagnetic simulations have shown the existence of two magnetic states, the 'onion' state, characterised by the presence of opposite head-

to-head and tail-to-tail domain walls, and the flux-closure 'vortex' state (Kläui *et al.*, 2003).

Nanoscale magnetic imaging and micromagnetic simulations have shown that the onion state exists with transverse or vortex walls as shown in the figure. The onion-to-vortex switching has been found to be a very fast nucleation-free domain wall propagation process with switching times of the order of 400 ps. It has been proposed for applications such as magnetic random access memory (MRAM) and bio-detection.

structures on micro- and nanoscales. Bland was a leader in the many aspects of this very rapidly developing field. By 1994, he and Heinrich had completed the editing of their two-volume monograph which spanned essentially all aspects of nanomagnetism, in which Bland was recognised as a world authority (Bland and Heinrich, 1994). He was a dynamic and charismatic leader who attracted a large body of graduate and postdoctoral students. The range of his interests was well summarised by Peter Littlewood, following Bland's tragic suicide in 2009:

> Tony's group is recognised as a world leader in the field of nanomagnetism, focusing on the atomic-scale origins of magnetism, spintronics, magnetic nanostructures and materials. He was held in high esteem and those who met him were stimulated by his enthusiasm and charisma.[6]

From his earliest researches, he fully appreciated the potential of *spintronics*, the use of the spin of the electron rather than its charge as the basic entity in nano- and micro-circuitry. The energy involved in switching spin states is very much less than that for electronic states and it can also take place much faster (Box 20.9). Consequently there would be enormous advantages in terms of speed and energy consumption if spintronics could become a reality. But the technical challenges are enormous. One of Bland's challenges was to devise techniques for injecting spin-polarised electrons into semiconductor materials. He also fully appreciated that magnetic nanostrutures could be used as bio-markers.

Following Bland's death, the management of the Thin Film Magnetism Group was taken over by Crispin Barnes, and in 2010 Russell Cowburn joined the Laboratory, enhancing the strength of the activity in nanomagnetism and spintronics (Section 22.9.4).

20.9 Theory of condensed matter

The Edwards era saw a dramatic expansion of theoretical activities, particularly in the area of of soft matter. The appointments of Robin Ball (1983), Michael Cates (1989) and Mark Warner (1989) had a large impact on the direction of theoretical research in condensed matter physics, and yet it was built into the philosophy of the group that common threads ran through the whole of the programme from fundamental issues in strongly correlated electron physics to biological physics. As Ball has written in his description of the nature of condensed matter physics:

> From the theoretical point of view, the chief characteristic of condensed matter is the presence of a very large (effectively infinite) number of strongly interacting particles, whether in the quantum mechanics of electrons or the classical mechanics of the motion of polymer chains. Fortunately, these many-body problems often reduce to the motion of nearly independent excitations called quasiparticles. Examples [include] the reptation of a polymer chain snaking its way in a polymer melt, the screened d-shell electrons causing magnetism in transition metals and the ordering modes which organise the arrangement of atoms in a mineral as it cools from the melt. However, the many body nature is essential in some research problems such as the fractional quantised quantum Hall effect, the localisation of electrons, polymer entanglements and aggregation.
>
> . . . similar mathematical ideas run through a wide range of topics on the atomic and electronic structure of various intrinsically disordered systems, including composite fibre materials, colloids, polymer melts, glasses, rubbers, amorphous semiconductors, spin glasses, fractals of aggregation, magnets disordered at high temperature. Aggregation theory has important applications in oil reservoir simulation.[7]

These efforts strongly benefitted from the presence of David Khmelnitskii, who had carried out pioneering research in many areas of many-body theory, having been trained at the Landau Institute of Theoretical Physics in Moscow. As is stated on his personal page at the Cavendish website,

> [Khmelnitskii's] accomplishments are associated with the renormalization-group theory of critical phenomena, effects of disorder on phase transitions, the Quantum Hall Effect and coherent phenomena in disordered conductors, including weak localisation, anomalous magnetoresistance and mesoscopic fluctuations.

Khmelnitskii was awarded a Senior Research Fellowship at Trinity College in January 1991, which he held until his retirement. He chose to make the TCM Group his academic base, where his inspiration and experience was of great benefit to the group.

Ball was a pioneer in the area of diffusion-limited aggregation, addressing issues such as how objects such as snow crystals and soot particles form. Once a nucleus is formed, individual molecules or other specks of soot stick to it, and then more and more are added to the cluster. They are most likely to first hit one of the outer parts of the incipient cluster and so it grows in a fluffy form. To put it more technically, Ball extended the concept of reptation to non-linear molecules and compatible blends using scattering techniques applied to diffusion in solution.

Box 20.10   **How to tie a tie**

One of the more amusing outcomes of the study of the topology of polymers was a remarkable little book by Thomas Fink and Yong Mao entitled *The 85 Ways to Tie a Tie* (Fink and Mao, 1999). They used their understanding of the ways in which polymers become entangled to work out topologically all the 85 ways in which the crossings could take place. It turned out that only five of the topological structures resulted in a genuine knot, one of them actually being a new way of tying a necktie.

As part of that programme he developed a computer code that was vastly superior to others available at the time. As a result, he could simulate what actually happens under different assumptions as well as trying to explain it. Two further examples of computational investigations are aggregation models and the motion of polymer melts. Clusters formed in kinematically controlled aggregation, such as colloid flocculations, develop spatially self-similar internal structures which are not yet fundamentally understood. Large computer simulations of polymer melts enabled the theories to be tested at the molecular level and allowed details of the way chains really move to be seen. In 1998, Ball accepted a professorship in theoretical physics at Warwick University.

Initially, Warner was appointed to the Polymer and Colloids Group, working directly with Edwards, as described in Section 18.5. He discovered many new phenomena in 'quasi-solids' after creating the first theories of liquid crystal elastomers.

Cates asked whether chains composed of assemblies of soap molecules can break and reform rather than wiggling out of entanglements, a special form of reptation. He became a specialist in complex phases made up of these amphiphiles to form onions, worms and flapping sheets. He was appointed to the Tait Professorship of Physics in Edinburgh in 1994 and returned to Cambridge in 2015 as Lucasian Professor of Mathematics, in succession to Stephen Hawking and Michael Green.

A surprising example of the extensive nature of the study of polymer topology is a book by Thomas Fink and Young Mao, *The 85 Ways to Tie a Tie* (Box 20.10). It became a best-seller.

The importance of computation for theoretical studies continued to form a major part of the programme of the group. The appointment first of Richard Needs in 1983 and then of Michael Payne in 1991 provided an extremely powerful nucleus of expertise and innovation in first-principles total energy pseudopotential computation. These innovative codes were of very wide use by the worldwide community of computational theorists. The CASTEP code was commercialised and marketed by Accelrys, which has a been a major commercial success. This was followed by the ONETEP code which enabled molecules containing thousands of atoms to be treated exactly quantum mechanically.

The appointment of Ben Simons in 1995 brought yet another dimension to the work of the group. An expert on many-body theory, on which topic he wrote a basic text with Alexander Altland (2010), his interests evolved towards the application of these techniques in biology and medicine, with notable success. In due course he was appointed as Herchel Smith Professor of the Physics of Medicine and Director of the Physics of Medicine Building.

The Edwards era: high-energy physics and radio astronomy

21.1 High-energy physics: the LEP era

The 1976 Report to the CERN Council recommended the construction of a Large Electron–Positron (LEP) collider capable of accelerating electrons and positrons to energies at which the W and Z bosons could be created in large numbers. The Council approved the project in 1981, the civil engineering work beginning in 1983. Excavation of the 27-kilometre circumference tunnel began in 1985 and was completed three years later – this excavation was Europe's largest civil-engineering project prior to the Channel Tunnel (see Figure 17.5). To measure the many products of electron–positron collisions, four huge detectors, ALEPH, DELPHI, L3 and OPAL, were installed at the experimental stations around the ring. As described in Section 17.3.1, CERN had already agreed a fast-track route to the discovery of the W and Z bosons and this was achieved in 1982–83. Now the task was to make precision measurements of their properties.

The LEP collider was commissioned in July 1989, its initial energy being about 91 GeV, at which energy Z bosons were created in huge numbers. LEP operated at about 100 GeV for seven years, a total of about 17 million Z particles being produced during that period. In 1995 LEP was upgraded to roughly twice that energy so that pairs of W bosons would be created as well – by 2000, the collider's maximum energy exceeded 209 GeV. During its 11 years of operation, the LEP experiments provided detailed information about the nature of the electroweak interaction as well as key tests of the standard model of particle physics. The Cavendish High Energy Physics Group became a member of the OPAL collaboration, which involved about 200 physicists from 34 institutes in Canada, Germany, Hungary, Italy, Israel, Japan, the United Kingdom and the United States – the acronym stands for Omni-Purpose Apparatus for LEP.

Janet Carter led the Cavendish participation in the OPAL experiment from about 1984. The group went through a further period of considerable crisis with the departure of Bill Neale to the Rutherford–Appleton Laboratory in 1988 and, even more seriously, the departure of the most senior scientist and head of the group, John Rushbrooke, to Australia in the same year. Although holding only the relatively junior post of Senior Assistant in Research, Carter became head of the group in 1988 and did a very impressive job of reinvigorating the research activity. Over the following decade, she brought the group to a position of real distinction in the UK high-energy physics community and internationally. She was promoted to an ADR post in 1990 and then to a readership in 1995. Her promotion to a professorship took place in 2003. The involvement of the group in the OPAL experiment

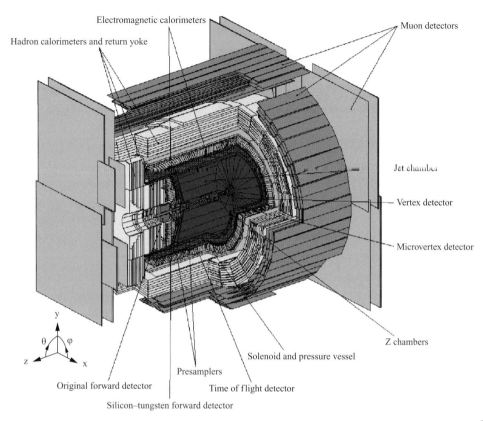

Electromagnetic calorimeters

Muon detectors

Hadron calorimeters and return yoke

Jet chamber

Vertex detector

Microvertex detector

Z chambers

y

θ φ

z x

Solenoid and pressure vessel

Presamplers

Original forward detector

Time of flight detector

Silicon–tungsten forward detector

A diagram showing the layout of the elements of the OPAL detector (copyright CERN).

Fig. 21.1

was further supported by the appointments of Richard Batley and Andy Parker as assistant lecturers in 1989 and of Val Gibson in 1994.

The Cambridge HEP Group was responsible for contributions to several components of the OPAL experiment: the Vertex Drift Chamber, the Track Trigger, the Silicon Microvertex Detector and the Endcap Electromagnetic Calorimeter, as well as providing Monte Carlo simulations of the products of electron–positron collisions and a graphical interface for displaying events (Figure 21.1). When she joined the OPAL project, Carter fully appreciated that accurate determination of the particle tracks as close to the interaction point as possible was essential and, in collaboration with Canadian and the other UK groups, designed and built a vertex detector based on the drift chamber principle, which achieved all its design parameters (Box 21.1). The instrument achieved a positional accuracy of about 50 μm.

Once the OPAL detector began taking data in 1989, Carter was one of the first to recognise that, as a result of advances in technology while the drift chamber was being constructed, greater accuracy could be achieved by using silicon microstrip detectors even closer to the intersection point. She needed additional experimental expertise in silicon detector technology and this was achieved with the appointment of Parker as an assistant

Box 21.1   **The microvertex detector of the OPAL experiment**

Half of a microvertex detector. The Cavendish High Energy Physics Group played a leading role in its design and construction. Two such hemicylinders were joined to form a cylindrical detector about the beryllium beam pipe inside which the beams collide. The actual detector consisted of two such barrels of silicon microstrip detectors, 250 or 300 μm thick. When charged particles traverse the microstrip detectors, electron–hole pairs are created, the electrons drifting to implanted strips of width 25 μm, enabling the track position to be accurately determined (copyright High Energy Physics Group, Cavendish Laboratory).

The original OPAL silicon detector was installed in 1991, and consisted of 25 ladders, each of three detectors, arranged in two cylindrical barrels at radii 6.1 and 7.5 cm about the thin beryllium beam pipe. The intrinsic resolution of the detectors was about 5 μm, with an effective precision of about 10 μm achieved within OPAL. In the original detector, only two-dimensional readouts were performed, but an updated detector installed in 1993 consisted of pairs of detectors with orthogonal strips placed back-to-back, in order to provide a three-dimensional readout. A further upgrade took place in 1995–96, when the geometry was modified to eliminate some of the gaps between adjacent detectors and to lengthen the detector acceptance along the axis of the detector.

lecturer. She set up a silicon detector research and development facility in the Laboratory and, with others in the UK, persuaded the collaboration that such a microstrip silicon detector could be inserted between the original gas vertex chamber and the beam pipe. The project eventually accepted the proposal, and the silicon detector was constructed very rapidly and inserted into the OPAL experiment in 1991. With the inclusion of the silicon microstrip detectors, the accuracy of particle track determinations improved to 5–10 μm. These developments were essential for the remarkable success of the LEP project.

Carter, David Ward (1978) and the HEP team were deeply involved in the analysis of the data and in deriving precision parameters which confirmed the success of the standard model. The quality of the data from the LEP-1 and LEP-2 phases of the project is illustrated by the data shown in Figure 21.2. The measured cross-section as a function of energy for the production of Z bosons was so precisely known that it was demonstrated that there are only three neutrino species (Box 21.2). The LEP experiment was closed down on 2 November 2000 to make way for the construction of the Large Hadron Collider (LHC) in the same tunnel. By that time, the HEP Group was fully involved in the LHC programme, group members playing prominent roles in the ATLAS and LHCb experiments (see Sections 22.6.2 and Box 22.2).

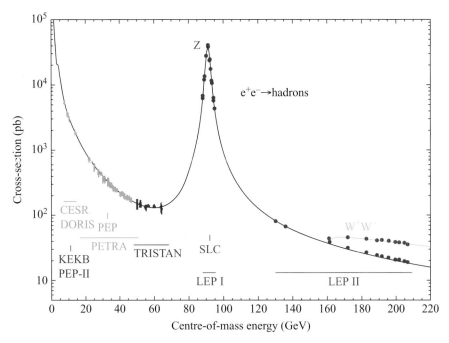

The cross-section for electron–positron collisions, showing the resonance associated with the formation of Z bosons and the contribution of W bosons at high energies in the LEP-2 experiments (copyright CERN).

Fig. 21.2

Limits to the number of neutrino species from the OPAL experiment

Box 21.2

The shape and amplitude of the Z resonance is determined by the number of neutrino species. With very high significance, the experiment demonstrated that there are only three generations of neutrinos. The lines show the expected resonance curves for two, three and four neutrino species (copyright CERN). Interestingly, a similar conclusion had been reached earlier from studies of the primordial synthesis of the light elements, hydrogen, helium, deuterium and lithium in the standard Hot Big Bang model of the early Universe (see, for example, Longair, 2008a).

21.2 Radio astronomy: new initiatives

With Martin Ryle's retirement in 1982, new directions were pursued to exploit the Radio Astronomy Group's expertise in aperture synthesis, at the same time broadening its astronomical and technical range of interests. Antony Hewish and then John Baldwin took over as Head of Group. The new areas of interest were optical and infrared aperture synthesis, the first interferometric studies of spatial fluctuations in the cosmic microwave background radiation and millimetre and submillimetre astronomy. At the same time, the group's traditional interest in radio source studies continued, as well as in deep radio sky surveys at low radio frequencies, making full use of the exponentially increasing power of computers.

21.2.1 Optical and infrared aperture synthesis

Baldwin took up the challenge of applying the concepts of aperture synthesis to high-resolution imaging at optical and near-infrared wavelengths. At first sight, the problems appeared to be formidable with the need to preserve phase stability at wavelengths 10^4–10^5 times shorter than radio wavelengths. Following the precepts set out in his pioneering papers with Warner on phaseless aperture synthesis, however, he and his colleagues solved the problem using the principles of closure phase at optical wavelengths (Baldwin and Warner, 1976, 1978).

The long-standing problem with optical astronomical imaging was that the angular resolution of long-exposure optical images taken with large telescopes on the best astronomical sites is limited by fluctuations in the refractive index of the atmosphere which degrade the angular size of the images to about 0.5–1 arcsec, the phenomenon known as 'seeing'. This angular resolution is some 10 to 50 times poorer than the theoretical Rayleigh diffraction limit of a 4-metre class telescope. In the first successful experiments, they adopted the approach of taking very short exposures which 'froze' the fluctuations. The observations were made through an aperture mask consisting of a non-redundant array of three or more holes in the optical train of the telescope. The use of circular apertures with diameters less than the scale r_0 of the irregularities in the atmosphere above the telescope ensured that there was little phase variation across each of the small apertures and so the wavefronts from the different apertures were undistorted on reaching the focus of the telescope. They then measured the visibilities and closure phases of the fringe patterns observed in each short-exposure image. In their paper of 1986 they reported the first successful application of the technique in observations made with the 2.5-metre Isaac Newton Telescope on La Palma, and demonstrated that it could be used to image objects as faint as 15th magnitude (Baldwin *et al.*, 1986).

The first optical images were obtained in the following year and consisted of images of a previously unresolved star, λ Peg, and the very clear separation of the two stars in the binary star ϕ And, which have separation 0.46 arcsec (Haniff *et al.*, 1987). The upper limit to the angular size of λ Peg was 50 milliarcsec, essentially the diffraction limit of a 2.5-metre telescope. The images had good dynamic range, even when the number of

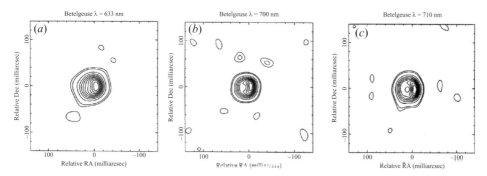

Observations of Betelgeuse made at the ground-based high-resolution imaging laboratory (GHRIL) at the Nasmyth focus of the William Herschel 4.2-m Telescope on La Palma. The reconstructed images are at wavelengths of 633, 700 and 710 nm with angular resolutions of 48, 38 and 39 milliarcsec. The 'hot-spot' on the surface of this supergiant star is present on all three images (Buscher *et al.*, 1990).

Fig. 21.3

photons in each short exposure was small, meaning less than 100 photons per exposure. The success of these observations was followed by high-resolution imaging using the 4.2-metre William Herschel Telescope, also on La Palma. Among the highlights of these pioneering observations was the imaging of structures on the surface of the supergiant star Betelgeuse (Figure 21.3; see also Box 22.1(*b*)), associated with 'hot-spots' in its photosphere (Buscher *et al.*, 1990).

With the technical understanding of how the problems of astronomical seeing could be overcome, Baldwin and his colleagues received funding to build a prototype optical interferometer, the Cambridge Optical Aperture Synthesis Telescope (COAST), at the Lord's Bridge Observatory (Baldwin and Mackay, 1988). The instrument eventually consisted of four telescopes with a maximum baseline of about 100 m. The signals were combined in an underground bunker containing four rail tracks with precisely moveable trolleys that provided the path compensation as the telescopes tracked the target over the sky. This was a demanding experimental challenge, since the separate beams from the four telescopes had to be combined in phase. As Baldwin remarked, the problem of carrying out optical aperture synthesis was not one of optics but rather of the construction and operation of ultra-high-precision control systems and path compensation instrumentation. The project proved to be technically demanding but the problems were solved and the first images published in 1996 in a paper with the title *The first images from an optical aperture synthesis array: mapping of Capella with COAST at two epochs* (Baldwin *et al.*, 1996). The success of these observations was to lead to the group's involvement in the large multi-telescope optical interferometer at the Magdalena Ridge Observatory in New Mexico. Chris Haniff and David Buscher were appointed project architects for that project.

21.2.2 The Cosmic Anisotropy Telescope (CAT)

The discovery in 1965 of the cosmic microwave background radiation by Penzias and Wilson offered new challenges for observational cosmology (Section 14.5). Of particular

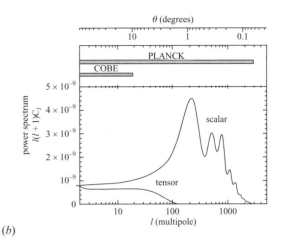

(a)   (b)

Fig. 21.4 (a) The Cosmic Anisotropy Telescope (CAT) at the Lord's Bridge Observatory, showing the three horn antennae within the protecting ground screen. (b) The predicted spectrum of fluctuations in the cosmic microwave background radiation shown as a spatial power spectrum. The angular scale is shown along the upper horizontal axis, as well as the angular scales observed by the COBE and Planck satellites (courtesy of the European Space Agency).

importance were searches for spatial anisotropies in the distribution of the radiation over the sky. The radiation is the thermal remnant left over from the very hot early phases of the Universe and it originates from the *last scattering surface* at the *epoch of recombination* of the primordial plasma at a redshift of about 1500, corresponding to an epoch about 380,000 years after the Big Bang. The theory of the origin of structure in the Universe makes strong predictions about the spatial distribution of the radiation resulting from the early phases of collapse of the structures out of which the large-scale distribution of matter in the Universe formed.[1]

From the early 1970s onwards, major efforts had been made to detect these tiny anisotropies in the background radiation, but they proved remarkably elusive.[2] By the early 1980s, the limits to the small-scale fluctuations in the radiation were sufficiently low to rule out the simplest model of structure formation, in which it was assumed that the matter content of the Universe was in baryonic form. To get round this difficulty, James Peebles at Princeton showed that the problem could be alleviated if most of the matter in the Universe was in some form of non-baryonic *dark matter* (Peebles, 1982). But even then, there still had to be intensity fluctuations in the background radiation at a level of about one part in 10^4–10^5 of the total intensity.

Members of the Radio Astronomy Group set about designing and building an interferometer with the sensitivity to detect the tiny predicted fluctuations. Paul Scott was the senior member of the group leading the project along with Peter Duffett-Smith, and they were supported by Anthony Lasenby, who had joined the group in 1984, and Richard Saunders, who joined in 1991. The radio astronomers constructed the Cosmic Anisotropy Telescope (CAT), a three-element interferometer operating at frequencies between 13 and 17 GHz with a large observing bandwidth and a system temperature of 50 K (Figure 21.4(a)). The baselines could be varied between 1 and 5 m, resulting in a synthesised beam of

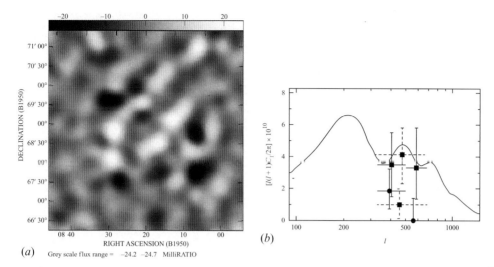

(a) Grey scale flux range = −24.2 −24.7 MilliRATIO

(a) The map of a region of sky $2° \times 2°$ of the fluctuations in the cosmic microwave background radiation observed by the CAT radio telescope (Scott et al., 1996; Baker et al., 1999). (b) The observed power spectrum beyond the first acoustic peak from the first CAT experiments (Scott et al., 1996).

Fig. 21.5

approximately 1/2 degree in angular diameter. This beam size was chosen to detect fluctuations on an angular scale at which a large signal at roughly the first predicted peak in the angular power spectrum was expected (Figure 21.4(b)). The telescope was surrounded by a 5-metre-high earth bank lined with aluminium to form a ground shield, reducing the effects of spillover and terrestrial radio interference.

By the mid 1990s, when the CAT came into operation, fluctuations on angular scales greater than about $7°$ had been observed by the COBE satellite (Smoot et al., 1992), in agreement with those predicted by the cold dark matter (CDM) theory of the formation of cosmic structure. In addition, the predicted rise in the fluctuation power spectrum towards scales of $1°$ ($l \sim 200$ in Figure 21.4(b)) had been measured by the Saskatoon experiment (Netterfield et al., 1995). The CAT was the first interferometer to measure fluctuations in the cosmic microwave background and the results, published in 1996, were the highest angular resolution CMB detection at that time (Scott et al., 1996) (Figure 21.5). These observations showed the decline in power at smaller angular scales, $\sim 0.5°$, $l = 500–700$, as compared with the Saskatoon experiment, demonstrating the existence of the long-predicted acoustic peak in the CMB power spectrum (Figure 21.5(b)). Specifically, as Scott and his colleagues wrote in their paper,

> The broadband power, averaged over spherical harmonic multipole orders between 330 and 680, is $\Delta T/T - 2.0^{+0.4}_{-0.4} \times 10^{-5}$, which is consistent with the predictions of a standard COBE-normalized, cold dark matter model. (Scott et al., 1996)

Further results for a different field were published in 1999 with the result $\Delta T/T = 2.1^{+0.4}_{-0.4} \times 10^{-5}$. The CAT was the forerunner of the Very Small Array (VSA) and established many of the design features for that more ambitious project (see Box 22.1).

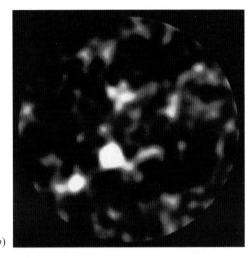

(a)   (b)

Fig. 21.6 (a) The James Clerk Maxwell Telescope (JCMT) on the summit of Mauna Kea in Hawaii (courtesy of the Royal Observatory, Edinburgh). (b) The Hubble Deep Field (HDF) as observed by the SCUBA submillimetre array camera on the JCMT at 850 μm (Hughes *et al.*, 1998). The brightest source close to the centre of the field is a distant powerful submillimetre source at the large redshift $z = 5.185$ (Walter *et al.*, 2012).

21.2.3 Millimetre astronomy

At the beginning of the Edwards era, Richard Hills and Anthony Lasenby were appointed to posts in support of millimetre and submillimetre astronomy (see Table 19.1). Hills had returned to the UK to take on the role of project scientist for the UK's millimetre telescope, later to become the James Clerk Maxwell Telescope. As described in Section 17.1.3, there was strong community support for the construction of a large 15-metre millimetre–submillimetre telescope. With support from the Canadian and Netherlands communities, the project went ahead to build the world's largest telescope for submillimetre observations. The original plan was that the telescope should be located at La Palma, but at an altitude of 2400 m, it was not as good a site for observations at submillimetre wavelengths as the Mauna Kea Observatory in Hawaii, at an altitude of 4200 m. The key parameter was the amount of precipitable water vapour over the telescope and that is a strong function of altitude. The decision was taken to locate the telescope on Mauna Kea, where, along with the UK Infrared Telescope (UKIRT), it established a strong UK presence in the Mauna Kea Observatory.

The telescope presented a number of major technical challenges, both in construction and in instrumentation. The parabolic 15-metre antenna was composed of 276 individually adjustable panels with a surface accuracy of better than 50 μm. The antenna and mountings were protected from the elements by a co-rotating carousel with a transparent membrane stretched across the carousel aperture (Figure 21.6(a)). Hills spent much of his time working at the Rutherford–Appleton Laboratory, which was responsible for the construction of the telescope, and in Hawaii where the local infrastructure to support the telescope in 'millimetre valley' had to be set up. Operational support was the responsibility of the Royal

(a) The HARP cryostat at one of the Nasmyth foci of the James Clerk Maxwell Telescope (courtesy of Richard Hills, Cavendish Laboratory). The detector elements of the submillimetre array spectrometer are maintained at 4 K. (b) The SCUBA cryostat mounted at one of the Nasmyth foci of the James Clerk Maxwell Telescope (courtesy of the Royal Observatory, Edinburgh).

Fig. 21.7

Observatory, Edinburgh, which was already operating UKIRT. I moved to Edinburgh in 1980 as Director of the Royal Observatory and was deeply involved in the negotiations with our international partners and the Institute of Astronomy in Hawaii through all phases of the project until I returned to Cambridge in 1991. Construction in Hawaii began in 1983 and 'first light' was obtained in 1987.

In 1991, Hills was appointed Professor of Radio Astronomy in the Cavendish Laboratory and in 1994 further support for millimetre astronomy was provided with the appointments of Stafford Withington and Rachel Padman. The major challenge was the construction of cutting-edge instrumentation for these wavelengths. Up until the 1990s, most of the science had been carried out using single-element detectors for both millimetre photometry and spectroscopy, mostly for studies of regions of star formation and active galaxies. Within the Laboratory, responsibility was taken for the construction of the common-user 16-element millimetre array spectrometer HARP, which was delivered to the JCMT in 2005 – HARP stands for Heterodyne Array Receiver Programme for millimetre and submillimetre wavelengths (Smith *et al.*, 2008) (Figure 21.7(a)). The principal areas of study were the chemistry and physics of interstellar gas and dust clouds.

While in Edinburgh, I had strongly advocated the construction of the world's first submillimetre common-user bolometer array (SCUBA) for the JCMT, and the project was approved by the JCMT board in 1987. SCUBA was a dual-camera system containing 91 pixels in a short-wavelength array optimised for observations at 450 μm and 37 pixels in a long-wavelength array optimised for the 850 μm waveband (Holland *et al.*, 1999). SCUBA combined high sensitivity with extensive wavelength coverage and wide field-of-view (Figure 21.7(b)). Its mapping speed was 10,000 times faster than the previous single-element detector, UKT14. It presented many demanding challenges, the detector array operating at a temperature of 0.1 K. It was commissioned on the JCMT in 1997.

On my return to Cambridge, one of my priority projects was to evaluate the role of millimetre and submillimetre astronomy for astrophysical cosmology. Unlike essentially all other wavebands, the spectra of dusty galaxies, in particular star-forming galaxies, are

Fig. 21.8   The ALMA submillimetre array on the Atacama plateau at 5000 m in Chile (courtesy of the European Southern Observatory).

'inverted' in the millimetre waveband because the dust has a temperature of about 30–60 K and so the millimetre waveband is in the modified Rayleigh–Jeans region of a body at that temperature. Andrew Blain and I showed that, as a result, galaxies at redshifts $z \geq 1$ do not decrease in observed intensity until very much larger redshifts. We predicted that a large population of such galaxies would be observable by the SCUBA bolometer array (Blain and Longair, 1993).[3] Blain went on to show that gravitational lensing of such submillimetre sources by clusters of galaxies would result in a large enhancement of their observed luminosities (Blain, 1997). Ian Smail, Rob Ivison and Blain made the first submillimetre observations of a rich cluster of galaxies with the SCUBA array and discovered this large population of distant submillimetre galaxies (Smail *et al.*, 1997). In fact, the number of observed sources exceeded even our most optimistic expectations.

This marked the beginning of a major international effort to understand this new population of far-infrared/submillimetre galaxies. The next target was to make submillimetre observations of the Hubble Deep Field (HDF), at that time the deepest optical image ever taken of the distant Universe. Blain and I were members of the consortium which carried out that survey, which found the expected number of faint submillimetre sources (Figure 21.6(*b*)) (Hughes *et al.*, 1998). Significantly, the brightest source in the centre of the field, HDF 850.1, could not be securely identified. It was only in 2012 that millimetre spectroscopic observations of the source with the ALMA submillimetre array determined a redshift $z = 5.185$ for the source (Walter *et al.*, 2012), although the optical counterpart has still not been seen.

The reason for telling this story in some detail is that these observations were one of the major motivations for the construction of the ALMA submillimetre array on the Atacama desert at 5000 m in Chile; submillimetre astronomy was now important for astrophysical cosmology as well as for studies of the interstellar gas and of star and planet formation. This major international project involved the construction of 66 submillimetre antennae to form an aperture synthesis array with quite remarkable sensitivity (Figure 21.8). Richard Hills was appointed project scientist in Chile for the construction and commissioning phases and John Richer was the UK project scientist for this $1.5 billion multinational project.

21.2.4 Radio source physics

The traditional interests of the Radio Astronomy Group in radio surveys and the physics of radio sources continued throughout the Edwards era. Special mention should be made of John Baldwin's ingenuity in using the increasing power of computers to enable all-sky surveys to be carried out using simple inexpensive antennae which were combined using computing techniques; these enabled all the baselines needed to produce a proper sampled radio image of the sky. This resulted in the sixth Cambridge (6C) survey of radio sources in the northern sky at declinations $\delta \geq 30°$ at a frequency of 151 MHz.[4] The success of the 6C programme led to an enhancement of the telescope system: the baselines were increased, more telescopes were built and now motorised so that a yet higher-resolution catalogue could be created. The 7C array, named the Cambridge Low-Frequency Synthesis Telescope (CLFST), resulted in a catalogue containing 43,683 sources over an area of about 1.7 steradians (Hales *et al.*, 2007). The final survey using these techniques was the 8C survey, carried out using the same principles by Nicholas Rees at 38 MHz of the region of the northern sky north of declination 60°, containing 5000 sources. The resolution and sensitivity were both improved by nearly an order of magnitude over previous surveys (Rees, 1990).

These surveys were important precursors of the current initiatives in low-frequency radio astronomy, including the LOFAR project in the Netherlands and the Square Kilometre Array (SKA) project. As George Miley wrote to me,

> The 38 MHz 8C survey is still the best survey below 50 MHz and was an important stimulus for the next-generation low-frequency arrays such as LOFAR. (George Miley, personal communication, 2010)

Support for these traditional areas of radio source physics was continued with the appointment of Paul Alexander as an assistant lecturer. Many of the lessons learned from these experiments have now been built into the Square Kilometre Array project. Alexander, now Professor and Head of the Cavendish Astrophysics Group, is UK Science Director of the SKA organisation and Lead of the Science Data Processor.

Let me end this chapter on a personal note, following my return to Cambridge in January 1991. A further project in which I had been involved since 1977 was the Hubble Space Telescope (HST). I had been fortunate enough to be selected as an interdisciplinary scientist on the NASA/ESA Science Working Group which entitled me to guaranteed time on the telescope. My project concerned HST imaging of the radio galaxies which James Gunn and I had discovered in the late 1970s (Gunn *et al.*, 1981). The HST project was delayed when major cost overruns occurred and then by the tragedy of the *Challenger* disaster which put the Space Shuttle programme on hold. When the telescope was finally launched in 1990, the problem of the telescope optics meant that my programme was not feasible until a solution to the spherical aberration problem was implemented during the first servicing mission in 1993. My first observations were only taken in 1994–95, but the wait was worth it.

The first three images obtained are shown in Figure 21.9 and came as a total surprise (Longair *et al.*, 1995). I had expected to see distant massive galaxies, but instead strange structures were observed aligned with the axes of the radio sources. The theory of these sources involves the outer radio source components being continuously supplied with relativistic material from a massive black hole in the nucleus of the galaxy. In the cases of 3C 324 (*b*) and 368 (*c*), it is apparent that the invisible jets had interacted strongly with

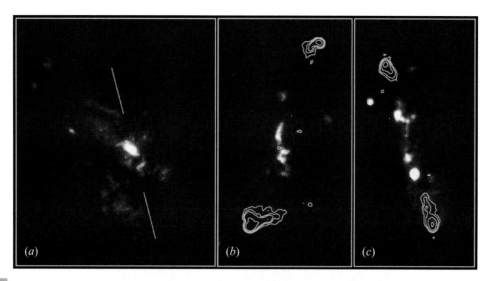

Fig. 21.9 Hubble Space Telescope images of the radio sources 3C 265, 3C 324 and 3C 368. Superimposed on the images are radio maps from the Very Large Array, shown as blue contour lines. In all three cases, it was known that the optical emission was aligned with the radio axis of the double radio source. (*a*) In 3C 265; the line indicates the axis of extended radio emissions. (*b*) 3C 324; the optical emission lies along the radio axis the double source. (*c*) 3C 368; the optical emission lies all the way along the radio axis of the source (Longair *et al.*, 1995).

the environment, resulting in the appearance of bright optical knots along the radio source axis. In the case of the much older source 3C 265 (*a*), the radio jet passed through this region of the galaxy long ago – the nature of the debris in this case is not so clear.

We obtained images of 28 3CR radio sources altogether, providing essentially a complete sample of these extremely powerful radio sources at redshifts between 0.6 and 1.8. It turned out that the strongest alignments occurred in the younger radio sources, while the jets were penetrating the interstellar gas within the parent galaxies, which were all massive giant elliptical galaxies at the centres of clusters of galaxies (Best *et al.*, 1996). My colleagues Philip Best and Huub Röttgering did a superb job in analysing the images and then in obtaining the high-resolution optical spectra as well as infrared images taken with UKIRT to understand the nature of the aligned optical emission (Best *et al.*, 1997, 1998). We concluded that in the younger sources, which displayed the most prominent aligned emissions, the excitation of the optical structures was due to shock waves associated with the jets of relativistic material ejected from the active galactic nucleus, while in the older sources the regions were excited by photoionisation (Best *et al.*, 2000).

PART X

1995 to present

Richard Henry Friend (b. 1953)

22.1 The end of history?

The Regius Professor of History, on being asked, 'When does history end?', responded, 'When they can no longer sue!' More pertinently, I was deeply involved in managing the Laboratory from the time of my return to Cambridge in 1991 until the present. Any pretense at objectivity would scarcely be credible. As a consequence, this chapter is more personal than the rest of the text, but I will endeavour to stick to the facts.

The changes in the named professorships through this period were as follows:

- Richard Friend appointed Cavendish Professor in 1995 at the age of 42 in succession to Sam Edwards
- Peter Littlewood appointed to the 1966 Professorship of Theoretical Physics in succession to Volker Heine in 1997; in 2013, this chair was transferred to Didier Queloz
- Henning Sirringhaus appointed to the Hitachi Professorship of Electron Device Physics in 2005
- Ullrich Steiner appointed John Humphrey Plummer Professor of Physics in 2005
- Jeremy Baumberg appointed Professor of Nanoelectronics, proleptically filling the Chair of Physics held by Michael Pepper in 2007
- James Stirling appointed Jacksonian Professor of Natural Philosophy following my retirement from the chair in 2008
- Ben Simons appointed to the Herchel Smith Professorship of the Physics of Medicine in 2011
- Roberto Maiolino appointed Professor of Astrophysics in succession to Richard Hills in 2012.

The Heads of Department during this period were:

- Archie Howie: 1990–97
- Malcolm Longair: 1997–2005
- Peter Littlewood: 2005–11
- James Stirling: 2011–13
- Andy Parker: 2013–present.

The days when Rutherford could manage the Cavendish Laboratory as a single dominating presence or Thomson manage the finances of the Laboratory from his personal cheque-book were long gone. The role of the Head of Department was much more that of the chief executive officer of a middle-sized company. By 2015 the number of persons on site was approaching 1000 and the turnover about £30 million. All those who have held the post of Head of Department since 1995 have found that, in the research funding and educational climate of the late twentieth and early twenty-first centuries, the role has become very much more demanding and time-consuming than might be expected of an academic institution. How did this come about?

22.2 Management, administration, responsibility and accountability

At the beginning of the Friend era, the University continued to be operated as a purely academic institution. Up until 1989, the role of Vice-Chancellor had rotated every two years among the Heads of the Colleges. In 1989, David Williams was appointed as the first full-time Vice-Chancellor, with a seven-year term of office. For the period 1996–2003, Alec Broers held the post and made strenuous efforts to reform the governance structure of the University, but was frequently frustrated by its antiquated procedures. The collegiate nature of the University, with the University and the Colleges being independent legal entities, resulted in a cumbersome administrative and management structure which was ill-adapted to the requirements of speed and managerial efficiency, particularly in the sciences. The ultimate approval for any decisions of the University Council lay with the Regent House, the body consisting of all the academic staff of University. The Vice-Chancellor had remarkably limited power to make changes to reflect the increasing pressures from the government for accountability and good governance. During this period, the funding of the universities in real terms was decreasing, the government wishing them to be more selective in research funding while also seeking to ensure that 50% of young people should benefit from a University education.

It is revealing to read the Council's recommendations of 26 June 2002 concerning amendments to the University's constitution, which were only partially implemented:

> Ideas for change were first published earlier this year. Extensive internal consultation has taken place over the last few months, leading to today's refined proposals.
>
> In the report, there is explicit recognition for the first time that the Vice-Chancellor is the principal executive officer of the University, responsible to the Council, with overall responsibility for the executive management of the University and its finances, and for its direction within the framework of agreed and approved policies.
>
> Significant revision is proposed to the composition of the University Council: three external members will join the Council, also for the first time. They will play a fundamentally important role in the University's future governance. One will chair the Council and another, the Audit Committee.
>
> Changes are also outlined in respect of the Regent House, the ultimate governing body of the University, both by extending membership to a wider academic community and also by increasing the number of signatories required to call for a ballot or a 'Discussion' on a topic of concern.
>
> Taken together, these proposals represent a significant modernisation of the governance arrangements.

But much more than the constitution of the University was involved. The University's financial management was scrutinised and found to be inadequate for the commitment accounting practices and auditing procedures required by government funding agencies. These were far from trivial changes and to implement them a commitment accounting software system project (CAPSA) was set up to introduce a University-wide unified accounting system. Introduction of the new system was not helped by the fact that most of the senior finance officers resigned or retired before the beginning of the project's implementation phase. The new accounting system had far-reaching consequences for the administration of the finances of the University and its departments, and a large programme of change management and additional expert finance staff were needed.[1] From the perspective of the Laboratory, there was a considerable turnover of financial staff, many of whom had provided dedicated service for many years. Matt Burgess was appointed Finance Manager and was thoroughly familiar with the contemporary accounting procedures and processes which had to be implemented.

The process of Research Assessment, instituted during the mid 1980s to evaluate the quality of University research funded by the Research Councils and other organisations, continued and intensified, with reviews taking place in 1996, 2001, 2008 and 2014. Success in these exercises was crucial for the funding of the departments – each academic staff member was expected to produce four pieces of internationally recognised outstanding research within a review period of typically five to six years. In the 2014 reincarnation, known as the Research Excellence Framework, 'impact' also played a significant part in the assessment, the term being interpreted as meaning the impact of the research outside the academic world. Whilst a case can be made that these reviews helped improve the quality of research in UK universities, the burden on those involved in the peer-review process was very large, as was the effort needed to put together the materials for the assessments.

Teaching quality was also subject to review, the largest exercise taking place in the 1998 Subject Review. Panels were set up to review the quality of all aspects of teaching by all Cambridge departments. This involved the assembly of vast amounts of data, and the toll on staff was considerable. The positive side of the exercise was that many aspects of the teaching provision for students were very significantly enhanced, and the department achieved the very high rating of 23/24.[2] The Quality Assurance Agency (QAA) of the Higher Education Funding Council for England (HEFCE) subsequently lightened the burden by devolving most of the responsibility to the University with the requirement that a programme of reviews spanning all the University departments be set up. Spot-checks by the QAA were undertaken in a number of departments.

Health and Safety played a much more prominent role in all aspects of the work of the Laboratory. We were fortunate to appoint Jane Blunt as a full-time Departmental Safety Officer and members of the department fully respected the need for compliance with the requirements of the government's Health and Safety Executive. Necessarily, this was a highly significant activity in an experimental physics laboratory, where dangerous gases, high-vacuum techniques, low temperatures and ultra-high pressures are part of the normal business of the Laboratory.

During the 2000s, Pro-Vice-Chancellors were appointed to relieve the pressure on the Vice-Chancellor who, in the case of Broers' successor from 2003 to 2010, Alison Richard, took a much more outwardly facing view of the role and left the Pro-Vice-Chancellors to get on with the day-to-day running of the University. Financial management of the academic programme was devolved onto the Schools, which consist of clusters of cognate disciplines – the Schools of Arts and Humanities, Biological Sciences, Clinical Medicine, Humanities and Social Sciences, Physical Sciences, and Technology. The Head of the Laboratory was necessarily a member of the School of Physical Sciences and a number of its subcommittees.

The University published a formal set of responsibilities for the Head of Department and, to cut a long story short, the document states that everything which happens within the department is the responsibility of the Head and that person is responsible to the Vice-Chancellor and the University for good governance of the department. Given the way in which the funding of higher education and its management within the UK have evolved, it is evident that the managerial load on heads of departments is very large and makes it a challenge for the incumbent to remain an active scholar. The problem is exacerbated by the fact that staff members were not appointed on the basis of their managerial abilities but because of their distinction as pure academics. Fortunately, within the Laboratory, a number of our colleagues have risen to this challenge.[3]

At the same time that the task of managing and administering the Laboratory increased considerably, the burden necessarily had to be communicated down the management chain to groups and individual staff members, who had to follow best practice in the pursuit of their research and teaching goals. This was a far cry from the string-and-sealing-wax image of physics research and teaching which was exploited with dramatic success by a Rutherford or a Ryle, but which would scarcely be allowed by the Health and Safety Executive in the twenty-first century.

It was also a very different administrative environment from the time when Bragg introduced the post of Secretary of the Department in 1948. In 1960, Ian Nicol was appointed Departmental Secretary on the retirement of Kenneth Dibden, and he was succeeded by John Deakin who held the post from 1966 to 1994. He was succeeded by Hilary Coote. On her retirement in 2001, Robert Hay was appointed Departmental Secretary with an emphasis upon academic matters, and in the following year I appointed David Peet as a second Departmental Secretary with an emphasis upon administrative matters. Even with this additional senior administrative support, the increase in administrative and accountability requirements by the University and the government remained a very large and often stressful challenge for a large department with huge amounts of expensive experimental facilities.

22.3　The evolution of the staff profile

In parallel with the gradual changes in governance, the distribution of academic staff changed markedly through the period 1995 to 2005, which happened to be the period during which I was Head of Department. This is illustrated by the distribution among various classes of academic staff shown in Figure 22.1. These data illustrate a number of important

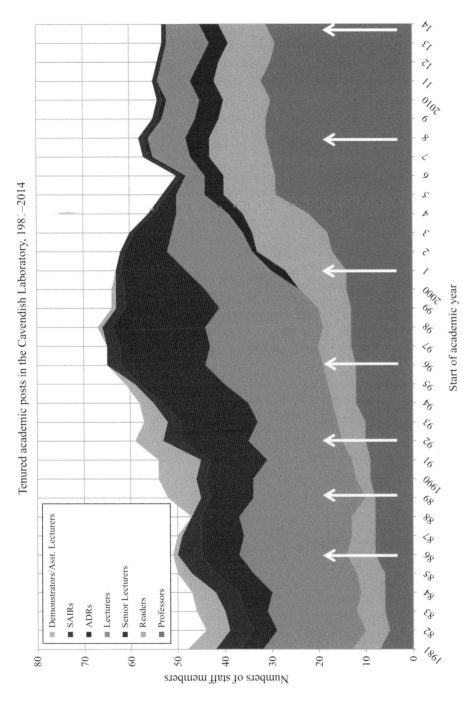

Tenured academic posts in the Cavendish Laboratory, 198⁻–2014

Numbers of staff members

Start of academic year

Legend: Demonstrators/Asst. Lecturers, SAIRs, ADRs, Lecturers, Senior Lecturers, Readers, Professors

Fig. 22.1 The diagram shows the changes in numbers of tenured academic staff from 1981 to 2014, spanning the Edwards and Frie¬d eras. The data are taken from the list of staff members published in the Cambridge University *Reporter* at the beginning of each academic year, generally in the first week of October. The white arrows indicate the years of the Research Assessment Exercises (RAE)/Research Excellence Frameworks (REF).

changes in the academic make-up of the department, and, by extension, of the University as a whole, as a result of the changing academic environment discussed in Section 22.2. The histogram has been extended back to the end of the Pippard era and the beginning of the Edwards era to illustrate how significant the changes were. Underlying the statistics of Figure 22.1 was the fact that the resources allocated to the universities were in decline through the period, and this had a severe impact upon physics departments throughout the UK. As the Gatsby Report of 2006 (Smithers and Robinson, 2006) states,

> Between 1994 and 2004, 17 university physics departments admitting more than ten students per year closed. Counting all departments with any physics students, the loss has been 24. At the same time first-year full-time UK-domiciled students reading physics fell by 905 (28.9%), including 166 from the 26 top-rated departments in the 2001 RAE (8.2%). A rank order correlation by region showed the drop in student numbers to be significantly correlated with departmental closures.

The closure of physics departments was largely associated with the fact that training students in such a heavily experimental discipline is expensive and the various formulae used to calculate the costs of such training did not take account of the real costs involved. Because of the breadth of the discipline, physics departments with small staff and student numbers could not survive in a stringent funding environment. As the Gatsby Report indicates, the knock-on effect was that smaller numbers of UK students were being trained in the physical sciences.

Returning to the distribution of staff among academic grades, Figure 22.1 illustrates the impact of a number of governmental and University policies upon the career development of Cavendish physicists.

- At the beginning of the period illustrated in Figure 22.1, the academic structure of the Physics Department was what it had always been. There were a few professors who held named chairs and a similar number of readers, while the majority of academic staff were University lecturers, these posts being regarded as career grades within the University. The post of Assistant Director of Research (ADR) had been introduced in the 1950s to accommodate experimental subjects such as radio astronomy which required more academic research support. As new research disciplines developed during the Edwards era, the ADR posts were an effective way of fostering these activities, often supported by fixed-term research contracts. The Senior Assistants in Research (SAIR) carried out a similar role but at a lower level. Finally, the normal entry grade to an academic career remained the post of University Demonstrator, harking back to the pioneering days of Maxwell. These were fixed-term posts, initially for three years extendible to five, after which the post-holder was either promoted to Lecturer or their University employment was discontinued. As can be seen from Figure 22.1, in 1988 the number of demonstrators fell to zero, after which time the post was renamed Assistant Lecturer, more in accord with the practice in other departments.[4] But Assistant Lectureships were to disappear by 2000 and the entry grade became the University Lecturer, with very different tenure arrangements. In accordance with European and UK law, the probation period as a new lecturer was reduced to six months, with the result that the appointment procedures for new staff had to be much more rigorous – mistakes could not be made or the department would be lumbered with staff who had peaked early in their careers.

- There was increasing recognition that the rarity of promotions to readerships and professorships was unfair to distinguished staff who would easily have satisfied the criteria for promotion to these grades at other UK universities. Many of the best lecturers understood that the possibilities for promotion were much greater at other universities, and this undoubtedly contributed to the departure of a number of outstanding lecturers and demonstrators/assistant lecturers. Recognising this inequality, the University decided that staff should be judged for promotion by the normal standards for distinguished universities in the UK. As a result, the number of promotions accelerated through the period 1997 to 2005 when I was Head of Department. By 2005, a new equilibrium was established, readers and professors now outnumbering lecturers and senior lecturers.

- The University also wished to phase out the post of ADR, which had always been a somewhat anomalous position. The numbers of these had increased during the Edwards era, with the funding generally being provided by research grants. Impending changes to European employment legislation meant that these posts would become permanent University positions after five years and the full costs of these staff would fall on the department rather than central University funds. This was a strong incentive to reduce the number of these staff, often by promotion to lectureships.

- In accordance with the practice at other universities, the post of Senior Lecturer was introduced in 2000. The intention was that staff would normally proceed through the sequence of lecturer, senior lecturer, reader, professor. The salaries of senior lecturers and readers were the same, but there was an emphasis upon outstanding research in promotion to a readership, whereas equal weight was given to research, teaching and management contributions in promotion to senior lecturer. In fact, many excellent staff took advantage of the senior lecturer route to promotion, which is now firmly embedded in Cambridge's academic career progression.

- Another feature of the staff profile is apparent from Figure 22.1. Throughout the period 1981 to 2015, there was a continuous atmosphere of financial stringency which meant that posts were often frozen when the incumbents retired and, in addition, all departments were given savings targets in successive financial years. During my spell as Head of Department, these savings amounted to typically £200,000 to £500,000 per year, which required careful management. At the same time, the RAE and REF exercises required the department to continue carrying out innovative science and employing new young staff. As a result, after periods when few appointment were made, there would be sudden bursts of new appointments in the years leading up to the RAEs. This is apparent from the strong correlation between increases in staff numbers, particularly lectureships, leading up to the assessment dates. The department made full use of proleptic appointments, meaning the early filling of posts which would become vacant in the future.

- The total number of staff members shown in Figure 16.5, which includes support staff, postdoctoral workers and research students, can be compared with the numbers of tenured academic staff members in Figure 22.1. Although the balance between different grades of tenured academic staff changed dramatically, the total number of tenured staff increased only modestly from 1981 to 2012 compared with the major increase in total staff. The large increase in staff numbers between 1981 and 2001 was associated with research fellows/postdoctoral research assistants (RF/PDRA) and graduate students. The numbers fluctuate from year to year, but representative figures from 1982 show that there were about 66 RF/PDRAs, as compared with 163 in 2003; for graduate students, the corresponding figures were about 132 and 281. This growth in numbers of non-tenured research staff is closely related to the dramatic increase in research grant income, also shown in Figure 16.5. Much of that resource was associated with research grants for the support of postdoctoral research associates, who were essential for the expanding research programme; these grants could also provide project studentships. Both sets of figures are a tribute to the success of academic staff members in winning grants for their research programmes from governmental and non-governmental agencies, including, increasingly, the European Research Council.

22.4 New areas of research

The development of the research programme was largely determined by the availability of tenured academic posts to attract the best physicists available. As staff members retired or moved to other universities, strategic decisions could be made about promising research areas, while maintaining the strength of dynamic existing programmes. As discussed in Section 22.3, the major additions to the tenured staff tended to occur in the years leading up to the Research Assessment Exercises, as illustrated by the arrival and departure data in Table 22.1. Many of these appointments strengthened the research activities of the existing groups, but in notable cases there were strategic decisions to develop new areas of research.

22.4.1 The physics of medicine and biology

The physics of soft condensed matter and biology had already been fostered during the Edwards era, led by the experimental activities of Athene Donald, as discussed in Section 20.3. The next stage in the advance of physics into the biomedical area took place with the arrival of Chris Dobson as John Humphrey Plummer Professor of Chemical and Structural Biology in 2001. He came with a vision of pulling together the activities of a number of cognate departments in the areas of biology and medicine to develop new approaches to interdisciplinary research. This initiative was taken up by the Regius Professor of Physic, Keith Peters, who was Head of the Clinical School, and me. Over the succeeding years, we put together a programme in the Physics of Medicine, to bring the culture of physics to appropriate areas of clinical medicine and also to bid for a new building to house these collaborative interdisciplinary activities. This building was also the first stage of the redevelopment of the Cavendish Laboratory (Section 22.10). The Department had received a

Table 22.1 Arrivals and departures of tenured academic staff, 1994–2015					
Academic year	Staff member	Research group	Notes	Staff member[a]	Research group[b]	Notes[c]

Academic year	Staff member	Research group	Notes	Staff member[a]	Research group[b]	Notes[c]
	Departures			Arrivals		
1994–95	S. Edwards	PC	Retirement	*W.Y. Liang*	HTS	Prof. Superconduct.
	M.E. Cates	TCM	To U. Edinburgh	S. Withington	RA	ADR
				V. Gibson	HEP	UAL
				D. Green	RA	UAL
1995–96	S. Kenderdine	AP	Retirement	*R.H. Friend*	OE	Cavendish Prof.
				S.R. Julian	LTP	UL
				B.D. Simons	TCM	UL
				C.G. Smith	SP	UL
				R. Padman	AP	ADR
1996–97	P.A.G. Scheuer	AP	Retirement	D.J.C. MacKay	AP/Inf	UL
	J.R. Shakeshaft	AP	Retirement	D.R. Ward (1978)	HEP	UL
				N.C. Greenham	OE	SAIR
1997–98	R.C. Ball	TCM	To U. Warwick			
	R.A.C. Jones	PC	To U. Sheffield			
1998–99	J.E. Baldwin	AP	Retirement	E.M. Terentjev	PC	UL
	C.J. Adkins	LTP	Retirement	C.A. Haniff	AP	UL
	T.O. White	HEP	Retirement			
1999–2000						
2000–01	L.M. Brown	MP	Retirement	N.C. Greenham	OE	UAL
	A. Howie	MP	Retirement			
2001–02	J.R. Waldram	LTP	Retirement	N.R. Cooper	TCM	UL
	P.F. Scott	AP	Retirement	M.P. Hobson	AP	UL
				H. Sirringhaus	OE/ME	UL
				M.A. Thomson	HEP	UL
				J. Ellis	SFM	ADR
2002–03	H. Ahmed	ME	Retirement	D.F. Buscher	AP	UL
	J.E. Field	PCS	Retirement	T.A.J. Duke	BSS	UL
				C.E. MacPhee	BSS	UL
2003–04	S.R. Julian	LTP	To U. Toronto			
2004–05	E. Marseglia	OE	Retirement	*H. Sirringhaus*	OE/ME	Hitachi Prof.
2005–06	C.E. MacPhee	BSS	To U. Edinburgh	U. Steiner	OE	Plummer Prof.
2006–07	T.A.J. Duke	BSS	To UCL			
2007–08	J.A.C. Bland	TFM	Died	*J.J. Baumberg*	NP	Prof. Nanophotonics
	M.S. Longair	AP	Retirement	M.K. Köhl	AMOP	R
				M. Atatüre	AMOP	UL
				P. Cicuta	BSS	UL
				J.R. Guck	PoM	UL
				Z. Hadzibabic	AMOP	UL
				C.G. Lester	HEP	UL

(*cont.*)

Academic year	Staff member	Research group	Notes	Staff member[a]	Research group[b]	Notes[c]
	Departures			Arrivals		
2008–09	J.R. Carter	HEP	Retirement	W.J. Stirling	HEP	Jacksonian Prof.
	M. Pepper	SP	To UC London	F.M. Grosche	QM	R
				C.H.W. Barnes	TFM	SUL
				E. Eiser	BSS	UL
2009–10	W.Y. Liang	QM	Retirement			
2010–11	J.R. Cooper	QM	Retirement	U.F. Keyser	PoM	UL
				J.D. Wells	HEP	UL, left after one year
2011–12	R.E. Hills	AP	Retirement	E. Artacho	TCM	From Earth Sciences
	P.B. Littlewood	TCM	To Argonne Nat. Lab.	B.M. Gripaios	HEP	UL
	G.G. Lonzarich	QM	Retirement	*B.D. Simons*	PoM	Herchel Smith Prof.
	R.E. Ansorge	PoM	Retirement			
	J. Guck	PoM	To Dresden U.			
2012–13	M.K. Köhl	AMOP	To U Bonn	R. Maiolino	AP	Prof. Astrophysics
	W.J. Stirling	HEP	To Imperial College	C. Castelnovo	TCM	UL
				J.L. Huppert	PoM	Member of Parliament
				A. Lamacraft	TCM	UL
2013–14	U. Steiner	OE	To ETH Zurich	D.P. Queloz	AP	Prof. Physics
	P.J. Duffett-Smith	AP	Retirement	S.E. Bohndiek	PoM	UL
				A.D. Mitov	HEP	UL
				S.E. Sebastian	QM	UL
2014–15				*R.P. Cowburn*	TFM	Prof. Physics[d]

[a] Names in italic font are elections to named Professorships, the appointee already being a member of the Laboratory.

[b] Interdisciplinary Centre for High Temperature Superconductivity.

[c] Notation: Prof. = Professor, R = Reader, SUL = Lecturer, UL = Lecturer, ADR = Assistant Director of Research, UAL = University Assistant Lecturer, SAIR = Senior Assistant in Research.

[d] Russell Cowburn transferred his research programme in Thin Film Magnetism from Imperial College, London to the Laboratory in 2010, where he was designated a Director of Research. He was formally appointed to the professorship in 2014.

major philanthropic gift from the Herchel Smith bequest,[5] which I decided to donate to the Physics of Medicine programme.

These proposals received strong support from the Vice-Chancellor, Alison Richard, and a successful bid was made to the Wolfson Foundation for the resources to construct the Physics of Medicine Building. In fact, the resources that were made available enabled us to realise half of the necessary space, the emphasis being upon the provision of experimental laboratories for physics, chemistry, biochemistry and biology. The appointments of Cate

McPhee and Tom Duke were made in support of this programme. On their departure to other universities, Jochen Guck, Pietro Cicuta, Erica Eiser, Ulrich Keyser, Julian Huppert and Sarah Bohndiek were appointed to support this activity. In 2011, Ben Simons was appointed to the Herchel Smith Chair of the Physics of Medicine. The concept was that other departments would carry out collaborative programmes in the new building, and this concept has been fostered by Simons, the activity being managed and operated by the Cavendish Laboratory. There is close collaboration with the activities of the Biological and Soft Systems Group in the Laboratory. The Physics of Medicine Building acts as a hub for the Cambridge Centre for the Physics of Medicine (CCPoM).

22.4.2 Atomic, mesoscopic and optical physics

When my term as Head of Department ended in September 2005, I bequeathed to my successor Peter Littlewood eight lectureship posts which could be filled in anticipation of the 2008 Research Assessment Exercise. He took the key decision to make a major investment in the area of atomic, mesoscopic and optical physics, a burgeoning area in which the Laboratory was lagging far behind. The Laboratory was able to attract an outstanding group of experimental physicists to lead these activities: Michael Köhl, Mete Atatüre and Zoran Hadzibabic were joined by Richard Phillips to form the new Atomic, Mesoscopic and Optical Physics (AMOP) Group. This involved not only considerable investment in experimental facilities for the new areas, but a major upgrade in the workshop facilities needed for the research. This had the beneficial knock-on effect of enhancing the workshop facilities and associated expertise for all areas of the Laboratory. Very quickly, the group developed state-of-the-art facilities for the study of quantum optics and mesoscopic systems (Atatüre), quantum gases and collective phenomena (Hadzibabic), quantum optoelectronics (Phillips) and, most recently, many-body quantum dynamics (Ulrich Schneider).

22.4.3 Physics of sustainability

In 2007, after my return from a two-year sabbatical leave, Peter Littlewood asked me to be the Director of Development for the Laboratory. This involved putting together a coherent strategy and programme of fund-raising for all aspects of the Laboratory's programme, including the reconstruction of the whole Laboratory. One of the first initiatives was a breakfast event at the Royal Society of London, hosted by Cavendish alumnus Humphrey Battcock[6] to which potential benefactors were invited. As a result of this event, David Harding, founder, Chairman and Head of Research of Winton Capital Management and also a Cambridge alumnus, pledged £20 million to the Cavendish Laboratory to set up and fund the Winton Programme for the Physics of Sustainability. His gift, the largest donation to the Laboratory since its creation in 1874, has created a new programme in the physics of sustainability, applying physics to meet the growing demands on our natural resources. The vision is well encapsulated in the words of Peter Littlewood at the formal opening of the programme in March 2011 (Figure 22.2):

Fig. 22.2 David Harding, Peter Littlewood and the Rt. Hon. David Willetts MP, the Minister of State for Universities and Science, viewing posters illustrating the diversity of basic physics to be supported by the Winton Programme for the Physics of Sustainability, at the opening of the programme in March 2011.

In 2100, the sources of energy on this planet will be either solar or fusion, and the preferred means to transport and use that energy will be electrical. The 'magic' technologies needed to deliver this new age and make them available to societies world-wide are: photovoltaics, electrical storage, refrigeration and lighting. These technologies are particularly important for use in the tropical, developing world. There are no basic physical principles preventing breakthroughs in all these areas. Today, solid state lighting is the closest to the appropriate performance. The discovery of new materials and the development of new concepts in physics are needed to bring this vision to fruition and to make the resulting technologies available to the worldwide community.

The programme supports research in the basic science expected to generate the new technologies and industries needed to meet the demands of a growing population with already strained natural resources. Richard Friend is the programme's Director. The programme provides PhD studentships, research fellowships and support for new academic staff as well as investment in research infrastructure at the highest level, pump-priming for novel research projects, support for collaborations within the University and outside, and sponsorship for outreach activities. The emphasis of the fellowship programme is on exciting and novel ideas that bring new activities to the Laboratory and which would be difficult to support through the normal Research Council routes.

'Top-down' activities involve introducing new areas of research to the Cavendish research programme. In the materials area, programmes have been generated in which the physics is supported by the complementary materials science and materials chemistry, enabling new directions to be explored. Two outstandingly successful Winton fellows, Suchitra Sebastian and Sian Dutton, have recently been appointed University lecturers in the Laboratory.

22.4.4 Nanophotonics

With his arrival at the Laboratory from Southampton University in 2007, Jeremy Baumberg set up the Cambridge Nanophotonics Centre. With funding from the EPSRC and the EU, as well as a number of industrial partners and collaborators, a new soft nanophotonics collaboration was set up with Ullrich Steiner in the Biological and Soft Systems (BSS) Group and colleagues in the Chemistry, Chemical Engineering and Engineering departments. The objective of the programme is to combine soft and photonic materials in unusual ways, for example in making flexible and stretchable nanomaterials with new properties for both functional and biological applications.

One of the objectives of the research programme in the area of nanophotonics is to design materials which assemble themselves on the nanoscale. The objective would be to replace the expensive facilities needed to create traditional semiconductor devices with materials which are designed to build themselves. The challenge is to combine nanoscience and nanotechnology with photonics. This necessarily involves an interdisciplinary approach, bringing together physics, chemistry, materials, engineering, biology and other sciences.

The group also supports the EPSRC Centre for Doctoral Training in Nanoscience and Nanotechnology. This involves a new type of PhD course in this interdisciplinary area. The programme is based on lecture courses, practicals and projects in the first year, and design of an interdisciplinary PhD programme, jointly with physics, chemistry, engineering, materials or another department. The programme also includes Entrepreneurial Learning at the Judge Business School.

22.4.5 Astrophysics

The Cavendish Astrophysics Group has considerably expanded its range of interests with the appointment of two new professors in the general area of experimental astrophysics, but now outside its traditional radio/millimetre/submillimetre wavebands. Roberto Maiolino came from Rome to lead observational and instrumentation projects in the infrared and submillimetre wavebands. His projects involve all the major astronomy facilities which will come into operation over the next ten years, including advanced instrumentation for the ESO Very Large Telescope, guaranteed time on the James Webb Space Telescope (JWST), the infrared successor to the Hubble Space Telescope and the European Extremely Large Telescope (E-ELT).

In 2013, Didier Queloz joined the Astrophysics Group. With his PhD supervisor, Michel Mayor, they have discovered extra–solar system planets or exoplanets, a completely new area of research for the group. This involved a considerable investment, including delivering his vision that the various Cambridge astronomy groups be collocated on the same site to foster collaboration. This was achieved with the construction of the Battcock Centre for Experimental Astrophysics, which was opened in 2013 (see Box 22.17).

The significance of these changes for the future of astronomical research in Cambridge is considerable. There has been significantly more collaboration between the Cavendish Astrophysics Group and the Institute of Astronomy. The expertise of the two groups is

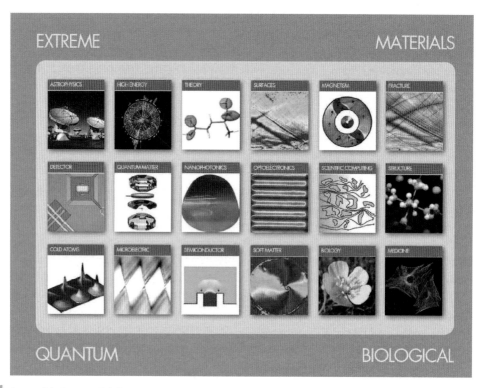

Fig. 22.3 A map of the 'universes' of physics as currently being pursued in the Cavendish Laboratory. The four universes – extreme, quantum, biological and materials – overlap considerably. This portrayal of the diversity of research in the Laboratory was developed by Peter Littlewood and me.

complementary, the emphasis in the Cavendish group being on the experimental and technological aspects of the astronomy programme.

22.5 The Cavendish research programme in 2016

It is not feasible to summarise briefly the full scope of the Laboratory's current research programme. To bring some coherence to the multi-faceted nature of the programme, Peter Littlewood and I developed the concept of the 'Universes of Physics' – the extreme, the biological, the quantum and the materials universes (Figure 22.3). The icons in Figure 22.3 refer to research groups or activities within the Laboratory, the location on the diagram giving a rough indication of synergies with nearby groups. With 18 icons, each group consists of about 50–100 persons, the canonical Bragg number for group size.

But there is more to the map than this. Many of the most important developments take place at the interfaces between disciplines, for example, between astrophysics and particle physics, between biophysics and soft matter research, between optoelectronics and nanophotonics. This is to be strongly welcomed and encouraged. The second aspect is that it is convenient to group the activities into these larger 'universes' for administrative efficiency. Each of the larger groupings needs professional expertise in dealing with research grants and all the University and government accounting requirements.

To give a flavour of the current research programme, the following sections summarise the principal research topics and facilities of each group. This is necessarily a very broad-brush picture, but the main themes and sub-themes are taken directly from the most up-to-date information on the groups' own websites. For each of them, selected highlights of the programmes through the Friend era are briefly outlined in boxes. I don't pretend that the coverage is complete, but these accomplishments are my personal selection of some of the more significant research topics of recent years.

22.6 The extreme universe

The extreme universe involves the physics of the very large and the very small – the physics of the Universe and of Particle Physics. As our understanding of the origin and evolution of our Universe has matured, it is apparent that many of the key cosmological and astrophysical problems are closely related to issues and concerns in high-energy particle physics. The discovery of the Higgs boson has demonstrated that scalar fields are present in the Universe and fields of a similar nature are needed to account for the inflationary expansion of the early Universe.

22.6.1 Astrophysics

The Cavendish Astrophysics Group carries out a wide range of research programmes centred on four major areas of astrophysics and cosmology, each linked to experimental and instrumentation programmes (Box 22.1). In summary, the areas and the associated facilities are:

- The formation of stars and planets
  - The Atacama Large Millimetre Array (ALMA)
  - The James Clerk Maxwell Telescope (JCMT)
- Observational cosmology of the microwave background radiation
  - The Arcminute Microkelvin Imager (AMI)
  - The ESA Planck Surveyor satellite
- The formation and evolution of galaxies
  - The Low Frequency Array (LOFAR)
  - The Square Kilometre Array (SKA)
- High-resolution imaging of stellar systems and active galactic nuclei
  - The Magdalena Ridge Observatory Interferometer (MROI).

These activities are supported by theoretical studies into fundamental physics applications in relativity and cosmology, and modelling and simulation of astrophysical phenomena. With the arrival of Roberto Maiolino and Didier Queloz, the instrumentation programme has expanded into new project areas in the optical, infrared, millimetre and submillimetre wavebands.

Website: http://www.mrao.cam.ac.uk

Box 22.1   **Astrophysics programme**

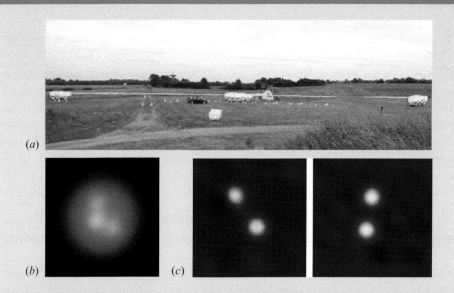

(a) The Cambridge Optical Aperture Synthesis Telescope (COAST), consisting of four telescopes with a baseline of up to 100 m. (b) A COAST image of structure on the surface of the red giant star Betelgeuse. The bright spots are due to huge 'holes' in the star's upper atmosphere (Young *et al.*, 2000). (c) Images of the binary star Capella between 13 and 28 September 1995 (Baldwin *et al.*, 1996). The separation of the stars is about 40 marcsec.

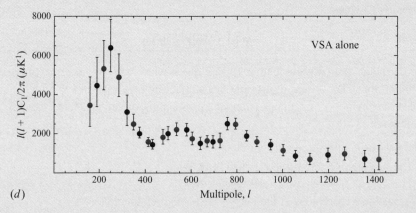

(d) The Very Small Array (VSA) is located on Tenerife at a high, dry site. It measured the first four peaks in the power spectrum of the cosmic microwave background radiation (Dickinson *et al.*, 2004).

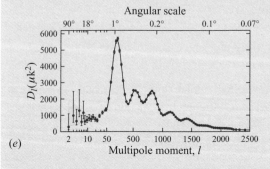

The Astrophysics Group participated in the analysis of the ESA Planck satellite data to determine precision cosmological parameters and to carry out tests of the ΛCDM model. Figure (e) shows a comparison between the observed power spectrum (red dots) and the prediction of a six-parameter fit to the observations (Ade *et al.*, 2014). This fit shows that the Universe is dominated by dark energy (68.3%) and dark matter (26.8%), ordinary baryonic matter making up only 4.9% of the total.

22.6.2 High-energy physics

The High Energy Physics Group's research is based upon experiments at high-energy particle accelerators, group members making up part of international collaborations working on experiments at CERN, Geneva; at Fermilab, Chicago; and on research and development activities for a future linear collider (Box 22.2). The group also studies a range of theoretical problems with a phenomenological emphasis. The major projects of the HEP Group at the Laboratory are:

- ATLAS: a general purpose detector for the Large Hadron Collider (LHC) at CERN. This was one of the two experiments at the LHC which discovered the Higgs boson.
- The Large Hadron Collider beauty experiment (LHCb): This experiment involves studies of CP violation and b-quark physics at the LHC.
- The Main Injector Neutrino Oscillation Search (MINOS) at Fermilab. This experiment concerns the detailed physics of neutrino oscillations.
- Research and Development for Future Collider Experiments.

The Cavendish HEP Theory Group supports HEP's experimental activities.
   Website: http://www.hep.phy.cam.ac.uk

22.7 The biological universe

Biological physics has been part of the Laboratory's programme since the pioneering work of Bernal and his colleagues in the 1930s. Bragg fully supported the development of the X-ray cystallographic analyses of biological molecules and this led to Watson and Crick's epochal paper in which the double-helix structure of the DNA molecule was determined (Section 12.6.2). The immense implications of this discovery led to the formation of the Laboratory of Molecular Biology on the Addenbrooke's Hospital site. After some years in which only modest interest was taken in physics and the life sciences, there has been a remarkable renaissance, with the physics of biology and medicine now playing a major role in the Laboratory's long-term strategy. The Biological and Soft Systems (BSS) sector, which is an association of groups working in soft condensed matter physics and the physics of biology, and the Physics of Medicine programme are at the heart of this activity, but there are strong links with many other groups in the Laboratory, University departments and other UK universities.

22.7.1 Biological and soft systems

The twenty-first century promises a major expansion at the interface of physics with the biological sciences and nanotechnology. These areas fall outside the conventional boundaries of chemistry, physics and biology, requiring a collaborative, multidisciplinary

Box 22.2   **High-energy physics**

(*a*) High Energy Physics Group members participate in the MINOS (Main Injector Neutrino Oscillation Search) experiment. The MINOS experiment is a long-baseline neutrino experiment designed to observe the phenomena of neutrino oscillations. MINOS uses two detectors, one at Fermilab at the source of the neutrinos and the other 450 miles away in northern Minnesota, at the Soudan Underground Mine State Park in Tower–Soudan. The image shows the front

(*a*)

face of the MINOS far detector. The control room is to the left and on the right is a mural by Joseph Giannetti (copyright Fermilab).

(*b*)

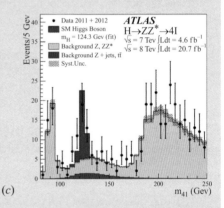

(*c*)

In 2012, the High Energy Physics Group participated in the discovery of the Higgs boson at CERN. (*b*) A four-muon decay of a Higgs boson detected in the ATLAS experiment at the LHC. (*c*) ATLAS evidence for the existence of the Higgs boson (red histogram) superimposed upon background events from other decays (copyright CERN).

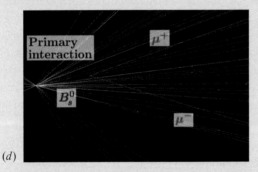

(*d*)

(*d*) An example of the rare decay of a neutral B_s particle into a pair of oppositely charged muons, observed in the LHCb experiment at CERN in 2013. The B_s particle decays within 10^{-12} s after its production but, during this short time, it travels far enough, about 1 cm, to be observed by the LHCb detector. Remarkably, the measured cross-section for this process is exactly as predicted by the standard

model of particle physics. This is a setback for the proponents of the simplest model of supersymmetry, since the new physics has failed to show up where it arguably had the best opportunity of being observed (copyright CERN).

(a) Athene Donald (right) with Deborah Stokes (left) at the console of the FEI Philips Dualbeam Quanta 3D ESEM. This dual-beam instrument permits *in situ* sectioning through samples using a focussed ion beam. The advantage of the ESEM is that biological specimens can be kept 'wet' during the imaging (see Figure 20.4).

(a)

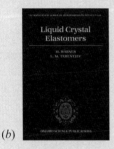

(b) In 2003, Mark Warner and Eugene Terentjev published their pioneering book on the physics of liquid crystal elastomers, describing the theory behind their extraordinary properties (Warner and Terentjev, 2003). These materials can change their lengths by factors of four or five on slight heating or exposure to light, change their shape without energy cost, distinguish between right and left, act as a laser when lightly pumped and change emission colour on stretching.

(b)

Recently appointed lecturer Sarah Bohndiek has used the technique of optoacoustic imaging to detect the impact of angiogenesis, the growth of new blood vessels, in cancer. This advanced imaging technique exploits the thermoelastic response of tissue to pulsed laser light in generating an ultrasound wave, combining an optical excitation with ultrasound detection. Significant advantages of the optoacoustic imaging method are high spatial resolution and penetration depth, high contrast due to the optical absorption of blood, and non-invasive application. Clinical trials in breast cancer detection are expected to commence in 2016.

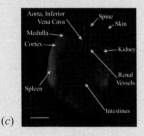

(c) An example of a blood haemoglobin oxygenation image acquired using multispectral optoacoustic tomography (MSOT). MSOT excites the distinct optical absorption profile of haemoglobin using a pulsed multiwavelength laser source and detects the acoustic wave resulting from the transient expansion of the tissue. MSOT is an emerging low-cost tool for clinical imaging and does not require injected contrast. The scale bar is 5 mm.

(c)

approach. The Biological and Soft Systems Sector (BSS) is pursuing such multidisciplinary research (Box 22.3). Techniques and inspirations from classical polymer physics, soft matter physics and the physics of condensed matter provide a foundation for research activities, with important progress being made in protein folding, biomaterials, cell biophysics and nanoscience, using theoretical, computational and experimental techniques. Much of the research takes place within the Cambridge Centre for the Physics of Medicine (CCPoM), with key strengths in:

- properties of soft matter
- polymers at surfaces and interfaces
- protein aggregation and folding
- cellular biophysics
- modelling of biological systems
- medical imaging and biophotonics
- nanoscience
- environmental scanning electron microscopy (ESEM).

Website: www.bss.phy.cam.ac.uk

22.7.2 Physics of medicine

From the structure of DNA to the development of magnetic resonance imaging, many major advances in biology and medicine have come about through collaborations between physical scientists and biologists. The Cambridge Centre for the Physics of Medicine (CCPoM) is a major new expansion of research activity in these areas in the University. With a hub in the Physics of Medicine (PoM) building and a focus on biomedical research, it aims to nucleate interactions between different disciplines as well as technology development at the interface of the physical, life and clinical sciences (Box 22.4). A major theme of the project is the fostering of new methods and concepts to understand the organisation and function of cells and their assembly into tissues and organs.

In addition to its function as a hub, CCPoM aims to identify talented scientists with multidisciplinary skills and interests at the early stages of their careers. PoM supports applications for independent career fellowships and provides space and research infrastructure.

Website: www.pom.cam.ac.uk

22.8 The quantum universe

Quantum phenomena are the subjects of study by many of the experimental groups and by the Theory of Condensed Matter Group. The distinction between the 'quantum universe' and the succeeding 'materials universe' is blurred and constructively overlapping. For convenience we group under this heading the activities of the following groups: Atomic, Mesoscopic and Optical Physics (AMOP), Microelectronics (ME) and the Hitachi Cambridge Laboratory (HCL), Nanophotonics (NP), Quantum Matter (QM), Semiconductor Physics (SP), Theory of Condensed Matter (TCM) and Detector Physics (DP).

22.8.1 Atomic, mesoscopic and optical physics

The research of the Atomic, Mesoscopic and Optical Physics (AMOP) Group is centred on the understanding of quantum aspects of condensed matter, from atomic Bose–Einstein condensates to semiconductor quantum dots (Box 22.5). The group has a high

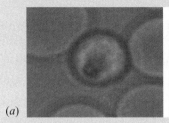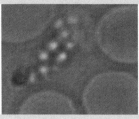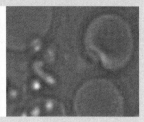

(a)

The unique automated imaging platform developed by Pietro Cicuta and his colleagues allows high-frame-rate videos of rare processes to be obtained with no human intervention. This technique has been used to observe how malaria parasites attack cells. Figure (a) shows: (left) a pre-rupture infected cell (schizont), (centre) the explosive egress of the parasites, or merozoite dispersal; (right) first deformation response on merozoite–red blood cell contact. The piercing of the red-cell wall results in the penetration of the healthy cell by the malarial infection (Crick *et al.*, 2013).

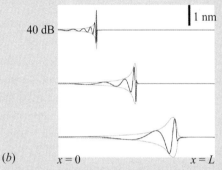

(b)

Duke and Jülicher (2003) developed a quantitative theory of hearing using a model of non-linear travelling waves in the basilar membrane (BM) within the cochlea of the ear. Figure (b) shows active travelling non-linear waves on the BM within the cochlea at three different frequencies ($f = 370$ Hz, 1.3 kHz, and 4.6 kHz). Each frequency reaches maximum amplitude at a different characteristic location: $x_r/L = 0.25$ (top), 0.5 (centre) and 0.75 (bottom). The frequencies of the strongly amplified waves are localised at different positions along the membrane.

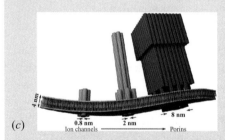

(c)

Structures at the molecular level can be created using DNA molecules as building blocks, referred to as 'DNA origami'. Figure (c) is a sketch of versatile DNA channel architectures in a lipid membrane. Channel diameters mimic the diversity of natural membrane components, from ion channels to large porins (Burns *et al.*, 2013) (image courtesy of Ulrich Keyser, Kerstin Göpfrich and Karl Goedel).

(d)

Stem cells contribute to both development and maintenance of multi-cellular organisms. The therapeutic potential of these extraordinary 'pluripotent' cells has created interest from both inside and outside the academic community. It has become possible to resolve the fate of individual labelled stem cells and their progeny, termed over the lifetime of an organism. Figure (d) shows genetic labelling of the small intestine. Following induction, stem cells express one of four fluorescent proteins. The figure shows ribbons of differentiating cells migrating from monoclonal crypts (bottom) onto villi (top) at 4 months post-induction (Simons and Clevers, 2011).

Box 22.5   **Atomic, mesoscopic and optical physics**

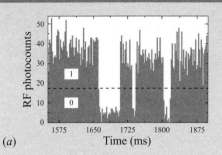

(a)

Members of the AMOP Group engineered a system comprising two separate quantum dots in close proximity. The optical transitions of the additional quantum dot act as a local sensor for the nearby qubit. When probed resonantly, these transitions generate a stream of single photons conditional on the electron spin state. The figure shows a continuous measurement of the state of the qubit as it goes through spin jumps in real time (Vamivakas *et al.*, 2010).

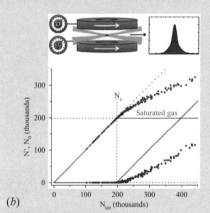

(b)

Even a weakly interacting gas, millions of times thinner than air, deviates strongly from Einstein's ideal-gas saturation picture for a Bose–Einstein gas, represented by the solid red and blue lines (lower panel). The upper panel shows how a combination of lasers and magnetic fields can be used to trap an atomic gas, cool it to a temperature T less than a micro-kelvin and vary the strength of the interactions. The number of atoms in the condensate (N_0, blue) and the thermal gas (N, red) as a function of the total atom number N_{tot} are compared with the ideal Bose–Einstein prediction. By extrapolating the

behaviour of the gas to low temperatures, Einstein's prediction of a saturated state is observed (Tammuz *et al.*, 2011).

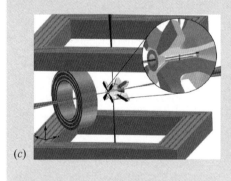

(c)

The diagram at left shows an ultra-high-vacuum chamber with an electrodynamic Paul trap (shown in grey) in which a single ion can be trapped in the small horizontal gap, about 0.6 mm width, between the electrodes. The neutral atoms are introduced into the trap and evaporatively cooled to the temperatures at which Bose–Einstein condensation occurs (Zipkes *et al.*, 2010). In these experiments, a hybrid system was created for the first time in which the trapped ion was immersed in a Bose–Einstein condensate.

turnover of projects and experiments in response to rapid advances in these areas. There are four subgroups, each led by a group member:

- Quantum gases and collective phenomena: superfluidity, quantum magnetism, two-dimensional systems, non-equilibrium phenomena

Microelectronics   **Box 22.6**

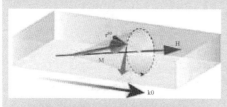

Spin electronics or 'spintronics' aims to understand how one can make use of the spin degree of freedom of electrons in order to realise electronic devices and functionalities which cannot be realised using the electron's charge degree of freedom alone.

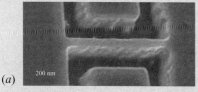

(*a*) As microwave frequency electricity is passed through our ferromagnets, a torque is applied to the magnetisation, causing it to orbit about its equilibrium position. (*b*) A micrograph showing one of the nanoscale ferromagnetic wires through which the electrical current is passed. This example is etched from a crystal of Ga(Mn,As) grown on an epitaxial substrate (Fang *et al.*, 2011).

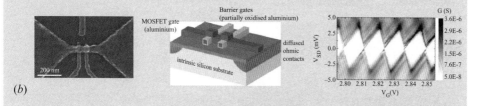

(*b*)

(left) Scanning electron micrograph of a silicon quantum dot. (centre) Schematic showing the device structure including the electrostatic gates, ohmic contacts and intrinsic substrate. (right) Coulomb diamonds measured for a single quantum dot device at mK temperatures.

- Quantum optics and cold atoms: quantum gases, optical lattices, single atoms, cavity QED
- Quantum optics and mesoscopic systems: quantum dot spins, cavity QED, plasmonics and diamond colour centres
- Quantum optoelectronics: Quantum dot spins, polymer semiconductors, coherence in semiconductors.

  Website: www-amop.phy.cam.ac.uk

22.8.2 Microelectronics

Members of the Microelectronics (ME) Group investigate electron spin physics in nanoscale electronic devices (Box 22.6). Ultra-sensitive electrical measurement techniques, at both low and microwave frequencies, are applied for sample characterisation. The group aims to explore both classical and quantum effects in spintronics that are relevant for future technologies.

Box 22.7   **Nanophotonics**

Flexible, polymer-based colloidal-sphere photonic crystals can be fabricated by self-assembly. Under strain, the crystal plane spacing changes. This results in a dynamic change in reflectivity and transmission, as seen in the picture. The colour changes results from the size of the polymer spheres used in fabrication and the electric field distribution in the close-packed structure. Incorporating different absorbing materials into the structure allows control of the absorption in the layers, opening up new sensor applications and novel filter designs.

Two CW pump lasers focussed 20 μm apart onto the microcavity (black holes on the right-hand image) create a trap for the polariton condensate in between. Photon emission shows the spontaneously formed $n = 3$ quantum harmonic oscillator state, but now on the macroscopic scale (Tosi *et al.*, 2012).

The first cohort of interdisciplinary PhD students in the Nano Doctoral Training Centre in Cambridge in 2009. Students attend courses and carry out practical experiments and projects across physics, chemistry, materials science and engineering. The Director of the programme, Jeremy Baumberg, is third from right in the front row.

There is close collaboration with the Hitachi Cambridge Laboratory (HCL), established in 1989 as an 'embedded laboratory' within the Cavendish. The aim is to create new concepts for advanced electronic/optoelectronic devices. The HCL specialises in advanced measurement and characterisation techniques, while the ME Group specialises in nanofabrication techniques. Collaborative research topics include:

- phase-state low electron (hole)-number drive memory (PLEDM)
- single-electron devices and nanoelectronics
- quantum information processing
- nanospintronics
- spin transport in carbon-based, organic semiconductors.

Website: www.me.phy.cam.ac.uk

22.8.3 Nanophotonics

In nanophotonics, new materials are constructed in which atoms are arranged in sophisticated ways on the nanometre scale. The Nanophotonics (NP) Group is one of the most recent groups in the Laboratory and is part of the EPSRC-funded UK Nanophotonics Portfolio at Cambridge (Box 22.7). Research conducted at the Laboratory includes:

- nanoplasmonics
- polymer photonic crystals
- semiconductor microcavities
- metamaterials
- nano-self-assembly.

Assembling nanoscale pieces of matter into sophisticated structures creates nanomaterials with emergent properties not found in their constituents, but is a major technological challenge. One of the goals is moving from expensive fabrication of devices to elegant nano-assembly, in which materials build themselves.

This convergence of nanoscience and nanotechnology with photonics is highly interdisciplinary across all physical sciences and beyond, including chemistry, engineering, biology, healthcare, materials and physics.

Website: www.np.phy.cam.ac.uk

22.8.4 Quantum matter

In the Quantum Matter (QM) Group, matter is studied under extreme conditions of very low temperatures, high magnetic fields and high pressures, using advanced experimental techniques (Box 22.8). The goal of this research is to understand new forms of magnetism and superconductivity and to find electrically conducting materials with physical properties not described within the standard models of solid state physics. Work is focussed on two major topics:

- the nature of quantum order in itinerant-electron systems on the border of magnetism at low temperatures
- the physics of novel superconducting materials such as high-T_c superconductors, graphite intercalates and the ruthenates.

Research interests include anisotropic superconductivity, the development of new cryogenic equipment, the electronic structure of correlated electron materials, exotic states of matter, high-T_c materials, hydrostatic pressure, novel superconductors and quantum ferroelectrics.

Website: www.qm.phy.cam.ac.uk

22.8.5 Theory of condensed matter

The Theory of Condensed Matter (TCM) Group carries out fronter research under three general headings:

- collective quantum phenomena
- first-principles quantum mechanical methods
- soft matter.

Box 22.8   **Quantum matter**

(a) Suchitra Sebastian working with the world's highest-field hybrid magnet, combining resistive and superconducting technologies to reach continuous magnetic fields of 45 T at the National High Magnetic Field Laboratory, Tallahassee, Florida. This facility has been used to discover quantum oscillations in $YBa_2Cu_3O_y$ with $y = 6.56$, at magnetic flux densities between 22 and 64 T through magnetic torque and electrical resistivity measurements (Sebastian et al., 2011) (courtesy of Suchitra Sebastian).

(a)

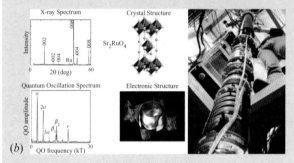

(b)

(b) A comparison of the roles of (upper left) X-ray diffraction in determining the crystal structure and (lower left) quantum oscillations from de Haas–van Alphen or Shubnikov–de Haas measurements which determine the electronic structure. The low-temperature insert of the QM Group's 18 Tesla, 1 mK cryomagnet is shown at right.

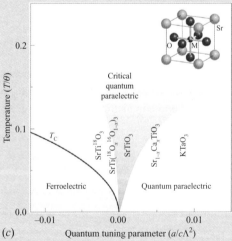

(c)

Ferroelectrics are materials comprising electrical dipoles in the unit cells of the crystal lattice. Due to the interactions between them, these dipoles may line up, resulting in ordered electric fields permeating the crystal. By using pressure, or chemical or isotopic substitution, ferroelectrics can be tuned to the quantum critical regime. Amazing transformations of the crystal properties take place as a ferroelectric approaches its quantum critical point. The theory of these processes includes the effects of coupled fluctuating polarisation and quantum strain fields on the measured dielectric susceptibility below a temperature of 4 K. The phase diagram for quantum critical ferroelectrics is shown in Figure (c) (Rowley et al., 2014).

The research interests of the group constantly evolve to address new theoretical challenges, some arising from experiments performed at the Laboratory and others from elsewhere (Box 22.9). For instance, Michael Payne and his colleagues have developed ONETEP, a quantum mechanical code that offers state-of-the-art accuracy, the cost scaling only linearly with system size. This allows quantum mechanical simulations of biological systems. The group has an outstanding record of training young researchers, 28 members having taken up permanent academic appointments within the last nine years and 14 within the last five.

Website: www.tcm.phy.cam.ac.uk

22.8.6 Semiconductor physics

The Semiconductor Physics (SP) Group uses semiconductor devices to investigate phenomena in fundamental physics. These investigations often involve the observation of effects which can only be explained by the laws of quantum mechanics and often require the control of small numbers of quantum particles such as electrons and photons. The aim is to develop semiconductor devices where quantum effects dominate the device operation, an area of work which has many future applications in science, technology and medicine, including quantum information processing, sensing and imaging technologies (Box 22.10).

To achieve these aims the group has developed a range of sophisticated technologies for the fabrication of high-quality devices, much of the group's research relying on the technique of molecular beam epitaxy (MBE). Recent work has seen the start of a number of projects based on carbon electronics. These rely on using either graphene, a single layer of carbon atoms with very unusual properties, or carbon nanotubes, a single layer of carbon atoms rolled up into a tube structure.

The group also works in collaboration with a number of other researchers at the Cavendish Laboratory, the Materials Science and Engineering departments, as well as with many other universities and industrial research laboratories such as Toshiba, TeraView and the National Physical Laboratory (NPL).

Examples of the research undertaken in the group are:

- single and entangled photon sources
- coupled electron-hole gases
- terahertz technology
- quantum computing using surface acoustic waves
- carbon electronics
- gate-defined quantum dots.

Website: www.sp.phy.cam.ac.uk/spweb/index,php

22.8.7 Quantum sensors

The Quantum Sensor (QS) Group operates a major facility for investigating, manufacturing and testing a new generation of superconducting detectors and sensors for astrophysics and the applied sciences. Much of the work relates to large-format superconducting

Box 22.9   **Theory of condensed matter**

David MacKay's influential book *Sustainable Energy – Without the Hot Air* (courtesy of UIT Publications). In 2008, David MacKay, who was Chief Scientific Advisory to the Department of Energy from 2010 to 2014, wrote this influential book, in which he made simple calculations using elementary physical principles to make reliable estimates of the UK's energy needs and what is likely to be available from all types of renewable resources.

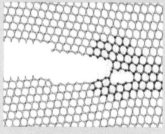

Development of the CASTEP and ONETEP programs has allowed quantum mechanical computations of systems containing thousands of atoms to be simulated with high precision. The red bonds in the figure at left show the region in which a full quantum mechanical computation of the propagation of a crack through a regular lattice has been carried out. CASTEP has been commercialised and has already generated over \$30 million in sales.

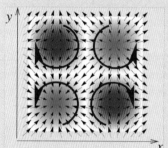

The standing wave pattern of a laser can lead to a lattice of vortices and anti-vortices in the phase of the light field (black arrows), corresponding to circulating photon angular momentum (blue arrows). An atom moving between regions in which its state is A (red) or B (green) exchanges momentum with the light in such a way that it experiences a Lorentz force equivalent to a magnetic field of fixed sign (Cooper, 2011). This is a means of creating enormously strong magnetic fields, millions of times stronger than in the strongest conventional magnets. Experiments are underway to test these theoretical predictions.

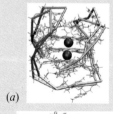

(*a*)

(*b*)

In collaboration with Julian Huppert of the Biological and Soft Systems Group, Payne and Cole have studied regions of DNA molecules called G-quadruplexes which form four-stranded structures, instead of the more familiar double-stranded helices. The images show (*a*) a snapshot of a G-quadruplex structure with metal ions (blue) running down the central channel. The original crystal structure is superposed in silver. (*b*) Electron transfer to a central K^+ ion in a small region of the quadruplex. This simulation uses the first-principles ONETEP program.

Box 22.10

Semiconductor physics

(a)

The figure on the left shows a molecular beam epitaxy (MBE) system for the growth of crystals of ultra-pure (1 part in 10^9) III-V semiconductors such as gallium arsenide (GaAs) and aluminium gallium arsenide (AlGaAs) with control close to the level of a single atomic layer. These semiconductors are processed into device structures using a range of techniques to etch small patterns as well as deposit patterns of metal on the surface to form electrical contacts and gates.

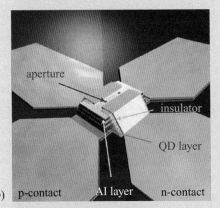

(b)

The collaboration between the Semiconductor Physics Group and Toshiba Research Europe has developed devices which produce a stream of single photons. This is achieved by embedding a nanoscale 'self-assembled' quantum dot of indium antimonide (InAs) in a slab of GaAs semiconductor, itself placed between regions of p- and n-type GaAs. If pulses of current are passed through this device, electrons and holes recombine in the quantum dot to produce a stream of single photons of very well defined wavelength (Yuan *et al.*, 2002) (diagram courtesy of Toshiba Ltd).

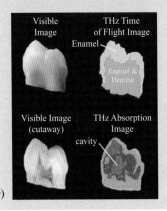

(c)

The Semiconductor Physics Group pioneered the use of THz radiation for imaging in biomedical and dental applications. In this example, THz imaging is used to locate a cavity in a tooth through a combination of time-of-flight imaging and absorption. The TeraView Company was spun out of Toshiba Research Europe in 2001 by its co-founders, Michael Pepper and Don Arnone. TeraView maintains close links with the Laboratory, where modern THz technology was pioneered in conjunction with TeraView (Townsend, 2013).

cameras and receivers for millimetre, submillimetre and far-infrared wavelengths. The group runs one of the most advanced fabrication facilities worldwide for superconducting microcircuits, and explores the limits of quantum detection and sensing (Box 22.11). Topics include:

- heat flow and thermal fluctuation noise in low-dimensional structures at 100 mK
- non-equilibrium superconductivity through above-gap and sub-gap photon and phonon excitation

Quantum sensors

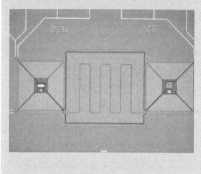

A 'lab-on-a-chip' to develop the ultra-low-noise detectors used for the next generation of astronomical space-borne observatories such as SPICA. The device here is an ultra-low-noise transition edge sensor (TES) on a suspended SiNx membrane; the device in the centre is a meandering super-conducting transmission line; the device on the right is a blackbody load, with thin-film resistor for radiometric ther-mometry at 100 mK. The whole device measures just 2×1 mm.

UHV sputtering with $<10^{-10}$ torr base pressure and sub-strate temperatures of $-200°C$ to $900°C$ allows a range of materials to be deposited with different stoichiometries. Materials used include Nb, Ta, Al, NbN, Mo, Hf, Ir, Cu, Au, SiO_2, SiO. Reactive ion etching (RIE) with laser endpoint detection is used for pattern definition. A complete end-to-end 4-inch process is available for large-format imaging arrays.

- partially coherent optics and electromagnetism
- detector materials science, mesoscopic device processing and quantum measurement techniques
- transition edge sensors, kinetic inductance Detectors, >1 THz SIS mixers, SQUIDs, and quantum qubits
- energy absorption interferometry for measuring coherent phenomena in energy-absorbing structures.

The group works with many international agencies, and under a European Space Agency (ESA) contract is currently developing 200–20 μm ultra-low-noise imaging arrays for the Japanese cooled-aperture space telescope, SPICA.

Website: www.phy.cam.ac.uk/research/research_groups/qs

22.9 The materials universe

The 'materials universe' is concerned with the discovery of new materials which are of scientific, industrial and social importance. The Winton Programme for the Physics of Sustainability is a prime example of the importance of these researches for society at large. The materials universe has many synergies with the quantum and biological universes. The

Akshay Rao and his colleagues discovered that, in organic semiconductors, electrons and holes that encounter each other form intermediate species, so-called charge transfer (CT) states, 75% of which have triplet (spin 1) and 25% singlet (spin 0) character (Rao *et al.*, 2013). The ground state of the system has singlet (spin 0) character and thus the triplet CT states cannot recombine to the ground state. By morphological control of the heterojunction, the delocalisation of electronic excited states could be such that the triplet CT states are re dissociated and recycled back as free charges. The images show the structure of such an organic solar cell: (top) the active layer of the device (green) with hole and electron transporting layers (blue and yellow) all sandwiched between electrodes (silver); (bottom) the structure of the active layer with an interpenetrating network of donor and acceptor semiconductors.

Device schematic and energy diagram for the proposed working mechanism of organic solar cells (Ehrler *et al.*, 2012). The excited singlet state in the pentacene layer converts to two triplet states after about 80 fs. These triplets are dissociated at the pentacene/PbS nanocrystal interface. At the same time the PbS nanocrystals absorb the infrared portion of the incident light.

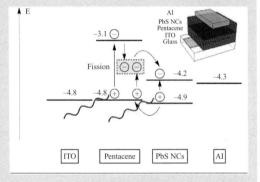

An advantage of organic semiconductor materials for photovoltaics is that large areas of solar cells can be produced by inkjet-printing techniques. This concept has involved a challenging programme in fluid mechanics to devise means by which the narrow channels for thin conducting wires can be manufactured by fluid processes. The image shows a plastic roll of solar cells made by this technique.

activities of the following groups are described under this heading: Optoelectronics (OE), Molecular Engineering (MEng), Surface, Microstructure and Fracture (SMF), Thin Film Magnetism (TFM) and Scientific Computing (SC).

22.9.1 Optoelectronics

Members of the Optoelectronics (OE) Group carry out fundamental physics studies of different aspects of organic semiconductor materials (Box 22.12). There is a particular focus

Box 22.13 **Molecular engineering**

Crystals of ammonium dihydrogenphosphate exhibit frequency doubling of light, in this case from infrared to blue wavelengths. This is dependent upon ionic displacement or molecular charge-transfer processes. Single-crystal X-ray and neutron diffraction studies are being conducted on organic materials that exhibit second harmonic generation (SHG) – the frequency doubling of light. Hydrogen bonding, thermal motion and charge-density analyses are employed to understand the electronic charge-transfer processes that dictate this phenomenon.

upon the physics of semiconducting conjugated polymers, which are long-chain organic molecules made from conjugated units such as benzene. They are inherently quantum mechanical objects with nanometer-sized dimensions.

In the late 1980s, the group discovered that these conjugated polymers behave in many respects like inorganic semiconductors and can be used in a number of semiconducting devices such as field-effect transistors, light-emitting diodes and solar cells (see Section 20.4). These pioneering discoveries were important milestones for the field of organic electronics, which has now developed into a large international research endeavour with many diverse academic and industrial activities. The current main applications of the OE Group's research interests are:

- light-emitting diodes
- solar cells
- transistors.

Website: www.oe.phy.cam.ac.uk

22.9.2 Molecular Engineering

The Molecular Engineering (MEng) Group investigates structure–property relationships in optoelectronic and energy-related materials. Armed with this knowledge, new materials can be structurally engineered to suit a given device application. Primarily interests are in materials which have potential use for dye-sensitised solar cells, optical telecommunications, nuclear waste storage and nano-electronic insulation (Box 22.13). A wide variety of techniques are used to realise this goal:

- diffraction and spectroscopy
- optoelectronics measurements
- sample fabrication and device testing
- non-linear optical devices fabricated by embedding them in a polymer

The helium spin-echo (HeSE) spectrometer consists of a long V-shaped vacuum system. The high-intensity helium source is at the rear centre of the picture, the ultra-high-vacuum sample chamber on the left and the detector on the right. Development of the HeSE technique was a major undertaking, made possible by substantial support from the Cavendish workshop facilities. The system has been used to measure the motions and dynamics of molecules on surfaces.

- Computational techniques are seminal in predicting or rationalising our experimental work.

Website: www.mole.phy.cam.ac.uk

22.9.3 Surfaces, microstructure and fracture

Activities in the Surface, Microstructure and Fracture (SMF) Group are broadly divided into surface phenomena and dynamic material processes (Box 22.14). The group has two main focus areas:

- fracture and shock physics
- surface physics.

The Fracture and Shock Physics subgroup has an international reputation in the areas of material fracture, shock physics and the study of energetic materials, and has pioneered the use of high-speed diagnostics. In the area of surface physics, fundamental research into surface structure and processes specialises in the experimental technique of helium atom scattering (HAS), complemented by more traditional surface techniques. Some of the most exciting work is in the development of new instrumentation.

Thin-film magnetism

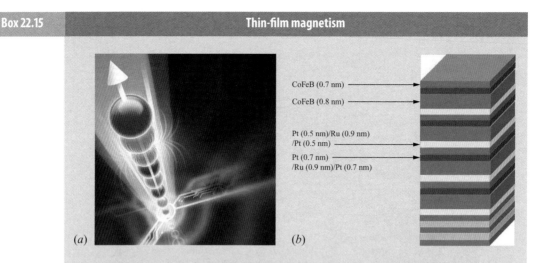

CoFeB (0.7 nm)

CoFeB (0.8 nm)

Pt (0.5 nm)/Ru (0.9 nm)
/Pt (0.5 nm)

Pt (0.7 nm)
/Ru (0.9 nm)/Pt (0.7 nm)

(*a*)   (*b*)

A shift register is a device which passes data along a chain of memory cells. It has only one input and one output but there can be several memory cells connected in series between these two points. (*a*) A schematic diagram of a spintronic shift register (image by C. van der Linden). (*b*) A schematic diagram of a three-dimensional spintronic shift register, made of layers of magnetic and non-magnetic materials only a few atoms thick stacked on top of each other (courtesy of R. Cowburn).

Current high-profile projects include using helium-3 spin-echo techniques to study dynamics on atomic length- and time-scales and the development of helium atom microscopy as an imaging technique with a uniquely delicate probe.

Website: www.smf.phy.cam.ac.uk

22.9.4 Thin-film magnetism

The Thin Film Magnetism (TFM) Group is at the forefront of research on ultra-thin film and magnetic nanostructures. Major research areas include magnetic mesostructures and spin transport (Box 22.15).

Novel magnetic properties and spin-polarised electron transport phenomena in molecular beam epitaxy (MBE) grown magnetic film structures are investigated, including ferromagnet–semiconductor hybrid structures. In particular, the fundamental electron spin-dependent transport processes which underpin the emerging field of spintronics are investigated, as well as nanomagnetism for biomedical applications.

The group is well equipped to characterise and fabricate nanomagnetic structures. Magnetic materials can be fabricated using an ultra-high-vacuum (UHV) multiple technique chamber, electro-chemical deposition and MBE. Characterisation facilities including superconducting quantum interference devices (SQUIDs), low-temperature magneto-optical Kerr effect (MOKE) apparatus and spin detector chambers are essential for these activities.

Website: www.tfm.phy.cam.ac.uk

Sample calculations of stratosphere–troposphere exchange during an event that took place over the North Atlantic in 1996. The technique captures explicitly a number of essentially three-dimensional features such as filaments and tubes in the tropopause region. These integrations use a chemistry and transport model forced by meteorological analyses.

22.9.5 Scientific computing

The niche of the Scientific Computing (SC) Group is the application of contemporary, cutting-edge research from the physical sciences (including applied mathematics, numerical analysis and fundamental physics) and contemporary high-performance computing methodologies to technology, engineering and applied science (Box 22.16). Our methodologies include:

- high-resolution, shock-capturing Riemann problem-based numerical schemes
- mesh generation and moving boundaries using Cartesian cut-cell and ghost-fluid approaches
- hierarchical, structured adaptive mesh refinement
- parallel computing using a message passing interface (MPI) and algorithm/code implementations on graphical processing units.

Research is funded mainly by companies including Orica Mining Services, Schlumberger Cambridge Research, Boeing Research and Technology, Jaguar Land Rover, AWE Aldermaston and BAE Systems. A wide range of projects in pure and applied physics include the determination of equations of state for hydrocodes by means of ab initio atomic-level modelling, atmospheric dispersion of pollutants, anti-icing of aircraft, heavy-oil recovery, coupled reactive two-phase flow and elastoplastic material algorithms and advanced vehicle simulation.

The group also supports a MPhil degree in Scientific Computing and an advanced training programme with short courses and summer schools at a national and international level, including the EPSRC Autumn Academy on High Performance Computing and the NCAS (National Centre for Atmospheric Science) Climate Modelling Summer School.

Website: www.lsc.phy.cam.ac.uk

22.10 The Cavendish III project

Soon after I took over as Head of Department in 1997–98, it became apparent that the operating costs of the Laboratory and the need for continual refurbishment of the space could not be sustained into the longer-term future. The specific problems were:

- The reconstruction of the Laboratory on the West Cambridge site was completed in 1974 with a design lifetime of about 25 years; it was constructed in an era when energy was cheap and there was little concern about the environment. For example, with single glazing there are severe heat losses.
- The buildings were far outside space and environmental standards and there were numerous deleterious design features which would be very expensive to mitigate. The flat roofs were a continuing nightmare and very vulnerable to water damage. This resulted in accidents and serious health and safety issues. Asbestos was a major problem which afflicted every building in the Laboratory. The internal construction of the buildings was wasteful of space which could not be reclaimed at economic cost. The top floor of the Mott Building was barely inhabitable during the summer months.
- The buildings proved to be functional, but were not of high quality, reflecting Pippard's goal of providing the maximum useable space using available resources. By the turn of the millennium, there was very significant overcrowding in many areas of the Laboratory with the growth in staff numbers shown in Figure 16.5.

The consequence of all these factors was that the operating costs of the Laboratory were very large compared with those of a building built to modern design norms.

There were, in addition, important scientific and strategic aims for the reconstruction programme:

- The Laboratory has a key role to play in the national interest, as one of the world's top physics departments. Large departments like the Cavendish Laboratory need to maintain a large output of highly qualified graduate students and undergraduates from the UK and abroad. Physics is still considered by HEFCE as a 'strategically important and vulnerable' (SIV) subject.
- Changes to higher education funding may hit SIVs particularly hard since Vice-Chancellors may close physics and chemistry departments due to lack of qualified undergraduates. In addition, with present tuition fees at £9000 per year, undergraduates may choose to study more applied/vocational subjects. There is a need to continue to enhance our extensive programme of public and schools education, which has been demonstrably successful and effective.
- The Laboratory has an even more important role in protecting the health of the subject. The pressing need continues to be for state-of-the art laboratories, office and supporting infrastructure and services to the highest level of modern design. These are essential if the Laboratory is to recruit and retain the very best physicists from the worldwide pool of outstanding individuals.

- The facilities need to be flexible to accommodate future developments and collaborations. Expansions of the programme may take place through strategic collaborations with other departments and universities and with industry.

In 2002, the University agreed that there was an urgent need to rebuild the Laboratory, and I was charged with developing a phased redevelopment plan on its existing site. The Laboratory's development programme formally began in October 2007 when I was appointed Director of Development with the charge of setting up and implementing a development and fund-raising programme. The activity spans all aspects of the Laboratory's programme and can be conveniently summarised under four headings:

- Support for people and their programmes, including outreach, student and postdoctoral support
- Endowment of lectureships and professorships
- Equipment
- Buildings and infrastructure.

These are summarised in the Cavendish Laboratory's Development Portfolio, which is regularly updated on our website.[7] Cooperation with the Cambridge University Development and Alumni Relations (CUDAR) programme has been essential in promoting the programme and achieving success.

The development plan involved raising the resources from benefactors and philanthropic bodies necessary to secure matched funding by the University and various government initiatives. During the period 2005–16, four new buildings were constructed, as illustrated in Box 22.17. Including the Winton donation (Section 22.4.3), over £80 million has been raised in support of the development plan from philanthropic gifts, grants from charities and matching funds from the government and the University.

The rebuilding of the rest of the Laboratory is a top priority for the University. The new buildings will occupy the neighbouring Paddocks site, and this development is embedded in the latest versions of the Master Plan for West Cambridge development. As a greenfield site, there are relatively few constraints, providing the opportunity to create a facility ideally matched to our requirements (Figure 22.4). The new buildings will house many of the groups working in the areas described in Section 22.8 as the 'quantum universe' and in Section 22.9 as the 'materials universe', as well as the High Energy Physics Group and the large experimental laboratories to be used by the Astrophysics Group. In addition, all the support services, including the Administration and Finance division, main workshop, stores, lecture halls and seminar rooms, the common room, library, exhibition space and the Physics Centre for Public Education will be located in the new buildings.

The absolute requirement is that the new buildings accommodate the very diverse needs of cutting-edge physics research and its cognate disciplines. For example, the physics of biology and of medicine require somewhat different facilities than a traditional physics laboratory. Many disciplines now place very severe requirements on vibrational stability and the use of clean-rooms. These are challenges for the architects and physicists who have worked closely together to achieve a balance between functionality, interaction, flexibility

New buildings 2008–2016 as part of the redevelopment of the Cavendish Laboratory

The Physics of Medicine Building was opened in December 2008. It was sponsored by the Wolfson Foundation and the University. Ben Simons was appointed Herchel Smith Professor of the Physics of Medicine in 2011.

The Kavli Institute for Cosmology is a joint programme between the Cavendish Astrophysics Group and the Institute of Astronomy. It was supported by a gift from the Kavli Foundation and by matched funding from the University. It was opened in November 2009.

The Battcock Centre for Experimental Astrophysics was made possible by a gift from Humphrey Battcock, a grant from the Wolfson Foundation and the support of the University. It was opened in October 2013.

The Maxwell Centre for collaboration between the physical sciences and industry has been funded by the Higher Education Funding Council for England, through its UK Research Partnership Infrastructure Fund (UKRPIF), and the University. The building was handed over to the Cavendish Laboratory in late 2015 (visualisation courtesy of BDP Architects/TIP).

and environmental friendliness. These new buildings and associated state-of-the-art infrastructure will place the Laboratory in a strong position to attract the best research physicists and maintain it at the forefront of physics on worldwide basis.

The synergy between the science and the buildings required to house the programme is well illustrated by the forthcoming completion of the Maxwell Centre. This will be the

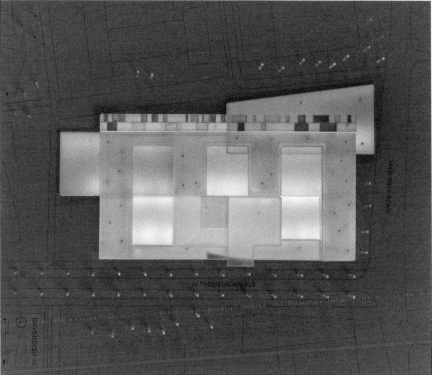

A schematic model by architects Jestico and Whiles of the new buildings to complete the redevelopment of the Cavendish Laboratory. The buildings will occupy the Paddocks site to the west of J.J. Thomson Avenue, subject to planning permission (model and photograph courtesy of Jestico and Whiles).

Fig. 22.4

centrepiece for industrial engagement with physical scientists working on the West Cambridge Physical Science and Technology Campus. The scale of industrial involvement is already substantial and our target is to double industrial funding through a combination of activities in the new building, in collaborating departments and in commercial space on the West Cambridge site. This will also be a central component of the Cavendish III project since we will offer use of the advanced technology facilities to our industrial partners. There will also be many opportunities for collaborative use of the instruments and facilities by other departments in Cambridge and by our partners in other universities.

In December 2015 we were delighted to learn that, in the Chancellor of the Exchequer's Autumn Statement on government funding for the next five years, £75 million has been allocated to the Cavendish III project, to be matched by a further £75 million from the University of Cambridge. We are optimistic that the rebuilding of the Cavendish Laboratory will be completed by about 2020. Maxwell's legacy will certainly continue to thrive for the rest of the 21st century.

Appendix The evolution of the New Museums site

These plans (Figures A.1–A.9) and the accompanying captions are taken from the series of papers prepared in 1936 by G.S. Graham-Smith of the Department of Pathology and are now housed in the Cambridge Collection in the Cambridge City Library. The originals contain information about the development of the Downing site, but these are not included in this summary, which concentrates upon the development of what became known as the New Museums site and which was to be the site of the Cavendish Laboratory until the move to West Cambridge in 1974. All references to buildings at the New Museums site are included, illustrating the continual construction and movements of departments on the site. Each map shows the site in a particular year, the captions listing the significant changes since the previous map.

The buildings associated with the Cavendish Laboratory and the Jacksonian Professor are indicated with blue shading. Salvin's building, which occupied the central region of the site, is shown in red, and Fawcett's building in green. Graham-Smith's original lettering has been preserved. The subsequent development of the site from the point of view of the Laboratory is summarised in Figure 5.3.

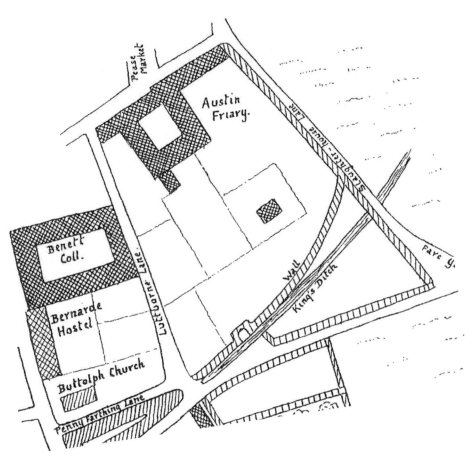

Fig. A.1   **1574:** What is now the New Museums site belonged to the Augustinian, or Austin, Friary. With the Dissolution of the Monasteries, the Friary was surrendered in 1538. In 1545, John Hatcher, Regius Professor of Physic and Vice-Chancellor in 1579, bought the grounds and lived there until his death in 1587. He left the house and grounds by his will to the University but this part of the will never came into effect. The Austin Friary passed to Stephen Perse, Senior Fellow of Gonville and Caius College, who left part of the site for a free grammar school, hence the name of Free School Lane, and part for alms-houses, which were erected soon after his death.

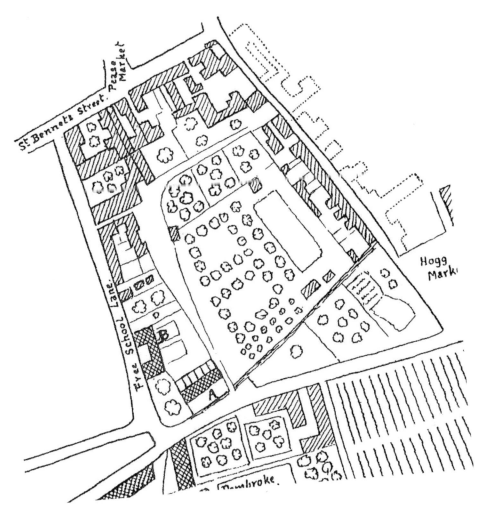

Fig. A.2

1688: Houses, including the Perse alms-houses (A) and the Free School (B) and their gardens occupy the west, north and east margins of the Friary grounds, but the south side is open. About three acres of the centre are occupied by a garden, with an entrance from Free School Lane, a little to the north of the present entrance to the Laboratory, and an entrance on the south, east of the alms-houses. Buildings have been erected along the east side of Slaughterhouse Lane.

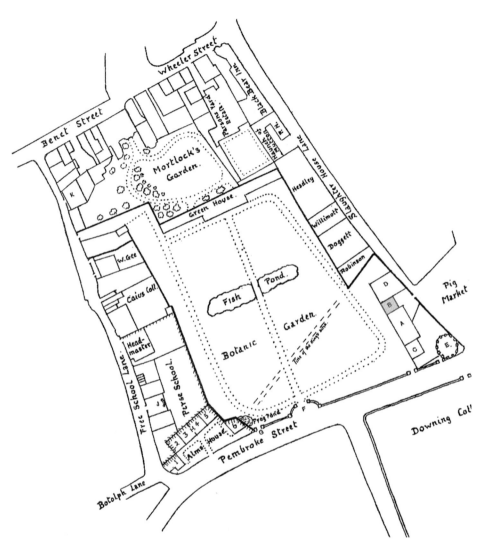

Fig. A.3 **1850:** In 1720, most of the Friary buildings had become ruinous and were pulled down. In 1760, the 'Great House' (K) in Free School Lane and about 5 acres of land were bought for £1600 and in 1762 conveyed to the University in trust as a public botanic garden by the enthusiastic botanist, the Rev. Richard Walker, Vice-Master of Trinity. In 1783, the Great House, which had become ruinous, and some land was sold by the University to John Mortlock who, 'from 1784 to his death in 1816 was undisputed dictator of the town'. In 1784, a lecture room (A) and two private rooms for the Jacksonian Professor (B) and the Professor of Botany (C) were built. These were the first buildings for scientific purposes erected by the University. In 1833 two other rooms were added (D) and a 'Round House' (E), with a dissecting room, a lecture room and a private room adjacent to it, was built. The site of the New Botanic Gardens was conveyed to the University in 1831, and the contents of the Old Botanic Gardens were transferred there between 1848 and 1852. The site of the Old Botanic Gardens was bought by the University from the trustees in 1853.

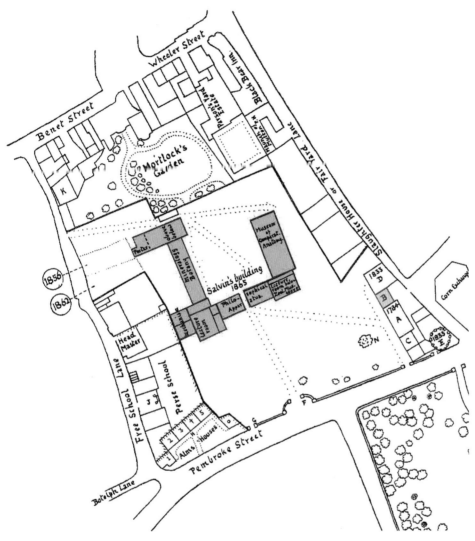

1865: In 1856, the University bought offices in Free School Lane, where the old part of the Cavendish Laboratory now stands, and in 1862 a house to the south. In 1853 a Museums and Lecture Room Syndicate was appointed 'to consider whether any and what steps should be taken for appropriating to the University the site of the Old Botanic Gardens and to consider whether any and what steps should be taken for erecting additional lecture rooms and museums'. In 1865, Salvin's two-storied building, erected to accommodate all the mathematical and scientific professors, except those of Anatomy and Chemistry and the Jacksonian Professor and the Professor of Geology, who remained in the old building, was completed. The north wing accommodated mineralogy above and botany on the ground floor, with mechanism and a lecture theatre at an angle. The south wing contained two rooms with an archway between them for philosophical apparatus and rooms for the mathematical professors above. To the east of these rooms was a lecture theatre for comparative anatomy below and the zoological museum above. The comparative anatomy museum extended to the north of the lecture room for that subject. The figures in circles show the dates on which the University purchased properties indicated by dotted lines.

Fig. A.4

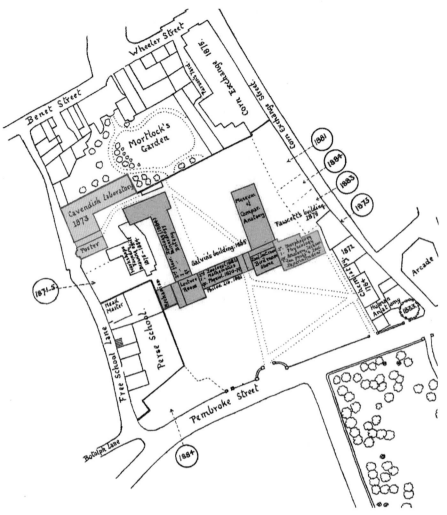

Fig. A.5   **1884:** In 1870 Foster was appointed Trinity praelector in physiology and the rooms on the ground floor on each side of the archway in Salvin's building were assigned to him. In 1871, the University bought two houses in Free School Lane. In 1872 a student's laboratory and private rooms were added to the Chemical Laboratory. In 1873 the Cavendish Laboratory 'was built and stocked with apparatus . . . '. Part of the building stands on the site of the Botanic Garden Curator's house. The entrance was shifted a little way to the south. The site in Corn Exchange Street east of Salvin's building was bought. In 1874–75 two houses north of the Perse School were bought. In 1875 the first workshop for mechanism was erected. The Corn Exchange was built. In 1878 further workshops for mechanism were built. In 1879 Fawcett's three-storied building on a site, sold in 1786 and repurchased in 1873, between the east end of Salvin's building and Corn Exchange Street, was completed. On the ground floor were rooms for the Human Anatomy and Jacksonian professors, the Superintendent of the Museum of Comparative Anatomy and store rooms. The first floor was assigned to Physiology and the second to Zoology. A new dissecting room was built over the room assigned to the Professor of Botany in the 1784 building and an adjoining private room built in 1833. In 1880 the archway of Salvin's building was moved to the east, so as to make two rooms, occupied by physiology, into one. In 1881, the Philosophical Library, which had been housed in a small room above, was moved into this room. A drawing office was built for mechanism and warehouses, rented to Messrs. Headley, were bought. . . . In 1882 a room 94 feet long was built for zoology over the south wing of Salvin's building, and workshops were built for mechanism. In 1883 houses in Corn Exchange Street just north of Fawcett's building were bought. In 1884, the roof of Salvin's building over the mineralogy laboratory was bodily lifted and walls built under it to give further accommodation for zoology. . . . A second floor was added to the annexe of the Herbarium and a foundry built for mechanism. The Perse alms-houses were bought for £2675, and a strip of land in Corn Exchange Street.

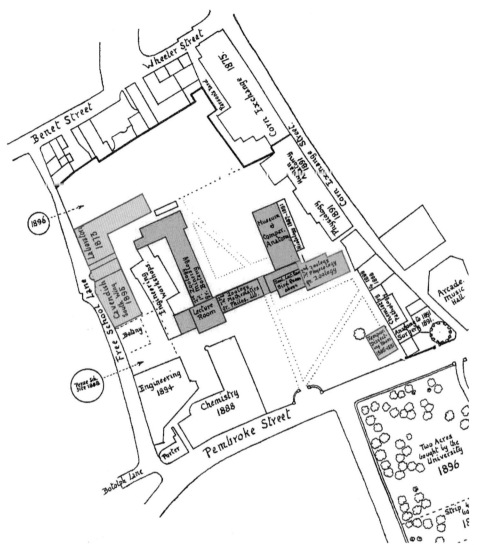

1897: In 1885, a temporary lecture room and museum for mechanism were completed. In 1886, a temporary corrugated iron dissecting room was built, which was used for teaching elementary physics after 1891. In 1887 a classroom for histology was built along the east side of the Museum of Comparative Anatomy. In 1888, the new Chemical Laboratory was completed, and Pathology, after large new windows had been inserted in the upper rooms, moved into the Old Chemical Laboratory. The Perse School site was purchased for £12,000. In 1890 the Perse School site was assigned to Engineering and the Headmaster's house to Botany. In 1891 the new laboratories for Human Anatomy and Physiology, with a common lecture room intervening between them were completed, and the Histology classroom, with a second floor added, became an annexe of the Museum of Comparative Anatomy. In 1894, the Engineering building, begun in 1893, was completed by adapting the Perse School buildings and adding a South Wing. In 1895, the South Wing of the Cavendish Laboratory was completed. In 1896 Mortlock's garden, to a line 80 feet south of Benet Street, was bought by the University, and also two acres of the Downing College grounds south of Pembroke Street. In 1897 a strip 40 feet wide, south of the two acres, was bought.

Fig. A.6

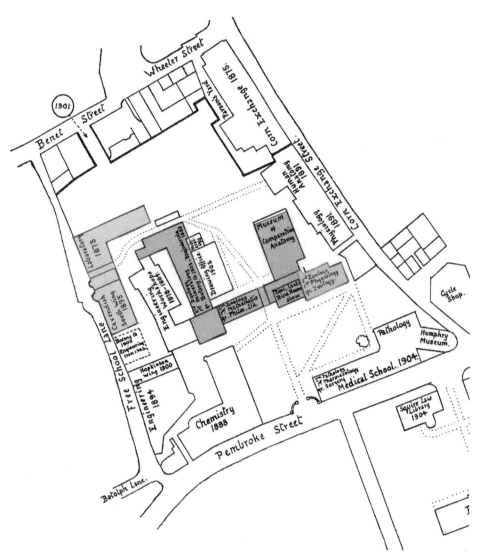

1904: In 1899 a house and four tenements belonging to the University in St Andrews Hill, opposite Fawcett's building, being ruinous, were pulled down, and the south end of Corn Exchange Street widened by four feet six inches. In 1900, the Hopkinson wing of the Engineering Department was completed and sites assigned to Archaeology, Botany, Geology and the Law Library on the Downing Site. In 1901, the Round House, up to and including the Jacksonian lecture theatre, was pulled down to make way for the Medical School, the museum specimens, etc. being stored at 16 Mill Lane. A house close to Barclay's Bank was bought, its demolition affording a northern entrance to the Old Botanic Garden site. In 1902 Downing College began to lay out the ground south of that which had been acquired by the University as a building site. This induced the University to buy six and a quarter acres. In 1903, a new drawing office for Engineering and a small lecture room for Mineralogy east of the north portion of Salvin's building were built. Botany moved into its new building on the Downing site. The greater part of the old botanical laboratory and the Perse Headmaster's house was also assigned to Engineering. The new department of Agriculture was housed in a basement of the Chemical Laboratory. On 1 March 1904, the Medical School, Botanical and Geological Laboratories and the Squire Law Library were opened by King Edward VII. The geological museum, begun with the purchase of Woodward's collection in 1827, was housed from 1842 to 1904 on the ground floor of Cockerill's building.

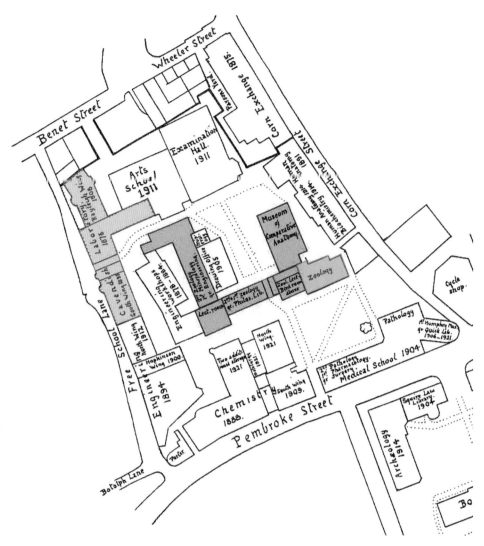

1922: In 1908, the north (Rayleigh) Wing of the Cavendish Laboratory was completed. The staircase in Salvin's building Fig. A.8 next to the archway was removed and three fair-sized rooms constructed. In 1909 an extension of the Chemical Laboratory to the west end of the Medical School, having under it an archway giving access to the court, was finished. Rooms in Salvin's building hitherto used by the Mathematical Professors was assigned to Zoology. In 1910, the new agricultural Laboratory on the Downing site was opened and Metallurgy occupied the old Agricultural Laboratory. In 1911, the Arts School and the Examination Hall on the Mortlock's garden site, commenced in 1909, were completed. In 1912 the north wing of the Engineering Laboratory was built on the site of the Perse School Headmaster's house. In 1913 the Psychological Laboratory and a new wing to the Agricultural Laboratory were completed. In 1914 the new laboratories for Archaeology, Forestry and Physiology were completed. Part of the old physiological laboratory (1891) was assigned to Human Anatomy and part to the new department of Biochemistry. In 1921 a north-east wing was added to Chemistry and storeys added to other parts, and to the south wing of the Cavendish Laboratory. In 1922, the Department of Engineering moved to the Scroope House site and the buildings vacated were handed over to Physics and to Chemistry. The Low Temperature Station was completed and the Molteno Institute for Parasitology opened (both on the Downing site).

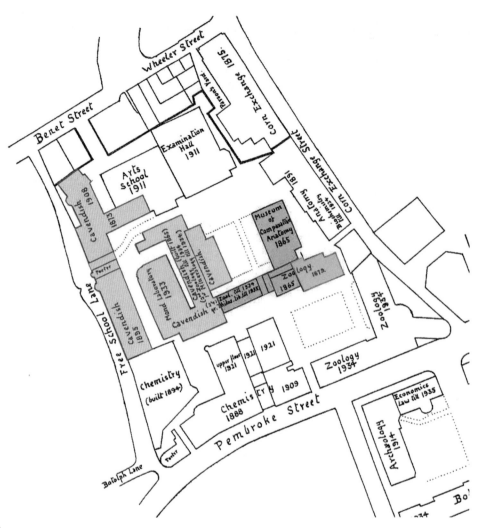

Fig. A.9   **1936:** In 1928 the Pathological Laboratory (on the Downing site) was completed and contained rooms on the ground floor for Animal Pathology. In the Medical School the rooms on the top floor, vacated by Pathology, were assigned to Zoology. In 1932, the Medical School was assigned to Zoology and its demolition commenced. Pharmacology was accommodated in the Physiological Laboratory and Surgery in the Pathological Laboratory. In 1933, The Mond Laboratory, on the site of the Engineering Workshops, was opened. In 1935, the western part of Salvin's building was assigned to Physics. In 1935, the new portion of the Zoology laboratory on the site of the Medical School . . . was finished. . . . The School of Forestry was closed and the Philosophical Library moved to the Arts School. In about 1936, the decision was taken to move many of the departments shown in the diagram to the Downing site. Those which remained on the Old Botanic Garden site included the numerous buildings of the Cavendish Laboratory which occupied much of the north-west quadrant of the site, the collection of buildings which formed the Chemistry Department in the south-west quadrant, and Zoology in the south-east quadrant. Those which moved to the Downing site included: Agriculture, Anatomy, Biochemistry, Botany, Medicine, Mineralogy, Parasitology, Pathology, Pharmacology and Surgery. Engineering had already moved to the Scroope House site.

Notes

Chapter 1

1. There is a wonderful literature on Tycho Brahe and his achievements. I have summarised his achievement in my book *Theoretical Concepts in Physics* (Longair, 2003), which includes references to some of the books on Tycho and his times.
2. *Theoretical Concepts in Physics* (Longair, 2003), Section 11.2.
3. Throughout this text, I translate all the formulae and expressions into their modern guise in SI units for the convenience of the reader.
4. See Harman (1998), page 74.
5. See Maxwell (1873), volume 1, page ix.
6. I have derived the expression for the Faraday rotation of linearly polarised radiation propagating along the magnetic field direction in Section 12.3.4 of my book *High Energy Astrophysics* (Longair, 2011a).
7. The history of Maxwell's discovery of the laws of electromagnetism is described in my book *Theoretical Concepts in Physics*, Chapter 4 (Longair, 2003).
8. To celebrate the 150th anniversary of the publication of Maxwell's paper of 1865 in the *Philosophical Transactions of the Royal Society of London*, I published a commentary on it in which I translated Maxwell's 20 equations into modern unrationalised notation and showed how easily they reduce to their modern form (Longair, 2015).
9. The relation between the electrostatic (esu) and electromagnetic (emu) systems of units of electric charge and the speed of light may be appreciated from the following considerations. For the sake of clarity, this calculation is carried out in SI units. We repeat the expressions for Coulomb's laws of electrostatics and magnetostatics:

$$f = \frac{q_1 q_2}{4\pi \epsilon_0 r^2} \, i_r \quad \text{and} \quad f = \frac{\mu_0 p_1 p_2}{4\pi r^2} \, i_r,$$

where i_r is the unit vector directed radially *away* from either electrostatic charge q_1, q_2 or magnetostatic monopole with pole strengths p_1, p_2 in the direction of the other. To find the ratio of the units of electric charge in the electrostatic and electromagnetic systems, we set the charges, pole strengths, forces and distances equal to unity. In addition, we need to relate the pole strength of a magnetic monopole to the electromagnetic unit of electric charge. The latter is most simply found by using the equivalence of a current loop carrying unit current and the unit magnetic dipole, $m = iA$. The current i is defined as the quantity of charge per unit time and so for unit current is $i(\text{unit}) = q_{\text{em}}(\text{unit})/\Delta t$, where $\Delta t = 1$. The unit magnetic pole strength is embedded in the expression of the unit dipole moment $m = p\Delta l$, where for unit values we take $\Delta l = 1$. It follows that $|p_1(\text{unit})| = |q_{\text{em}}(\text{unit})|$. Thus, for unit electrostatic charges and unit electromagnetic charges we find

$$1 = \frac{q_{\text{es}}^2(\text{unit})}{4\pi \epsilon_0} \quad \text{and} \quad 1 = \frac{\mu_0 q_{\text{em}}^2(\text{unit})}{4\pi}.$$

Hence,

$$\frac{q_{es}(\text{unit})}{q_{em}(\text{unit})} = (\epsilon_0 \mu_0)^{1/2} = \frac{1}{c}.$$

Although this calculation has been carried out in SI units, inspection of the expressions for the inverse-square laws shows that the ratio of unit electromagnetic and electrostatic charge would be the speed of light in any system of units.

These considerations refer to an era when the systems of units were defined in terms of separate physical standards of length, time, mass and current. In the SI system, the speed of light is defined to be exactly $c = 2.99792458 \times 10^8$ m s^{-1} and the constants μ_0 and ϵ_0 are related by $c = (\epsilon_0 \mu_0)^{-1/2}$, with μ_0 being defined to have the particular value $\mu_0 = 4\pi \times 10^{-7}$ H m^{-1}.

For reference, one esu unit of charge is 3.336×10^{-10} C; one emu of charge is 10 C. As can be seen, their ratio is $1/c$ in cgs units. For those who would enjoy learning more about the issue of units, the website http://info.ee.surrey.ac.uk/Workshop/advice/coils/unit_systems/ provides more enlightenment.

10. The complex history of the laying of the transatlantic cable is splendidly told in Smith and Wise's outstanding biography of William Thomson (Smith and Wise, 1989).

11. Many more details of these arguments are given in Sections 9.4 to 9.8 of my book *Theoretical Concepts in Physics* (Longair, 2003).

12. John Michell was a remarkable pioneer of many aspects of physics, astronomy and geology. A short biography was published long after his death by the geologist Sir Archibald Geikie (Geikie, 1918). More recently a biography has been published by Russell McCormmach (McCormmach, 2012).

Chapter 2

1. The contents of the *Principia* and its geometrical approach to the working out of the implications of the laws of motion are splendidly recounted in Subrahmanyan Chandrasekhar's book *Newton's Principia for the Common Reader* (Chandrasekhar, 1995). Be warned that Chandrasekhar's 'common reader' is expected to be pretty strong in geometry.

2. The term 'Tripos' is unique to the Cambridge examination system. According to the University of Cambridge website,

> The term 'Tripos' goes back to the seventeenth century when verses would be read out by someone sat on a three-legged stool (or Tripos) at graduation ceremonies. These became known as the Tripos verses. Eventually 'Tripos' was used to refer to courses offered by the University, when the lists of graduating students for each subject were written on the back of the Tripos verses.

> All Cambridge courses are assessed through examinations in broad subject areas called Triposes. Each Tripos is divided into one or more Parts and you need to complete a number of Parts in one or more Triposes to qualify for the B.A. degree.

The Natural Sciences and Mathematical Triposes nowadays have the structure shown in Table N.1. The Tripos system was only formalised in 1827; before that date the examinations were known as the Senate House Examinations. The original Tripos course consisted of the first three years shown in Table N.1, the first two being designated Part I and the third year Part II. In recent years, the course has been extended to four years for most physical science subjects, with the final year (the fourth year) being known as Part III.

3. Becher's paper gives a detailed history of the way in which the Cambridge Mathematical Tripos was reformed up to about 1850 and is centred on the role played by William Whewell (Becher, 1980).

Table N.1 The Natural Sciences and Mathematical Triposes					
Tripos	Year 1	Year 2	Year 3	Year 4	Degrees
Natural Sciences	Part IA	Part IB	Part II	Part III	BA
					MSci MASt
Mathematics	Part IA	Part IB	Part II	Part III	BA
					MMath MASt

4. A number of important figures in this history held 1851 Fellowships. The Royal Commission for the Exhibition of 1851 is a flourishing organisation today, offering major awards to scientists and engineers for research, development and design. Its aims are to support education for the benefit of productive industry.

5. The tortuous history of the introduction of physics into the Natural Sciences Tripos is brilliantly told in Wilson's paper *Experimentalists among the Mathematicians: Physics in the Cambridge Natural Sciences Tripos, 1851–1900* (Wilson, 1982).

6. For many more details of the events leading up to the decision to found the Cavendish Laboratory and the Cavendish Professorship of Experimental Physics, J.G. Crowther's *The Cavendish Laboratory 1874–1974* (Crowther, 1974) provides an excellent account of the internal workings of the University.

7. Henry Cavendish (1731–1810) was the first son of Lord Charles Cavendish, who was in turn the third son of William Cavendish, the 2nd Duke of Devonshire. William Cavendish, the 7th Duke, succeeded his cousin, the 6th Duke, known as the 'Bachelor Duke', who remained unmarried and had no family.

8. Matters were not to improve under Willis's successor, James Dewar, who was appointed to the Jacksonian Chair in 1875. In 1877, Dewar was appointed simultaneously to the Fullerian Professorship of Chemistry at the Royal Institution, London, where he was to spend most of his time. He was the inventor of the Dewar flask for containing liquids within a vacuum flask. He refused to resign from the Fullerian and Jacksonian chairs, which only became vacant on his death in 1923.

Chapter 3

1. A brief discussion of Maxwell's viscosity experiment and a sketch and photograph of the apparatus is contained in my book *Quantum Concepts in Physics* (Longair, 2013).

2. It is a great joy to appreciate Maxwell's imagination in both theory and experiment through reading his letters and other writings. These are beautifully presented in Peter Harman's monumental edition of Maxwell's letters and writings (Harman, 1990, 1995, 2002).

3. The endowment for the Clarendon Laboratory of Oxford University came from sales of the *History of the Great Rebellion* by Edward Hyde, 1st Earl of Clarendon, which had been accumulating income and interest since 1751. The original goal of the Trust had been to establish a riding academy.

4. In 1970, when I was first appointed a demonstrator in the Laboratory, I gave my first course of lectures on mechanics and electromagnetism in the Maxwell Lecture Theatre and fully appreciated all these splendid facilities, including the preparation room, almost a century after the building had been constructed. It is still one of the best lecture theatres for physics demonstrations that I know.

5. This remark is recorded by J.G. Crowther on page 74 of his book *The Cavendish Laboratory 1874–1974* (Crowther, 1974).

6. Schuster's chapter on the Clerk Maxwell period in *A History of the Cavendish Laboratory, 1871–1910* (Fitzpatrick *et al.*, 1910) provides an excellent summary of the research activities of the Laboratory.

7. The classing of students who took the first two sets of papers for the Mathematical Tripos is described by Wilson (1982).

> The top group were called 'wranglers,' and its best performer 'senior wrangler.' The middle and bottom groups were called 'senior optimes' and 'junior optimes,' respectively.

8. In fact, Maxwell had already described the problems of measuring the motion of the Earth through the aether in his article *Ether*, a contribution to the 9th edition of the *Encyclopaedia Britannica* (Maxwell, 1878). The brief paper in the *Proceedings of the Royal Society* was communicated by Stokes, who had received from Mr D.P. Todd of the Nautical Almanac Office, Washington, a copy of a letter sent to him by Maxwell dated 19 March 1979. In it Maxwell states:

> Even if we were sure of the theory of aberration, we can only get differences of position of stars, and in the terrestrial methods of determining the velocity of light, the light comes back along the same path again, so that the velocity of the earth with respect to the ether would alter the time of the double passage by a quantity depending on the square of the ratio of the earth's velocity to that of light, and this is quite too small to be observed.

In his book *Studies in Optics*, Michelson states,

> Maxwell was the first to point out that while it must be admitted that there can be no first-order effect which can be brought to light by experiment, this need not necessarily follow for effects depending on the second order. He expressed the doubt, however, as to the possibility to detecting such exceedingly small quantities, which may be expected to be of the order of the square of the aberration, i.e., one part in one hundred million. (Michelson, 1927)

In his earlier book *Light Waves and their Uses*, he stated,

> Maxwell considered it possible, theoretically at least, to deal with the square of the ratio of two velocities; that is the square of 1/10,000 or 1/100,000,000. He further indicated that if we made two measurements of the velocity of light, one in the direction in which the earth is travelling, and one in a direction at right angles to this, then the time it takes light to pass over the same length of path is greater in the first case than in the second. (Michelson, 1903)

Maxwell described the principles of optical interferometry in his *Encyclopaedia Britannica* article, but he does not seem to have been aware of its potential for measuring such tiny path differences.

9. Wilson's splendid paper entitled *Experimentalists among the Mathematicians: Physics in the Cambridge Natural Sciences Tripos, 1851–1900* (Wilson, 1982) provides a detailed and enlightening description of the problems of developing a balanced course of physics in the NST, combining the need to integrate both experiment and theory into a satisfactory course within the Natural Sciences Tripos.

10. I am grateful to Dr Isobel Falconer for pointing out that the dates stated in Wilson's paper should be for 1877–78 rather than 1878–79.

11. I am most grateful to Dr Isobel Falconer for allowing me to use these data, which are taken from her article about Maxwell's Cavendish Laboratory (Falconer, 2014).

12. Table N.2 shows in summary the career destinations of students registered for the Natural Sciences Tripos who took courses in physics for the period 1871 to 1900 (Wilson, 1982).

Table N.2 Careers of Natural Sciences Tripos physicists, 1871–1900[a]			
		Numbers of students	
Career Destination	1871–1881	1882–1889	1890–1900
Total numbers	34	46	39
Cambridge career	3	7	3.5
Higher education elsewhere	3	8.5	9
Secondary education	10	17.5	12
Medicine	4	1	2
Law	3	1.5	1
Engineering	3	1.5	1
Church	2	3	2
Other	2	1.5	2.5
Unknown	4	5	6

[a] Wilson, 1982.

Chapter 4

1. Biographies of Rayleigh include that by his son, Robert John Strutt, the 4th Lord Rayleigh, entitled *John William Strutt, Third Baron Rayleigh* (Rayleigh, 1924), and that by R.B. Lindsay, *Lord Rayleigh, the Man and His Works* (Lindsay, 1970). Rayleigh's years as Cavendish Professor are described by Glazebrook in his contribution to *A History of the Cavendish Laboratory, 1871–1910* (Glazebrook, 1910).

2. Henry and Nora Sidgwick were deeply involved in the struggle to gain the admission of women to University examinations, which they won, but they lost the battle to allow their degrees to be conferred. Instead, women only received a certificate by post confirming their success in the examinations. The admission of women to degrees only began in 1948. In 1998, degrees were awarded retrospectively to 900 women who had passed the examination before 1948. I was present at this rather moving and happy occasion.

3. A detailed description of what Pickering achieved at Harvard during the 42 years he held the position of Director of the Observatory is contained in my history of astrophysics and cosmology, *The Cosmic Century* (Longair, 2006).

4. I am indebted to Dr Isobel Falconer for providing this information about the numbers of students who attended the physics courses (private communication, 2014). She has described the problems of interpreting the information provided by Glazebrook (Glazebrook, 1910), who was inconsistent in how he counted the numbers of students.

5. Rayleigh's account of his discovery is described in his Nobel Prize lecture (Rayleigh, 1912).

6. Excellent photographs of some of the instruments built by the Cambridge Scientific Instrument Company are included in the booklet *Empires of Physics* (Bennett *et al.*, 1993).

Chapter 5

1. It is intriguing that Maxwell designed a zoetrope strip to illustrated the stability of three interacting frictionless vortices as a model for a molecule consisting of three vortex atoms. This is on display in the Cavendish Collection of Historic Scientific Instruments.

2. See the quotation of R. Steven Turner in the first paragraph of Section 3.1.

3. These data are taken from the information provided in the appendices of the volume *A History of the Cavendish Laboratory, 1871–1910* (Fitzpatrick *et al.*, 1910).

4. More details of the research carried out during this period are contained in the essays by Thomson, Newall, Rutherford and Wilson in *A History of the Cavendish Laboratory, 1871–1910* (Fitzpatrick *et al.*, 1910). Crowther (1974) provides a summary of the topics described in this history.

5. The Cavendish Physical Society continues to this day, an enticement to attend the lectures now being provided by wine and canapés after the lecture.

Chapter 6

1. I have given a demonstration of the essence of Maxwell's argument in Chapter 4 of my book *Theoretical Concepts in Physics* (Longair, 2003).

2. See http://isobelf.files.wordpress.com/2013/08/falconer_corpusclestoelectrons_preprint.pdf for an electronic version of Falconer's paper.

3. Thomson's struggles with vortex models of atoms and molecules and his other theoretical interests up to the discovery of the electron in 1897 are splendidly described in the book *J.J. Thomson and the Discovery of the Electron* by E.A. Davis and I.L. Falconer (Davis and Falconer, 1997) and by J.L. Heilbron in his biography of Thomson in the *Dictionary of Scientific Biography* (Heilbron, 1970).

4. The cause of the colours of the Brocken spectre is Mie scattering in which the size of the droplets is of the same order as the wavelength of light. The droplets are consequently very much smaller than the raindrops which are responsible for rainbows.

5. The book *J.J. Thomson and the Discovery of the Electron* by Davis and Falconer gives an excellent, detailed account of Thomson's experiments and the thinking behind them, as well as including reprints of many of the key papers or excerpts from them (Davis and Falconer, 1997).

6. Had the lines been observed with higher spectral resolution, the anomalous splitting of the D lines of sodium would have been observed and this could not have been explained by Lorentz's purely classical calculation.

7. In fact, most of the ionisation of air at ground level is a result of radioactive decays of minerals in the rocks. This ionisation drops of quite rapidly with height but, above about 1 km altitude, the degree of ionisation *increases* with increasing height. This is associated with the decay products of high-energy cosmic rays entering the top of the atmosphere (see, for example, Longair, 2011a).

8. The tradition of writing comic songs to well-known tunes for the Cavendish dinners is immortalised in the booklet entitled the *Post-Prandial Proceedings of the Cavendish Society* (published by Bowes and Bowes, Cambridge), which appeared in numerous editions until the sixth and final edition in 1926.

Chapter 7

1. Poynting and Thomson co-authored a multi-volume undergraduate *Textbook of Physics*, which was in print for about 50 years. According to Davis and Falconer (1997), Poynting wrote most of it.

2. See Section 9.2.1 of *High Energy Astrophysics* (Longair, 2011a).

3. Thomson's formula is similar to that derived for the interaction of cosmic ray electrons with thermal electrons. The formula is derived in Section 6.1. of *High Energy Astrophysics* (Longair, 2011a).

4. I describe the discovery of the Fermi–Dirac distribution in Sections 16.3 and 16.4 of my book *Quantum Concepts in Physics* (Longair, 2013).

5. The list of Aston's publications can be found in Hevesy's obituary of Aston in the *Obituary Notices of Fellows of the Royal Society* (Hevesy, 1948).

6. I have traced in detail the development of the old quantum theory and the quantum revolution of 1925 to 1930 in my book *Quantum Concepts in Physics* (Longair, 2013).

7. Taylor was a keen sailor and he stated that during the three-month exposure he had time to sail his new sailboat on a month-long cruise round the coast.

8. Nebulium was identified as a forbidden transition of doubly ionised oxygen O^{2+} by Ira Bowen (1927). In the late 1930s Walter Grotrian (1939) and Bengt Edlén (1941) identified the coronium line with highly ionised iron (Fe^{13+}).

9. This quotation appears in the introduction to Leon Rosenberg's 1963 edition of Bohr's unpublished memorandum for Rutherford, *On the Constitution of Atoms and Molecules*, ed. Rosenfeld, L. Copenhagen: Munksgaard.

10. How Bohr arrived at his model and its subsequent refinement by Sommerfeld, Born and many others is described in my book *Quantum Concepts in Physics* (Longair, 2013).

11. A useful survey of the range of research activities being carried out in the period 1900 to 1909 can be found in a list of memoirs containing an account of work done in the Cavendish Laboratory, in *A History of the Cavendish Laboratory, 1871–1910*, pp. 281–323 (Fitzpatrick *et al.*, 1910).

Chapter 8

1. This is a slightly revised version of the list which I presented in the last section of Chapter 9 of my book *Quantum Concepts in Physics* (Longair, 2013).

2. Langevin, himself a physicist of remarkable energy, was quoted later as stating that he and Rutherford had had an excellent relationship as research students, but that one could scarcely be 'friendly' with a 'force of nature' (Crowther, 1974).

3. In the language of heraldry: '**Crest**, a baron's coronet. On a helm wreathed of the Colors, a kiwi Proper. **Escutcheon**: Per saltire arched Gules and Or, two inescutcheons voided of the first in fess, within each a martlet Sable. **Supporters**: Dexter, Hermes Trismegistus (mythological patron of knowledge and alchemists). Sinister, a Maori warrior. **Motto**: Primordia Quaerere Rerum ("To seek the first principles of things". Lucretius).' The shield with the martlets, birds similar to house martins, was derived from the coat of arms of the Clan Rutherford.

4. In modern terms, the thorium derivatives identified by Rutherford and Soddy turned out to be: thorium $\equiv {}^{232}_{90}Th$, thorium X $\equiv {}^{224}_{88}Ra$, thorium emanation $\equiv$ radon $\equiv {}^{220}_{86}Rn$, thorium A $\equiv$ polonium $\equiv {}^{216}_{84}Po$, thorium B $\equiv$ lead $\equiv {}^{212}_{82}Pb$, thorium C $\equiv$ bismuth $\equiv {}^{212}_{83}Bi$.

5. The modern version of the radioactive decay chain of thorium is shown in Figure N.1. This can be compared with the simpler version shown in Table 8.1 which was presented in Rutherford's Bakerian Lecture of 1904 (Rutherford, 1905a).

6. Rutherford's struggles to increase the accuracy of these measurements are beautifully told in Chapter 2 of his book with Chadwick and Ellis, *Radiations from Radioactive Substances* (Rutherford *et al.*, 1930a).

7. Some examples of radiocarbon dating are given in Section 10.4.2 of my book *High Energy Astrophysics* (Longair, 2011a). Applications of radioactivity to estimate the confinement time of cosmic rays in the Galaxy are given in Section 15.6, which also includes the use of the transfer equation for radioactive species.

8. The passage of an energetic charged particle through the sensitive volume of the detector results in the ionisation of the atoms of the gas because of the electrostatic forces between the particle and the electrons in the atoms, a process known as *ionisation losses*. These 'collisions' result in the liberation of large numbers of electron–ion pairs, which cause the breakdown of the gas in the presence of a strong electric field. (For details of the process of ionisation losses, see Sections 5.2–5.5 of *High Energy Astrophysics* (Longair, 2011a).)

9. A complete list of 1851 Exhibition science Research Scholars and Senior Research Students in experimental physics who joined the Cavendish Laboratory in the period 1919 to 1936 is given in Appendix 1 of Jeffrey Hughes' PhD dissertation (Hughes, 1993). The list also includes their future career destinations.

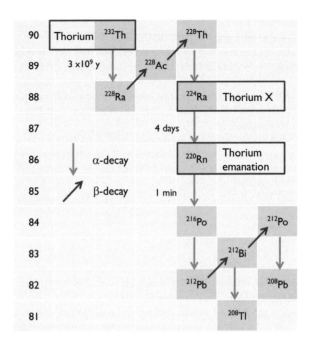

The radioactive decay chain of thorium.

10. A complete list of Cambridge PhD dissertations in experimental physics originating in the Cavendish Laboratory in the period 1919 to 1936 is given in Appendix 2 of Jeffrey Hughes' PhD dissertation (Hughes, 1993).
11. I have on my shelves the beautifully transcribed notes of the late Professor John F. Coales, Professor of Engineering at the University of Cambridge. He studied Part I and Part II Physics during the period 1926 to 1929, gaining a first in Part I of the Mathematical Tripos in 1927 and graduating with a second-class degree in physics in 1929.
12. See John Baker's Austinmemories website, http://www.austinmemories.com.
13. Crowther notes that, although Cockeroft carried out one of the most important experiments in nuclear physics, he was primarily an engineer and mathematician rather than a physicist. He published only 16 scientific papers (Crowther, 1974).
14. The 377th meeting of the Kapitsa Club, the last at which Kapitsa was present before his detention in the Soviet Union, was held on 21 August 1934. The series was continued under the leadership of David Shoenberg until the last regular meeting, the 675th, on 4 March 1958. By then, Shoenberg considered that the club had had its day with the increased specialisation of physics. In 1966, Kapitsa was at last allowed to return to Cambridge and a special meeting of the club was held on 10 May, with a number of the original members from 30 years earlier present (Shoenberg, 1985).

Chapter 9

1. The capability of carrying out research in radioactivity depended upon having access to sources of radioactive materials. Thus, when Marsden went to New Zealand, he was cut off from supplies of radioactive minerals. This theme of the difficulty of acquiring samples of radioactive materials is a recurrent one throughout this chapter.

2. For many more details of these calculations for the non-relativistic and relativistic cases, see my book *High Energy Astrophysics*, Chapter 5 (Longair, 2011a). Note that I have slightly altered the notation in Box 9.1 as compared with the book so that the α-particle properties are all lower-case letters and the atoms of the medium are upper-case.

3. It turned out that ^{17}O is a very rare isotope of oxygen, amounting to only about 0.04% of the oxygen in seawater and atmospheric oxygen.

4. Hughes (1993) gives a detailed description of the protocols and many precautions which they devised in order to obtain reliable results using the scintillation method.

5. More details of Dirac's achievement in predicting the properties of the electron and the existence of the positron are given in my book *Quantum Concepts in Physics* (Longair, 2013).

6. Wynn-Williams gives a delightful account of the invention of the scale-of-two counter in his contribution to the book *Cambridge Physics in the Thirties* (Wynn Williams, 1984).

7. During the war Wynn-Williams was brought in by Max Newman to utilise his electronic skills to speed up the code-breaking efforts at Bletchley Park. He designed and developed the counters of Heath Robinson, a machine which was a direct precursor of Colossus, the world's first programmable digital electronic computer.

8. Isobel Falconer (private communication) has drawn my attention to the long interval between the use of the first Geiger counters by Rutherford and Geiger in 1908 (Section 7.2.2) and the perfected Geiger–Müller detector of 1928. This may have been partly due to Rutherford's strong attachment to and expertise in the techniques of scintillation counting and partly to a mistrust of instruments which removed the experimenter from direct experience of the counting process.

9. This result is a pleasant exercise in the use of four-vectors in special relativity. In Rindler's notation (2001), the four-vectors of the photon and the proton are, respectively,

$$P_\gamma \equiv \left[\frac{h\nu}{c}, \frac{h\nu}{c} i_x \right] \quad P_p \equiv [m_p c, 0]. \tag{N.1}$$

Transforming to the centre of momentum frame of reference of the collision in which the total momentum of the photon plus proton is zero, the velocity of that frame is

$$V = c \frac{h\nu}{m_p c^2 + h\nu}. \tag{N.2}$$

In a head-on collision, the proton is returned along its incoming path with momentum $\gamma_V m_p V$ and total energy $\gamma_V m_p c^2$, where $\gamma_V = (1 - V^2/c^2)^{-1/2}$. Carrying out an inverse Lorentz transformation into the laboratory frame of reference, the total energy of the proton is

$$E_{tot} = \gamma_V^2 m_p c^2 \left[1 + \frac{V^2}{c^2} \right]. \tag{N.3}$$

Subtracting the rest mass energy of the proton from the total energy E_{tot}, its kinetic energy is

$$E = 2\gamma_V^2 m_p V^2. \tag{N.4}$$

Inserting the value of V from (7.18) into (7.20) and recalling that $\gamma_V = (1 - V^2/c^2)^{-1/2}$, we find the exact result

$$E = \frac{2h\nu}{2 + \dfrac{m_p c^2}{h\nu}}. \tag{N.5}$$

10. The concept of *spin* was introduced to understand the splittings of lines in atomic spectra. The details of the discovery of spin and its consequences are described in my book *Quantum Concepts in Physics* (Longair, 2013). Particles with fractional spin, 1/2, 3/2, etc. obey Fermi–Dirac statistics and are known as *fermions*. Particles with integral spin, 0, 1, 2, etc. obey Bose–Einstein statistics and are known as *bosons*. Spin is a fundamental property of all particles and combines

with the orbital angular momentum, for example, as if it were the intrinsic angular momentum of the particle.

11. A delightful account of the history of the Cockcroft and Walton experiment is contained in the book by Brian Cathcart, *The Fly in the Cathedral: How a Small Group of Cambridge Scientists Won the Race to Split the Atom* (Cathcart, 2005).

12. There is a delightful reminiscence by Oliphant of Rutherford's continuing enthusiasm for nuclear physics at the age of 63:

> We found a group of particles which clearly carried a double charge and appeared to be α-particles, in numbers equal to the protons and tritons. Rutherford produced hypothesis after hypothesis . . . I went to bed tired out. At 3.00 am the telephone rang. . . . I heard an apologetic voice express sorrow for waking me up, then excitedly say: 'I've got it. Those short-range particles are helium of mass three.' Shocked into attention, I asked on what possible grounds could he conclude that this was so, as no possible combination of twice two could give two particles of mass three and one of unity. Rutherford roared: 'Reasons, reasons! I feel it in my water!' (Oliphant, 1984)

13. Chapter 12 of the book by Rutherford, Chadwick and Ellis (1930a) contains many tables of the results of Ellis's careful experiments.

14. This translation is taken from Pais's book *Inward Bound*, in which many more details of this complex story are recounted (Pais, 1985).

Chapter 10

1. I have told the story of the discovery of quantum mechanics in some detail in my book *Quantum Concepts in Physics* (Longair, 2013).

2. Van der Pol spent two years in the Cavendish Laboratory, from 1917 to 1919. In his paper on relaxation oscillations, he introduced the *van der Pol equation* for a non-linear oscillator,

$$\frac{d^2x}{dt^2} - \mu(1 - x^2)\frac{dx}{dt} + x = 0. \tag{N.6}$$

Besides resulting in limit-cycle behaviour, the dynamics of the oscillator become chaotic for certain ranges of the damping term μ. Such chaotic behaviour had been observed in the performance of the triode oscillator, but its significance for the understanding of chaos was not appreciated at the time (Gleick, 1988).

3. The values of the rotation measure for an ionised plasma are worked out in my book *High Energy Astrophysics* (Longair, 2011a).

4. As Shoenberg remarks, a more fanciful version of the origin of the nickname 'Crocodile' relates to the crocodile in J.M. Barrie's play *Peter Pan*, which had swallowed an alarm clock and thus gave warning of its approach. Rutherford's heavy tread and loud voice, including his rendition of the hymn 'Onward Christian Soldiers', gave Kapitsa early warning of the Crocodile's approach.

5. A detailed account of the events of this period is contained in Chapter 16 of David Wilson's biography of Rutherford (Wilson, 1983).

6. A discussion of the circumstances surrounding the award of the Nobel Prize to Kapitsa is contained in the article by Allan Griffin (Griffin, 2008).

7. George Batchelor's biography of G.I. Taylor, *The Life and Legacy of G.I. Taylor*, does full justice to Taylor's remarkable contributions (Batchelor, 1996).

8. The order of the names is in decreasing order of golfing ability. Taylor wrote that

> Though Fowler was a scratch golfer and Aston fairly good, Rutherford was not good and I was just awful. (Batchelor, 1996)

Chapter 11

1. The problem also occurs in radio and optical interferometry. In a discussion of this issue with my colleague, the late Leslie Little, he made the percipient remark, 'the amplitude tells you how much there is, but the phase tells you where you are'. In radio astronomy, the problem was solved by John Baldwin and Peter Warner in their important paper on phaseless aperture synthesis (Baldwin and Warner, 1978) (Section 17.1.3).

2. Bernal was a convinced Marxist and member of the Communist Party from his student days, travelling regularly to the Soviet Union during the years of Stalin's dictatorship. His lifestyle was bohemian and his interests in science and politics very extensive. These attributes were the antithesis of Rutherford's approach to life and science. Rutherford did not like Bernal, but there is no evidence that he actually obstructed Bernal's research activities. These issues are elaborated in Brown's book, *J.D. Bernal: The Sage of Science* (Brown, 2005).

3. Many more details of Bernal's researches at Cambridge are contained in Dorothy Hodgkin's obituary of Bernal in the *Biographical Memoirs of the Fellows of the Royal Society* (Hodgkin, 1980).

4. Blackett, who was now at Manchester, was to play a major role in the war effort through his introduction of the techniques of operational research to military operations.

5. In May 1946, Nunn May was convicted of espionage for passing nuclear secrets to the Soviet Union.

6. This research was the subject of classified reports, which were only declassified after the war.

7. Nicholas Kemmer, who was part of the Cavendish group working on nuclear reactions involving slow neutrons, suggested the names neptunium and plutonium following the sequence of the outer planets from uranium, which had been named after the planet Uranus. Fortunately, the American nuclear physicists came up with the same suggestion and the names stuck.

8. This paper was submitted to the *Physical Review* in May 1941, but was not published until the material was declassified in 1946.

9. I am indebted to Dr Sanjoy Mahajan for introducing me to the remarkable power of this theorem.

10. For many more details of the story of radar and the Second World War, see L. Brown, *A Radar History of World War II* (Brown, 1999).

11. Francis Graham Smith was always known as Graham Smith. When he was knighted in 1986, he changed his surname to Graham-Smith. To avoid confusion, reference to his works even before 1986 are listed under Graham-Smith, F.

12. Anthony Kinloch has written a splendid brief biography of de Bruyne in the *Biographical Memoirs of Fellows of the Royal Society* (Kinloch, 2000).

Chapter 12

1. G.K. Hunter's biography of Lawrence Bragg can be warmly recommended, providing a much more complete picture of Bragg's life and period as Cavendish Professor (Hunter, 2004).

2. In Hoyle's novel *The Black Cloud* (Hoyle, 1957), the thinly disguised Mathematical Laboratory is described as the hero carries out overnight computations in a dingy room full of 1950s computing equipment.

3. The history of Lawrence's development of the cyclotron is described in *Lawrence and his Laboratory: A History of the Lawrence Radiation Laboratory* (Heilbron and Seidel, 1989).

4. Edward Shire carried out research work in atomic and low-temperature physics at the Laboratory until the outbreak of the Second World War. In 1939 Cockcroft invited him to take part in secret training in air defence, and at the end of that Shire year took up a position at the Air Defence Experimental Establishment, working on research into radar. In 1940 he was one of two inventors of the proximity fuse. Shire returned to Cambridge in 1944. In 1947 he became Reader in Physics in the Cavendish Laboratory and a senior member of the High Energy Physics Group.

5. Perutz worked on the Habakkuk project in which ice was strengthened by the addition of wood pulp and cotton wool. The objective turned out to be to build floating icefloe airfields to extend the range of land-based planes (Blow, 2004). The project came to nothing.

6. The remarkable story of the unravelling of the structures of the DNA molecules, myoglobin and haemoglobin has been told in a number of books, some of them by the participants themselves, including Watson's notorious *Double Helix* (1968), Crick's *What Mad Pursuit* (1988) and Perutz's essays in *Science is Not a Quiet Life: Unravelling the Atomic Mechanism of Haemoglobin* (Perutz, 1997).

7. This procedure had already been used by William Cochran in his analysis of the much simpler sucrose molecule (Beevers and Cochran, 1946; Cochran, 1946).

8. Michael Woolfson's sympathetic biographical memoir of Cochran makes clear Cochran's modest view of his own achievements. As Woolfson remarks:

> Cochran's scientific achievements were considerable. He made important contributions in three areas that eventually earned Nobel Prizes for others – in understanding the helical structure of DNA (Crick and Watson), in direct methods in crystallography (Karle and Hauptman) and in lattice dynamics (Brockhouse). (Woolfson, 2005)

9. Crick told his wife later that evening about their discovery. Years later, she told him she had not believed a word of it, remarking, 'You were always coming home and saying things like that'.

10. The early history of radio astronomy is splendidly told in the book by Sullivan (2009), *Cosmic Noise: A History of Early Radio Astronomy*. Sullivan includes interviews with many of the participants in the events summarised in this book and many other details of the technical, scientific and social background of the early development of radio astronomy in Cambridge and elsewhere.

11. The early history of electron microscopy in Germany is described in the paper by Niedrig, *The Early History of Electron Microscopy in Germany* (Niedrig, 1996).

12. Experimental investigation of the intermediate state began in Moscow and elsewhere immediately after the war. In the Mond Laboratory, Désirant and Shoenberg (1948) made careful measurements of the magnetisation of cylinders of different dimensions in a transverse field, and the theory was developed by Andrew (1948) and Kuper (1951) to explain anomalies in their results. Andrew and Lock (1950) made observations on the restoration of resistance in thin plates in a transverse field, and Tom Faber (1949, 1952) made an extensive study of supercooling and the dynamics of phase movement.

13. The coherence length ξ must be distinguished from Pippard's later electromagnetic coherence length ξ_0, which does not diverge at T_c. He thought that the two lengths might be related and of similar magnitudes at low temperatures, which later proved to be correct in pure materials.

14. Their paper had been published in Russian in 1950, but not properly appreciated in the west until Shoenberg circulated a translation some time later.

Chapter 13

1. The complexity of the system arises from the legal independence of the University and the Colleges. The General Board of the Faculties deals with the needs of the University departments to carry out their research and teaching functions. It does not, however, represent the Colleges. The only formal means of taking the views of the Colleges into account is through the University Council, which includes representatives of both the University and the Colleges. Even then, a decision of the Council can be overturned by a vote of the Regent House, which is 'the governing body and principal electoral constituency of the University'. Nowadays (2016), the Regent House has more than 3800 members, including University officers, and Heads and Fellows of Colleges. It makes and amends the regulations that govern the University. A vote on decisions of the Council, published as Graces, can be called if 25 or more members of the Regent House request it. This typically takes place when contentious issues are published as

Graces. The cumbersome nature of the decision-making process is one of the frustrations of the College–University structure, which otherwise has admirable qualities.

2. See Note 2 of Chapter 2.

3. Crowther provides an extended summary of Pippard's speech and its arguments in Chapter 32 of his book *The Cavendish Laboratory 1874–1974* (Crowther, 1974).

Chapter 14

1. The Radio Astronomy Group members included Francis Graham-Smith, Antony Hewish, John Baldwin, John Shakeshaft, Bruce Elsmore, Peter Scheuer, Paul Scott and Sidney Kenderdine.

2. $1 \, \text{Jy} = 1 \, \text{jansky} = 10^{-26} \, \text{W m}^{-2} \, \text{Hz}^{-1}$.

3. To express this result more technically, the comoving space density of the sources had to increase with distance. The term 'comoving' means the adoption of a frame of reference which expands with the Universe. If the number of sources were conserved as the Universe expands and their properties were unchanged, their comoving number density would be the same at all times.

4. Redshift is the best measure of distance in cosmology. It is defined by the ratio of the increase of the wavelength of the observed lines in the spectrum, $\lambda - \lambda_0$, divided by the rest wavelength λ_0, $z = (\lambda - \lambda_0)/\lambda_0$. It is sometimes converted into a 'recessional velocity' by multiplying by the speed of light, but this is only meaningful at small redshifts and should be avoided. Roughly speaking, objects with redshift $z \sim 1$ are observed at cosmic epochs when the Universe was less than half its present age. The best interpretation of the redshift is that $1/(1+z)$ is the average distance between galaxies in the expanding Universe when the radiation was emitted. Thus, at a redshift $z = 1$, the galaxies were on average half as far apart as they are today.

5. It was a great sadness that relations between Hoyle and Ryle were so soured by the controversy over the radio source counts. In 1965, Peter Scheuer and I attempted a reconciliation between them when the four of us got together in Hoyle's newly founded Institute of Theoretical Astronomy in Cambridge to try to understand their different positions. Sadly, there was no longer any common ground and each simply repeated their entrenched views. It was one of the saddest events of my career.

6. Over the subsequent years, improved flux densities and radio structures became available for essentially all the bright sources in the northern sky and it became possible to create a yet more complete sample of sources at 178 MHz. What became known as the revised 3CR catalogue (3CRR) was published in 1983 by Robert Laing, Julia Riley and me (Laing *et al.*, 1983). This catalogue includes all extragalactic radio sources in the northern hemisphere with flux densities $S \geq 10 \times 10^{-26} \, \text{W m}^{-2} \, \text{Hz}^{-1}$ at 178 MHz on the flux density scale of Kellermann, Pauliny-Toth and Williams with declinations greater than 10 degrees, and galactic latitude greater than $10°$ or less than $-10°$. The catalogue forms a complete sample of 173 radio galaxies and radio loud quasars.

7. In fact, the angular resolution was 23 arcsec in right ascension and 23 cosec δ arcsec in declination δ because the telescopes were located along an east–west baseline. Thus the full resolution in both coordinates was only obtained for sources located at the celestial pole, $\delta = 0°$.

8. For many more details of the astrophysical significance of these observations for relativistic astrophysics and the origin of cosmic rays, see my book *High Energy Astrophysics* (Longair, 2011a).

9. I have given a summary of the evidence for the evolution of source populations in all observable wavebands in my book *Galaxy Formation* (Longair, 2008a).

10. The Science Research Council (SRC) replaced the Department of Scientific and Industrial Research in 1965 as the principal governmental organisation in the UK with the responsibilities listed in Section 13.2. In 1981, the SRC changed its name to the Science and Engineering Research Council (SERC). In 1994, the Director General of the Research Councils reorganised the

Table N.3 Contributions to the total measured radio background signal in experiments of Penzias and Wilson (1965) at 4.08 GHz (7.35 cm)	
Signal	Noise signal T/K
Total zenith noise temperature	6.7 ± 0.3
Atmospheric emission	2.3 ± 0.3
Ohmic losses	0.8 ± 0.4
Backlobe response	≤ 0.1
Cosmic background radiation	3.5 ± 1.0

research council structure with the formation of the Particle Physics and Astronomy Research Council (PPARC) and the Engineering and Physics Sciences Research Council (EPSRC).

11. I have described in detail the events leading up to the discovery of the cosmic microwave background radiation in my book *The Cosmic Century* (Longair, 2006).

12. The various contributions to the observed background signal are listed in Table N.3.

13. One of Frisch's earliest graduate students was Allan Cormack, who joined him as a graduate student to study problems connected with ^{6}He. He did not complete the degree but accepted a lectureship in the Physics Department at Cape Town. As the local expert in nuclear physics, he became involved in nuclear medicine and, in 1956, by chance became interested in X-ray computer tomography, which led to practical CAT scanners. He was awarded the 1979 Nobel Prize in Physiology or Medicine jointly with Godfrey Hounsfield 'for the development of computer assisted tomography'.

Chapter 15

1. It is also possible, by measuring how the amplitude of the oscillations varies with temperature, to measure the electron relaxation time, and to map it out over the Fermi surface.

2. These developments are described in Fritz London's *Superfluids*, Vol. I on superconductivity (London, 1950) and Vol. II on superfluid helium (London, 1954).

3. The Magnus effect is the force acting on a rotating sphere passing through a fluid due to the pressure difference at the surface of the body, in accordance with Bernoulli's equation. The effect gives rise to the swerve of a spinning football, cricket ball or golf ball.

4. Section 11.2.4 of Leggett's history of superfluidity and superconductivity contains an excellent, accessible description of how the theory came together through the joint efforts of Bardeen, Cooper and Schreiffer (Leggett, 1995).

5. The microbeam technique involves scanning the solid with a very fine X-ray beam and so splitting up the diffraction rings into spots which provide much more information about the crystal structure. Kelly provides an illuminating history of development of microbeam techniques during the Bragg era (Kelly, 2012).

6. See Section 12.3 for a description of the origin and crucial role of EDSAC in Cambridge research.

7. The section became known as the Section of Tribophysics in 1946 and then the Division of Tribophysics in 1948. Tabor invented the designations *tribophysics* and *tribology* from the Greek word *tribos*, meaning rubbing.

8. Tabor gives a clear discussion of these calculation in his classic textbook *Gases, Liquids and Solids* (Tabor, 1991).

9. Heine's *TCM Group History* (Heine, 2015) is an informal set of notes on the development of condensed matter theory in the Cavendish Laboratory, from the time of the formation of the

Solid State Physics Group when Mott arrived and Ziman was appointed as a new lecturer to lead the theoretical work in solid state physics. I am most grateful to Volker Heine for letting me have a copy of this 'living' history.

10. In the 1970s, I was the head of this set of examples classes and benefitted enormously from Volker Heine's insight into the mathematics students really needed to understand in some depth. I still have his marvellous list of the mathematical topics we aimed to cover in the classes, and by and large we managed to incorporate most of them into the ten two-hour sessions.

Chapter 16

1. At that time, the University Grants Committee (UGC) presented to the UK government the funding requirements of the universities for the next five-year period or quinquennium. Pippard and Mott succeeded in obtaining the funding for the new Laboratory within a single quinquennium. As Pippard told me, it was a good thing they did this since the funds for such major developments were not available in subsequent quinquennia.

2. These booklets were in-house publications produced yearly during the period 1975 to 1983. They are similar in content to Mott's more formal reports published in 1963, 1965, 1967 and 1971.

3. Gordon Squires was one of the pioneers of neutron scattering in the post-war period. His Cavendish group established itself at Harwell, long before any formal user programme existed. They shared a time-of-flight beam line, originally on the world's first liquid hydrogen cold source at the BEPO reactor and later on the cold source at the DIDO reactor. Squires made a number of original scientific contributions, including measurements of the ortho- and para-hydrogen cross-sections (Stewart and Squires, 1953), the first observation of critical scattering with neutrons (Squires, 1954) and a series of studies of the lattice dynamics of elements together with his students. He is best remembered for his textbook *Introduction to the Theory of Thermal Neutron Scattering* (Squires, 2012), which quickly became the definitive text on this topic. He was an excellent lecturer and teacher, remembered fondly by all who benefitted from his dedicated teaching at Trinity College.

4. Tom Faber's academic research was in three areas: superconductivity, liquid metals and liquid crystals. In superconductivity, he investigated how super-states coexisted with, and were transformed from, the normal states of a metal. Liquid metals did not apparently fall within the experimental and theoretical framework that could, by the end of the 1950s, explain with quantum mechanics the flow of current and other phenomena in solid metals. With John Ziman, Faber developed a theory of liquid metals. This research concluded with an authoritative monograph, *Introduction to the Theory of Liquid Metals* (Faber, 1972), in which he gathered together published and unpublished experiments, at the same time clarifying the essence of its complex theoretical content. Liquid crystals flow like liquids but are elastic-like solids. Faber carried out experiments and constructed the theory of the optics, elasticity and ordering necessary to understand how they function. He was passionate about the teaching of classical physics, in particular fluid mechanics. The remarkable course of undergraduate lectures he gave on fluids led to *Fluid Dynamics for Physicists* (Faber, 1995), one of the finest expositions of the physics of fluids.

   Faber's father was Sir Geoffrey Faber, co-founder of the publishing house that would become Faber and Faber. T.S. Eliot, in the preface to *Old Possum's Book of Practical Cats*, acknowledged 'Mr T.E. Faber', who was 12 at the time. Tom Faber became chairman of Geoffrey Faber Holdings Ltd, and reorganised the finances of the company to ensure its survival.

Chapter 17

1. The radio source Cygnus A is abnormally close to the Earth for a source of its radio luminosity. Once the luminosity function of radio sources was known, it became clear that it is about ten times closer to the Earth than it should be if the sources were randomly distributed throughout the Universe. Consequently, it is 100 times brighter as a radio source than it should be and

sources of comparable luminosity are at cosmological distances with redshifts $z \sim 1$ (Stockton and Ridgway, 1996).

2. Reputedly, PHUEI was an acronym for Palomar–Hale Universal Extragalactic Instrument.

3. I gave a summary of the history of the identification of the sources in the 3CRR sample in Spinrad's festschrift (Longair, 1999).

4. Many more details of these processes and the interpretation of the radio data are contained in my book *High Energy Astrophysics* (Longair, 2011a).

5. The Half-Mile Telescope was a four-element aperture synthesis radio telescope built under the leadership of John Baldwin which used the 800-metre track of the One-Mile Telescope to map the velocity distribution of neutral hydrogen in nearby galaxies. It carried out such observations between 1967 and 1974.

6. The founder countries of CERN were Belgium, Denmark, France, Germany, Greece, Italy, Netherlands, Norway, Sweden, Switzerland and the United Kingdom.

7. Much of the material in this subsection is taken from the Cavendish High Energy Physics Group website (http://www.hep.phy.cam.ac.uk/history/), the authors being as follows. For UA2: Richard Ansorge, Bob deWolf, Sven Katvars, Michel Lefebvre, David Munday, Andy Parker, Mike Pentney, John Rushbrooke, Simon Singh, Wai Tsang, Pippa Wells, Tom White and Steve Wotton. For UA5: Richard Ansorge, Chris Booth, Bob deWolf, David Munday, Jonathon Ovens, John Rushbrooke, Pat Ward, David Ward, Chris Webber, Tony Weidberg and Tom White. The material is used with their kind permission.

8. The members of the group involved in the UA2 project are listed in Note 7.

9. A hodoscope is a detector which can determine the trajectories of charged particles emitted in particle interactions. The instrument consists of many individual segmented detectors which define the path of the charged particle through the detector.

10. The pseudorapidity, η, is a convenient spatial coordinate describing the angle of a particle relative to the beam axis. It is defined as $\eta \equiv -\ln\left[\tan\left(\frac{\theta}{2}\right)\right]$, where θ is the angle between the particle three-momentum $\boldsymbol{p}$ and the positive direction of the beam axis.

11. A review of measurements of the absolute value of the gravitational constant through the period of Cook's tenure of the Jacksonian Chair and into the mid 1990s is contained in Gillies' review article (Gillies, 1997).

Chapter 18

1. Hermann Hauser completed his PhD dissertation on *Mechanically Activated Chemical Reactions* in 1977 under the supervision of John Field in the PCS Group; his research involved computer analysis of calorimetric data from chemical reactions. After a year as a postdoctoral fellow, his interest in the potential of microprocessors for scientific and domestic computing led him and his collaborator Chris Curry to set up the Acorn Computer Company in 1978. In 1981, they designed and marketed the BBC Microcomputer, which eventually sold more than 1.5 million units. The machine was very influential in making microcomputing cheap and available to everyone, particularly to school pupils. Acorn next developed Reduced Instruction Set Computing (RISC) for their next-generation Archimedes microcomputer. In turn, this led to the formation of the company ARM (Advanced RISC Machine), which rapidly became a very successful international venture, billions of ARM chips being used in mobile devices such as mobile telephones.

2. I am most grateful to Volker Heine for letting me have a copy of his history of the first 50 years of the Theory of Condensed Matter Group (Heine, 2015).

3. The EPSRC website describes the Industrial CASE Award system as follows:

> Industrial *Cooperative Awards in Science & Technology* (CASE) provides funding for PhD studentships where businesses take the lead in arranging projects with an academic partner of their choice. The aim of these awards is to provide PhD students with a first-rate, challenging research training experience, within the context of a mutually beneficial

research collaboration between academic and partner organisations, for example industry and policy-making bodies.

4. The government's 'New Blood' scheme ran for three years, 1983, 1984 and 1985. According to the figures provided by David Smith (1985), a total of 938.5 new lectureships were provided for all subject areas, compared with a loss of about 5600 posts as a result of public expenditure cuts during the early 1980s. The physical sciences, engineering and technology fared best – 21.2% of the new posts went to all areas of the physical sciences. In the physical sciences there was an upper age limit of 35 on appointment.

Chapter 19

1. Part of the cause of this problem was the decline in the numbers of school teachers with physics qualifications in the schools. It was of concern that in many schools physics was being taught by teachers without qualifications in physics. In the early 1990s, the Institute of Physics reported that only about a quarter of the physics school teachers needed to replace those who were retiring were being trained.
2. Table 19.3 is a mildly edited version of the entry in the Lecture-List for the year 1998–99 published in the Special Edition of the *Reporter* of Friday 2 October 1998, Vol. CXXIX, 198–199.

Chapter 20

1. A Veeco mod GEN II system was installed in 2002, capable of uniform growth on 3-inch substrates. It is used mainly for the growth of THz quantum cascade lasers as well as samples in the GaInAs-AlInAs-InP system. A GEN Veeco III system was installed in 2007 and took over much of the InGaAs-AlGaAs-GaAs work from the 'A' chamber. This system, which has 12 source ports, is capable of highly uniform growth over 3-inch substrates and is connected to a surface decontamination chamber for regrowth on patterned substrates.
2. In her blog on the *Guardian's* Occam's Typewriter website, she recalled the difficult dilemma facing her at the time:

> There was one absolutely vital statement [Sam Edwards] made to me that meant that I stayed in science and had the confidence to attempt to combine motherhood and a career. At the time that it became clear a lectureship was going to open up in polymer physics at the Cavendish I went to see him to discuss the situation. I pointed out that I wanted to start a family. I have never forgotten his response: 'Intelligent women should have families', he said, making it clear that were I pregnant at interview he, for one, would not hold this against me . . . Remember this was a very long time ago when such attitudes could not be counted on; perhaps they still cannot. (Donald, 2015)

3. Following the reorganisation of the Research Council structure in 1993, the Agriculture and Food Research Council (AFRC) became the Biotechnology and Biological Sciences Research Council (BBSRC).
4. This citation, quoted in https://en.wikipedia.org/wiki/Richard_Friend, is from the Royal Society's Library and Archive catalogue: EC/1993/12 Friend, Richard Henry. London: the Royal Society.
5. Specifically, the crystals used by Bednorz and Müller (1986) had composition $Ba_xLa_{5-x}Cu_5O_{5(3-y)}$, where their samples had $x = 1$ and $y > 0$.
6. http://physicsworld.com/cws/article/news/2007/dec/13/tony-bland-1958-to-2007
7. This statement was prepared by Robin Ball on behalf of his colleagues in the Theory of Condensed Matter group during the Edwards era as a contribution to the work of the University's Industrial Liaison Office, emphasising the broad range of applicability of theoretical techniques in condensed matter physics.

Chapter 21

1. For details of the origin of structure in the Universe and the formation of galaxies in the standard Hot Big Bang model of the Universe, see my book *Galaxy Formation* (Longair, 2008a).

2. A splendid history of measurements of the cosmic microwave background is contained in the book *Finding the Big Bang*, edited by Peebles, Page and Partridge (Peebles *et al.*, 2009). The appendices showing the increasingly strict limits to the fluctuation spectrum per decade are of particular interest.

3. The detailed quantitative predictions of a large population of submillimetre sources was the result of computer simulations (Blain and Longair, 1993). Just before I returned to Cambridge in 1990, however, I presented an analytic version of the arguments to the JCMT board as supporting evidence for the prediction of the existence of such a population of sources. This presentation helped convince the board of the importance of building the ambitious SCUBA submillimetre camera.

4. References for the six parts of the 6C survey published over the period 1985 to 1993 can be found in my Biographical Memoir of John Evan Baldwin written for the Royal Society (Longair, 2011b).

Chapter 22

1. I was chair of the CAPSA Advisory Group, which was a pretty major poisoned chalice. The scale of the problem can be appreciated from the fact that the 150 departments or institutions of the University had all developed their own accounting procedures and had been promised that these could be maintained. In fact, this was not feasible with the adopted Oracle financial package – rather, the departments had to change their procedures to match the Oracle model. There was considerable anger among Cambridge academics about the complexity, delays and initial inefficiencies of CAPSA implementation, which are recorded in the Shattock and Finkelstein reports (Finkelstein and Shattock, 2001). What must be borne in mind is that these changes had to be implemented. After a difficult couple of years, the system settled down and now meets proper financial management and government requirements.

2. To improve the facilities and teaching environment for students, it was necessary to spend about £750,000 of the department's own resources to cope with the refurbishments which should have been carried out years ago.

3. I certainly benefitted from the fact that for 11 years I was Director of the Royal Observatory, Edinburgh, where I managed the organisation according to strict Civil Service rules. As a result, I had had experience of how to cope with essentially any managerial or personnel problems which were likely to happen in an organisation as large as the Cavendish Laboratory.

4. A number of us who were demonstrators were sad that the term 'demonstrator' disappeared, since it corresponded more accurately of the needs of an experimental physics department and was part of the original conception of Maxwell's Laboratory.

5. Herchel Smith was an alumnus of Emmanuel College, where he studied Natural Sciences with a particular interest in organic chemistry. His important inventions were oral and injectable contraceptives, from which he made a fortune which he used for philanthropic purposes. During his lifetime and after his death, Smith donated over $200 million to Cambridge University.

6. Humphrey Battcock is a Cavendish Laboratory and Downing College alumnus who has been an enthusiastic supporter of the University's initiatives since the beginning of the present Cavendish redevelopment programme. He is a managing partner of Advent International, one of the largest and best-established global private equity groups.

7. Details of the portfolio can be found at http://www.phy.cam.ac.uk/development/Development_Portfolio_Final.pdf

References

Adair, G. S. (1925). The haemoglobin system, *Journal of Biological Chemistry*, **63**, 493–545. This paper consists of six parts, covering many aspects of Adair's pioneering studies of the heamoglobin molecule.

Adams, A. (2009). Walter Eric Spear, *Biographical Memoirs of Fellows of the Royal Society*, **55**, 267–289.

Ade, P. A. R., Aghanim, N., Armitage-Caplan, C. *et al.* Planck 2013 results, XV: CMB power spectra and likelihood, *Astronomy and Astrophysics*, **571**, A15.

Adkins, C. J. (1961). The Cavendish high-fields laboratory, in *High Magnetic Fields*, eds Kolm, H., Lax, B., Bitter, F., and Mills, R., pp. 393–397. Cambridge, MA: MIT Press and New York: John Wiley & Sons.

Adkins, C. J. (1962). On the energy gap in superconducting tin, *Proceedings of the Royal Society of London*, **A268**, 276–289.

Adkins, C. J. and Phillips, W. A. (1985). Inelastic electron tunnelling spectroscopy, *Journal of Physics C: Solid State Physics*, **18**, 1313–1346.

Ahmed, H. (2013). *Cambridge Computing: The First 75 Years*. London: Third Millennium Information.

Ahn, C. C. and Krivanek, O. L. (1983). *EELS Atlas: A Reference Collection of Electron Energy Loss Spectra Covering All Stable Elements*. Warrendale: Gatan.

Alexander, P. (2006). Models of young powerful radio sources, *Monthly Notices of the Royal Astronomical Society*, **368**, 1404–1410.

Alexandrowicz, G. and Jardine, A. P. (2007). Helium spin-echo spectroscopy: Studying surface dynamics with ultra-high-energy resolution, *Journal of Physics: Condensed Matter*, **19**, D5001.

Alfvén, H. and Herlofson, N. (1950). Cosmic radiation and radio stars, *Physical Review*, **78**, 616.

Allen, J. F. and Jones, H. (1938). New phenomena connected with heat flow in helium II, *Nature*, **141**, 243–244.

Allen, J. F. and Misener, A. D. (1938). Flow of liquid helium II, *Nature*, **141**, 75.

Allen, J. F., Peierls, R., and Zaki Uddin, M. (1937). Heat conduction in liquid helium, *Nature*, **140**, 62–63.

Alpher, R. A., Follin, J. W., and Herman, R. C. (1953). Physical conditions in the initial stages of the expanding universe, *Physical Review*, **92**, 1347–1361.

Alpher, R. A., Herman, R. C., and Gamow, G. A. (1948). Thermonuclear reactions in the expanding universe, *Physical Review*, **74**, 1198–1199.

Altland, A. and Simons, B. (2010). *Condensed Matter Field Theory*. Cambridge: Cambridge University Press.

Ampère, A.-M. (1826). *Théorie des Phénomènes Électro-Dynamique, Uniquement Déduite de l'Expérience*. Paris: Méquingon-Marvis.

Anderson, C. D. (1932). The apparent existence of easily deflected positives, *Science*, **76**, 238–239.

Anderson, P. W. (1958). Absence of diffusion in certain random lattices, *Physical Review*, **109**, 1492–1505.

Anderson, P. W. (1992). Local moments and localized states: 1977 Nobel Prize lecture, in *Nobel Lectures, Physics 1971–1980*, ed. Lundqvist, S., pp. 376–398. Singapore: World Scientific.

Anderson, P. W. and Rowell, J. M. (1963). Probable observation of the Josephson superconducting tunneling effect, *Physical Review Letters*, **10**, 230–232.

Andersson, S. and Pendry, J. B. (1972). Surface structures from low energy electron diffraction, *Journal of Physics C: Solid State Physics*, **5**, L41–L45.

Andrade, E. N. da C. (1964). *Rutherford and the Nature of the Atom*. New York: Doubleday. The quotation is on page 111.

Andrew, E. R. (1948). The intermediate state of superconductors, III: Theory of behaviour of superconducting cylinders in transverse magnetic fields, *Royal Society of London Proceedings, Series A*, **194**, 98–112.

Andrew, E. R. and Lock, J. M. (1950). The magnetization of superconducting plates in transverse magnetic fields, *Proceedings of the Physical Society A*, **63**, 13–25.

Appleton, E. V. (1925). Geophysical influences on the transmission of wireless waves, *Proceedings of the Physical Society of London*, **37**, 16D–22D.

Appleton, E. V. (1927). The existence of more than one ionised layer in the upper atmosphere, *Nature*, **120**, 330.

Appleton, E. V. (1931). *Thermionic Vacuum Tubes*. London: Metheun and Co.

Appleton, E. V. (1932). Wireless studies of the ionosphere, *Journal of the Institution of Electrical Engineers*, **71**, 642–650.

Appleton, E. V. and Barnett, M. A. F. (1925a). Local reflection of wireless waves from the upper atmosphere, *Nature*, **115**, 333–334.

Appleton, E. V. and Barnett, M. A. F. (1925b). A note on wireless signal strength measurements made during the solar eclipse of 24 January 1925, *Proceeedings of the Cambridge Philosophical Society*, **22**, 672–675.

Appleton, E. V. and Barnett, M. A. F. (1925c). On some direct evidence for downward atmospheric reflection of electric rays, *Proceeedings of the Royal Society of London*, **A109**, 621–641.

Appleton, E. V. and Barnett, M. A. F. (1925d). Wirelss wave propagation, the magneto-ionic theory: The part played by the atmosphere; the effect of diurnal variations, *Electrician*, **94**, 398.

Appleton, E. V. and van der Pol, B. (1921). On the form of free triode vibrations, *Philosophical Magazine (6)*, **42**, 201–221.

Appleton, E. V. and van der Pol, B. (1922). On a type of oscillation-hysteresis in a simple triode generator, *Philosophical Magazine (6)*, **43**, 177–193.

Archard, G. D. (1956). Electron optical properties of electrode systems of four- and eight-fold symmetry, in *Proceedings of the International Conference on Electron Microscopy, 1954*, ed. Ross, R., pp. 97–105. London: Royal Microscopical Society.

Ashmead, J. (1950). A Joule–Thomson cascade liquefier for helium, *Proceedings of the Physical Society*, **63**, 504–508.

Atkins, K. R. and Osborne, D. V. (1950). The velocity of second sound below 1 K, *Philosophical Magazine (7)*, **41**, 1078–1081.

Avery, O. T., MacLeod, C. M., and McCarty, M. (1944). Studies on the chemical nature of the substance inducing transformation of pneumococcal types: Induction of transformation by desoxyribonucleic, *Journal of Experimental Medicine*, **79**, 137–158.

Baade, W. and Minkowski, R. (1954). Identification of the radio sources in Cassiopeia, Cygnus A, and Puppis A, *Astrophysical Journal*, **119**, 206–214.

Bacher, R. F. and Condon, E. U. (1932). The spin of the neutron, *Physical Review*, **41**, 683–685.

Baeyer, O. von and Hahn, O. (1910). Magnetische lininenspektren von β-strahlen, *Physicalische Zeitung*, **11**, 488–493.

Baeyer, O. von, Hahn, O., and Meitner, L. (1911a). Nachwies von β-strahlen bei radium D, *Physicalische Zeitung*, **12**, 378–379.

Baeyer, O. von, Hahn, O., and Meitner, L. (1911b). Über die β-strahlen des aktiven neiderschlags des radiums, *Physicalische Zeitung*, **12**, 273–279.

Baeyer, O. von, Hahn, O., and Meitner, L. (1912). Das magnetische spektren der β-strahlen des thoriums, *Physicalische Zeitung*, **13**, 264–266.

Baker, J. C., Grainge, K., Hobson, M. P. *et al.* (1999). Detection of cosmic microwave background structure in a second field with the Cosmic Anisotropy Telescope, *Monthly Notices of the Royal Astronomical Society*, **308**, 1173–1178.

Baldwin, J. E. (1967). The non-thermal radio-emission from the Galaxy (Introductory Report), in *Radio Astronomy and the Galactic System*, ed. van Woerden, H., volume 31 of *IAU Symposium*, pp. 337–354. London: Academic Press.

Baldwin, J. E., Harris, C. S., and Ryle, M. (1973). 5 GHz observations of the infrared star MWC 349, and the HII condensation W3(OH), *Nature*, **241**, 38–39.

Baldwin, J. E. and Mackay, C. D. (1988). The COAST interferometer project, in *High-Resolution Imaging by Interferometry*, ed. Merkle, F., volume 29 of *European Southern Observatory Conference and Workshop Proceedings*, pp. 935–938. Garching: European Southern Observatory.

Baldwin, J. E. and Warner, P. J. (1976). Aperture synthesis without phase, *Monthly Notices of the Royal Astronomical Society*, **175**, 345–353.

Baldwin, J. E. and Warner, P. J. (1978). Phaseless aperture synthesis, *Monthly Notices of the Royal Astronomical Society*, **182**, 411–422.

Baldwin, J. E., Beckett, M. G., Boysen, R. C. *et al.* (1996). The first images from an optical aperture synthesis array: Mapping of Capella with COAST at two epochs, *Astronomy and Astrophysics*, **306**, L13–L16.

Baldwin, J. E., Haniff, C. A., Mackay, C. D., and Warner, P. J. (1986). Closure phase in high-resolution optical imaging, *Nature*, **320**, 595–597.

Balibar, S. (2010). The enigma of supersolidity, *Nature*, **464**, 176–182.

Bardeen, J., Cooper, L. N., and Schreiffer, J. R. (1957). Theory of superconductivity, *Physical Review*, **108**, 1175–1204.

Barišić, N. (2013). Universal sheet resistance and revised phase diagram of the cuprate high-temperature superconductors, *Proceedings of the National Academy of Sciencess of the United States of America*, **110**, 12235–12240.

Barkla, C. G. (1903). Secondary radiation from gases subject to X-rays, *Philosophical Magazine (6)*, **5**, 685–698.

Barkla, C. G. (1904). Polarisation in Röntgen rays, *Nature*, **19**, 463.

Barkla, C. G. (1906). Polarisation of secondary Röntgen radiation, *Proceedings of the Royal Society of London*, **A77**, 247–255.

Barkla, C. G. (1911a). Note on the energy of scattered X-radiation, *Philosophical Magazine (6)*, **21**, 648–652.

Barkla, C. G. (1911b). The spectra of the fluorescent Röntgen radiations, *Philosophical Magazine (6)*, **22**, 396–412.

Barkla, C. G. and Sadler, C. A. (1908). Homogeneous secondary Röntgen radiations, *Philosophical Magazine (6)*, **16**, 550–584.

Batchelor, G. (1953). *The Theory of Homogeneous Turbulence*. Cambridge: Cambridge University Press.

Batchelor, G. (1996). *The Life and Legacy of G. I. Taylor*. Cambridge: Cambridge University Press.

Beauchamp, K. (1997). *Exhibiting Electricity: IEE History of Technology, Series 21*. London: The Institution of Electrical Engineers.

Becher, H. (1980). William Whewell and Cambridge mathematics, *Historical Studies in the Physical Sciences*, **11**, 1–48.

Becquerel, H. (1896). Sur les radiations invisibles émises par les corps phosphorescents (On the invisible radiation emitted by phosphorescent bodies), *Comptes Rendus de l'Academie des Sciences*, **122**, 501–503.

Bednorz, J. G. and Müller, K. A. (1986). Possible high T_c superconductivity in the Ba-La-Cu-O system, *Zeitschrift fur Physik B: Condensed Matter*, **64**, 189–193.

Beevers, C. A. and Cochran, W. (1946). The crystal structure of sucrose sodium bromide dihydrate, *Proceedings of the Royal Society of London*, **A190**, 257–272.

Bell, A. R. (1978a). The acceleration of cosmic rays in shock fronts I, *Monthly Notices of the Royal Astronomical Society*, **182**, 147–156.

Bell, A. R. (1978b). The acceleration of cosmic rays in shock fronts II, *Monthly Notices of the Royal Astronomical Society*, **182**, 443–455.

Bennett, A. (1962). The revised 3C catalogue of radio sources, *Memoirs of the Royal Astronomical Society*, **67**, 163–172.

Bennett, F., Brain, R., Bycroft, K. *et al.* (1993). *Empires of Physics: A Guide to the Exhibition*. Cambridge: Whipple Museum of the History of Science.

Bergemann, C., Julian, S. R., MacKenzie, A. P., Nishizaki, S., and Maeno, Y. (2000). Detailed topography of the Fermi surface of Sr_2RuO_4, *Physical Review Letters*, **84**, 2662.

Bergemann, C., MacKenzie, A. P., Julian, S. R., Forsythe, D., and Ohmichi, E. (2003). Quasi-two-dimensional Fermi liquid properties of the unconventional superconductor Sr_2RuO_4, *Advances in Physics*, **52**, 639–725.

Berggren, K.-F., Thornton, T. J., Newson, D. J., and Pepper, M. (1986). Magnetic depopulation of 1D subbands in a narrow 2D electron gas in a GaAs:AlGaAs heterojunction, *Physical Review Letters*, **57**, 1769–1772.

Bernal, J. D. (1924). The structure of graphite, *Proceedings of the Royal Society of London*, **A106**, 749–773.

Bernal, J. D. (1926). On the interpretation of X-ray single rotation photographs, *Proceedings of the Royal Society of London*, **A113**, 117–160.

Bernal, J. D. (1929). A universal X-ray photogoniometer, *Journal of Scientific Instruments*, **6**, 343–353.

Bernal, J. D. and Crowfoot, D. (1934). X-ray photographs of crystalline pepsin, *Nature*, **133**, 794–795.

Bernal, J. D., Fankuchen, I., and Perutz, M. (1938). An X-ray study of chymotrypsin and hæmoglobin, *Nature*, **141**, 523–524.

Bernhoeft, N. R., Lonzarich, G. G., Mitchell, P. W., and Paul, D. M. (1983). Magnetic excitations in Ni_3Al at low energies and long wavelengths, *Physical Review B*, **28**, 422–424.

Best, P. N., Longair, M. S., and Röttgering, H. J. A. (1996). Evolution of the aligned structures in $z \sim 1$ radio galaxies, *Monthly Notices of the Royal Astronomical Society*, **280**, L9–L12.

Best, P. N., Longair, M. S., and Röttgering, H. J. A. (1997). HST, radio and infrared observations of 28 3CR radio galaxies at redshift $z \sim 1$, I: The observations, *Monthly Notices of the Royal Astronomical Society*, **292**, 758.

Best, P. N., Longair, M. S., and Röttgering, H. J. A. (1998). HST, radio and infrared observations of 28 3CR radio galaxies at redshift $z \sim 1$, II: Old stellar populations in central cluster galaxies, *Monthly Notices of the Royal Astronomical Society*, **295**, 549.

Best, P. N., Röttgering, H. J. A., and Longair, M. S. (2000). Ionization, shocks and evolution of the emission-line gas of distant 3CR radio galaxies, *Monthly Notices of the Royal Astronomical Society*, **311**, 23–36.

Biondi, M. A., Garfunkel, M. P., and McCoubrey, A. O. (1956). Millimeter wave absorption in superconducting aluminum, *Physical Review*, **101**, 1427–1429.

Biot, J. B. and Savart, F. (1820). Note sur le magnétisme de al pile de Volta, *Annales de Chimie et de Physique*, **15**, 222–223.

Birtwistle, G. (1926). *The Quantum Theory of the Atom*. Cambridge: Cambridge University Press.

Birtwistle, G. (1928). *The New Quantum Mechanics*. Cambridge: Cambridge University Press.

Blackett, P. M. S. (1925). The ejection of protons from nitrogen nuclei, photographed by the Wilson method, *Proceedings of the Royal Society of London*, **A107**, 349–360.

Blackett, P. M. S. (1960). Charles Thomson Rees Wilson, *Biographical Memoirs of the Fellows of the Royal Society*, **6**, 268–295.

Blackett, P. M. S. (1964). Cloud chamber researches in nuclear physics and cosmic Radiation: 1948 Nobel Prize lecture, in *Nobel Lectures, Physics 1942–1962*, pp. 97–119. Amsterdam: Elsevier.

Blackett, P. M. S. and Champion, F. C. (1931). The scattering of slow alpha-particles by helium, *Proceedings of the Royal Society of London*, **A130**, 380–388.

Blackett, P. M. S. and Occhialini, G. P. S. (1933). Some photographs of the tracks of penetrating radiation, *Proceedings of the Royal Society of London*, **A139**, 699–722.

Blackler, J. M. (1958). Models for main sequence stars, *Monthly Notices of the Royal Astronomical Society*, **118**, 38–44.

Blain, A. W. (1997). Gravitational lensing by clusters of galaxies in the millimetre/submillimetre waveband, *Monthly Notices of the Royal Astronomical Society*, **290**, 553–565.

Blain, A. W. and Longair, M. S. (1993). Submillimetre cosmology, *Monthly Notices of the Royal Astronomical Society*, **264**, 509–521.

Bland, J. A. C. and Heinrich, B., eds (1994). *Ultrathin Magnetic Structures*, vol. I and II. Berlin: Springer. The second edition was published in 2005, by which date it had expanded to four volumes.

Blow, D. M. (2004). Max Ferdinand Perutz, *Biographical Memoirs of Fellows of the Royal Society*, **50**, 227–256.

Blythe, J. H. (1957a). A new type of pencil beam aerial for radio astronomy, *Monthly Notices of the Royal Astronomical Society*, **117**, 644–651.

Blythe, J. H. (1957b). Results of a survey of galactic radiation at 38 Mc/s, *Monthly Notices of the Royal Astronomical Society*, **117**, 652–662.

Bohr, N. (1912). Unpublished memorandum for Ernest Rutherford, in *On the Constitution of Atoms and Molecules*, ed. Rosenfeld, L. Copenhagen: Munksgaard.

Bohr, N. (1913a). On the constitution of atoms and molecules (Part I), *Philosophical Magazine (6)*, **26**, 1–25.

Bohr, N. (1913b). On the constitution of atoms and molecules, Part II: Systems containing only a single electron, *Philosophical Magazine (6)*, **26**, 476–502.

Bohr, N. (1913c). On the constitution of atoms and molecules, Part III: Systems containing several nuclei, *Philosophical Magazine (6)*, **26**, 857–875.

Bohr, N. (1913d). On the theory of the decrease of velocity of moving electrified particles on passing through matter, *Philosophical Magazine (6)*, **25**, 10–31.

Bohr, N. (1918). On the quantum theory of line spectra, Part II: On the hydrogen spectrum, *Mathematisk-Fysiske Meddelelser, Det Kgl. Danske Videnskabernes Selskab: Skrifter 8*, **4.1**, 37–100. Reprinted in *Collected works*, **3**, 103–166.

Bohr, N. (1922). The structure of the atom: 1922 Nobel Prize lecture, in *Nobel Lectures 1922–1941*, pp. 7–43. Amsterdam: Elsevier. This volume was published in 1965.

Bohr, N., Kramers, H. A., and Slater, J. C. (1924). The quantum theory of radiation, *Philosophical Magazine (6)*, **47**, 785–822.

Bolton, J. G., Stanley, G. J., and Slee, O. B. (1949). Positions of three discrete sources of galactic radio-frequency radiation, *Nature*, **164**, 101–102.

Boltwood, B. B. and Rutherford, E. (1911). Production of helium by radium, *Philosophical Magazine (6)*, **22**, 586–604.

Bose, S. N. (1924). Planck's gesetz und lichtquantenhypothese (Planck's law and the hypothesis of light quanta), *Zeitschrift für Physik*, **26**, 178–181.

Cook, A. H. (1975). On the structure of hydroxyl maser sources, *Monthly Notices of the Royal Astronomical Society*, **171**, 605–618.

Cook, A. H. (1977). *Celestial Masers*. Cambridge: Cambridge University Press.

Cook, A. H. (1978). Physics of celestial masers, *Quarterly Journal of the Royal Astronomical Society*, **19**, 255–268.

Cook, A. H. (1980). *Interiors of the Planets*. Cambridge: Cambridge University Press.

Cook, A. H. (1981). Molecular spectroscopy prompted by astrophysical observations, *Philosophical Transactions of the Royal Society of London*, **A303**, 551–563.

Cook, A. H. (1988). *The Motion of the Moon*. Bristol: Adam Hilger, 1988.

Cook, A. H. and Chen, Y. T. (1982). On the significance of the radial Newtonian gravitational force of the finite cylinder, *Journal of Physics A: Mathematical and General*, **15**, 1591–1597.

Cooley, J. W. and Tukey, J. W. (1965). An algorithm for the machine calculation of complex Fourier series, *Mathematics of Computation*, **19**, 297–301.

Cooper, N. R. (2011). Optical flux lattices for ultracold atomic gases, *Physical Review Letters*, **106**, 175301.

Cosslett, V. E. and Duncumb, P. (1956). Micro-analysis by a flying-spot X-ray method, *Nature*, **177**, 1172–1173.

Cosslett, V. E. and Nixon, W. C. (1950). *X-Ray Microscopy*. Cambridge: Cambridge University Press.

Cosslett, V. E. and Nixon, W. C. (1951). X-ray shadow microscope, *Nature*, **168**, 24–25.

Cosslett, V. E., Camps, R. A., Saxton, W. O. *et al.* (1979). Atomic resolution with a 600-kV electron microscope, *Nature*, **281**, 49–51.

Courtney-Pratt, J. S. (1957). A review of the methods of high-speed photography, *Reports on Progress in Physics*, **20**, 379–432.

Crick, A., Tiffert, T., Shah, S. *et al.* (2013). An automated live imaging platform for studying merozoite egress-invasion in malaria cultures, *Biophysical Journal*, **104**, 997–1005.

Crick, F. (1988). *What Mad Pursuit: A Personal View of Scientific Discovery*. New York: Basic Books.

Crowther, J. G. (1974). *The Cavendish Laboratory, 1874–1974*. London: Macmillan.

Curie, I. (1931). Sur le rayonnement γ nucléaire excité dans le glucinium et dans le lithium par les rayons α du polonium, *Comptes Rendus (Paris)*, **193**, 1412.

Curie, I. and Joliot, F. (1932). The emission of high energy protons from hydrogenous substances irradiated with very penetrating γ-rays, *Comptes Rendus (Paris)*, **194**, 273–275.

Curie, M. P. and Skłodowska-Curie, M. (1898). On a new radioactive substance contained in pitchblende, *Comptes Rendus*, **127**, 175–178.

Curie, M. P., Skłodowska-Curie, M., and Démont, G. (1898). On a new, strongly radioactive substance, contained in pitchblende, *Comptes Rendus*, **127**, 1215–1217.

Dahl, P. F. (2002). *From Nuclear Transmutation to Nuclear Fission, 1932–1939*. Bristol: IOP Publishing.

Dalton, J. (1808). *A New System of Chemical Philosophy*. Manchester: R. Bickerstaff.

Darwin, C. G. (1912). A theory of the absorption and scattering of the α-rays, *Philosophical Magazine (6)*, **23**, 901–920.

Darwin, C. G. (1958). Douglas Rayner Hartree, 1897–1958, *Biographical Memoirs of Fellows of the Royal Society*, **4**, 102–116.

Darwin, G. and Darwin, H. (1881). Lunar disturbance of gravity: Report of the committee, *Nature*, **25**, 20–21.

Davies, R. A., Uren, M., and Pepper, M. (1981). Magnetic separation of localisation and interaction effects in a two-dimensional electron gas at low temperatures, *Journal of Physics C: Solid State Physics*, **14**, 5737–5762.

Davis, B. and Barnes, A. H. (1929). Capture of electrons by swiftly-moving alpha-particles, *Physical Review*, **34**, 152–156.

Davis, E. A. and Falconer, I. J. (1997). *J.J. Thomson and the Discovery of the Electron*. London: Taylor and Francis.

de Boer, J. H. and Verwey, E. J. W. (1937). Semi-conductors with partially and with completely filled ₃d-lattice bands, *Proceedings of the Physical Society*, **49**, 59–71.

de Broglie, L. (1923a). Les quanta, la théorie cinétique de gaz et la principe et le principe de Fermat, *Comptes Rendus (Paris)*, **177**, 630–632.

de Broglie, L. (1923b). Ondes et quanta, *Comptes Rendus (Paris)*, **177**, 507–510.

de Broglie, L. (1923c). Quanta de lumière, diffraction et interférence, *Comptes Rendus (Paris)*, **177**, 548–550.

de Bruyne, N. A. (1928). The action of strong electric fields on the current from a thermionic cathode, *Proceedings of the Royal Society of London*, **A120**, 423–437.

Debye, P. (1912). Zur theorie der spezifischen wärme (On the theory of specific heats), *Annalen der Physik (4)*, **39**, 789–839. English translation: *Collected Papers of Peter J. W. Debye*, 1954, pp. 650–696. New York: Interscience.

Dee, P. I. and Walton, E. T. S. (1933). A photographic investigation of the transmutation of lithium and boron by protons and of lithium by ions of the heavy isotope of hydrogen, *Proceedings of the Royal Society of London*, **A141**, 733–742.

De Forest, L. (1906). The audion: A new receiver for wireless telegraphy, *Transactions of the American Institute of Electrical and Electronic Engineers*, **25**, 735–763.

de Gennes, P.-G. (1966). *Superconductivity of Metals and Alloys*. New York: Addison-Wesley.

de Gennes, P.-G. (1971). Reptation of a polymer chain in the presence of fixed obstacles, *Journal of Chemical Physics*, **55**, 572–579.

de Haas, W. J. and van Alphen, P. M. (1931). The dependence of the susceptibility of diamagnetic metals upon the field, *Leiden Communications*, **212a**, 3–16.

Deltrap, J. H. M. (1964a). Correction of spherical aberration of electron lenses. PhD dissertation, University of Cambridge.

Deltrap, J. H. M. (1964b). Correction of spherical aberration with combined quadrupole–octopole units, in *Proceedings of the 3rd European Conference on Electron Microscopy, Prague, Czech Republic*, ed. Titlbach, M., volume A, pp. 45–46. Prague: Czechoslovak Academy of Sciences.

Désirant, M. and Shoenberg, D. (1948). The intermediate state of superconductors, I: Magnetization of superconducting cylinders in transverse magnetic fields, *Royal Society of London Proceedings, Series A*, **194**, 63–79.

De Young, D. S. and Axford, W. I. (1967). Inertial confinement of extended radio sources, *Nature*, **216**, 129–131.

Dickenson, H. W. (1958). The steam engine to 1830, in *A History of Technology, Vol. IV*, eds Singer, L. C., Holmyard, E. J., Hall, A., and Williams, T. I., pp. 168–198. Oxford: Clarendon Press. The quotation from Henderson appears on page 165.

Dickinson, C., Battye, R. A., Carreira, P. *et al.* (2004). High-sensitivity measurements of the cosmic microwave background power spectrum with the extended Very Small Array, *Monthly Notices of the Royal Astronomical Society*, **353**, 732–746.

Dirac, P. A. M. (1925). The fundamental equations of quantum mechanics, *Proceedings of the Royal Society of London*, **A109**, 642–653.

Dirac, P. A. M. (1928a). The quantum theory of the electron, *Proceedings of the Royal Society of London*, **A117**, 610–624.

Dirac, P. A. M. (1928b). The quantum theory of the electron, *Proceedings of the Royal Society of London*, **A118**, 351–361.

Dirac, P. A. M. (1930). *Principles of Quantum Mechanics*. Oxford: Clarendon Press.

Dirac, P. A. M. (1931). Quantized singularities in the electromagnetic field, *Proceedings of the Royal Society of London*, **A133**, 60–72.

Dirac, P. A. M. (1939). A new notation for quantum mechanics, *Proceedings of the Cambridge Philosophical Society*, **35**, 416–418.

Dirac, P. A. M. (1984). Blackett and the positron, in *Cambridge Physics in the Thirties*, ed. Hendry, J., pp. 61–62. Bristol: Adam Hilger.

Dodson, G. (2002). Dorothy Mary Crowfoot Hodgkin, *Biographical Memoirs of Fellows of the Royal Society*, **48**, 179–219.

Doi, M. and Edwards, S. F. (1978a). Dynamics of concentrated polymer systems, I: Brownian motion in equilibrim state, *Journal of the Chemical Society, Faraday Transactions II*, **74**, 1789–1801.

Doi, M. and Edwards, S. F. (1978b). Dynamics of concentrated polymer systems, II: Molecular motion under flow, *Journal of the Chemical Society, Faraday Transactions II*, **74**, 1802–1817.

Doi, M. and Edwards, S. F. (1978c). Dynamics of concentrated polymer systems, III: The constitutive equation, *Journal of the Chemical Society, Faraday Transactions II*, **74**, 1818–1832.

Doi, M. and Edwards, S. F. (1979). Dynamics of concentrated polymer systems, IV: Rheological properties, *Journal of the Chemical Society, Faraday Transactions II*, **75**, 38–54.

Doi, M. and Edwards, S. F. (1986). *The Theory of Polymer Dynamics*. Oxford: Clarendon Press.

Donald, A. M. (1994). Physics of foodstuffs, *Reports on Progress in Physics*, **57**, 1081–1135.

Donald, A. M. (2015). On the Loss of a Giant. http://occamstypewriter.org/athenedonald/2015/05/09/on-the-loss-of-a-giant/.

Duke, T. and Jülicher, F. (2003). Active traveling wave in the cochlea, *Physical Review Letters*, **90**(15), 158101 1–4.

Dyson, F. (1999). Why is Maxwell's theory so hard to understand?, in *James Clerk Maxwell Commemorative Booklet*, pp. 8–13. Edinburgh: James Clerk Maxwell Foundation.

Eddington, A. (1920). The internal constitution of the stars, *Observatory*, **43**, 341–358.

Eden, R. J. (2012). *Sometimes in Cambridge: Memoirs*. Cambridge: Clare Hall in the University of Cambridge.

Eden, R. J., Landshoff, P. V., Olive, D. I., and Polkinghorne, J. C. (1966). *The Analytic S-Matrix*. Cambridge: Cambridge University Press.

Edge, D. O., Shakeshaft, J. R., McAdam, W. B., Baldwin, J. E., and Archer, S. (1959). A survey of radio sources at a frequency of 159 Mc/s, *Memoirs of the Royal Astronomical Society*, **68**, 37–60.

Edlén, B. (1941). An attempt to identify the emission lines in the spectrum of the solar corona, *Arkiv för Matematik, Astronomi och Fysik*, **28B**, 1–4. Also published as The interpretation of the emission line spectrum of the solar corona, *Zeitschrift für Astrophysik*, **22**, 30–64, 1942.

Edwards, S. F. (1958). A new method for the evaluation of electric conductivity in metals, *Philosophical Magazine*, **3**, 1020–1031.

Edwards, S. F. (1964). The statistical dynamics of homogeneous turbulence, *Journal of Fluid Dynamics*, **18**, 239–273.

Edwards, S. F. (1965). The statistical mechanics of polymers with excluded volume, *Proceedings of the Physical Society of London*, **85**, 613–624.

Edwards, S. F. (1966). The theory of polymer solutions at intermediate concentration, *Proceedings of the Physical Society of London*, **88**, 265–280.

Edwards, S. F. (1968). Statistical mechanics with topological constraints, II, *Journal of Physics*, **A1**, 15–28.

Edwards, S. F. and Anderson, P. W. (1975). Theory of spin glasses, *Journal of Physics F: Metal Physics*, **5**, 965–974.

Edwards, S. F. and Anderson, P. W. (1976). Theory of spin glasses, II, *Journal of Physics F: Metal Physics*, **6**, 1927–1937.

Ehrler, B., Wilson, M. W. B., Rao, A., H., Friend, R. H., and Greenham, N. C. (2012). Singlet exciton fission-sensitized infrared quantum dot solar cells, *Nano Letters*, **12**, 1053–1057.

Einstein, A. (1905). Über einen die erzeugung und verwandlung des lichtes betreffenden heuristischen gesichtspunkt (On a heuristic point of view concerning the production and transformation of light), *Annalen der Physik (4)*, **17**, 132–148.

Einstein, A. (1906). Die plancksche theorie der strahlung und die theorie der spezifischen wärme, *Annalen der Physik (4)*, **22**, 180–190.

Einstein, A. (1924). Quantentheorie des einatomigen idealen gases, I, *Sitzungberichte der (Kgl.) Preussischen Akademie der Wissenschaften (Berlin)*, pp. 261–267.

Einstein, A. (1925). Quantentheorie des einatomigen idealen gases, II, *Sitzungberichte der (Kgl.) Preussischen Akademie der Wissenschaften (Berlin)*, pp. 3–14.

Einstein, A. (1931). Maxwell's influence on the development of the conception of physical reality, in *James Clerk Maxwell, A Commemorative Volume 1831–1931*, pp. 66–73. Cambridge: Cambridge University Press.

Eldredge, K. R. and Tabor, D. (1955). The mechanism of rolling friction, I: The plastic range, *Proceedings of the Royal Society of London*, **A229**, 181–198.

Ellinger, S., Graham, K. R., Shi, P. *et al.* (2011). Donor–acceptor–donor-based π–conjugated oligomers for nonlinear optics and near-IR emission, *Chemistry of Materials*, **23**(17), 3805–3817.

Ellis, C. D. and Mott, N. F. (1933). Energy relations in the β-type of radioactive disintegration, *Proceedings of the Royal Society of London*, **A141**, 502–511.

Ellis, C. D. and Wooster, W. A. (1927). The average energy of disintegration of radium E, *Proceedings of the Royal Society of London*, **A117**, 109–123.

Elsmore, B. and Mackay, C. D. (1969). Observations of the structure of radio sources in the 3C catalogue, III: The absolute determination of positions of 78 compact sources, *Monthly Notices of the Royal Astronomical Society*, **146**, 361–379.

Epstein, P. S. (1916). Zur theorie des Starkeffektes (On the theory of the Stark effect), *Annalen der Physik (4)*, **50**, 489–520.

Eve, A. S. (1939). *Rutherford: Being the Life and Letters of the Rt. Hon. Lord Rutherford, O. M.* Cambridge: Cambridge University Press.

Eve, A. S. and Chadwick, J. (1938). Lord Rutherford, 1871–1937, *Obituary Notices of Fellows of the Royal Society, Vol. 2*, **6**, 394–423.

Ewald, P. P. (1962). *Fifty Years of X-ray Diffraction*. Utrecht: N.V. A. Oosthoek's Uitgeversmaatschappij. This volume was edited by Ewald. It includes articles by and biographies of many of the pioneers of X-ray spectroscopy.

Faber, T. E. (1949). Creation and growth of superconducting nuclei, *Nature*, **164**, 277–278.

Faber, T. E. (1952). The phase transition in superconductors, I: Nucleation, *Royal Society of London Proceedings, Series A*, **214**, 392–412.

Faber, T. E. (1972). *Introduction to the Theory of Liquid Metals*. Cambridge: Cambridge University Press.

Faber, T. E. (1995). *Fluid Dynamics for Physicists*. Cambridge: Cambridge University Press.

Fabian, A. C., Sanders, J. S., Ettori, S. *et al.* (2000). Chandra imaging of the complex X-ray core of the Perseus cluster, *Monthly Notices of the Royal Astronomical Society*, **318**, L65–L68.

Falconer, I. (1988). J.J. Thomson's work on positive rays, 1906–1914, *Historical Studies in the Physical and Biological Sciences*, **18**, 265–310.

Falconer, I. (1989). J.J. Thomson and 'Cavendish physics', in *The Development of the Laboratory*, ed. James, F. A. J. L., pp. 104–117. London: Macmillan.

Falconer, I. (2014). Maxwell and building the Cavendish Laboratory, in *James Clerk Maxwell*, eds Flood, R., McCartney, M., and Whitaker, A., pp. 67–98. Oxford: Oxford University Press.

Fanaroff, B. L. and Riley, J. M. (1974). The morphology of extragalactic radio sources of high and low luminosity, *Monthly Notices of the Royal Astronomical Society*, **167**, 31P–36P.

Fang, D., Kurebayashi, H., Wunderlich, J. *et al.* (2011). Spin-orbit-driven ferromagnetic resonance, *Nature Nanotechnology*, **6**, 413–417.

Faraday, M. (1846). Thoughts on ray-vibrations, *Faraday's Researches in Electricity*, **3**, 447–452. This letter was first published in the *Philosophical Magazine (3)*, **53**, 345–350, 1846, following his lecture at the Royal Institution in April 1846.

Feather, N. (1932). The collisions of neutrons with nitrogen nuclei, *Proceedings of the Royal Society of London*, **A136**, 709–727.

Feather, N. (1984). The experimental discovery of the neutron, in *Cambridge Physics in the Thirties*, ed. Hendry, J., pp. 31–41. Bristol: Adam Hilger.

Fermi, E. (1934). Versuch einer theorie der β-strahlen, I, *Zeitschrift für Physik*, **88**, 161–177.

Fermi, E., Amaldi, E., D'Agostino, O., Rasetti, F., and Segré, E. (1934). Radioattività provocata da bombardamento di neutroni, III, *La Recerca Scientifica*, **5**, 452–453.

Feynman, R. (1955). Applications of quantum mechanics to liquid helium, *Progress in Low Temperature Physics*, **1**, 17–53.

Field, J. E. (1971). Brittle fracture: Its study and application, *Contemporary Physics*, **12**, 1–31.

Field, J. E. (1983). High-speed photography, *Contemporary Physics*, **24**, 439–459.

Fink, T. and Mao, Y. (1999). *The 85 Ways to Tie a Tie: The Science and Aesthetics of Tie Knots*. London: Fourth Estate.

Finkelstein, A. and Shattock, M. (2001). CAPSA and its implementation: Report to the Audit Committee and the Board of Scrutiny, *Cambridge University Reporter*, **132**, 153–208.

Fitzpatrick, T. C. and Whetham, W. C. D. (1910). The building of the Laboratory, in *A History of the Cavendish Laboratory, 1871–1910*, pp. 1–13. London: Longmans, Green and Co.

Fitzpatrick, T. C., Whetham, W. C. D., Schuster, A. *et al.*, eds (1910). *A History of the Cavendish Laboratory, 1871–1910*. London: Longmans, Green and Co.

Forbes, R. J. (1958). Power to 1850, in *A History of Technology, Vol. 4*, eds Singer, C., Holmyard, E. J., Hall, A. R., and Williams, T. I., pp. 148–167. Oxford: Clarendon Press. The quotation is on page 165.

Forfar, D. (1995). What became of the Senior Wranglers?, *Mathematical Spectrum*, **29**, 1–4.

Foucault, L. (1849). Lumiére électrique (Electric light), *L'Institut, Journal Universal des Sciences*, **17**, 44–46.

Fourier, J. B. J. (1822). *Théorie Analytique de la Chaleur (Analytical Theory of Heat)*. Paris: Firmin Didot Pre et fils.

Franklin, R. E. and Gosling, R. G. (1953a). Evidence for 2-chain helix in crystalline structure of sodium desoxyribonucleate, *Nature*, **172**, 156–157.

Franklin, R. E. and Gosling, R. G. (1953b). Molecular configuration in sodium thymonucleate, *Nature*, **171**, 740–741.

Franklin, R. E. and Gosling, R. G. (1953c). The structure of sodium thymonucleate fibres, I: The influence of water content, *Acta Crystallographica*, **6**, 673–677.

Franklin, R. E. and Gosling, R. G. (1953d). The structure of sodium thymonucleate fibres, II: The cylindrically symmetrical Patterson function, *Acta Crystallographica*, **6**, 678–685.

Franklin, R. E. and Gosling, R. G. (1953e). The structure of sodium thymonucleate fibres, III: The three-dimensional Patterson function, *Acta Crystallographica*, **8**, 151–156.

Fraunhofer, J. (1817a). Bestimmung des brechungs- und des farbenzerstreuungs-vermögens verschiedener glasarten, in bezug auf die vervollkommnung achromatischer fernröhre (On the refractive and dispersive power of different species of glass in reference to the improvement of achromatic telescopes, with an account of the lines or streaks which cross the spectrum), *Denkschriften der Königlichen Akademie der Wissenschaften zu München*, **5**, 193–226. Translation: *Edinburgh Philosophical Journal*, **9**, pp. 288–299, 1823; **10**, pp. 26-40, 1824.

Fraunhofer, J. (1817b). Bestimmung des brechungs- und des farbenzerstreuungs-vermögens verschiedener glasarten, in bezug auf die vervollkommnung achromatis-cher fernröhre (On the refractive and dispersive power of different species of glass in reference to the improvement of achromatic telescopes, with an account of the lines or streaks which cross the spectrum), *Gilberts Annalen der Physik (1)*, **56**, 264–313.

Fraunhofer, J. (1821). Neue modifikation des lichtes durch gegenseitige einwirkung und beugung der strahlen, und gesetze derselben, *Denkschriften der Königlichen Akademie der Wissenschaften zu München*, **8**, 1–76.

Friedrich, W., Knipping, P., and Laue, M. von. (1912). Interferenz-erscheinungen bei Röntgenstrahlen (Interference effects with Röntgen rays), *Sitzberichte der Königlich Bayerischen Akademie der Wissenschaften*, pp. 303–312.

Friend, R. H. (2008). Richard Friend interviewed by Alan Macfarlane. http://www.sms.cam.ac.uk/media/1116837.

Frisch, O. (1974). Nuclear physics in the modern Cavendish, in *A Hundred Years of Cambridge Physics*, ed. Moralee, D., pp. 39–41. Cambridge: Cavendish Laboratory. The second and third editions were published in 1980 (ed. Parker, J.) and 1995 (eds. Jacques, G. and Bache, I.) with the title *A Hundred Years and More of Cambridge Physics*. See also http://www.phy.cam.ac.uk/history/years/nuclearphys.

Frisch, O., ed. (1980). *What Little I Remember*. Cambridge: Cambridge University Press.

Fröhlich, H. (1950). Theory of the superconducting state, I: The ground state at the absolute zero of temperature, *Physical Review*, **79**, 845–856.

Fujii, Y. (1971). Dilaton and possible non-Newtonian gravity, *Nature Physical Sciences*, **234**, 5–7.

Fujii, Y. (1972). Scale invariance and gravity of hadrons, *Annals of Physics*, **69**, 494–521.

Galvani, L. (1791). De viribus electricitatis in motu musculari commentarius, *De Bononiensi Scientiarum et Artium Institute atque Academia Commentarii*, **7**, 363–418.

Gamow, G. (1928). Zur quantentheorie des atomkernes, *Zeitschrift für Physik*, **51**, 204–212.

Geiger, H. and Marsden, E. (1909). On a diffuse reflection of the α-particles, *Proceedings of the Royal Society of London*, **A82**, 495–500.

Geiger, H. and Marsden, E. (1913). The laws of deflexion of α-particles through large angles, *Philosophical Magazine (6)*, **25**, 604–623.

Geiger, H. and Müller, W. (1928). Das electronenzählrohr (The electron-counting tube), *Physikalische Zeitschrift*, **29**, 839–841.

Geiger, H. and Müller, W. (1929). Technische bermerkungen zum electronenzählrohr (Technical remarks on the electron-counting tube), *Physikalische Zeitschrift*, **30**, 489–493.

Geiger, H. and Nuttall, J. M. (1911). The ranges of the α-particles from various radioactive substances and a relation between range and period of transformation, *Philosophical Magazine (6)*, **22**, 613–621.

Geikie, A. (1918). *Memoir of John Michell*. Cambridge: Cambridge University Press.

Gerchberg, R. W. and Saxton, W. O. (1972). A practical algorithm for the determination of the phase from image and diffraction plane pictures, *Optik*, **35**, 237–246.

Gerlach, W. and Stern, O. (1922). Der experimentelle nachweis der richtungsquantelung im magnetfeld, *Zeitschrift für Physik*, **9**, 349–352.

Giaever, I. (1960a). Electron tunneling between two superconductors, *Physical Review Letters*, **5**, 464–466.

Giaever, I. (1960b). Energy gap in superconductors measured by electron tunneling, *Physical Review Letters*, **5**, 147–148.

Gillies, G. T. (1997). The Newtonian gravitational constant: Recent measurements and related studies, *Reports on Progress in Physics*, **60(2)**, 151–225.

Ginzburg, V. L. (1951). Cosmic rays as a source of galactic radio-radiation, *Doklady Akademiya Nauk SSSR*, **76**, 377–380.

Ginzburg, V. L. and Landau, L. D. (1950). On the theory of superconductivity (in Russian), *Zhurnal Experimentalnoi i Teoreticheskikh Fizika*, **20**, 1064–1082.

Glazebrook, R. T. (1882). On the refraction of plane polarised light at the surface of a uniaxial crystal, *Philosophical Transactions of the Royal Society of London*, **173**, 595–620.

Glazebrook, R. T. (1910). Lord Rayleigh's professorship, in *A History of the Cavendish Laboratory, 1871–1910*, pp. 40–74. London: Longmans, Green and Co.

Glazebrook, R. T. and Shaw, W. (1885). *Practical Physics*. London: Longmans, Green and Co. This volume ran to many subsequent editions.

Glazebrook, R. T., Dodds, J. M., and Sargant, E. B. (1883). IV. Experiments on the value of the British Association unit of resistance, *Philosophical Transactions of the Royal Society of London*, **174**, 223–268.

Gleick, J. (1988). *Chaos: Making a New Science*. London: William Heinemann.

Gold, T. (1968). Rotating neutron stars as the origin of the pulsating radio sources, *Nature*, **218**, 731–732.

Goldbart, P., Goldenfeld, N., and Sherrington, D., eds (2004). *Stealing the Gold: A Celebration of the Pioneering Physics of Sam Edwards*, volume 126 of International Series of Monographs on Physics. Oxford: Clarendon Press.

Goldstein, E. (1886). Über eine noch nicht untersuchte strahlungsform an der kathodeinducirter entladungen, *Berlin Monatsberichte II*, 691–699.

Gorter, C. J. and Casimir, H. (1934). On supraconductivity, I, *Physica*, **1**, 306–320. Also published in *Physikalische Zeitung*, **35**, 963, 1934.

Gower, J. F. R. (1966). The source counts from the 4C survey, *Monthly Notices of the Royal Astronomical Society*, **133**, 151–161.

Gower, J. F. R., Scott, P. F., and Wills, D. (1967). A survey of radio sources in the declination ranges −07° to 20° and 40° to 80°, *Memoirs of the Royal Astronomical Society*, **71**, 49–144.

Gowing, M. (1964). *Britain and Atomic Energy, 1935–1945*. London: Macmillan.

Graham-Smith, F. (1951). An accurate determination of the positions of four radio stars, *Nature*, **168**, 555.

Graham-Smith, F. (1986). Martin Ryle, *Biographical Memoirs of Fellow of The Royal Society of London*, **32**, 496–524.

Gray, K. E., Long, A. R., and Adkins, C. J. (1969). Measurements of the lifetime of excitations in superconducting aluminium, *Philosophical Magazine*, **20**, 273–278.

Greenstein, J. L. and Matthews, T. A. (1963). Redshift of the radio source 3C 48, *Astronomical Journal*, **68**, 279.

Greenwood, J. A., Minshall, H., and Tabor, D. (1960). Hysteresis losses in rolling and sliding friction, *Proceedings of the Royal Society of London*, **A259**, 480–507.

Greinacher, H. (1926). Eine neue methode zur messung elemenarstrahlen, *Zeitschrift für Physik*, **36**, 364–373.

Greinacher, H. (1927). Über die registrierung von α- und H-strahlen nach der neuen elektrischen zählmethode, *Zeitschrift für Physik*, **44**, 319–325.

Griffin, A. (2008). Superfluidity: Three people, two papers, one prize, *Physics World*, **August 2008**, 27–30.

Grotrian, W. (1939). Zur frage der deutung der linien im spektrum der sonnenkorona, *Naturwissenschaften*, **27**, 214.

Gull, S. F. and Daniell, G. J. (1978). Image reconstruction from incomplete and noisy data, *Nature*, **272**, 686–690.

Gull, S. F. and Northover, K. J. E. (1973). Bubble model of extragalactic radio sources, *Nature*, **244**, 80–83.

Gunn, J. E. and Ostriker, J. P. (1969). Acceleration of high-energy cosmic rays by pulsars, *Physical Review Letters*, **22**, 728–731.

Gunn, J. E., Hoessel, J. G., Westphal, J. A., Perryman, M. A. C., and Longair, M. S. (1981). Investigations of the optical fields of 3CR radio sources to faint limiting magnitudes, IV, *Monthly Notices of the Royal Astronomical Society*, **194**, 111–123.

Gurney, R. W. and Condon, E. U. (1928). Quantum mechanics and radioactive disintegration, *Nature*, **122**, 439.

Gurney, R. W. and Condon, E. U. (1929). Quantum mechanics and radioactive disintegration, *Physical Review*, **33**, 127–140.

Haas, A. E. (1910a). Der zusammenhang des Planckschen elementaren wirkungsquantums mit den grundgrössen der elektronentherorie, *Jahrbuch der Radioaktivität und Elektronik*, **7**, 261–268.

Haas, A. E. (1910b). Über die electrodynamische bedeutung des Planckschen strahlungsgesetzes und über eine neue bestimmung des elektrischen elementarquantums unde der dimensionen des wasserstoffatoms, *Sitzberichte der Kaiserlichen Akademie der Wissenschaften (Wien). Abteilung II*, **119**, 119–144.

Haas, A. E. (1910c). Über eine neue theoretische methode zur bestimmung des elektrischen elementarquantums unde des halbmessers des wasserstoffatoms, *Physikalische Zeitschrift*, **11**, 537–538.

Hahn, O. and Strassmann, F. (1939). Über den nachweis und das verhalten der bei der bestrahlung des urans mittels neutronen entstehenden erdalkalimetalle, *Naturwissenschaften*, **27**, 11–15.

Hales, S. E. G., Riley, J. M., Waldram, E. M., Warner, P. J., and Baldwin, J. E. (2007). A final non-redundant catalogue for the 7C 151-MHz survey, *Monthly Notices of the Royal Astronomical Society*, **382**, 1639–1642.

Hall, H. E. (1957). The angular acceleration of liquid helium II, *Philosophical Transactions of the Royal Society of London*, **A250**, 359–385.

Hall, H. E. and Vinen, W. F. (1956a). The rotation of liquid helium II, I: Experimnets on the propagation of second sound in uniformly rotating helium II, *Proceedings of the Royal Society of London*, **A238**, 204–214.

Hall, H. E. and Vinen, W. F. (1956b). The rotation of liquid helium II, II: The theory of mutual friction in uniformly rotating helium II, *Proceedings of the Royal Society of London*, **A238**, 215–234.

Haniff, C. A., Mackay, C. D., Titterington, D. J., Sivia, D., and Baldwin, J. E. (1987). The first images from optical aperture synthesis, *Nature*, **328**, 694–696.

Hargrave, P. J. and Ryle, M. (1974). Observations of Cygnus A with the 5 km radio telescope, *Monthly Notices of the Royal Astronomical Society*, **166**, 305–327.

Harman, P. (1990). *The Scientific Letters and Papers of James Clerk Maxwell. Volume I, 1846-1862*. Cambridge: Cambridge University Press.

Harman, P. (1995). *The Scientific Letters and Papers of James Clerk Maxwell. Volume II, 1862-1873*. Cambridge: Cambridge University Press.

Harman, P. (1998). *The Natural Philosophy of James Clerk Maxwell*. Cambridge: Cambridge University Press.

Harman, P. (2002). *The Scientific Letters and Papers of James Clerk Maxwell. Volume III, 1874-1879*. Cambridge: Cambridge University Press.

Harrison, W. A. (1958). Cellular method for wave functions in imperfect metal lattices, *Physical Review*, **110**, 14–25.

Hartree, D. R. (1929). The propagation of electromagnetic waves in stratified media, *Mathematical Proceedings of the Cambridge Philosophical Society*, **25**, 97–120.

Hartree, D. R. (1957). *The Calculation of Atomic Structures*. New York: John Wiley & Sons.

Haselgrove, C. B. and Hoyle, F. (1956). A mathematical discussion of the problem of stellar evolution, with reference to the use of an automatic digital computer, *Monthly Notices of the Royal Astronomical Society*, **116**, 515–526.

Hawkes, P. W. (2009). Aberration corrections past and present, *Philosophical Transactions of the Royal Society of London*, **A367**, 3637–3664.

Haydock, R. and Nex, C. M. M. (1984). Comparison of quadrature and termination for estimating the density of states within the recursion method, *Journal of Physics C: Solid State Physics*, **17**, 4783–4789.

Hazard, C., Mackey, M. B., and Shimmins, A. J. (1963). Investigation of the radio source 3C 273 by the method of lunar occultations, *Nature*, **197**, 1037–1039.

Heaviside, O. (1902). The theory of electric telegraphy, *Encyclopaedia Britannica*, tenth edition. This article was commissioned by the Encyclopaedia Britannica for its tenth edition. The article was reprinted in volume 3 of Heaviside's *Electromagnetic Theory*, pp. 331–346 (London: 'The Electrician' Printing and Publishing Company).

Hedgeland, H., Fouquet, P., Jardine, A. P. *et al.* (2009). Measurement of single-molecule frictional dissipation in a prototypical nanoscale system, *Nature Physics*, **5**, 561–564.

Heilbron, J. L. (1970). Joseph John Thomson, in *Dictionary of Scientific Biography*, volume 13, pp. 362–372. New York: Charles Scribner's Sons.

Heilbron, J. L. (1977). Lectures on the history of atomic physics 1900–1922, in *History of Twentieth Century Physics: 57th Varenna International School of Physics, 'Enrico Fermi'*, ed. Weiner, C., pp. 40–108. New York and London: Academic Press.

Heilbron, J. L. and Seidel, R. W. (1989). *Lawrence and His Laboratory: A History of the Lawrence Radiation Laboratory*. Berkeley: University of California Press.

Heine, V. (2015). TCM Group History. http://www.tcm.phy.cam.ac.uk/about/history.

Heine, V. (1970). The pseudopotential concept, *Solid State Physics*, **24**, 1–36.

Heisenberg, W. (1925). Über quanten theoretische umdeutung kinematischer und mechanischer beziehungen (Quantum-theoretical re-interpretation of kinematic and mechanical relations), *Zeitschrift für Physik*, **33**, 879–893.

Heisenberg, W. (1927). Über den anschaulichen Inhalt der quantentheoretischen kinematik und mechanik, *Zeitschrift für Physik*, **43**, 172–198. An English translation is included in the volume *Quantum Theory and Measurement*, eds. Wheeler, J. A. and Zurek, W. H., pp. 62–84, Princeton: Princeton University Press, 1983.

Hendry, J. (1984). Introduction to Part 3, Underlying themes, in *Cambridge Physics in the Thirties*, ed. Hendry, J., pp. 103–124. Bristol: Adam Hilger.

Henyey, L. G. and Keenan, P. C. (1940). Interstellar radiation from free electrons and hydrogen atoms, *Astrophysical Journal*, **91**, 625–630.

Hertz, H. (1893). *Electric Waves*. London: Macmillan. The original book, *Untersuchungen über die Ausbreitung der elektrischen Kraft*, was published by Johann Ambrosius Barth in Leipzig in 1892.

Hess, V. F. (1913). Über beobachtungen der durchdringenden strahlung bei sieben freiballonfahrten (Concerning observations of penetrating radiation on seven free balloon flights), *Physikalische Zeitschrift*, **13**, 1084–1091.

Hevesy, G. (1948). Francis William Aston, 1877–1945, *Obituary Notices of Fellows of the Royal Society, Vol. 5*, **5**, 634–650.

Hewish, A. (1951). The diffraction of radio waves in passing through a phase-changing ionosphere, *Proceedings of the Royal Society of London*, **A209**, 81–96.

Hewish, A. (1955). The irregular structure of the outer regions of the solar corona, *Proceedings of the Royal Society of London*, **A228**, 238–251.

Hewish, A., Bell, S., Pilkington, J., Scott, P., and Collins, R. (1968). Observations of a rapidly pulsating radio source, *Nature*, **217**, 709–713.

Hewish, A., Scott, P. F., and Wills, D. (1964). Interplanetary scintillation of small diameter radio sources, *Nature*, **203**, 1214–1217.

Hey, J. S. (1946). Solar radiations in the 4–6 metre radio wave-length band, *Nature*, **157**, 47–48.

Hey, J. S., Parsons, S. J., and Phillips, J. W. (1946). Fluctuations in cosmic radiation at radio-frequencies, *Nature*, **158**, 234.

Hirsch, P. B., Horne, R. W., and Whelan, M. J. (1956). LXVIII. Direct observations of the arrangement and motion of dislocations in aluminium, *Philosophical Magazine*, **1**, 677–684.

Hirsch, P. B., Howie, A., Nicholson, R., Pashley, D. W., and Whelan, M. J. (1965). *Electron Microscopy of Thin Crystals*. London and Malabar, FL: Butterworths/Krieger. Second edition 1977.

Hirsch, P. B., Howie, A., and Whelan, M. J. (1960). A kinematical theory of diffraction contrast of electron transmission microscope images of dislocations and other defects, *Philosophical Transactions of the Royal Society of London*, **A252**, 499–529.

Hodgkin, D. M. C. (1980). John Desmond Bernal, *Biographical Memoirs of Fellows of the Royal Society*, **26**, 17–84.

Holland, W. S., Robson, E. I., Gear, W. K. *et al.* (1999). SCUBA: A common-user submillimetre camera operating on the James Clerk Maxwell Telescope, *Monthly Notices of the Royal Astronomical Society*, **303**, 659–672.

Howell, T. F. and Shakeshaft, J. R. (1966). Measurement of the minimum cosmic background radiation at 20.7-cm wave-length, *Nature*, **210**, 1318–1319.

Howell, T. F. and Shakeshaft, J. R. (1967). Spectrum of the 3° K cosmic microwave radiation, *Nature*, **216**, 753–754.

Howie, A. (1960). The electrical resistivity of stacking faults, *Philosophical Magazine*, **5**, 251–271.

Howie, A. (1979). Image contrast and localized signal selection techniques, *Journal of Microscopy*, **11–23**.

Howie, A. and Whelan, M. J. (1961). Diffraction contrast of electron microscope images of crystal lattice defects, II: The development of a dynamical theory, *Proceedings of the Royal Society of London*, **A263**, 217–237.

Howie, A. and Whelan, M. J. (1962). Diffraction contrast of electron microscope images of crystal lattice defects, III: Results and experimental confirmation of the dynamical theory of dislocation image contrast, *Proceedings of the Royal Society of London*, **A267**, 206–230.

Hoyle, F. (1957). *The Black Cloud*. London: William Heinemann.

Huggins, M. L. (1943). The structure of fibrous proteins, *Chemical Review*, **32**, 195–218.

Hughes, D. H., Serjeant, S., Dunlop, J. *et al.* (1998). High-redshift star formation in the Hubble Deep Field revealed by a submillimetre-wavelength survey, *Nature*, **394**, 241–247.

Hughes, J. A. (1993). The radioactivitists: Community, controversy and the rise of nuclear physics. PhD dissertation, University of Cambridge.

Hunter, G. (2004). *Light is a Messenger: The Life and Science of William Lawrence Bragg*. Oxford: Oxford University Press.

Huygens, C. (1690). *Treatise on Light*. Leiden.

Israelachvili, J. N. and Tabor, D. (1972). The measurement of van Der Waals dispersion forces in the range 1.5 to 130 nm, *Proceedings of the Royal Society of London*, **A331**, 19–38.

Jaklevic, R. C., Lambe, J., Silver, A. H., and Mercereau, J. E. (1964). Quantum interference effects in Josephson tunneling, *Physical Review Letters*, **12**, 159–160.

James, F. (2000). *Guides to the Royal Institution of Great Britain, I: History*. London: Royal Institution of Great Britain.

Jammer, M. (1989). *The Conceptual Development of Quantum Mechanics*, 2nd edition. New York: American Institute of Physics and Tomash Publishers. The first edition of Jammer's important book was published in 1966 by the McGraw-Hill Book Company in its International Series of Pure and Applied Physics. The 1989 edition is an enlarged and revised version of the first edition and I use this as the primary reference to Jammer's history.

Jansky, K. G. (1933). Electrical disturbances apparently of extraterrestrial origin, *Proceedings of the Institution of Radio Engineers*, **21**, 1387–1398.

Jardine, A. P., Hedgeland, H., Alexandrowicz, G., Allison, W., and Ellis, J. (2009). Helium-3 spin-echo: Principles and application to dynamics at surfaces, *Progress in Surface Science*, **84**, 323–379.

Jenkins, C. J., Pooley, G. G., and Riley, J. M. (1977). Observations of 104 extragalactic radio sources with the Cambridge 5-km telescope at 5 GHz, *Memoirs of the Royal Astronomical Society*, **84**, 61–99.

Jennison, R. C. and Das Gupta, M. K. (1953). Fine structure of the extra-terrestrial radio source Cygnus 1, *Nature*, **172**, 996–997.

Jones, R. V. (1978). *Most Secret War: British Scientific Intelligence 1939–1945*. London: Hamish Hamilton. The revised edition has been published in paperback by Penguin.

Jones, R. V. (1981). Some consequences of physics, *Nature*, **293**, 23–25.

Josephson, B. D. (1960). Temperature-dependent shift of γ rays emitted by a solid, *Physical Review Letters*, **4**, 341–342.

Josephson, B. D. (1962). Possible new effects in superconductive tunnelling, *Physics Letters*, **1**, 251–253.

Jungnickel, C. and McCormmach, R. (1986a). *The Intellectual Mastery of Nature: Theoretical Physics from Ohm to Einstein, Vol. 1. The Torch of Mathematics, 1800 to 1870*. Chicago: University of Chicago Press.

Jungnickel, C. and McCormmach, R. (1986b). *The Intellectual Mastery of Nature: Theoretical Physics from Ohm to Einstein, Vol. 2. The Now Mighty Theoretical Physics, 1870 to 1925*. Chicago: University of Chicago Press.

Kamerlingh Onnes, H. (1911a). *On the Change of Resistance of Pure Metals at Very Low Temperatures, etc. III. The Resistance of Platinum at Helium Temperatures*. Communications of the Physical Laboratory of the University of Leiden, No. 119b.

Kamerlingh Onnes, H. (1911b). *Further Experiments with Liquid Helium. D. On the Change of Electric Resistance of Pure Metals at Very Low Temperatures, etc. V. The Disappearance of the Resistance of Mercury*. Communications of the Physical Laboratory of the University of Leiden, No. 122b.

Kapitsa, P. L. (1922). The loss of energy of an α-ray beam in its passage through matter, I: Passage through air and CO_2, *Proceedings of the Royal Society of London*, **A102**, 48–71.

Kapitsa, P. L. (1923). Some observations on α-particle tracks in a magnetic field, *Proceedings of the Cambridge Philosophical Society*, **21**, 511–516.

Kapitsa, P. L. (1924a). α-ray tracks in a strong magnetic field, *Proceedings of the Royal Society of London*, **A106**, 602–622.

Kapitsa, P. L. (1924b). A method of producing strong magnetic fields, *Proceedings of the Royal Society of London*, **A105**, 691–710.

Kapitsa, P. L. (1927). Further developments of the method of obtaining strong magnetic fields, *Proceedings of the Royal Society of London*, **A115**, 658–683.

Kapitsa, P. L. (1938). Viscosity of liquid helium below the λ-point, *Nature*, **141**, 74.

Kapitsa, P. L. and Cockcroft, J. D. (1932). Hydrogen liquefaction plant at the Royal Society Mond Laboratory, *Nature*, **129**, 224–226.

Kaufmann, W. (1902). Die elektromagnetische masse des elektrons, *Physikalische Zeitschrift*, **4**, 54–56.

Keesom, W. H. and Keesom, A. P. (1936). On the heat conductivity of liquid helium, *Physica*, **3**, 359–360.

Keesom, W. H. and van den Ende, J. N. (1930). The specific heats of solid substances at the temperatures attainable with the aid of liquid helium II: Measurements of the atomic heats of lead and of bismiuth, *Proceedings of the Koniklijke Akadamie van Wetenschappen te Amsterdam*, **33**, 243–254. Leiden Communications No. 203.

Kelly, A. (2012). Lawrence Bragg's interest in the deformation of metals and 1950–1953 in the Cavendish: A worm's eye view, *Acta Crystallographica*, **A69**, 16–24.

Kendrew, J. C. (1990). Bragg's broomstick and the structure of proteins, in *Selections and Reflections: The Legacy of Sir Lawrence Bragg*, eds Thomas, J. M. and Phillips, D., pp. 88–91. Northwood, UK: Science Reviews.

Kendrew, J. C., Bodo, G., Dintzis, H. M., Parrish, R. G., and Wyckoff, H. (1958). A three-dimensional model of the myoglobin molecule obtained by X-ray analysis, *Nature*, **181**, 662–666.

Kendrew, J. C., Dickerson, R. E., Strandberg, B. *et al.* (1960). Structure of myoglobin: 3-dimensional Fourier synthesis at 2Å resolution, *Nature*, **185**, 422–427.

Kennedy, J. W., Seaborg, G. T., Segrè, E., and Wahl, A. C. (1946). Properties of 94(239), *Physical Review*, **70**, 555–556.

Kennelly, A. E. (1902). On the elevation of the electrically-conducting strata of the Earth's atmosphere, *Electrical World and Engineer*, **39**, 473.

Khmelnitskii, D. (2004). Impurity diagrammatics and the physics of disordered metals, in *Stealing the Gold: A Celebration of the Pioneering Physics of Sam Edwards*, eds Goldbart, P., Goldenfeld, N., and Sherrington, D., volume 126 of International Series of Monographs on Physics: pp. 23–29. Oxford: Clarendon Press.

Kiepenheuer, K. (1950). Cosmic rays as the source of general galactic radio emission, *Physical Review*, **79**, 738–739.

Kinloch, A. J. (2000). Norman Adrian de Bruyne, *Biographical Memoirs of Fellows of the Royal Society*, **46**, 127–143.

Kirchhoff, G. (1859). Ueber den zusammenhang zwischen emission und absorption von licht und wärme (On the connection between emission and absorption of light and heat), *Berlin Monatsberichte*, pp. 783–787.

Kirchhoff, G. (1861). Untersuchungen über das sonnenspektrum und die spectren der chemischen elemente (Investigations of the solar spectrum and the spectra of the chemical elements), Part 1, *Abhandlungen der Königlich Preussischen Akademie der Wissenschaften zu Berlin*, pp. 62–95.

Kirchhoff, G. (1862). Untersuchungen über das sonnenspektrum und die spectren der chemischen elemente (Investigations of the solar spectrum and the spectra of the chemical elements), Part 1 (continued), *Abhandlungen der Königlich Preussischen Akademie der Wissenschaften zu Berlin*, pp. 227–240.

Kirchhoff, G. (1863). Untersuchungen über das sonnenspektrum und die spectren der chemischen elemente (Investigations of the solar Spectrum and the spectra of the chemical elements), Part 2, *Abhandlungen der Königlich Preussischen Akademie der Wissenschaften zu Berlin*, pp. 225–240.

Kirsch, G. and Pettersson, H. (1923). Long-range particles from radium active deposit, *Nature*, **112**, 394–395.

Kirsch, G. and Pettersson, H. (1926). *Atomzertrümmerung (Atomic Fragmentation)*. Leipzig: Akademische Verlagsgesellschaft.

Kläui, M., Vaz, C. A. F., Lopez-Diaz, L., and Bland, J. A. C. (2003). Vortex formation in narrow ferromagnetic rings, *Journal of Physics: Condensed Matter*, **15**, R985–R1023.

Klitzing, K. V., Dorda, G., and Pepper, M. (1980). New method for high-accuracy determination of the fine-structure constant based on quantized Hall resistance, *Physical Review Letters*, **45**, 494–497.

Knoll, M. and Ruska, E. (1932a). Beitrag zur geometrischen Elektronenoptik I and II, *Annalen der Physik (4)*, **12**, 607–640 and 641–661.

Knoll, M. and Ruska, E. (1932b). Das elektronenmikroskop (The electron microscope), *Zeitschrift für Physik*, **78**, 318–339.

Kohlrausch, W. (1870). *Leitfaden der praktischen physik zunächst für das physikalische prakticum in Göttingen (An introduction to physical measurements)*. Leipzig: Teubner.

Kolhörster, W. (1913). Messungen der durchdringenden strahlung im freiballon in grösseren höhen (Measurements of penetrating radiation in free balloon flights at great altitudes), *Physikalische Zeitschrift*, **14**, 1153–1156.

Kragh, H. (1996). *Cosmology and Controversy: The Historical Development of Two Theories of the Universe*. Princeton: Princeton University Press.

Krivanek, O. L., Dellby, N., and Brown, L. M. (1996). Spherical aberration corrector for a dedicated STEM, in *Proceedings of the 11th European EM Congress, Dublin 1996*, pp. 352–353. Brussels: CESEM.

Krivanek, O. L., Dellby, N., and Lupini, A. R. (1999). Towards sub-Å electron beams, *Ultramicroscopy*, **78**, 1–11.

Krivanek, O. L., Dellby, N., Spence, A. J., Camps, R. A., and Brown, L. M. (1997a). Aberration correction in the STEM, in *Institute of Physics Conference Series 153 (Proceedings of the 1997 EMAG Meeting)*, ed. Rodenburg, J., pp. 35–39. Bristol: IOP Publishing.

Krivanek, O. L., Dellby, N., Spence, A. J., Camps, R. A., and Brown, L. M. (1997b). On-line aberration measurement and correction in STEM, in *Microscopy and Microanalysis: Proceedings of the 55th MSA Meeting*, **3**, 1171–1172. This paper appears in Suppl. 2 of this volume of *Microscopy and Microanalysis*.

Kuper, C. G. (1951). An unbranched laminar model of the intermediate state of superconductors, *Philosophical Magazine (7)*, **42**, 961–977.

Lagendijk, A., van Tiggelen, B., and Wiersma, D. S. (2009). Fifty years of Anderson localization, *Physics Today*, **62**, 24–29.

Laing, R. A., Riley, J. M., and Longair, M. S. (1983). Bright radio sources at 178 MHz: Flux densities, optical identifications and the cosmological evolution of powerful radio galaxies, *Monthly Notices of the Royal Astronomical Society*, **204**, 151–187.

Landau, L. D. (1930). Diamagnetismus der metalle, *Zeitschrift für Physik*, **64**, 629–637.

Landau, L. D. (1941). The theory of superfluidity of helium II, *Journal of Physics (Moscow)*, **5**, 71–90.

Landé, A. (1919). Eine quantenregel für die räumliche orientierung von elektron-ringen, *Verhandlungen der Deutschen Physikalischen Gesellschaft*, **21**, 585–588.

Langevin, P. and de Broglie, M. (1912). *La Théorie du Rayonnement et les Quanta: Rapports et Discussions de la Réunion Tenue à Bruxelles, du 30 Octobre au 3 Novembre 1911*. Paris: Gautier-Villars.

Laue, M. von (1912). Eine quantative prüfung der theorie für die interferenzerscheinungen bei Röntgenstrahlung (A quantitative test of the theory of X-ray interference phenomena), *Sitzberichte der Königlich Bayerischen Akademie der Wissenschaften*, pp. 363–373.

Lauritsen, C. C. and Bennett, R. D. (1928). A new high potential X-ray tube, *Physical Review*, **32**, 850–857.

Lawrence, E. O. and Livingston, M. S. (1931). The production of high speed protons without the use of high voltages, *Physical Review*, **38**, 834.

Lawrence, E. O. and Sloan, D. H. (1931). The production of high speed canal rays without the use of high voltages, *Proceedings of the National Academy of Sciences of the United States of America*, **17**, 64–70.

Lawrence, E. O., Alvarez, L. W., Brobeck, W. M. *et al.* (1939). Initial performance of the 60-inch cyclotron of the William H. Crocker Radiation Laboratory, University of California, *Physical Review*, **56**, 124.

Lechner, B. A. J., de Wijn, A. S., Hedgeland, H. *et al.* (2013). Atomic scale friction of molecular adsorbates during diffusion, *Journal of Chemical Physics*, **138**, 194710.

Leggett, A. J. (1995). Superfluids and superconductors, in *Twentieth Century Physics, Vol. II*, eds Brown, L. M., Pais, A., and Pippard, A. B., pp. 913–966. Bristol and Philadelphia: Institute of Physics Publishing, and New York: American Institute of Physics Press.

Lenz, E. (1834). Über die bestimmung der richtung durch elektodyanamische vertheilung erregten galvanischen ströme, *Annalen der Physik (2)*, **31**, 483–483.

Lewis, W. B. (1984). The development of electrical counting methods in the Cavendish, in *Cambridge Physics in the Thirties*, ed. Hendry, J., pp. 133–136. Bristol: Adam Hilger.

Lilly, S. J. and Longair, M. S. (1984). Stellar populations in distant radio galaxies, *Monthly Notices of the Royal Astronomical Society*, **211**, 833–855.

Lindsay, R. B. (1970). *Lord Rayleigh, the Man and His Works*. Oxford and New York: Pergamon Press.

Little, L. T. and Hewish, A. (1966). Interplanetary scintillation and its relation to the angular structure of radio sources, *Monthly Notices of the Royal Astronomical Society*, **134**, 221–237.

Lockyer, N. (1874a). Editorial, *Nature*, **9**, 298.

Lockyer, N. (1874b). The new physical laboratory of the University of Cambridge, *Nature*, **10**, 139–142.

London, F. (1938). On the Bose–Einstein condensation, *Physical Review*, **54**, 947–954.

London, F. (1950). *Superfluids, I; Macroscopic Theory of Superconductivity*. New York: John Wiley & Sons; London: Chapman & Hall.

London, F. (1954). *Superfluids, II: Macroscopic Theory of Superfluid Helium*. New York: John Wiley & Sons.

London, F. and London, H. (1935a). The electromagnetic equations of the supraconductor, *Proceedings of the Royal Society of London*, **A149**, 71–88.

London, F. and London, H. (1935b). Supraleitung und diamagnetismus, *Physica*, **2**, 341–354.

London, H. (1940). The high-frequency resistance of superconducting tin, *Nature*, **141**, 643–644.

Long, A. and Adkins, C. (1973). Transfer characteristics of phonon-coupled superconducting tunnel junctions, *Philosophical Magazine*, **27**, 865–882.

Longair, M. S. (1965). Objects in the fields of 88 radio sources, *Monthly Notices of the Royal Astronomical Society*, **129**, 419–436.

Longair, M. S. (1966). On the interpretation of radio source counts, *Monthly Notices of the Royal Astronomical Society*, **133**, 421–436.

Longair, M. S. (1971). Observational cosmology, *Reports of Progress in Physics*, **34**, 1125–1248.

Longair, M. S. (1974). The counts of radio sources, in *Confrontation of Cosmological Theories with Observational Data*, ed. Longair, M. S., volume 63 of *IAU Symposium*, pp. 93–108. Dordrecht: D. Reidel.

Longair, M. S. (1999). The 3CR sample: 1962–1999, in *The Hy-Redshift Universe: Galaxy Formation and Evolution at High Redshift*, eds Bunker, A. J. and van Breugel, W. J. M., volume 193 of Astronomical Society of the Pacific Conference Series, pp. 11–22. San Francisco: Astronomical Society of the Pacific Publications.

Longair, M. S. (2003). *Theoretical Concepts in Physics: An Alternative View of Theoretical Reasoning in Physics*. Cambridge: Cambridge University Press.

Longair, M. S. (2006). *The Cosmic Century: A History of Astrophysics and Cosmology*. Cambridge: Cambridge University Press. A paperback edition of this book with a few small corrections was published by Cambridge University Press in 2013.

Longair, M. S. (2008a). *Galaxy Formation*, 2nd edition. Berlin: Springer.

Longair, M. S. (2008b). Maxwell and the science of colour, *Philosophical Transactions of the Royal Society of London*, **A366**, 1685–1696.

Longair, M. S. (2011a). *High Energy Astrophysics*, 3rd edition. Cambridge: Cambridge University Press.

Longair, M. S. (2011b). John Evan Baldwin, *Biographical Memoirs of Fellows of the Royal Society*, **57**, 3–23.

Longair, M. S. (2013). *Quantum Concepts in Physics*. Cambridge: Cambridge University Press.

Longair, M. S. (2015). '. . . a paper . . . I hold to be great guns': A commentary on Maxwell (1865) 'A dynamical theory of the electromagnetic field', *Philosophical Transactions of the Royal Society of London*, **A373**, 20140473.

Longair, M. S. and Macdonald, G. H. (1969). Observations of the structure of radio sources in the 3C catalogue, IV: Correlation diagrams and the evolution of radio sources, *Monthly Notices of the Royal Astronomical Society*, **145**, 309–325.

Longair, M. S. and Waldram, J. R. (2009). Alfred Brian Pippard, *Biographical Memoirs of Fellows of the Royal Society*, **55**, 201–220.

Longair, M. S. and Willmore, A. P. (1974). The X-ray spectrum of Cygnus-A, *Monthly Notices of the Royal Astronomical Society*, **168**, 479–490.

Longair, M. S., Best, P. N., and Röttgering, H. J. A. (1995). HST observations of three radio galaxies at redshift $z \sim 1$, *Monthly Notices of the Royal Astronomical Society*, **275**, L47–L51.

Longair, M. S., Ryle, M., and Scheuer, P. A. G. (1973). Models of extended radiosources, *Monthly Notices of the Royal Astronomical Society*, **164**, 243–270.

Lonzarich, G. G. (1984). Band structure and magnetic fluctuations in ferromagnetic or nearly ferromagnetic metals, *Journal of Magnetism and Magnetic Materials*, **45**, 43–53.

Lonzarich, G. G. (1986). The magnetic equation of state and heat capacity in weak itinerant ferromagnets, *Journal of Magnetism and Magnetic Materials*, **54–57**, 612–616.

Lonzarich, G. G. (1987). Quasiparticles and magnetic fluctuations in metals with large magnetic susceptibilities, *Journal of Magnetism and Magnetic Materials*, **70**, 445–450.

Lonzarich, G. G. (1988). Magnetic oscillations and the quasiparticle bands of heavy electron systems, *Journal of Magnetism and Magnetic Materials*, **76–77**, 1–10.

Lonzarich, G. G. (1994). Magnetic phase transitions at low temperatures, in *Spring College in Condensed Matter on Quantum Phases*. Trieste: International Centre for Theoretical Physics, SMR. 758-21.

Lonzarich, G. G. (1997). The magnetic electron, in *Electron*, ed. Springford, M., pp. 109–147. Cambridge: Cambridge University Press.

Loram, J. W., Mirza, K. A., Cooper, J. R., and Liang, W. Y. (1993). Electronic specific heat of $YBa_2Cu_3O_{6+x}$ from 1.8 to 300 K, *Physical Review Letters*, **71**, 1740–1743.

Lovell, A. C. B. (1975). Patrick Maynard Stuart Blackett, *Biographical Memoirs of the Fellows of the Royal Society*, **21**, 1–115.

Lovell, A. C. B. (1981). The mood of research: Optimism into doubt, *New Scientist*, **92**, 490–495. This article appeared in the edition of *New Scientist* for 19 November 1981.

Lovell, A. C. B. (1985). Martin Ryle, *Quarterly Journal of the Royal Astronomical Society*, **26**, 358–368.

Lovell, A. C. B. (1987). The emergence of radio astronomy in the UK after World War II, *Quarterly Journal of the Royal Astronomical Society*, **28**, 1–9.

Macdonald, G. H., Kenderdine, S., and Neville, A. C. (1968). Observations of the structure of radio sources in the 3C catalogue, I, *Monthly Notices of the Royal Astronomical Society*, **138**, 259–311.

Mack, J. E. (1947). *Semi-Popular Motion Picture Record of the Trinity Explosion*. MDDC221: United States Atomic Energy Commission, Washington, DC.

Mackay, C. D. (1969). Observations of the structure of radio source in the 3C catalogue, II, *Monthly Notices of the Royal Astronomical Society*, **145**, 31–65.

MacKay, D. J. C. (2008). *Sustainable Energy: Without the Hot Air*. Cambridge: UIT Publications.

MacKenzie, A. P., Haselwimmer, R. K. W., and Taylor, A. W. (1998a). Extremely strong dependence of superconductivity on disorder in Sr_2RuO_4, *Physical Review Letters*, **80**, 161–164.

MacKenzie, A. P., Julian, S. R., Diver, A. J. *et al.* (1998b). Quantum oscillation in the layered perovskite superconductor Sr_2RuO_4, *Physical Review Letters*, **80**, 161–164.

MacKenzie, A. P., Julian, S. R., Lonzarich, G. G. *et al.* (1993). Resistive upper critical field of $Tl_2Ba_2CuO_6$ at low temperatures and high magnetic fields, *Physical Review Letters*, **71**, 1238–1241.

Mann, L. F. (1964). David Keilin, 1887–1963, *Biographical Memoirs of Fellows of the Royal Society*, **10**, 183–205.

Marcus, J. A. (1947). The de Haas–van Alphen effect in a single crystal of zinc, *Physical Review*, **71**, 559.

Marsden, E. (1914). The passage of α-particles through hydrogen, *Philosophical Magazine (6)*, **27**, 824–830.

Mather, J. C., Cheng, E. S., Eplee, Jr., R. E. *et al.* (1990). A preliminary measurement of the cosmic microwave background spectrum by the Cosmic Background Explorer (COBE) satellite, *Astrophysical Journal Letters*, **354**, L37–L40.

Mathur, N. D., Grosche, F. M., Julian, S. R. *et al.* (1998). Magnetically mediated superconductivity in heavy fermion compounds, *Nature*, **394**, 39–43.

Matthews, T. A. and Sandage, A. R. (1963). Optical identification of 3C 48, 3C 196 and 3C 286 with stellar objects, *Astrophysical Journal*, **138**, 30–56.

Mattis, D. C. and Bardeen, J. (1958). Theory of anomalous skin effect in normal and superconducting metals, *Physical Review*, **111**, 412–417.

Maxwell, J. C. (1853). On the equilibrium of elastic solids, *Transactions of the Royal Society of Edinburgh*, **20**, 87–120. This paper was read to the Royal Society of Edinburgh on 18 February 1850 and includes confirmation of the theory from his experiments on strained glasses.

Maxwell, J. C. (1856a). Analogies in nature, in *The Scientific Letters and Papers of James Clerk Maxwell*, ed. Harman, P. M., volume 1, p. 244. Cambridge: Cambridge University Press. This volume was published in 1990.

Maxwell, J. C. (1856b). On Faraday's lines of force, *Transactions of the Cambridge Philosophical Society*, **10**, 155–188.

Maxwell, J. C. (1860a). Illustrations of the dynamical theory of gases, Part 1: On the motions and collisions of perfectly elastic spheres, *Philosophical Magazine (4)*, **20**, 21–33.

Also published in *The Scientific Papers of James Clerk Maxwell* (ed. Niven, W.D.), 1890. Volume 1, pp. 377–391. Cambridge: Cambridge University Press.

Maxwell, J. C. (1860b). Illustrations of the dynamical theory of gases, Part 2: On the process of diffusion of two or more kinds of moving particles among one another, *Philosophical Magazine (4)*, **19**, 19–32. Also published in *The Scientific Papers of James Clerk Maxwell* (ed. Niven, W.D.), 1890. Volume 1, pp. 392–405. Cambridge: Cambridge University Press.

Maxwell, J. C. (1860c). Illustrations of the dynamical theory of gases, Part 3: On the collision of perfectly elastic bodies of any form, *Philosophical Magazine (4)*, **20**, 33–37. Also published in *The Scientific Papers of James Clerk Maxwell* (ed. Niven, W.D.), 1890. Volume 1, pp. 405–409. Cambridge: Cambridge University Press.

Maxwell, J. C. (1861a). On physical lines of force, I: The theory of molecular vortices applied to magnetic phenomena, *Philosophical Magazine (4)*, **21**, 161–175. Also published in *The Scientific Papers of James Clerk Maxwell* (ed. Niven, W.D.), 1890. Volume 1, pp. 451–466. Cambridge: Cambridge University Press.

Maxwell, J. C. (1861b). On physical lines of force, II: The theory of molecular vortices applied to electric currents, *Philosophical Magazine (4)*, **21**, 281–291; plate, 338–348. Also published in *The Scientific Papers of James Clerk Maxwell* (ed. Niven, W.D.), 1890. Volume 1, pp. 467–488. Cambridge: Cambridge University Press.

Maxwell, J. C. (1862a). On physical lines of force, III: The theory of molecular vortices applied to statical electricity, *Philosophical Magazine (4)*, **23**, 12–24. Also published in *The Scientific Papers of James Clerk Maxwell* (ed. Niven, W.D.), 1890. Volume 1, pp. 489–502. Cambridge: Cambridge University Press.

Maxwell, J. C. (1862b). On physical lines of force, IV: The theory of molecular vortices applied to the action of magnetism on polarised light, *Philosophical Magazine (4)*, **23**, 85–95. Also published in *The Scientific Papers of James Clerk Maxwell* (ed. Niven, W.D.), 1890. Volume 1, pp. 502–512. Cambridge: Cambridge University Press.

Maxwell, J. C. (1865). A dynamical theory of the electromagnetic field, *Philosophical Transactions of the Royal Society of London*, **155**, 459–512. Also published in *The Scientific Papers of James Clerk Maxwell* (ed. Niven, W.D.), 1890. Volume 1, pp. 526–597. Cambridge: Cambridge University Press.

Maxwell, J. C. (1867). On the dynamical theory of gases, *Philosophical Transactions of the Royal Society of London*, **157**, 49–88. Also published in *The Scientific Papers of James Clerk Maxwell* (ed. Niven, W.D.), 1890. Volume 2, pp. 26–78. Cambridge: Cambridge University Press.

Maxwell, J. C. (1870). *Theory of Heat*. London: Longmans and Co.

Maxwell, J. C. (1873). *A Treatise on Electricity and Magnetism* 2 volumes. Oxford: Clarendon Press. The second edition was published postmously in 1881 and contains a few additional notes.

Maxwell, J. C. (1874). Letter to Robert Dundas Cay, 12th May 1874, in *The Scientific Letters and Paper of James Clerk Maxwell*, ed. Harman, P. M., volume 3, p. 70. Cambridge: Cambridge University Press.

Maxwell, J. C. (1878). Ether, in *Encyclopaedia Britannica, 9th edition*, volume 8, pp. 568–572. Edinburgh: A. & C. Black.

Maxwell, J. C. (1879). *Electrical Researches of the Hon. Henry Cavendish*. Cambridge: Cambridge University Press.

Maxwell, J. C. (1880). On a possible mode of detecting a motion of the solar system through the luminiferous ether, *Proceedings of the Royal Society of London*, **30**, 109–110.

Maxwell, J. C. (1890). Introductory lecture on experimental physics, *Scientific Papers*, **2**, 241–255. This is Maxwell's inaugural lecture as first Cavendish Professor of Experimental Physics, delivered in October 1871. The quotation is on page 244.

Maxwell, J. C., Stewart, B., and Jenkin, F. (1863). *Description of an Experimental Measurement of Electrical Resistance, Made at King's College*, British Association Reports, pp. 140–158. London: British Association Reports.

Mayer, A. M. (1878). On the morphological laws of the configurations formed by magnets floating vertically and subjected to the attraction of a superposed magnet; with notes on some of the phenomena in molecular structure which these experiments may serve to explain and illustrate, *American Journal of Science*, **16**, 247–256.

McCormmach, R. (2012). *Weighing the World: The Reverend John Michell of Thornhill*. Berlin: Springer.

McMullan, G. J. and Lonzarich, G. G. (1997). The normal states of magnetic itinerant electron systems, in *Magnetism in Metals: Matematish-Fysiske Meddelelser 45*, eds McMorrow, D. F., Jensen, J., and Rønnow, H. M., pp. 247–257. Copenhagen: The Royal Danish Academy of Sciences and Letters.

Meadows, D. H., Meadows, D. L., Randers, J., and Behrens III, W. W. (1972). *The Limits to Growth: A Report for the Club of Rome's Project on the Predicament of Mankind*. New York: Universe Books.

Meissner, W. and Ochsenfeld, R. (1933). Ein neuer effekt bei eintritt der supraleitfhigkeit, *Naturwissenschaften*, **21**, 787–788.

Meitner, L. and Frisch, O. R. (1939). Disintegration of uranium by neutrons: A new type of nuclear reaction, *Nature*, **143**, 239–240.

Mendeleyev, D. I. (1869). On the relationship of the properties of the elements to their atomic weights, *Zhurnal Russkoe Fiziko-Khimicheskoe Obshchestvo*, **1**, 60–77.

Menter, J. W. (1956). The direct study by electron microscopy of crystal lattices and their imperfections, *Proceedings of the Royal Society of London*, **A236**, 119–135.

Michelson, A. A. (1903). *Light Waves and Their Uses*. Chicago: University of Chicago Press. This quotation appears on page 24.

Michelson, A. A. (1927). *Studies in Optics*. Chicago: University of Chicago Press.

Michelson, A. A. and Morley, E. W. (1887). On the relative motion of the Earth and the luminiferous ether, *American Journal of Science*, **34**, 333–345.

Millikan, R. A. (1913). On the elementary electric charge and the Avogadro constant, *Physical Review*, **2**, 109–143.

Millikan, R. A. (1916a). A direct photoelectric determination of Planck's h, *Physical Review*, **7**, 355–388.

Millikan, R. A. (1916b). Einstein's photoelectric equation and contact electromotive force, *Physical Review*, **7**, 18–32.

Mills, B. Y. and Slee, O. B. (1957). A preliminary survey of radio sources in a limited region of the sky at a wavelength of 3.5 m, *Australian Journal of Physics*, **10**, 162–194.

Minkowski, R. (1960). A new distant cluster of galaxies, *Astrophysical Journal*, **132**, 908–908.

Mitton, S. and Ryle, M. (1969). High resolution observations of Cygnus A at 2. 7 GHz and 5 GHz, *Monthly Notices of the Royal Astronomical Society*, **146**, 221–233.

Monthoux, P. and Lonzarich, G. G. (1999). p-wave and d-wave superconductivity in quasi-two-dimensional metals, *Physical Review B*, **59**, 14598–14605.

Moseley, H. G. J. (1913). The high frequency spectra of the elements, *Philosophical Magazine (6)*, **26**, 1024–1034.

Moseley, H. G. J. (1914). The high frequency spectra of the elements, Part II, *Philosophical Magazine (6)*, **27**, 703–713.

Mott, N. F. (1928). The solution of the wave equation for the scattering of particles by a Coulombian centre of force, *Proceedings of the Royal Society of London*, **A118**, 542–549.

Mott, N. F. (1930). *An Outline of Wave Mechanics*. Cambridge: Cambridge University Press.

Mott, N. F. (1949). The basis of electron theory of metals, with special reference to the transition metals, *Proceedings of the Physical Society of London*, **A62**, 416–422.

Mott, N. F. (1966). *The Cavendish Laboratory: The Need for a New Building*. Cambridge: Cavendish Laboratory.

Mott, N. F. (1984). Theory and experiment at the Cavendish circa 1932, in *Cambridge Physics in the Thirties*, ed. Hendry, J., pp. 125–132. Bristol: Adam Hilger.

Mott, N. F. (1986). *Sir Nevill Mott: A Life in Science*. London: Taylor and Francis.

Mott, N. F. (1990). Manchester and Cambridge, in *Selections and Reflections: The Legacy of Sir Lawence Bragg*, eds Thomas, J. M. and Phillips, D., pp. 96–97. Northwood, UK: Science Reviews.

Mott, N. F. (1992). Electrons in glass: 1977 Nobel Prize lecture, in *Nobel Lectures, Physics 1971–1980*, ed. Lundqvist, S., pp. 403–413. Singapore: World Scientific.

Mott, N. F. (1996). Dislocation in metal crystals, in *The Life and Legacy of G. I. Taylor*, ed. Batchelor, G., pp. 150–152. Cambridge: Cambridge University Press.

Mott, N. F. and Davis, E. A. (1971). *Electronic Processes in Non-Crystalline Materials*. Oxford: Oxford University Press.

Mott, N. F. and Davis, E. A. (1979). *Electronic Processes in Non-Crystalline Materials*, 2nd edition. Oxford: Oxford University Press.

Mott, N. F. and Gurney, R. W. (1940). *Electronic Processes in Ionic Crystals*. Oxford: Clarendon Press.

Mott, N. F. and Jones, H. (1936). *The Theory of the Properties of Metals and Alloys*. Oxford: Clarendon Press.

Mott, N. F. and Massey, H. S. W. (1934). *Theory of Atomic Collisions*. Oxford: Clarendon Press.

Mott, N. F., Pepper, M., Pollitt, S., Wallis, R. H., and Adkins, C. J. (1975). The Anderson transition, *Proceedings of the Royal Society of London*, **A345**, 169–205.

Mulvey, T. (1994). Vernon Ellis Cosslett, *Biographical Memoirs of the Fellows of the Royal Society*, **40**, 63–84.

Nabarro, F. R. N. and Argon, A. S. (1995). Egon Orowan, *Biographical Memoirs of Fellows of the Royal Society*, **41**, 317–340.

Nagaoka, H. (1904a). Kinematics of a system of particles illustrating the line and band spectrum and the phenomenon of radioactivity, *Philosophical Magazine (6)*, **7**, 445–455.

Nagaoka, H. (1904b). Kinematics of a system of particles illustrating the line and band spectrum and the phenomenon of radioactivity, *Nature*, **69**, 392–393.

Nakazato, K. and Ahmed, H. (1995). The multiple-tunnel junction and its application to single-electron memory and logic circuits, *Japanese Journal of Applied Physics*, **34**, 700–706.

Nakazato, K., Blaikie, R. J., Cleaver, J. R. A., and Ahmed, H. (1993). Single-electron memory, *Electronic Letters*, **29**, 384–385.

Navarro, J. (2005). J. J. Thomson on the nature of matter: Corpusles and the continuum, *Centaurus*, **47**, 259–282.

Neary, G. J. (1940). The β-ray spectrum of radium-E, *Proceedings of the Royal Society of London*, **A175**, 71–87.

Netterfield, C. B., Jarosik, N., Page, L., Wilkinson, D., and Wollack, E. (1995). The anisotropy in the cosmic microwave background at degree angular scales, *Astrophysical Journal Letters*, **445**, L69–L72.

Newall, H. F. (1910). 1885–1894, in *A History of the Cavendish Laboratory, 1871–1910*, pp. 102–158. London: Longmans, Green and Co.

Newton, I. (1687). *Philosophiæ Naturalis Principia Mathematica*. London: Royal Society of London.

Newton, I. (1704). *Optiks*. London: Royal Society of London.

Newton, (1736). *Method of Fluxions*. London: Henry Woodfall. Newton's book was completed in 1671, but published posthumously.

Nichols, H. W. and Schelleng, J. C. (1925a). Propagation of electric waves over the Earth, *Bell Systems Technical Journal*, **4**, 215–234.

Nichols, H. W. and Schelleng, J. C. (1925b). The propagation of radio waves over the Earth, *Nature*, **115**, 334.

Nicholson, J. W. (1911). A structural theory of chemical elements, *Philosophical Magazine (6)*, **22**, 864–889.

Nicholson, J. W. (1912). The spectrum of nebulium, *Monthly Notices of the Royal Astronomical Society*, **72**, 49–64.

Niedrig, H. (1996). The early history of electron microscopy in Germany, in *Advances in Imaging and Electron Physics, Vol. 96*, ed. Hawkes, P. W., pp. 131–148. London: Academic Press.

Occhialini, G. P. S. (1975). Memorial meeting for Lord Blackett, O.M., C.H., F.R.S. at the Royal Society on 31 October 1974, *Notes and Records of the Royal Society of London*, **29**, 144–146.

Ohm, G. S. (1826a). Versuch einer theorie der durch galvanishe kraft hervorgebrachten electroskopische erscheinungen, *Annalen der Physik und Chemie (2)*, **6**, 459–469.

Ohm, G. S. (1826b). Versuch einer theorie der durch galvanishe kraft hervorgebrachten electroskopische erscheinungen, *Annalen der Physik und Chemie (2)*, **7**, 45–54.

Ohm, G. S. (1827). *Die Galvanische Kette: Mathematisch Bearbeitet*. Berlin: Riemann.

Oliphant, M. L. E. (1984). Working with Rutherford, in *Cambridge Physics in the Thirties*, ed. Hendry, J., pp. 184–188. Bristol: Adam Hilger.

Oliphant, M. L. E. and Rutherford, E. (1933). Experiments on the transmutation of elements by protons, *Proceedings of the Royal Society of London*, **A141**, 259–281.

Oliphant, M. L. E., Harteck, P., and Rutherford, E. (1934). Transmutation effects observed with heavy hydrogen, *Proceedings of the Royal Society of London*, **A144**, 692–703.

Oliphant, M. L. E., Kinsey, B. B., and Rutherford, E. (1933). The transmutation of lithium by protons and by ions of the heavy isotope of hydrogen, *Proceedings of the Royal Society of London*, **A141**, 722–733.

Onsager, L. (1949). Remark following a paper by C. J. Gorter on the two fluid model of liquid helium, *Nuovo Cimento Supplement 2*, **6**, 249–450.

Onsager, L. (1952). Interpretation of the de Haas–van Alphen effect, *Philosophical Magazine (7)*, **43**, 1006–1008.

Ørsted, H. C. (1820). *Experimenta circum Effectum Conflictus Electrici in Acum Magneticam*. Copenhagen: pamphlet.

Orowan, E. (1934). Zur kristallplastizitat, I–III, *Zeitschrift für Physik*, **89**, 605–659.

Orowan, E. (1943). The calculation of roll pressure in hot and cold flat rolling, *Proceedings of the Institution of Mechanical Engineering*, **150**, 140–167.

Orowan, E., Nye, J. F., and Cairns, W. J. (1944). *Notch brittleness and ductile fracture in metals*, Theoretical Research Report, Ministry of Supply, Armament Research Department, England, **16/45**.

Owen, F. N. and Ledlow, M. J. (1994). The FRI/Il break and the bivariate luminosity function in Abell clusters of galaxies, in *The Physics of Active Galaxies*, eds Bicknell, G. V., Dopita, M. A., and Quinn, P. J., volume 54 of Astronomical Society of the Pacific Conference Series, pp. 319–323. San Francisco: Astronomical Society of the Pacific Publications.

Pais, A. (1982). *'Subtle is the Lord . . . : The Science and the Life of Albert Einstein*. Oxford: Clarendon Press.

Pais, A. (1985). *Inward Bound*. Oxford: Clarendon Press.

Pauli, W. (1925). Über den zussamenhang des abschlusses der elektronengruppen im atom mit der komplexstrucktur der spektren, *Zeitschrift für Physik*, **31**, 765–785.

Pauling, L., Corey, R. B., and Branson, H. R. (1951). The structure of proteins: Two hydrogen-bonded helical configurations of the polypeptide chain, *Proceedings of the National Academy of Sciences of the United States of America*, **37**, 205–211.

Payne, M. C., Levi, A. F. J., Phillips, W. A., Inkson, J. C., and Adkins, C. J. (1984). Phonon structure of amorphous germanium by inelastic electron tunnelling spectroscopy, *Journal of Physics C: Solid State Physics*, **17**, 1643–1653.

Peebles, P. J. E. (1982). Large-scale background temperature fluctuations due to scale-invariant primaeval perturbations, *Astrophysical Journal*, **263**, L1–L5.

Peebles, P. J. E., Page, Jr, L. A., and Partridge, R. B. (2009). *Finding the Big Bang.* Cambridge: Cambridge University Press.

Pendry, J. B. (1969a). The application of pseudopotentials to low energy electron diffraction. PhD dissertation, University of Cambridge.

Pendry, J. B. (1969b). The application of pseudopotentials to low-energy electron diffraction, I: Calculation of the potential and 'inner potential', *Journal of Physics C: Solid State Physics*, **2**, 1215–1221.

Pendry, J. B. (1969c). The application of pseudopotentials to low-energy electron diffraction, II: Calculation of the reflected intensities, *Journal of Physics C: Solid State Physics*, **2**, 2273–2282.

Pendry, J. B. (1969d). The application of pseudopotentials to low-energy electron diffraction, III: The simplifying effect of inelastic scattering, *Journal of Physics C: Solid State Physics*, **2**, 2283–2289.

Penzias, A. A. and Wilson, R. W. (1965). A measurement of excess antenna temperature at 4080 MHz, *Astrophysical Journal*, **142**, 419–421.

Pepper, M. (1998). Working with Nevill Mott, in *Nevill Mott: Reminiscences and Appreciations*, ed. Davis, E. A., pp. 211–218. London: Taylor and Francis.

Pepper, M., Pollitt, S., and Adkins, C. J. (1974a). Anderson localisation of holes in a Si inversion layer, *Physics Letters A*, **48**, 113–114.

Pepper, M., Pollitt, S., and Adkins, C. J. (1974b). The spatial extent of localized state wavefunctions in silicon inversion layers, *Journal of Physics C: Solid State Physics*, **7**, L273–L277.

Pepper, M., Pollitt, S., Adkins, C. J., and Stradling, R. (1975). Anderson localization in silicon inversion layers, *CRC Critical Reviews in Solid State Physics*, **5**, 375–384.

Perrin, J. B. (1896). New experiments on the kathode rays, *Nature*, **53**, 298–299.

Perrin, J. B. (1901). Les hypothèses moléculaires, *Revue Scientifique*, **15**, 449–461.

Perutz, M. (1970). Bragg, protein crystallography and the Cavendish Laboratory, *Acta Crystallographica*, **A26**, 183–185.

Perutz, M. (1993). Co-chairman's remarks: Before the double helix, *Gene*, **135**, 9–13.

Perutz, M. (1997). *Science is Not a Quiet Life: Unravelling the Atomic Mechanism of Haemoglobin.* London: Imperial College Press.

Perutz, M., Rossmann, M. G., Cullis, A. F. *et al.* (1960). Structure of haemoglobin: A three-dimensional Fourier synthesis at 5.5 Å resolution obtained by X-ray analysis, *Nature*, **185**, 416–422.

Peshkov, V. (1944). 'Second sound' in helium II, *Journal of Physics (Moscow)*, **8**, 381.

Pfleiderer, C., McMullan, G. J., Julian, S. R., and Lonzarich, G. G. (1997). Magnetic quantum phase transition in MnSi under hydrostatic pressures, *Physical Review B*, **55**, 8330–8338.

Phillips, D. C. (1979). William Lawrence Bragg, *Biographical Memoirs of Fellows of the Royal Society*, **25**, 74–143.

Phillips, J. C. (1958). Energy-band interpolation scheme based on a pseudopotential, *Physical Review*, **112**, 685–695.

Phillips, W. A. (1969). Ultrasonic attenuation at 500 MHz by superconducting tin, *Proceedings of the Royal Society of London*, **A309**, 259–280.

Phillips, W. A. (1972). Tunneling states in amorphous solids, *Journal of Low Temperature Physics*, **7**, 351–360.

Phillips, W. A. (1987). Two-level states in glasses, *Reports of Progress in Physics*, **50**, 1657–1708.

Phillips, W. A., Buchenau, U., Nücker, N., Dianoux, A.-J., and Petry, W. (1989). Dynamics of glassy and liquid selenium, *Physical Review Letters*, **63**, 2381–2384.

Pickering, E. C. (1873). *Elements of Physical Manipulation, Vol. 1*. New York: Hurd and Houghton.

Pickering, E. C. (1876). *Elements of Physical Manipulation, Vol. 2*. New York: Hurd and Houghton.

Pilkington, J. D. H. and Scott, J. F. (1965). A survey of radio sources between declinations 20° and 40°, *Memoirs of the Royal Astronomical Society*, **69**, 183–224.

Pilkington, J. D. H., Hewish, A., Bell, S. J., and Cole, T. W. (1968). Observations of some further pulsed radio sources, *Nature*, **218**, 126–129.

Pippard, A. B. (1953). An experimental and theoretical study of the relation between magnetic field and current in a superconductor, *Proceedings of the Royal Society of London*, **A216**, 547–568.

Pippard, A. B. (1957). Experimental determination of the Fermi surface in copper, *Philosophical Transactions of the Royal Society of London*, **A250**, 325–357.

Pippard, A. B. (1960). Theory of ultrasonic attenuation in metals, and magneto-acoustic oscillations, *Proceedings of the Royal Society of London*, **A257**, 165–193.

Pippard, A. B. (1962). Quantisation of coupled electron orbits in metals, *Proceedings of the Royal Society of London*, **A270**, 1–13.

Pippard, A. B. (1963). Commentary on a conjecture of Shoenberg's concerning the de Haas–van Alphen effect, *Proceedings of the Royal Society of London*, **A272**, 192–206.

Pippard, A. B. (1964). Quantization of coupled orbits in metals, II: The two-dimensional network, with special reference to the properties of zinc, *Philosophical Transactions of the Royal Society of London*, **A256**, 317–355.

Pippard, A. B. (1965). *The Dynamics of Conduction Electrons*. New York: Gordon and Breach.

Pippard, A. B. (1971). *Reconciling Physics with Reality*. Cambridge: Cambridge University Press.

Pippard, A. B. (1974). The move to West Cambridge, in *A Hundred Years of Cambridge Physics*, ed. Moralee, D., pp. 42–44. Cambridge: Cavendish Laboratory. The second and third editions were published in 1980 (ed. Parker, J.) and 1995 (eds. Jacques, G. and Bache, I.) with the title *A Hundred Years and More of Cambridge Physics*.

Pippard, A. B. (1978a). *The Physics of Vibration, Vol. 1: The Simple Classical Vibrator*. Cambridge: Cambridge University Press.

Pippard, A. B. (1978b). *The Physics of Vibration, Vol. 2: The Simple Vibrator in Quantum Mechanics*. Cambridge: Cambridge University Press.

Pippard, A. B. (1985). *Response and Stability: An Introduction to the Physical Theory*. Cambridge: Cambridge University Press.

Pippard, A. B. (1989). *Magnetoresistance in Metals*. Cambridge: Cambridge University Press.

Pippard, A. B. (1990). Bragg: The Cavendish Professor, in *Selections and Reflections: The Legacy of Sir Lawence Bragg*, eds Thomas, J. M. and Phillips, D., pp. 97–100. Northwood, UK: Science Reviews.

Pippard, A. B. (1995). Physics in 1900, in *Twentieth Century Physics*, eds Brown, L. M., Pais, A., and Pippard, A. B., pp. 1–41. Bristol and Philadelphia: Institute of Physics Publishing, and New York: American Institute of Physics Press.

Pippard, A. B. (1998). Sir Nevill Francis Mott C. H., *Biographical Memoirs of Fellows of the Royal Society*, **44**, 315–328. Mott's complete bibliography was originally available on microfiche. This has now been scanned and is available in the Royal Society's online version of the Biographical Memoir.

Pippard, A. B., Shepherd, J. G., and Tindall, D. A. (1971). Resistance of normal–superconducting interfaces, *Proceedings of the Royal Society of London*, **A324**, 17–35.

Planck, M. (1900). Zur theorie des gesetzes der energieverteilung im normalspektrum, *Verhandlungen der Deutschen Physikalischen Gesellschaft*, **2**, 237–245. Also published in Planck's collected papers: *Physikalische Abhandlungen und Vortrage*, **1**, pp. 698–706, Braunschweig: Vieweg. English translation: Hermann, A. (1971). *The Genesis of Quantum Theory (1899–1913)*, p. 10. Cambridge, MA: MIT Press.

Planck, M. (1902). Über die natur des weisen lichtes, *Annalen der Physik (4)*, **7**, 390–400. Also published in Planck's collected papers: *Physikalische Abhandlungen und Vortrage*, **1**, pp. 763–773, Braunschweig: Vieweg.

Poisson, S.-D. (1812). *Mémoire sur la Distribution de l'Électricité à la Surface des Corps Conducteurs*. Paris: Académie des Sciences de l'Institut de France.

Polanyi, M. (1934). Über eine art gitterstörung, die einen kristall plastisch machen könnte, *Zeitschrift für Physik*, **89**, 660–664.

Pollitt, S., Pepper, M., and Adkins, C. J. (1976). The Anderson transition in silicon inversion layers, *Surface Science*, **58**, 79–88.

Porch, A., Cheah, H. M., and Waldram, J. R. (1990). Microwave response of aligned $YBa_2Cu_3O_{7-\delta}$ powders, *Physica B: Condensed Matter*, **165**, 1197–1198.

Porch, A., Cooper, J. R., Zheng, D. N. *et al.* (1993). Temperature dependent magnetic penetration depth of Co and Zn doped $YBa_2Cu_3O_7$ obtained from the AC susceptibility of magnetically aligned powders, *Physica C: Superconductivity*, **214**, 350–358.

Proudman, J. (1916). On the motion of solids in a liquid possessing vorticity, *Proceeedings of the Royal Society of London*, **A92**, 408–424.

Quinn, T. (2005). Sir Alan Hugh Cook, *Biographical Memoirs of Fellows of the Royal Society*, **51**, 87–100.

Rao, A., Chow, P. C. Y., Gélinas, S. *et al.* (2013). The role of spin in the kinetic control of recombination in organic photovoltaics, *Nature*, **500**, 435–439.

Ratcliffe, J. A. (1966). Edward Victor Appleton, *Biographical Memoirs of Fellows of the Royal Society*, **12**, 1–21.

Ratcliffe, J. A. and Huxley, L. G. H. (1949). A survey of ionospheric cross modulation, *Proceedings of the Institution of Electrical Engineers*, **96**, 433–440.

Ratcliffe, J. A., Booker, H. G., and Shinn, D. H. (1950). Diffraction from an irregular screen with applications to ionospheric problems, *Philosophical Transactions of the Royal Society of London*, **A242**, 579–607.

Rayleigh, J. W. (1877). *The Theory of Sound, Vol. 1*. London: Macmillan.

Rayleigh, J. W. (1878). *The Theory of Sound, Vol. 2*. London: Macmillan.

Rayleigh, J. W. (1882a). Comparison of methods for the determination of resistances in absolute measure, *Philosophical Magazine (5)*, **14**, 329–346. Rayleigh's paper is also contained in *Scientific Papers by John William Strutt, Baron Rayleigh, 1881–1887*, Vol. 2. Cambridge: Cambridge University Press.

Rayleigh, J. W. (1882b). Experiments to determine the value of the British Association unit of resistance in absolute measure, *Philosophical Transactions of the Royal Society of London*, **173**, 661–697. Rayleigh's paper is also contained in *Scientific Papers by John William Strutt, Baron Rayleigh, 1881–1887*, Vol. 2. Cambridge: Cambridge University Press.

Rayleigh, J. W. (1912). *Scientific Papers by John William Strutt, Baron Rayleigh, 1902–1910*, volume 5. Cambridge: Cambridge University Press. Relevant excerpts from Rayleigh's Nobel Prize lecture are on pages 212–215.

Rayleigh, J. W. and Ramsay, W. (1895). Argon, a new constituent of the atmosphere, *Philosophical Transactions of the Royal Society of London*, **A186**, 187–241. Rayleigh and Ramsay's paper is also contained in *Scientific Papers by John William Strutt, Baron Rayleigh, 1892–1901*, Vol. 4. Cambridge: Cambridge University Press.

Rayleigh, J. W. and Schuster, A. (1881). On the determination of the ohm in absolute measure, *Proceedings of the Royal Society of London*, **32**, 104–141. Rayleigh and Schuster's paper is also contained in *Scientific Papers by John William Strutt, Baron Rayleigh, 1881–1887*, Vol. 2. Cambridge: Cambridge University Press.

Rayleigh, J. W. and Sidgwick, E. M. (1882). On the specific resistance of mercury, *Philosophical Transactions of the Royal Society of London*, **174**, 173–185. Rayleigh and Sidgwick's paper is also contained in *Scientific Papers by John William Strutt, Baron Rayleigh, 1881–1887*, Vol. 2. Cambridge: Cambridge University Press.

Rayleigh, J. W. and Sidgwick, E. M. (1883). Experiments, by the method of Lorentz, for the further determination in absolute value of the British Association unit of resistance, with an appendix on the determination of the pitch of a standard tuning fork, *Philosophical Transactions of the Royal Society of London*, **174**, 295–322. Rayleigh and Sidgwick's paper is also contained in *Scientific Papers by John William Strutt, Baron Rayleigh, 1881–1887*, Vol. 2. Cambridge: Cambridge University Press.

Rayleigh, J. W. and Sidgwick, E. M. (1884). On the electro-chemical equivalent of silver, and on the absolute electromotive force of Clark cells, *Philosophical Transactions of the Royal Society of London*, **175**, 411–460. Rayleigh and Sidgwick's paper is also contained in *Scientific Papers by John William Strutt, Baron Rayleigh, 1881–1887*, Vol. 2. Cambridge: Cambridge University Press.

Rayleigh, R. J. S. (1924). *John William Strutt, Third Baron Rayleigh*. London: Edward Arnold and Co.

Reber, G. (1940). Notes: Cosmic static, *Astrophysical Journal*, **91**, 621–624.

Reber, G. (1944). Cosmic static, *Astrophysical Journal*, **100**, 279–287.

Rees, M. J. (1971). New interpretation of extragalactic radio sources, *Nature*, **229**, 312–317.

Rees, N. (1990). A deep 38-MHz radio survey of the area delta greater than +60 degrees, *Monthly Notices of the Royal Astronomical Society*, **244**, 233–246.

Reines, F. and Cowan, C. L. (1956). The neutrino, *Nature*, **178**, 446–449.

Reuter, G. E. H. and Sondheimer, E. H. (1956). The theory of the anomalous skin effect in metals, *Proceeedings of the Royal Society of London*, **A195**, 336–364.

Rhodes, R. (1986). *The Making of the Atomic Bomb*. New York: Simon and Schuster.

Richardson, O. (1901). On the negative radiation from hot platinum, *Proceedings of the Cambridge Philosophical Society*, **11**, 286–295.

Richardson, O. and Brown, K. (1908). Kinetic energy of negative electrons emitted by hot bodies, *Philosophical Magazine (6)*, **16**, 353–376.

Riley, J. M. and Pooley, G. G. (1975). Observations of 31 extragalactic radio sources with the Cambridge 5-km telescope at 5 GHz, *Memoirs of the Royal Astronomical Society*, **80**, 105–137.

Riley, J. M. and Pooley, G. G. (1978). The radio structure of 3C 123 at 2.7 adn 15 GHz, *Monthly Notices of the Royal Astronomical Society*, pp. 245–255.

Rindler, W. (2001). *Relativity: Special, General and Cosmological*. Oxford: Oxford University Press.

Rodenburg, J. M. (1989). The phase problem, microdiffraction and wavelength-limited resolution, *Ultramicrosopy*, **27**, 413–422.

Roll, P. G. and Wilkinson, D. T. (1966). Cosmic background radiation at 3.2 cm: Support for cosmic black-body radiation, *Physical Review Letters*, **16**, 405–407.

Röntgen, W. C. (1895). Über eine neue art von strahlen. (On a new type of ray. Preliminary communication), *Erste Mittheilung: Sitzungsberichte der Physikalisch-Medizinische Gesellschaft, Würzburg*, p. 137. Röntgen's paper was published in December 1895. It was also published in English in 1896, Nature, **53**, 274.

Röntgen, W. R. (1901). Wilhelm Conrad Röntgen: Biographical. Nobel Media AB 2014. http://www.nobelprize.org/nobel_prizes/physics/laureates/1901/rontgen-bio.html.

Rowley, S. E., Spalek, L. J., Smith, R. P. *et al.* (2014). Ferroelectric quantum criticality, *Nature Physics*, **10**, 367–372.

Roy, M. (2004). *The Weathermen of Ben Nevis 1883–1904*. Edinburgh: Royal Meteorological Society.

Rubinowicz, A. (1918). Bohrsche frequenzbedingnug und erhaltung des impulsmomentes, I: Tiel, *Physikalische Zeitschrift*, **19**, 441–445.

Rutherford, E. (1899). Uranium radiation and the electrical conduction produced by it, *Philosophical Magazine (5)*, **47**, 109–163.

Rutherford, E. (1903). The electric and magnetic deviation of the easily absorbed rays from radium, *Philosophical Magazine (5)*, **5**, 177–187.

Rutherford, E. (1905a). Bakerian Lecture: The succession of changes in radioactive bodies, *Philosophical Transactions of the Royal Society of London*, **A204**, 169–219.

Rutherford, E. (1905b). The radium: The cause of the Earth's heat, *Harper's Magazine*, pp. 390–396.

Rutherford, E. (1907). Some cosmical aspects of radioactivity, *Journal of the Royal Society of Canada*, **1**, 145–165.

Rutherford, E. (1911). The scattering of α and β particles by matter and the structure of the atom, *Philosophical Magazine (6)*, **21**, 669–688.

Rutherford, E. (1913). The structure of the atom, *Nature*, **92**, 423.

Rutherford, E. (1919a). Collisions of α particles with light atoms, I: Hydrogen, *Philosophical Magazine (6)*, **37**, 537–561.

Rutherford, E. (1919b). Collisions of α particles with light atoms, II: Velocity of the hydrogen atom, *Philosophical Magazine (6)*, **37**, 562–571.

Rutherford, E. (1919c). Collisions of α particles with light atoms, III: Nitrogen and oxygen atoms, *Philosophical Magazine (6)*, **37**, 571–580.

Rutherford, E. (1919d). Collisions of α particles with light atoms, IV: An anomalous effect in nitrogen, *Philosophical Magazine (6)*, **37**, 581–587.

Rutherford, E. (1920). Bakerian Lecture: Nuclear constitution of atoms, *Proceedings of the Royal Society of London*, **A117**, 300–316.

Rutherford, E. (1928). Address of the President, Sir Ernest Rutherford, O.M., at the Anniversary Meeting, November 30, 1927, *Proceedings of the Royal Society of London*, **A117**, 300–316.

Rutherford, E. and Andrade, E. N. da C. (1913). The reflection of γ-rays from crystals, *Nature*, **92**, 267.

Rutherford, E. and Chadwick, J. (1921). The artificial disintegration of light elements, *Philosophical Magazine (6)*, **42**, 809–825.

Rutherford, E. and Chadwick, J. (1924a). The bombardment of elements by α-particles, *Nature*, **113**, 457.

Rutherford, E. and Chadwick, J. (1924b). Further experiments on the artificial disintegration of elements, *Proceedings of the Physical Society of London*, **36**, 417–422.

Rutherford, E. and Geiger, H. (1908a). The charge and nature of the alpha particle, *Proceeedings of the Royal Society of London*, **A81**, 162–173.

Rutherford, E. and Geiger, H. (1908b). An electrical method of counting the number of α-particles from radio-active substances, *Proceeedings of the Royal Society of London*, **A81**, 141–161.

Rutherford, E. and Robinson, H. R. (1913). The analysis of the β rays from radium B and radium C, *Philosophical Magazine (6)*, **26**, 717–729.

Rutherford, E. and Robinson, H. R. (1914). The mass and velocities of the alpha particles from radioactive substances, *Philosophical Magazine, (6)*, **28**, 552–572.

Rutherford, E. and Royds, T. (1909). The nature of the α particle from radioactive substances, *Philosophical Magazine (6)*, **17**, 281–286.

Rutherford, E. and Soddy, F. (1902a). The cause and nature of radioactivity, Part I, *Philosophical Magazine (6)*, **4**, 370–396.

Rutherford, E. and Soddy, F. (1902b). The cause and nature of radioactivity, Part II, *Philosophical Magazine (6)*, **4**, 569–585.

Rutherford, E. and Soddy, F. (1903). Radioactive change, *Philosophical Magazine (6)*, **5**, 576–591.

Rutherford, E., Chadwick, J., and Ellis, C. (1930a). *Radiations from Radioactive Substances*. Cambridge: Cambridge University Press.

Rutherford, E., Robinson, H. R., and Rawlinson, W. F. (1914). Spectrum of the β-rays excited by γ rays, *Philosophical Magazine (6)*, **28**, 281–286.

Rutherford, E., Ward, F. A. B., and Wynn-Williams, C. E. (1930b). A new method of analysis of groups of α-rays. (1) α-rays from radium C, thorium C and actinium C, *Proceedings of the Royal Society of London*, **A129**, 211–234.

Ryle, M. (1952). A new radio interferometer and its application to the observation of weak radio stars, *Proceedings of the Royal Society of London*, **A211**, 351–378.

Ryle, M. (1955). Radio stars and their cosmological significance, *The Observatory*, **75**, 137–147.

Ryle, M. (1972). The 5-km radio telescope at Cambridge, *Nature*, **239**, 435–438.

Ryle, M. (1975). Radio telescopes of large resolving power, *Reviews of Modern Physics*, **47**, 557–566.

Ryle, M. (1982). Wind power, in *Research at the Cavendish Laboratory*, ed. Cook, A. H., pp. 8–9. Cambridge: Cavendish Laboratory.

Ryle, M. and Graham-Smith, F. G. (1948). A new intense source of radio-frequency radiation in the constellation of Cassiopeia, *Nature*, **162**, 462–463.

Ryle, M. and Hewish, A. (1960). The synthesis of large radio telescopes, *Monthly Notices of the Royal Astronomical Society*, **120**, 220–230.

Ryle, M. and Neville, A. C. (1962). A radio survey of the north polar region with a 4.5 minute of arc pencil-beam system, *Monthly Notices of the Royal Astronomical Society*, **125**, 39–56.

Ryle, M. and Pooley, G. G. (1968). The extension of the number–flux density relation for radio sources to very small flux densities, *Monthly Notices of the Royal Astronomical Society*, **139**, 515–528.

Ryle, M. and Sandage, A. (1964). The optical identification of three new radio objects of the 3C 48 class, *Atrophysical Journal*, **139**, 419–421.

Ryle, M. and Vonberg, D. D. (1946). Solar radiation on 175 Mc/s, *Nature*, **158**, 339–340.

Ryle, M. and Vonberg, D. D. (1948). An investigation of radio-frequency radiation from the sun, *Proceedings of the Royal Society of London*, **A193**, 98–120.

Ryle, M., Elsmore, B., and Neville, A. C. (1965). High-resolution observations of the radio sources in Cygnus and Cassiopeia, *Nature*, **205**, 1259–1262.

Ryle, M., Graham-Smith, F. G., and Elsmore, B. (1950). A preliminary survey of the radio stars in the northern hemisphere, *Monthly Notices of the Royal Astronomical Society*, **110**, 508–523.

Sandage, A. R. (1965). The existence of a major new constituent of the universe: The quasistellar galaxies, *Astrophysical Journal*, **141**, 1560–1578.

Saxena, S. S., Agarwal, P., Ahilan, K. *et al.* (2000). Superconductivity on the border of itinerant-electron ferromagnetism in UGe_2, *Nature*, **406**, 587–592.

Scherzer, O. (1947). Sphärische und chromatische korrektur von elektronen-linsen, *Optik*, **2**, 114–132.

Scheuer, P. A. G. (1957). A statistical method for analysing observations of faint radio sources, *Proceedings of the Cambridge Philosophical Society*, **53**, 764–773.

Scheuer, P. A. G. (1974). Models of extragalactic radio sources with a continuous energy supply from a central object, *Monthly Notices of the Royal Astronomical Society*, **166**, 513–528.

Scheuer, P. A. G. (1975). Radio astronomy and cosmology, in *Galaxies and the Universe*, eds Sandage, A., Sandage, M., and Kristian, J., pp. 725–760. Chicago: University of Chicago Press.

Scheuer, P. A. G. (1984). The development of aperture synthesis at Cambridge, in *The Early Years of Radio Astronomy: Reflections Fifty Years after Jansky's Discovery*, ed. Sullivan III, W. T., pp. 249–265. Chicago and London: University of Chicago Press.

Schmidt, G. C. (1898). Ueber die von den thorvebindungen und einigen anderen substanzen ausgehende strahlung, *Annalen der Physik und Chemie (Wiedemanns Annalen) (3)*, **65**, 141–151.

Schmidt, M. (1963). 3C 273: A star-like object with large red-shift, *Nature*, **197**, 1040–1040.

Schmidt, M. and Matthews, T. A. (1964). Redshift of the quasi-stellar radio sources 3C 47 and 3C 147, *Astrophysical Journal*, **139**, 781–785.

Schuster, A. (1910). The Clerk–Maxwell Period, in *A History of the Cavendish Laboratory, 1871–1910*, pp. 14–39. London: Longmans, Green and Co.

Schwarzschild, K. (1916). Zur quanten hypothese, *Sitzungberichte der (Kgl.) Preussischen Akademie der Wissenschaften (Berlin)*, pp. 548–568.

Scott, P. F., Saunders, R., Pooley, G. *et al.* (1996). Measurements of structure in the cosmic background radiation with the Cambridge Cosmic Anistropy Telescope, *Astrophysical Journal Letters*, **461**, L1–L4.

Sebastian, S. E., Harrison, N., Altarawneh, M. M. *et al.* (2011). Chemical potential oscillations from nodal Fermi surface pocket in the underdoped high-temperature superconductor $YBa_2Cu_3O_{6+x}$, *Nature Communications*, **2**, 471.

Sekido, Y. and Elliot, H. (1985). *Early History of Cosmic Ray Studies*. Dordrecht: D. Reidel.

Shakeshaft, J. R., Ryle, M., Baldwin, J. E., Elsmore, B., and Thomson, J. (1955). A radio survey of radio sources between declinations −38 and +83, *Memoirs of the Royal Astronomical Society*, **67**, 106–154.

Shimizu, T. (1921a). A preliminary note on branched α-ray tracks, *Proceedings of the Royal Society of London*, **A99**, 432–435.

Shimizu, T. (1921b). A reciprocating expansion apparatus for detecting ionising rays, *Proceedings of the Royal Society of London*, **A99**, 425–431.

Shoenberg, D. (1939). The magnetic properties of bismuth, III: Further measurements on the de Haas–van Alphen effect, *Proceedings of the Royal Society of London*, **A170**, 341–364.

Shoenberg, D. (1940). Properties of superconducting colloids and emulsions, *Proceedings of the Royal Society of London*, **A175**, 49–70.

Shoenberg, D. (1952). The de Haas–van Alphen effect, *Philosophical Transactions of the Royal Society of London*, **A245**, 1–57.

Shoenberg, D. (1985). Piotr Leonidovich Kapitsa, *Biographical Memoirs of Fellows of the Royal Society*, **31**, 327–374.

Siegel, D. M. (1991). *Innovation in Maxwell's Electromagnetic Theory: Molecular Vortices, Displacement Current, and Light*. Cambridge: Cambridge University Press.

Simons, B. M. and Clevers, H. (2011). Strategies for homeostatic stem cell self-renewal in adult tissues, *Cell*, **145**, 851–862.

Sirringhaus, H., Kawase, T., Friend, R. *et al.* (2000). High-resolution inkjet printing of all-polymer transistor circuits, *Science*, **290**, 2123–2126.

Skilling, J. and Bryan, R. K. (1984). Maximum entropy image reconstruction: General algorithm, *Monthly Notices of the Royal Astronomical Society*, **211**, 111–124.

Skobeltsyn, D. (1929). Über eine neue art sehr schneller β strahlen (On a new type of very fast β-ray), *Zeitschrift für Physik*, **54**, 686–702.

Smail, I., Ivison, R. J., and Blain, A. W. (1997). A deep sub-millimeter survey of lensing clusters: A new window on galaxy formation and evolution, *Astrophysical Journal Letters*, **490**, L5–L8.

Smith, C. and Wise, N. M. (1989). *Energy and Empire: A Biographical Study of Lord Kelvin*. Cambridge: Cambridge University Press.

Smith, C. G., Pepper, M., Ahmed, H. *et al.* (1988). The transition from one- to zero-dimensional ballistic transport, *Journal of Physics C: Solid State Physics*, **21**, L893–L898.

Smith, D. M. (1985). The 'New Blood' scheme and its application to geography, *Area*, **17**, 237–243.

Smith, H., Buckle, J., Hills, R. *et al.* (2008). HARP: A submillimetre heterodyne array receiver operating on the James Clerk Maxwell Telescope, *Proceedings of the SPIE*, **7020**, 70200Z-1–70200Z-15.

Smith, H. E. and Spinrad, H. (1980). An update of the status of the revised 3C catalog of radio sources: 22 new galaxy redshifts, *Publications of the Astronomical Society of the Pacific*, **92**, 553–569.

Smith, H. J. and Hoffleit, D. (1963). Light variations in the superluminous radio galaxy 3C 273, *Nature*, **198**, 650–651.

Smithers, A. and Robinson, P. (2006). *Physics in Schools and Universities, II: Patterns and Policies*. Gatsby Charitable Foundation report. Buckingham: Carmichael Press.

Smoot, G. F., Bennett, C. L., Kogut, A. *et al.* (1992). Structure in the COBE differential microwave radiometer first-year maps, *Astrophysical Journal*, **396**, L1–L5.

Sommerfeld, A. (1915a). Die feinstruktur der wassenstoff- und wasserstoffähnlichen linien, *Münchener Berichte*, pp. 459–500.

Sommerfeld, A. (1915b). Zur theorie der Balmerschen serie, *Münchener Berichte*, pp. 425–458.

Sommerfeld, A. (1916). Zur quantentheorie der spektrallinien, *Annalen der Physik (4)*, **51**, 1–94, 125–167.

Sommerfeld, A. (1919). *Atombau und Spektrallinien*. Braunschweig: Vieweg. The third edition was published in English as *Atomic Spectra and Spectral Lines*, trans. Brose, H.L., 1923. London: Methuen.

Spear, W. E. and Le Comber, P. G. (1975). Substitutional doping of amorphous silicon, *Solid State Communications*, **17**, 1193–1196.

Squires, G. L. (1954). The scattering of slow neutrons by ferromagnetic crystals, *Proceedings of the Physical Society of London*, **A67**, 248–2XX.

Squires, G. L. (2012). *Introduction to the Theory of Thermal Neutron Scattering*. Cambridge: Cambridge University Press. This is a reprint of the original edition published in 1978.

Stewart, A. T. and Squires, G. L. (1953). The scattering of slow neutrons by ortho- and para-hydrogen, *Physical Review*, **90**, 1125–1125.

Stobbs, W. M. (1973). Dislocation interaction with irradiation damage in the high voltage electron microscope, *Philosophical Magazine*, **27(1)**, 257–263.

Stockton, A. and Ridgway, S. (1996). Optical and near IR observations of Cygnus A, in *Cygnus A: Study of a Radio Galaxy*, eds Carilli, C. L. and Harris, D. E., pp. 1–4. Cambridge: Cambridge University Press.

Stoner, E. C. (1924). The distribution of electrons among atomic levels, *Philosophical Magazine (6)*, **48**, 719–736.

Stoney, G. J. (1891). On the cause of double lines and of equidistant satellites in the spectra of gases, *Scientific Transactions of the Royal Dublin Society*, **4**, 563–608. The reference to the term 'electron' appears on page 583.

Storey, L. R. O. (1953). An investigation of whistling atmospherics, *Philosophical Transactions of the Royal Society of London*, **A246**, 113–141.

Sullivan III, W. T. (2009). *Cosmic Noise: A History of Early Radio Astronomy*. Cambridge: Cambridge University Press.

Sviedrys, R. (1976). The rise of physics laboratories in Britain, *Historical Studies in the Physical Sciences*, **7**, 405–436.

Szilard, L. and Zinn, W. H. (1939). Instantaneous emission of fast neutrons in the interaction of slow neutrons with uranium, *Physical Review*, **55**, 799–800.

Tabor, D. (1955). The mechanism of rolling friction, II: The elastic range, *Proceedings of the Royal Society of London*, **A229**, 199– 220.

Tabor, D. (1969). Frank Philip Bowden, *Biographical Memoirs of Fellows of the Royal Society*, **15**, 1–34.

Tabor, D. (1991). *Gases, Liquids and Solids and Other States of Matter*. Cambridge: Cambridge University Press. This is the third edition of Tabor's book, first published by Penguin in 1969.

Tabor, D. and Winterton, R. H. S. (1969). The direct measurement of normal and retarded van der Walls forces, *Proceedings of the Royal Society of London*, **A312**, 435–450.

Taillefer, L. and Lonzarich, G. G. (1985). Effect of spin fluctuations on the magnetic equation of state of ferromagnetic or nearly ferromagnetic metals, *Journal of Physics C: Solid State Physics*, **18**, 4339–4371.

Taillefer, L. and Lonzarich, G. G. (1988). Heavy-fermion quasiparticles in UPt_3, *Physical Review Letters*, **60**, 1570–1573.

Tammuz, N., Smith, R. P., Campbell, R. L. D. *et al.* (2011). Can a Bose gas be saturated?, *Physical Review Letters*, **106**, 230401.

Taylor, G. I. (1909). Interference fringe with feeble light, *Proceeedings of the Cambridge Philosophical Society*, **15**, 114–115.

Taylor, G. I. (1910). The conditions necessary for discontinuous motion in gases, *Proceeedings of the Royal Society of London*, **A84**, 371–377.

Taylor, G. I. (1915). Eddy motion in the atmosphere, *Philosophical Transactions of the Royal Society of London*, **A215**, 1–26.

Taylor, G. I. (1916). Pressure distribution over the wing of an aeroplane in flight, *Reports and Memoranda of the Advisory Committee of Aeronautics*, **287**.

Taylor, G. I. (1917a). Motion of solids in fluids when the flow is not irrotational, *Proceeedings of the Royal Society of London*, **A93**, 99–113.

Taylor, G. I. (1917b). Phenomena connected with turbulence in the lower atmosphere, *Proceeedings of the Royal Society of London*, **A94**, 137 155.

Taylor, G. I. (1919). Tidal friction in the Irish Sea, *Philosophical Transactions of the Royal Society of London*, **A220**, 1–33.

Taylor, G. I. (1920). Tidal friction and the secular acceleration of the Moon, *Monthly Notices of the Royal Astronomical Society*, **80**, 308–309.

Taylor, G. I. (1923a). Experiments on the motion of solid bodies in rotating fluids, *Proceeedings of the Royal Society of London*, **A104**, 213–218.

Taylor, G. I. (1923b). Stability of a viscous liquid contained between two rotating cylinders, *Philosophical Transactions of the Royal Society of London*, **A223**, 289–343.

Taylor, G. I. (1934). The mechanism of plastic deformation in crystals, I: Theoretical, *Proceeedings of the Royal Society of London*, **A145**, 362–387.

Taylor, G. I. (1935a). Statistical theory of turbulence I, *Proceeedings of the Royal Society of London*, **A151**, 421–444.

Taylor, G. I. (1935b). Statistical theory of turbulence II, *Proceeedings of the Royal Society of London*, **A151**, 444–454.

Taylor, G. I. (1935c). Statistical theory of turbulence, III: Distribution of dissipation of energy in a pipe over its cross-section, *Proceeedings of the Royal Society of London*, **A151**, 455–464.

Taylor, G. I. (1935d). Statistical theory of turbulence, IV: Diffusion in a turbulent air stream, *Proceeedings of the Royal Society of London*, **A151**, 465–478.

Taylor, G. I. (1938). The spectrum of turbulence, *Proceeedings of the Royal Society of London*, **A164**, 476–490.

Taylor, G. I. (1950a). The formation of a blast wave by a very intense explosion, I: Theoretical discussion, *Proceeedings of the Royal Society of London*, **A201**, 159–174.

Taylor, G. I. (1950b). The formation of a blast wave by a very intense explosion, II: The atomic explosion of 1945, *Proceeedings of the Royal Society of London*, **A201**, 175–186.

Taylor, G. I. (1958). *Scientific Papers I. Mechanics of Solids*, ed. Batchelor, G. K. Cambridge: Cambridge University Press.

Taylor, G. I. (1960). *Scientific Papers II. Meteorology, Oceanography and Turbulent Flow*, ed. Batchelor, G. K. Cambridge: Cambridge University Press.

Taylor, G. I. (1963). *Scientific Papers III. Aerodynamics and the Mechanics of Projectiles and Explosives*, ed. Batchelor, G. K. Cambridge: Cambridge University Press.

Taylor, G. I. (1971). *Scientific Papers IV. Mechanics of Fluids: Micellaneous Papers*, ed. Batchelor, G. K. Cambridge: Cambridge University Press.

Thomson, G. P. (1928). Experiments on the diffraction of cathode rays, *Proceedings of the Royal Society of London*, **A117**, 600–609.

Thomson, J. J. (1881). On the electric and magnetic effects produced by the motion of electrified bodies, *Philosophical Magazine (5)*, **11**, 229–249.

Thomson, J. J. (1882). The vibrations of a vortex ring, and the action upon each other of two vortices in a perfect fluid, *Philosophical Transactions of the Royal Society of London*, **173**, 493–521.

Thomson, J. J. (1883a). On the theory of electric discharge in gases, *Philosophical Magazine (5)*, **15**, 427–434.

Thomson, J. J. (1883b). *A Treatise on the Motion of Vortex Rings*. London: Macmillan.

Thomson, J. J. (1887). On the dissociation of some gases by the electric discharge, *Proceedings of the Royal Society of London*, **42**, 343–344.

Thomson, J. J. (1893a). *Notes on Recent Researches in Electricity and Magnetism*. Oxford: Clarendon Press.

Thomson, J. J. (1893b). On the effect of electrification and chemical action on a steam jet and of water vapour on the discharge of electricity through gases, *Philosophical Magazine (5)*, **36**, 313–327.

Thomson, J. J. (1895). On the electrolysis of gases, *Proceedings of the Royal Society of London*, **58**, 244–257.

Thomson, J. J. (1896a). On the discharge of electricity produced by Röntgen rays, *Proceedings of the Royal Society of London*, **59**, 274–276.

Thomson, J. J. (1896b). The Röntgen rays, *Nature*, **53**, 391–392.

Thomson, J. J. (1897a). Cathode rays, *Journal of the Royal Institution*, **15**, 1–14.

Thomson, J. J. (1897b). Cathode rays, *The Electrician*, **39**, 104–109.

Thomson, J. J. (1897c). Cathode rays, *Philosophical Magazine (5)*, **44**, 293–316.

Thomson, J. J. (1898). On the charge of electricity carried by the ions produced by Röntgen rays, *Philosophical Magazine (5)*, **46**, 528–545.

Thomson, J. J. (1899). On the masses of the ions in gases at low pressures, *Philosophical Magazine (5)*, **48**, 547–567.

Thomson, J. J. (1903a). *Conduction of Electricity through Gases*. Cambridge: Cambridge University Press.

Thomson, J. J. (1903b). On the charge of electricity carried by a gaseous ion, *Philosophical Magazine (6)*, **5**, 346–355.

Thomson, J. J. (1906a). On the number of corpuscles in an atom, *Philosophical Magazine (6)*, **11**, 769–781.

Thomson, J. J. (1906b). *Conduction of Electricity through Gases*. Cambridge: Cambridge University Press.

Thomson, J. J. (1910). Survey of the last twenty-five years, in *A History of the Cavendish Laboratory, 1871–1910*, pp. 75–101. London: Longmans, Green and Co.

Thomson, J. J. (1912a). Further experiments on positive rays, *Philosophical Magazine (6)*, **24**, 209–253.

Thomson, J. J. (1912b). Ionisation by moving electrified particles, *Philosophical Magazine (6)*, **23**, 449–457.

Thomson, J. J. (1913). Rays of positive electricity, *Proceedings of the Royal Society of London*, **A89**, 1–20.

Thomson, J. J. (1933). Mr. E. Everett, *Nature*, **132**, 774.

Thomson, J. J. (1936). *Recollections and Reminiscences*. London: G. Bell and Sons.

Thomson, J. J. and Newall, H. F. (1887). On the rate at which electricity leaks through liquids which are bad conductors of electricity, *Proceedings of the Royal Society of London*, **42**, 410–429.

Thomson, W. (1855). On the theory of the electric telegraph, *Proceedings of the Royal Society of London*, **7**, 382–399.

Thomson, W. and Tait, P. G. (1867). *Treatise on Natural Philosophy*. Oxford: Oxford University Press.

Timpe, A. (1905). Probleme der spannungsverteilung in ebenen systeme, einfach gelöst mit hilfe der Airyschen funktion (The problem of the stress distribution in a planar system, easily solved using Airy functions), *Zeitschrift für Mathematik und Physik*, **52**, 348–383.

Tosi, G., Christmann, G., Berloff, N. G. *et al.* (2012). Sculpting oscillators with light within a nonlinear quantum fluid, *Nature Physics*, **8**, 190–194.

Townsend, A. A. (1956). *The Structure of Turbulent Shear Flow*. Cambridge: Cambridge University Press.

Townsend, A. A. (1990). Early days of turbulence research in Cambridge, *Jounral of Fluid Mechanics*, **212**, 1–5.

Townsend, J. S. (1900). The diffusion of ions in gases, *Philosophical Transactions of the Royal Society of London*, **A193**, 129–158.

Townsend, P. (2013). *Fundamentals of Terahertz Radiation*. http://www.paultownsend.co.uk/research/fundamentals/terahertz-radiation/.

Tsui, D. C., Störmer, H. L., and Gossard, A. C. (1982). Two-dimensional magnetotransport in the extreme quantum limit, *Physical Review Letters*, **48**, 1559–1562.

Turner, R. S. (1970). Hermann von Helmholtz, in *Dictionary of Scientific Biography*, volume 6, pp. 241–253. New York: Charles Scribner's Sons.

Uhlenbeck, G. E. and Goudsmit, S. (1925). Ersetzung der hypothese vom unmechanischen zwang durch eine forderung bezüglich des inneren verhaltens jedes einzelnen elektrons, *Die Naturwissenschaften*, **13**, 953–954.

Vamivakas, A. N., Lu, C.-Y., Matthiesen, C. *et al.* (2010). Observation of spin-dependent quantum jumps via quantum dot resonance fluorescence, *Nature*, **467**, 297–300.

van de Graaff, R. J. (1931). A 1,500,000 volt electrostatic generator, *Physical Review*, **38**, 1919–1920.

van der Kloot, W. (2005). Lawrence Bragg's role in the development of sound-ranging in World War I, *Notes and Records of the Royal Society*, **59**, 273–284.

van der Pol, B. (1927). On relaxation-oscillations, *Philosophical Magazine (7)*, **2**, 978–992.

Varela, M., Findlay, S. D., Lupini, A. R. *et al.* (2004). Spectroscopic imaging of single atoms within a bulk solid, *Physical Review Letters*, **92**, 095502.

Villard, P. (1900a). Sur la réflection et la réfraction des rayons cathodique et les rayons déviables de radium (On the reflection and refraction of cathode rays and the deviable rays of radium), *Comptes Rendus de l'Academie des Sciences*, **130**, 1010–1012.

Villard, P. (1900b). Sur le rayonnement du radium (On the radiation of radium), *Comptes Rendus de l'Academie des Sciences*, **130**, 1178–1179.

Vinen, W. F. (1961). The detection of single quanta of circulation in liquid helium II, *Proceedings of the Royal Society of London*, **A260**, 218–236.

Volta, A. (1800). On the electricity excited by the mere contact of conducting substances of different kinds. In a letter from Mr. Alexander Volta, F.R.S. Professor of Natural Philosophy in the University of Pavia, to the Rt. Hon. Sir Joseph Banks, Bart. K.B. P.R.S., *Philosophical Transactions of the Royal Society of London*, **90**, 403–431.

Volterra, V. (1907). Sur l'équilibre des corps élastique multiplement connexes, *Annales Scientifique de l'École Normale Superieur, Suppl. (3)*, **24**, 401–517.

von Halban, H., Joliot, F., and Kowarski, L. (1939). Number of neutrons liberated in the nuclear fission of uranium, *Nature*, **143**, 680.

Wade, J. M., Loram, J. W., Mirza, K. A., Cooper, J. R., and Tallon, J. L. (1994). Electronic specific heat of $TL_2Ba_2Cu_{6+\delta}$ from 2 K to 300 K for $0 < \delta < 0.1$, *Journal of Superconductivity*, **7**, 261–264.

Waldram, J. R. (1964). The surface impedence of superconductors, *Advances in Physics*, **13**, 1–88.

Waldram, J. R. (1975). Chemical potential and boundary resistance at normal–superconducting interfaces, *Proceedings of the Royal Society of London*, **A345**, 231–249.

Waldram, J. R. (1996). *Superconductivity of Metals and Cuprates*. Bristol: IOP Publishing.

Waldram, J. R. and Battersby, S. J. (1992). Thermopower and resistance of superconducting–normal interfaces, I: Theory, *Journal of Low Temperature Physics*, **86**, 1–30.

Waldram, J. R. and Lumley, J. M. (1975). Direct measurement of the current-phase relation in superconducting weak links, *Revue de Physique Appliquée*, **10**, 7–10.

Waldram, J. R., Broun, D. M., Morgan, D. C., Ormeno, R., and Porch, A. (1999). Fluctuation effects in the microwave conductivity of cuprate superconductors, *Physical Review B*, **59**, 1528–1537.

Waldram, J. R., Pippard, A. B., and Clarke, J. (1970). Theory of the current–voltage characteristics of SNS junctions and other superconducting weak links, *Philosophical Transactions of the Royal Society of London*, **A268**, 265–287.

Waldram, J. R., Theopistou, P., Porch, A., and Cheah, H.-M. (1997). Two-fluid interpretation of the microwave conductivity of $YBa_2Cu_3O_{7-\delta}$, *Physical Review B*, **55**, 3222–3229.

Walter, F., Decarli, R., Carilli, C. *et al.* (2012). The intense starburst HDF 850.1 in a galaxy overdensity at $z \approx 5.2$ in the Hubble Deep Field, *Nature*, **486**, 233–236.

Ward, F. A. B., Wynn-Williams, C. E., and Cave, H. M. (1929). The rate of emission of α-particles from radium, *Proceedings of the Royal Society of London*, **A125**, 713–730.

Warner, M. and Terentjev, E. M. (2003). *Liquid Crystal Elastomers*. Oxford: Oxford University Press. The second edition was publihsed in 2007.

Watson, J. D. (1968). *The Double Helix: A Personal Account of the Discovery of the Structure of DNA*. New York: Athenium.

Watson, J. D. and Crick, F. H. C. (1953a). Genetical implication of the structure of deoxyribonucleic acid, *Nature*, **171**, 964–967.

Watson, J. D. and Crick, F. H. C. (1953b). A structure for deoxyribose nucleic acid, *Nature*, **171**, 737–738.

Weaver, H., Williams, D., Dieter, N., and Lum, W. (1965). Observations of a strong unidentified microwave line and of emission from the OH molecule, *Nature*, **208**, 29–31.

Weber, W. and Kohlrausch, R. (1856). Über die elektricitätsmenge, welche bei galvanischenen strömen durch den querschnitt der kette fleisst (On the amount of electricity which flows through the cross-section of the circuit in galvanic currents), *Annalen der Physik (2)*, **99**, 10–25.

Webster, H. C. (1932). The artificial production of nuclear γ-radiation, *Proceedings of the Royal Society of London*, **A136**, 428–453.

Weinreb, S., Meeks, M., Carter, J., Barrett, A., and Rogers, A. (1965). Observations of polarized OH emission, *Nature*, **208**, 440–441.

Wharam, D. A., Thornton, T. J., Newbury, R. *et al.* (1988). One-dimensional transport and the quantisation of the ballistic resistance, *Journal of Physics C: Solid State Physics*, **21**, L209–L214.

Wheeler, J. M. (1992). Applications of the EDSAC, *IEEE Annals of the History of Computing*, **14**, 27–33.

Whittaker, E. T. (1951). *A History of the Theories of Aether and Electricity*, Vols I and II. London: Thomas Nelson and Sons. The first edition was published in 1910 and the revised and enlarged edition in 1951.

Wideröe, R. (1928). Über ein neues prinzip zur herstellung hoher spannungen, *Archiv für Elektrotechnik*, **21(4)**, 387–406.

Wilhelm, J. O., Misener, A. D., and Clark, A. R. (1935). The viscosity of liquid helium, *Proceedings of the Royal Society of London*, **A151**, 342–347.

Wilkins, M. H. F., Stokes, A., and Wilson, H. R. (1953). Molecular structure of deoxypentose nucleic acids, *Nature*, **171**, 738–739.

Willis, R. (1841). *Principles of Mechanism*. London: John W. Parker.

Willis, R. (1851), *System of Apparatus for the Use of Lecturers and Experimenters in Mechanical Philosophy*. London: J. Weale.

Willis, R. and Clark, R. W. (1886). *Architectural History of the University of Cambridge and of the Colleges of Cambridge and Eton*. Cambridge: Cambridge University Press.

Wilson, A. (1984). Theoretical physics in Cambridge in the late 1920s and early 1930s, in *Cambridge Physics in the Thirties*, ed. Hendry, J., pp. 174–175. Bristol: Adam Hilger.

Wilson, C. T. R. (1897). Condensation of water vapour in the presence of dust-free air and other gases, *Philosophical Transactions of the Royal Society of London*, **A189**, 265–306.

Wilson, C. T. R. (1899a). On the comparative efficiency as condensation nuclei of particles and negatively charged ions, *Philosophical Transactions of the Royal Society of London*, **A193**, 289–317.

Wilson, C. T. R. (1899b). On the condensation nuclei produced in gases by the action of Röntgen rays, uranium rays, ultraviolet light, and other agents, *Philosophical Transactions of the Royal Society of London*, **A192**, 403–452.

Wilson, C. T. R. (1901). On the ionisation of atmospheric air, *Proceedings of the Royal Society of London*, **A68**, 151–161.

Wilson, C. T. R. (1912). On an expansion apparatus for making visible the tracks of ionising particles in gases and some results obtained by its use, *Proceedings of the Royal Society of London*, **A87**, 277–292.

Wilson, C. T. R. (1954). Ben Nevis 60 years ago, *Weather*, **9**, 309–311.

Wilson, C. T. R. (1960). Reminiscences of my early years, *Notes and Records of The Royal Society*, **14**, 163–173.

Wilson, D. (1982). Experimentalists among the mathematicians: Physics in the Cambridge Natural Sciences Tripos, 1851–1900, *Historical Studies in the Physical Sciences*, **12**, 325–371.

Wilson, D. (1983). *Rutherford: Simple Genius*. London: Hodder and Stoughton.

Wilson, H. A. (1903). A determination of the charge on the ions produced in air by Röntgen rays, *Philosophical Magazine (6)*, **6**, 429–441.

Wollaston, W. H. (1802). A method of examining refractive and dispersive powers, by prismatic reflection, *Philosophical Transactions of the Royal Society*, **92**, 365–380.

Woodward, R. M., Wallace, V. P., Pye, R. J. *et al.* (2003). Terahertz pulse imaging of ex vivo basal cell carcinoma, *Journal of Investigative Dermatology*, **120**, 72–78.

Woolfson, M. (2005). William Cochran 1922–2003, *Biographical Memoirs of Fellows of the Royal Society*, **51**, 67–85.

Wynn-Williams, C. E. (1931). The use of thyratrons for high speed automatic counting of physical phenomena, *Proceedings of the Royal Society of London*, **A132**, 295–310.

Wynn-Williams, C. E. (1932). A thyratron "scale of two" automatic counter, *Proceedings of the Royal Society of London*, **A136**, 312–324.

Wynn-Williams, C. E. (1984). The scale-of-two counter, in *Cambridge Physics in the Thirties*, ed. Hendry, J., pp. 141–149. Bristol: Adam Hilger.

Yoffe, A. B. (1998). NFM and PCS, in *Nevill Mott: Reminiscences and Appreciations*, ed. Davis, E. A., pp. 166–168. London: Taylor and Francis.

Young, J. S., Baldwin, J. E., Boysen, R. C. *et al.* (2000). New views of Betelgeuse: Multiwavelength surface imaging and implications for models of hotspot generation, *Monthly Notices of the Royal Astronomical Society*, **315**, 635–645.

Young, T. (1802). On the theory of light and colours, *Philosophical Transactions of the Royal Society*, **92**, 12–48.

Yuan, Z. L., Kardynal, B. E., Stevenson, R. M. *et al.* (2002). Electrically driven single-photon source, *Science*, **295**, 102–105.

Zeeman, P. (1896a). Over den invloed eener magnetisatie op den aard van het door eenstof uitgezonden licht (On the influence of magnetism on the nature of light emitted by a substance), *Verslag van de gewone Vergadering der Wis- en Natuurkundige Afdeeling, Koniklijke Akadamie van Wetenschappen te Amsterdam*, **5**, 181–185. English translations in *Philosophical Magazine (5)*, **43**, 226–239, 1897 and *Astrophysical Journal*, **5**, 332–347, 1897.

Zeeman, P. (1896b). Over den invloed eener magnetisatie op den aard van het door eenstof uitgezonden licht (On the influence of magnetism on the nature of light emitted by a substance), *Verslag van de gewone Vergadering der Wis- en Natuurkundige Afdeeling,*

Koniklijke Akadamie van Wetenschappen te Amsterdam, **5**, 242–248. English translations in *Philosophical Magazine (5)*, **43**, 226–239, 1897 and *Astrophysical Journal*, **5**, 332–347, 1897.

Zeleny, J. (1898). The ratio of velocities of the two ions produced in gases by Röntgen radiation, *Philosophical Magazine (5)*, **46**, 120–154.

Zeleny, J. (1900). The velocity of the ions produced in gases by Röntgen rays, *Philosophical Transactions of the Royal Society of London*, **A195**, 193–234.

Ziman, J. M. (1960). *Electrons and Phonons: The Theory of Transport in Solids*. Oxford: Clarendon Press.

Ziman, J. M. (1963) *Electrons in Metals: A Short Guide to the Fermi Surface*. London: Taylor and Francis.

Ziman, J. M. (1969). *Elements of Advanced Quantum Theory*. Cambridge: Cambridge University Press.

Ziman, J. M. (1972). *Principles of the Theory of Solids*. Cambridge: Cambridge University Press.

Zimmerman, J. and Silver, A. H. (1964). Quantum effects in type II superconductors, *Physics Letters*, **10**, 47–48.

Zipkes, C., Palzer, S., Sias, C., and Köhl, M. (2010). A trapped single ion inside a Bose–Einstein condensate, *Nature*, **464**, 388–391.

Author index

Subject index